AF323157

PLASMA SCIENCE AND TECHNOLOGY - PROGRESS IN PHYSICAL STATES AND CHEMICAL REACTIONS

Edited by **Tetsu Mieno**

Plasma Science and Technology - Progress in Physical States and Chemical Reactions

http://dx.doi.org/10.5772/60692
Edited by Tetsu Mieno

Contributors

Rodrigo Perito Cardoso, Silvio Francisco Brunatto, Marcio Mafra, Barbara Nebe, Claudia Bergemann, Cornelia Hoppe, Martin Eggert, Torsten Gerling, Maxi Höntsch, Maryna Karmazyna, Kashif Chaudhary, Syed Zuhaib Haider Rizivi, Jalil Ali, Marek Andrzej Rabiński, Krzysztof Zdunek, Jiayin Guo, Yoshihiro Deguchi, Zhenzhen Wang, Pieter Cools, Nathalie De Geyter, Rino Morent, Rouba Ghobeira, Stijn Van Vrekhem, Christoph Gerhard, Wolfgang Viöl, Stephan Wieneke, Mikhail Gitlin, Anton Nikiforov, Patrick Vanraes, Christophe Leys, Nikolay Britun, Stephanos Konstantinidis, Rony Snyders, Tiago Silva, Thomas Godfroid, Jacek Tyczkowski, Takeshi Higashiguchi, Padraig Dunne, Gerry O'Sullivan, Afsaneh Edrisy, Kora Farrokhzadeh, Federico Vazquez, Aldo Figueroa, Raúl Salgado, Farook Yousif Bashir, Marco A. Rivera, Shahidi, Jakub Wiener, Mahmood Ghoranneviss, Fatemeh Rezaei

CBS, Edition **2017**

Published by InTech

Janeza Trdine 9, 51000 Rijeka, Croatia

Notice

Statements and opinions expressed in the chapters are these of the individual contributors and not necessarily those of the editors or publisher. No responsibility is accepted for the accuracy of information contained in the published chapters. The publisher assumes no responsibility for any damage or injury to persons or property arising out of the use of any materials, instructions, methods or ideas contained in the book.

Publishing Process Manager Iva Lipovic

Technical Editor InTech DTP team

Cover Designer InTech Design Team

Additional hard copies can be obtained from orders@intechopen.com

Plasma Science and Technology - Progress in Physical States and Chemical Reactions, Edited by Tetsu Mieno
p. cm.
ISBN 978-953-51-2280-7

Contents

Preface IX

Section 1 **Plasma Processings 1**

Chapter 1 **Plasma-Enhanced Laser Materials Processing 3**
Christoph Gerhard, Wolfgang Viöl and Stephan Wieneke

Chapter 2 **Cold Plasma Produced Catalytic Materials 25**
Jacek Tyczkowski

Chapter 3 **Plasma Nitriding of Titanium Alloys 67**
Afsaneh Edrisy and Khorameh Farokhzadeh

Chapter 4 **Low-temperature Thermochemical Treatments of Stainless Steels – An Introduction 107**
Rodrigo P. Cardoso, Marcio Mafra and Silvio F. Brunatto

Chapter 5 **Computational Studies of the Impulse Plasma Deposition Method 131**
Marek Rabiński and Krzysztof Zdunek

Section 2 **Medical and Green Applications of Plasma 153**

Chapter 6 **Physicochemical Analysis of Argon Plasma-Treated Cell Culture Medium 155**
Claudia Bergemann, Cornelia Hoppe, Maryna Karmazyna, Maxi Höntsch, Martin Eggert, Torsten Gerling and Barbara Nebe

Chapter 7 **Non-thermal Plasma Technology for the Improvement of Scaffolds for Tissue Engineering and Regenerative Medicine - A Review 173**
Pieter Cools, Rouba Ghobeira, Stijn Van Vrekhem, Nathalie De Geyterand and Rino Morent

Contents

Chapter 8   **Plasma-Enhanced Vapor Deposition Process for the Modification of Textile Materials   213**
Sheila Shahidi, Jakub Wiener and Mahmood Ghoranneviss

Section 3   **Special Plasma Sources   239**

Chapter 9   **Laser-Produced Heavy Ion Plasmas as Efficient Soft X-Ray Sources   241**
Takeshi Higashiguchi, Padraig Dunne and Gerry O'Sullivan

Chapter 10   **Laser-Induced Plasma and its Applications   259**
Kashif Chaudhary, Syed Zuhaib Haider Rizvi and Jalil Ali

Section 4   **Plasma Diagnostics   293**

Chapter 11   **Diagnostics of Magnetron Sputtering Discharges by Resonant Absorption Spectroscopy   295**
Nikolay Britun, Stephanos Konstantinidis and Rony Snyders

Chapter 12   **A Technique for Time-Resolved Imaging of Millimeter Waves Based on Visible Continuum Radiation from a Cs-Xe DC Discharge — Fundamentals and Applications   329**
Mikhail S. Gitlin

Chapter 13   **Optically Thick Laser-Induced Plasmas in Spectroscopic Analysis   363**
Fatemeh Rezaei

Chapter 14   **Industrial Applications of Laser-Induced Breakdown Spectroscopy   401**
Yoshihiro Deguchi and Zhenzhen Wang

Section 5   **Plasmas for Environmental Technology   427**

Chapter 15   **Electrical Discharge in Water Treatment Technology for Micropollutant Decomposition   429**
Patrick Vanraes, Anton Y. Nikiforov and Christophe Leys

Chapter 16   **Study of CO2 Decomposition in Microwave Discharges by Optical Diagnostic Methods   479**
Tiago Silva, Nikolay Britun, Thomas Godfroid and Rony Snyders

Contents

Section 6   Basic Plasma Phenomena   505

Chapter 17   **Stochastic and Nonlinear Dynamics in Low-Temperature Plasmas   507**
Aldo Figueroa, Raúl Salgado-García, Jannet Rodríguez, Farook Yousif Bashir, Marco A. Rivera and Federico Vázquez

Preface

In the early twentieth century, Dr. Irving Langmuir actively studied plasma discharge and surface science. Since then, great progress has been made in the development of applications of discharges and plasmas such as discharge lamps, electric tubes, and arc welding. In relation to studies on space physics and controlled nuclear fusion, plasma physics has greatly advanced. Plasma chemistry has also progressed along with its applications in LSI fabrication technology, the chemical vapor deposition of functional films, and the production of nanomaterials.

In the twenty-first century, the further development of applications of plasma physics and plasma chemistry is certainly expected. In this book, 17 chapters on the recent progress in plasma science and technology have been written by active specialists worldwide.

In the section "Plasma Processing," reviews of the combination of plasma and laser processing for surface treatments, the creation of specially shaped catalytic materials by plasma deposition methods, the surface-hardening treatment of titanium alloys by plasma nitridation methods, and the thermochemical treatment of stainless steel by low-temperature plasma are presented. Finally, a computer simulation study on the use of the impulse plasma deposition (IPD) method for surface engineering is reported.

The recent progress in plasma applications in medical and green technologies has been unexpectedly rapid. In the section "Medical and Green Applications of Plasma," a physicochemical analysis of argon plasma–treated cells and plasma-treated scaffolds for tissue engineering are described. As a green technology, the plasma deposition on textile materials is also described.

Regarding laser technology, the development of new plasma source is described in the section "Special Plasma Sources." Reviews of efficient soft-X-ray sources using laser-produced heavy-ion plasmas and the principle and applications of intense laser-induced plasma are included in this section.

The large-scale application of plasma technology is expected to solve some of the world's environmental problems in our lifetime. In the section "Plasmas for Environmental Technology," waste water treatment and micropollutant decomposition by nonthermal plasmas and optical monitoring of the decomposition of exhaust CO_2 gas using microwave plasma are described.

We can produce unconventional states (thermal plasma and nonthermal plasma states) for specialized purposes, and these states should be monitored. In the section "Plasma Diagnostics," specially developed diagnostics methods are introduced, which are expected to be utilized in plasma-processing works. A method for the spectral analysis of optically thick plasmas and a review of resonant optical absorption spectroscopy (ROAS) and spatial imaging of millimeter wave propagation using Cs-Xe plasma, and a review of applications of laser-induced breakdown spectroscopy are given.

In the final section "Basic Plasma Phenomena," curious self-oscillating phenomena are described. As plasmas are not exactly in thermal equilibrium, they can exhibit distinct group motion. A theoretical and experimental study of the stochastic and nonlinear motion in plasma is described.

So we can produce plasmas in various states to observe physical and chemical phenomena, which will be used for many specialized applications. Thus, the progress in plasma science and technology is expected now and in future.

Prof. Tetsu Mieno
Graduate School of Science and Technology
Shizuoka University
Japan

Plasma Processings

Chapter 1

Plasma-Enhanced Laser Materials Processing

Christoph Gerhard, Wolfgang Viöl and Stephan Wieneke

Additional information is available at the end of the chapter

http://dx.doi.org/10.5772/61567

Abstract

In the last few years, the combination of laser irradiation with atmospheric pressure plasmas, also referred to as laser–plasma hybrid technology, turned out to be a powerful technique for different materials processing tasks. This chapter gives an overview on this novel approach. Two methods, simultaneous and sequential laser-plasma processing, are covered. In the first case, both the plasma and the laser irradiation are applied to the substrate at the same time. Depending on the process gas and the discharge type, the plasma provides a number of species that can contribute to the laser process plasma-physically or plasma-chemically. Sequential plasma-enhanced laser processing is based on a plasma-induced modification of essential material properties, thus improving the coupling of laser energy into the material during subsequent laser ablation. Simultaneous plasma-assisted laser processing allows increasing the efficiency of a number of different laser applications such as cleaning, microstructuring, or annealing processes. Sequential plasma-assisted laser processing is a powerful method for the processing of transparent media due to a reduction in the laser ablation threshold and an increase in the ablation rate at the same time. In this chapter, the possibilities, underlying mechanisms, performance, and limits of the introduced approaches are presented in detail.

Keywords: Laser-plasma-hybrid techniques, atmospheric pressure plasma, cleaning, microstructuring, modification

1. Introduction

During the last decades, both lasers and plasmas have become essential integral parts of modern manufacturing technology and surface engineering. The coupling of laser irradiation and plasmas, which can be realized either at low or atmospheric pressure, has turned out to be a powerful combination for a number of different applications. For example, plasma-assisted pulsed laser deposition (PLD), which is performed in vacuum environment, allows

the generation of special functional thin films [1,2]. In this context, laser–plasma coupling also offers the possibility of increasing the total efficiency of PLD processes by a reduction of the laser ablation threshold of some target materials such as tungsten due to the plasma perforation of the target surface and a subsequent laser-induced bursting of gas-filled pores [3].

Plasma-enhanced laser materials processing at atmospheric pressure is most commonly known from welding technology. Here, the laser-induced plasma results in the formation of a so-called keyhole, which allows an improved stirring of the molten material as well as a significant increase in welding depth and thus a larger and more homogenous welding seam [4,5]. In this context, laser irradiation can also be added to plasma arc welding processes in order to improve the process performances [6,7]. The plasmas used for this purpose are usually high-energetic arc discharges, driven at several amperes, where the metallic work piece itself acts as counter electrode. However, these techniques are not suitable for the machining of temperature-sensitive materials or for obtaining marginal and locally well-defined material removal as required in the case of cleaning or microstructuring.

For the latter task, the so-called laser-induced plasma-assisted ablation (LIPAA) method represents a promising approach for plasma-enhanced laser materials processing. LIPAA was developed in order to overcome the limits during laser ablation of glasses which are given by the high transparency of such media. Here, a focused laser beam is guided through the glass work piece onto a metallic target which is placed close to the rear side of the glass. As a result, a laser-induced plasma is ignited within the gap between the work piece and the target. Ablation of the transparent glass is then obtained due to a successive deposition of an absorbing metallic layer on its surface and a charge exchange amongst ions and electrons as well as a transfer of kinetic or potential energy, provided by radicals and ions [8,9].

In the last years, the simultaneous or sequential combination of low-energetic atmospheric pressure plasmas, mainly based on dielectric barrier discharges (DBD) and thus featuring low gas temperatures, and laser irradiation has turned out to be a powerful tool for a number of different applications. Such combination is also referred to as laser–plasma hybrid technology and represents another approach for improving laser-based materials processing techniques such as modification, cleaning, drilling, and cutting as well as structuring. In this context, auxiliary plasmas allow to increase the total process and energy efficiency and to improve the machining quality as presented in more detail in this chapter.

2. Laser–plasma coupling

The aim of plasma-enhanced laser materials processing is to benefit from interactions and synergies arising from the particular phenomena provided by laser irradiation and plasmas in order to improve the machining efficiency and quality of different processes. As shown in Figure 1, this technology is based on either simultaneous or subsequent coupling of both energy sources and the resulting impact on any work piece surface.

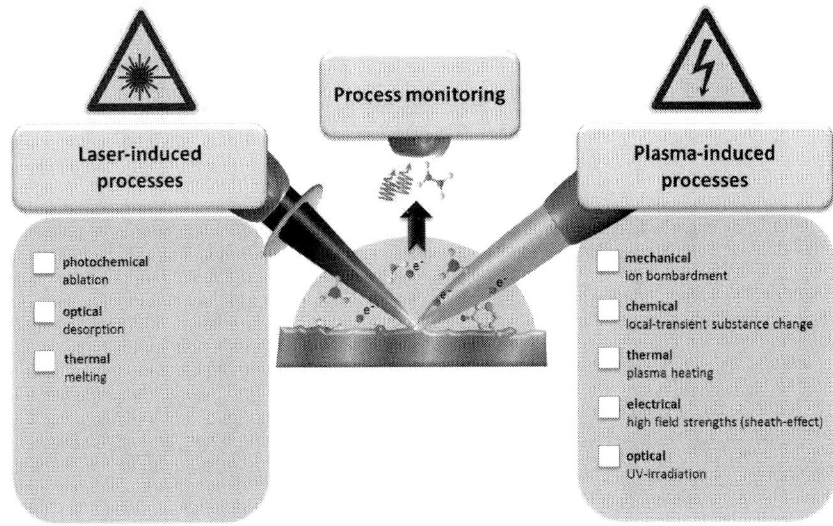

Figure 1. Schematic diagram of a hybrid laser–plasma machining process including particularly usable effects and mechanism.

Especially in the case of simultaneous processing, this approach results in a complex matrix laser-plasma-substrate. Thus, a proper inline process monitoring, for example, laser-induced breakdown spectroscopy (LIBS) or Rayleigh scattering analysis becomes crucial for a description and control of the process. This is of significant importance due to the number of different interacting parameters and mechanisms provided by laser irradiation and plasmas, which are introduced in more detail in the following sections.

2.1. Laser–matter interactions

Only three years after the presentation of the first operable laser source, its suitability for materials processing applications was demonstrated by the fabrication of diaphragms with an inner diameter of some microns [10]. Since that time, the laser has established as a state-of-the-art technology in different fields of materials processing and mainly material removal, commonly referred to as laser ablation. However, laser materials processing involves a number of different mechanisms such as photochemical, optical, and/or thermal processes, namely, ablation, desorption, or melting. Here, the particular process is set by the applied laser parameters.

The main influencing parameter during laser ablation is the wavelength-dependent absorption coefficient $\alpha(\lambda)$ of the particular substrate material, which determines the optical penetration depth of incoming laser irradiation with a certain wavelength λ. This material-specific property is given by

$$\alpha(\lambda) = \frac{4\pi k(\lambda) n(\lambda)}{\lambda} \tag{1}$$

with $k(\lambda)$ being the imaginary part of the complex index of refraction, i.e., the extinction coefficient, and $n(\lambda)$ being its real part, usually referred to as the refractive index. The optical penetration depth $d_p(\lambda)$ is then given by the reciprocal of the absorption coefficient according to

$$d_p(\lambda) = \frac{1}{\alpha(\lambda)} \tag{2}$$

It can easily be seen that surface absorption results from high absorption coefficients, whereas bulk absorption occurs in the case of low absorption coefficients. This fact is crucial for the choice of the laser wavelength for machining a given material with specific optical properties.

Another essential parameter for materials processing is the laser pulse duration τ. In the case of nanosecond laser ablation, material removal is mainly achieved by laser-induced heating and the subsequent desorption or melting and evaporation of material. Picosecond laser processing is based on the depletion of electrons, giving rise to a fast ionization within the substrate material, and thus resulting in material removal by Coulomb explosion due to the occurring repulsive forces of the remaining ions. Femtosecond laser machining processes are dominated by nonlinear effects induced by such ultrashort laser pulses such as temporally limited modifications in absorption within the processed material. Long-pulse and short-pulse laser materials processing are classified by the comparison of the optical penetration depth d_p (see Eq. (2)) and the thermal penetration depth d_{th} of laser irradiation. According to

$$d_{th} \approx \sqrt{2K\tau}, \tag{3}$$

the latter results from both thermodynamic material characteristics, expressed by the thermal conductivity K, and the laser pulse duration [11,12]. In the case of $d_p < d_{th}$, laser machining is referred to as long-pulse processing. In contrast, short-pulse processing is on hand when $d_p > d_{th}$.

The laser wavelength and its pulse duration further determine the laser fluence Φ (i.e., laser pulse energy E per illuminated area A) required for removing a given material. This parameter can also be expressed as dose D, given by the product of the laser pulse energy and the number of the applied laser pulses. Due to the thermal relaxation characteristics of solids, the laser pulse repetition rate f_{rep} finally has a considerable impact on multiple pulse laser machining processes given the fact that the average laser power P_{av} results from

$$P_{av} = Ef_{rep}. \tag{4}$$

To summarize, a number of different laser parameters have to be controlled in order to realize an optimized and high-quality machining process, depending on specific optical and thermo-dynamic characteristics of the work piece material. These laser parameters are wavelength, pulse duration, energy, and fluence as well as the repetition rate. In terms of plasma-enhanced laser materials processing, plasma–matter interactions have to be considered additionally.

2.2. Plasma–matter interactions

Similar to lasers, plasma sources have been used in manufacturing for several decades now, for example, for plasma cutting, arc welding, surface activation, or EUV lithography. This wide range of clearly different applications can be addressed, thanks to the versatility of technical plasmas. Plasma effects are generally influenced by the particularly applied operating parameters such as the high-voltage U, the high-voltage pulse frequency f, the composition of the process gas, the working distance, and the discharge type (direct or indirect). Due to this multiplicity of possible parameters and the resulting plasma–matter interactions, plasma technology is indeed a rapidly growing area of application but still represents a complex field of research.

Within a plasma volume, both neutral and charging electron collision-induced gas phase processes such as dissociation, excitation, ionization, and attachment can take place as visualized in Figure 2.

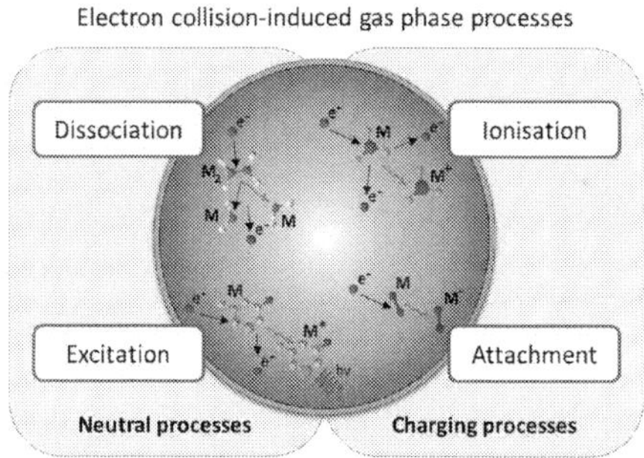

Figure 2. Schematic visualization of electron collision-induced gas phase processes in plasma bulks.

In terms of materials processing, plasmas thus potentially allow the following:

- The chemical reduction or oxidation of the work piece surface or adherents by chemically active plasma species (applicable for both direct and indirect discharges)

- An energy transfer to the work piece surface resulting from the de-excitation of metastable plasma species at walls (applicable for both direct and indirect discharges)

- A certain plasma heating of the work piece surface (applicable for both direct and indirect discharges)

- An acceleration of charge carriers due to a high electrical field strength in the sheath region on the work piece surface (applicable for direct discharges)

- The formation of UV radiation (applicable for direct discharges where UV radiation is generated close to the surface)

As listed above, some of these phenomena especially become of importance in direct plasma discharges where the work piece itself represents a component of the discharge geometry. The choice of the discharge type thus plays an essential role for any plasma-based materials processing application.

Plasma treatment is a state-of-the-art method for surface activation, mainly for improving the adhesion of lacquers or glues on synthetic materials by the formation of polar groups on the substrate surface [13,14]. Another established field of application is plasma sterilization [15] or plasma cleaning [16,17]. For the latter application, hydrocarbons ($-CH_2-$) represent the main surface pollutants. These pollutants can easily be removed by reactive oxygen species (ROS) such as atomic oxygen (O*), provided by air plasmas, according to

$$\left(-CH_2 - CH_2 -\right) + 6O^* \rightarrow 2CO_2 + 2H_2O. \tag{5}$$

However, an excitation of a carbon atom can also be achieved by collisions with electrons within the plasma as shown in Figure 3.

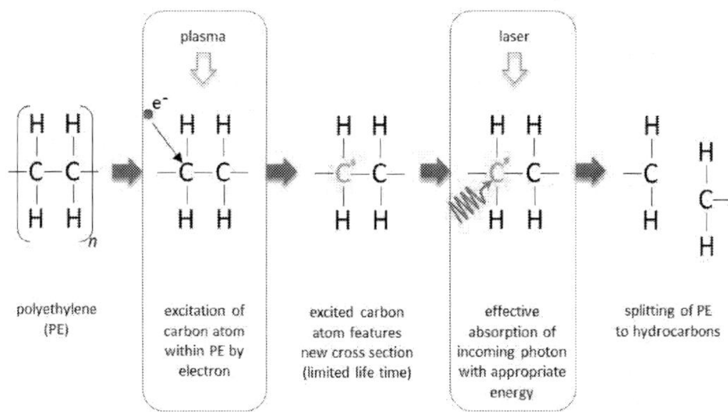

Figure 3. Splitting of polyethylene to hydrocarbons by electron-induced excitation of carbon atoms and subsequent absorption of incoming laser irradiation.

The carbon–hydrogen (C-H) bond features an original binding energy of approximately 4.4eV. Relating to laser irradiation, this energy corresponds to a laser wavelength of approximately 282 nm. Once a carbon atom within this bond is excited by such a collision, the energy required for breaking this bond is <4.4eV. The excitation of atoms by plasma thus offers new possibilities for laser processes and especially allows the use of energy- and cost-efficient standard laser sources. In the example given above, a 3rd-harmonic Nd:YAG laser instead of a 4th-harmonic one could be used for material removal, consequently improving the total energy efficiency of such a laser-based process.

Due to the possibility of generating reactive species, plasmas are further suitable for surface etching [18]. For instance, silicon-based materials such as silicon dioxide (SiO_2) can be processed using fluorochemical compounds as process gas. According to

$$SiO_2 + 4F \rightarrow SiF_4 + O_2, \tag{6}$$

etching is achieved by chemical reactions of both fluorine (F) provided by the plasma and silicon dioxide. As a result, gaseous silicon tetrafluoride (SiF_4) and gaseous dioxygen (O_2) are formed. Such plasma etching is also referred to as reactive atomic plasma technology (RAPT®).

3. Plasma-enhanced laser processes

3.1. Simultaneous laser–plasma processing

For simultaneous laser–plasma processing, suitable plasma and laser irradiation are applied to the work piece at the same time in order to generate synergetic effects between both sources of energy. Since plasma-induced chemical reactions require a certain length of time that is typically much longer than the temporal distance between the single pulses of pulsed laser irradiation (e.g., 100 µs for a laser pulse repetition rate of 1 kHz), this approach is rather based on the utilization of plasma-physical effects as also listed in the inlet in Figure 1. Here, the plasma discharge type plays a key role since some of these effects are only existent in either direct discharges or remote plasmas. In addition, plasma-chemical effects such as a constant chemical reduction or oxidation can contribute to the process as for example, in the case of laser–plasma cleaning.

3.1.1. Laser–plasma cleaning

Plasma-enhanced laser processing is suitable for cleaning contaminated and oxidized or corroded surfaces and can also be applied for the removal of coatings or lacquers. In the first case, the actual laser cleaning process, based on desorption or ablation of surface-adherent contaminants, is supported by a modification of such contaminants by suitable plasmas. Depending on the chemical composition of the contaminants, either chemically reducing or oxidizing plasmas are applied. As an example, the reduction of surface-adherent oxides is achieved by the use of nitrogenous or hydrogenous process gases, whereas oxygen plays a key role for the removal of carbon-based layers (compare Eq. 5). Employing both laser and plasma cleaning at the same time allows a significant increase in cleaning efficiency as exemplified in Figure 4.

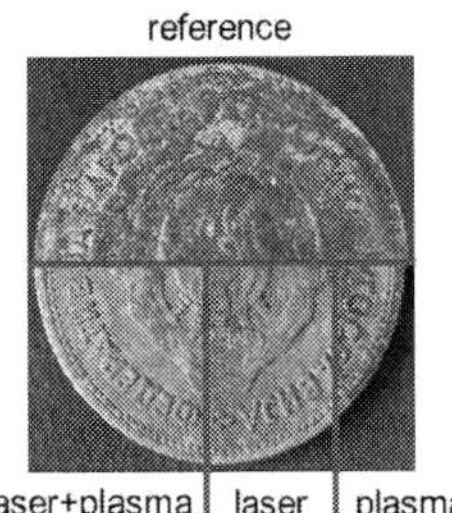

Figure 4. Photograph of a corroded coin (top) including plasma-cleaned (bottom right), laser-cleaned (bottom middle), and plasma-assisted laser-cleaned (bottom left) area

Here, laser–plasma cleaning of corroded coins was performed by introducing a plasma jet to Nd:YAG laser irradiation with a wavelength of 1064 nm. The plasma process gas was a mixture of argon (95%, carrier gas) and hydrogen (5%), leading to the formation of hydrogen radicals within the plasma [19,20]. Even though a pure plasma treatment features only a marginal cleaning efficiency, the combination of both laser-induced cleaning and plasma-induced reduction is a powerful tool for this task. In addition to an improvement of the cleaning efficiency, this hybrid approach also allows to reduce the laser fluence required for the cleaning process. In the present case shown in Figure 4, the laser fluence for cleaning was 0.6 J/cm² without any plasma and 0.4 J/cm² when applying the plasma simultaneously [19]. Thus, the required laser fluence is reduced by a factor of 1.5, which notably contributes to the overall energy efficiency of such hybrid cleaning processes.

As shown by Eq. 5 and illustrated in more detail in Figure 3, both plasma-chemical and plasma-physical processes can furthermore contribute to the laser removal of carbon-based coating materials such as parylene [21] by the excitation of carbon bonds or the oxidation of hydrocarbons. For this purpose, air can be used as process gas. Based on these effects, laser–plasma cleaning is thus a suitable technique for an efficient removal of graffiti sprays, spray lacquers, acryl paints, and acrylic and alkyd resins [22]. Since laser-based processes can be performed localized with a high accuracy, laser–plasma cleaning is thus of interest for the selective removal of protecting layers, e.g., for electrical contacting, the large-scale removal of bleached and aged lacquers, the restoration of artworks, etc.

3.1.2. Laser–plasma annealing

For the production of thin film transistor (TFT)-based flat screen displays, excimer laser annealing (ELA) represents an established method. Here, large-scale crystallization of amorphous silicon layers on glass substrates is performed by specifically shaped, i.e., laterally homogenously distributed laser beams [23]. The efficiency of this technique is currently limited by the laser energy of available sources. In order to overcome this restriction and to machine a large surface area, several excimer laser sources are coupled in order to achieve energy densities sufficient for large-scale annealing. In this context, adding appropriate plasma sources to the ELA process allows an increase in the process efficiency. Such increase can be achieved by simple plasma-induced heating of the silicon layer, which directly impacts the temperature-dependent bad gap $E_g(T)$ as expressed by the Varshni equation [24], given by

$$E_g\left(T\right) = E_g\left(T = 0K\right) - \frac{\alpha T_{max}^2}{T_{max} + \beta}.$$

(7)

Here, α, β, and E_g $(T = 0K)$ are material constants.

Another approach is to add cold, physically active plasmas to an ELA process. For example, direct DBD argon plasmas turned out to be suitable for increasing the crystallization efficiency by a factor of maximum 1.9 [25] as exemplified in Figure 5.

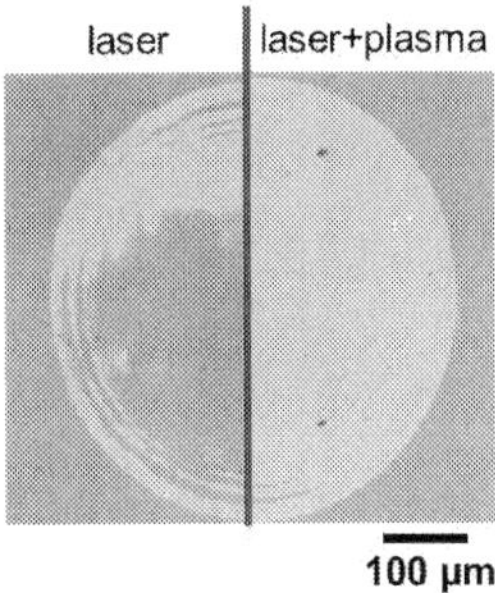

Figure 5. Microscopy image of silicon layer after pure excimer laser annealing (left) and plasma-assisted excimer laser annealing (right).

Even though the underlying mechanisms of this approach are not yet fully understood, the increase in crystallization efficiency is most likely due to a plasma-induced additional energy transfer by inelastic collisions of excited argon atoms and metastable argon species at the silicon layer surface as described in more detail in Section 3.1.4. Further, the formation of dipoles within the amorphous silicon layer due to the used direct discharge and the accompanying change in optical properties can contribute to an improved coupling of incoming laser energy into the layer. Even though laser–plasma annealing allows a significant enhancement of an ELA process in terms of its efficiency, it is subject to the restriction that the silicon-coated substrate necessarily has to be brought into a direct DBD plasma. The application of this method on an industrial scale, where typically laser line foci with a width of 1 to 1.5 m are used for ELA, is thus a challenging task to be solved in the future.

3.1.3. Laser–plasma engraving

Laser engraving is applied for the realization of microstructures for micromechanical devices and actuators or for marking print rolls. Here, the most disturbing effect is the formation of burrs and debris, i.e., resolidified material and condensed material vapor close to the ablation area. Typically, such burrs and debris droplets are of the same size as the actual laser-engraved structure and consequently limit the performance of any microstructured functional device. The use of assisting plasmas during laser engraving allows a reduction of these effects. As confirmed by X-ray photoelectron spectroscopic analysis of laser-induced debris on stainless steel, it is chemically reduced when applying appropriate plasma process gases. As a result, the debris features notably lower content of oxygen [26]. The additional (thermal) energy provided by the plasma further gives rise to a higher degree of vaporization of the laser-induced plume and thus inhibits the recondensation of material on the surface as shown by the comparison in Figure 6.

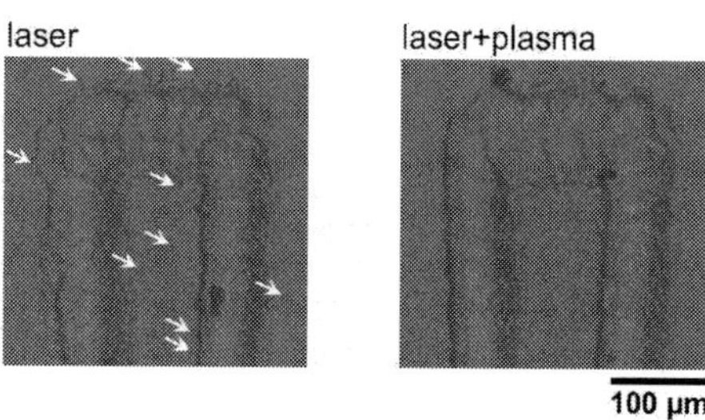

Figure 6. Scanning electron microscopy image of microstructures on stainless steel realized by pure (left) and plasma-assisted (right) laser engraving, considerable debris droplets are indicated by white arrows.

In principle, the formation of debris during laser engraving of different metals and alloys can be reduced by assisting plasmas [26]. However, this effect is strongly dependent on the particular chemical composition of the work piece and the plasma process gas mixture.

3.1.4. Laser–plasma drilling and cutting

Simultaneous plasma-enhanced laser processing can be applied for increasing the ablation rate during laser machining of various materials. As a consequence, the efficiency of laser drilling and cutting can be improved. For this purpose, a direct DBD-based argon plasma beam can be introduced to a focused laser beam, where both the plasma beam and the laser beam are guided coaxially as presented in more detail in [25,27–29]. As shown in Figure 7, this configuration allows increasing the ablation depth during laser drilling of optical glasses by a factor of approximately 2 in comparison to pure laser drilling without any assisting plasma [27].

It can also be applied to metals, where the maximum increase in volume ablation rate was even reported to be by a factor of 9.6 in the case of aluminum [28]. Principally, an increase in ablation rate was also achieved for aluminum oxide [29]. As observed by laser-induced breakdown spectroscopic measurements during both pure laser ablation and simultaneous plasma-assisted laser ablation of aluminum oxide, applying plasma additionally results in an intensification of characteristic spectral lines of aluminum ions (AlII) as shown in Figure 8.

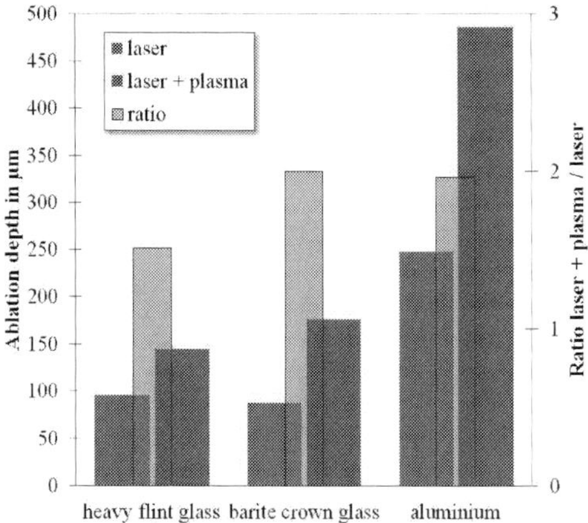

Figure 7. Comparison of the particular laser ablation depths of heavy flint glass, barite crown glass and aluminum obtained by 40 laser pulses at a fluence of 404.6 J/cm² without (blue columns) and with (red columns) assisting plasma including particular ratios (gray columns).

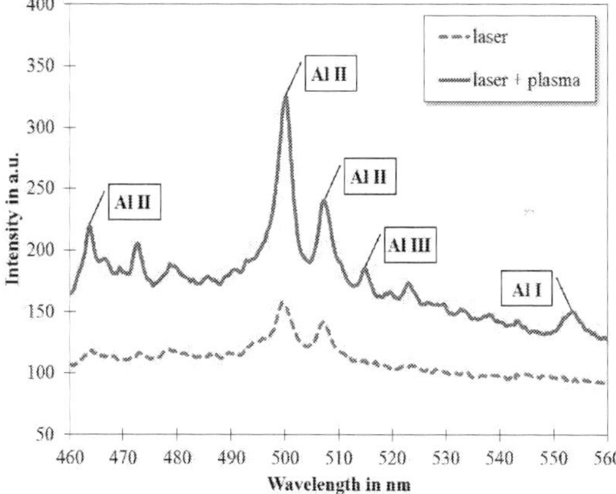

Figure 8. Optical emission spectra taken during pure laser ablation (blue dashed line) and plasma-assisted laser ablation (red solid line) of aluminum oxide ceramic.

Further, more spectral lines of neutral aluminum (AlI) and aluminum ions (AlII and AlIII) appear. This confirms the increase in material removal by a plasma-induced high-energetic process with respect to pure laser machining. However, this increase cannot be explained by a simple addition of the particular energies of the laser and the plasma. It is rather a synergetic effect, which can be described as follows: In the course of the laser evaporation process, the particles within the arising laser plume feature a considerably large surface for collision processes. Such collision results in a de-excitation of excited plasma species [30] and thus comes along with the disposal of energy, which amounts to 11.55 eV for the $^3P_2$ argon metastable (Ar^m) state and 11.72 eV for the $^3P_0$ Ar^m state, respectively. The formation of an ablation plume by the laser is thus of significant importance for energetic laser–plasma synergies that arise from this mechanism.

Since in the above-described setup, the work piece is placed within the discharge gap, the coupling of incoming laser energy can furthermore be supported by local-transient substance changes as, for example, a formation of dipoles and the accompanying modification of near-surface optical properties. Another — although marginal — effect during simultaneous plasma-enhanced laser machining is the influence of the plasma beam on the propagation characteristics of the laser beam. Here, the plasma acts as thermal gradient index lens, thus resulting in a certain focus shift of the passing focused laser beam. However, due to the relatively low gas temperature of DBD plasmas, this focus shift of 2 to 4 mm [31] is usually much lower than the Rayleigh length z_R of the focused laser beam which is given by

$$z_R = \frac{\pi w_0^2}{\lambda}. \tag{8}$$

For instance, z_R amounts to approximately 31.5 mm for a focused fundamental Nd:YAG laser beam with a beam waist radius w_0 of 100 μm, which is approximately 8-fold the focus shift induced by the plasma beam. This effect thus turns out to have a marginal influence during plasma-assisted laser drilling but may become of importance in the case of surface micro-structuring applications.

3.2. Sequential laser–plasma processing

Sequential laser–plasma processing represents a powerful tool for enhancing the laser machining quality and efficiency of transparent media such as glasses. This method is a combination of plasma pretreatment and subsequent laser ablation. Using hydrogenous process gases for plasma pretreatment, the energy coupling of incoming laser irradiation can be significantly improved due to a modification of the optical properties of the work piece. First, the process gas is dissociated as a result of collisions with electrons within the plasma volume. The atomic hydrogen generated in this vein can then induce several reactions at the surface of the work piece. In the case of glasses, the main mechanisms are the chemical reduction of glass network-forming oxides such as silicon dioxide (SiO$_2$) according to

$$\mathrm{SiO_2(s) + 2H(g) \rightarrow SiO(s) + H_2O(g)}, \tag{9}$$

resulting in the formation of a substoichiometric mixed phase at the surface, and the implantation of hydrogen into deeper regions of the glass bulk material. The substoichiometric mixed phase can be referred to as silicon suboxide (SiO$_x$, where $1 < x < 2$). Due to the comparatively low penetration depth of plasma-chemical reactions, the thickness of this reduced layer amounts to approximately 100 nm [32]. Silicon suboxide is dominated by oxygen deficiencies and E′ centers, which represent optically active defects and give rise to an increase in absorption in the wavelength range from 177 to 218 nm [33]. Moreover, the hydrogen implanted by the plasma induces a hydrolytic scission of the glass network, resulting in the formation of non-bridging oxygen (NBO) and H(I)-centers up to a penetration depth of several microns. These defects generate an increased absorption between 206 and 258 nm [33]. As shown in Figure 9, both effects (and further interacting phenomena such as a certain surface wrinkling [34]) result in a considerable increase in UV-absorption, depending on the plasma treatment duration.

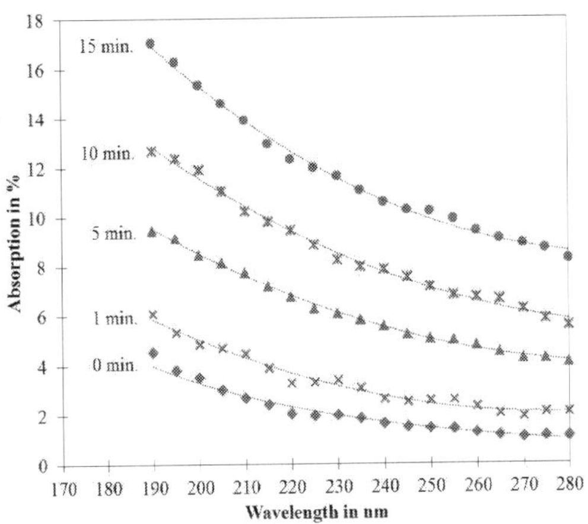

Figure 9. Absorption spectra of 5 mm thick fused silica before (0 minutes) and after plasma treatment for 1, 5, 10, and 15 minutes, respectively.

One has to notice that this increase in absorption is a reversible process. The initial transmission characteristics can nearly be reobtained by tempering the plasma-treated glass and thus removing the implanted hydrogen and reoxidizing the glass network [35].

In terms of laser ablation subsequent to the plasma pretreatment, the increase in absorption allows a notable reduction of the laser ablation threshold up to a factor of 4.6 when using an argon-fluoride excimer laser with an emission wavelength of 193 nm [36]. Due to the accompanying decrease in energy deposition into the glass, a higher contour accuracy and lower roughness of the ablated area is achieved as exemplified in Figure 10.

Thus, the sequential laser–plasma processing of glasses allows both a saving of laser energy and an improvement of the machining quality at the same time [37]. The implanted hydrogen further acts as reaction partner for oxygen within the SiO_2 network during multiple pulse laser ablation. This results in the formation of a new substoichiometric layer at the bottom of the ablated area as confirmed by secondary ion mass spectroscopic investigations. As a result of

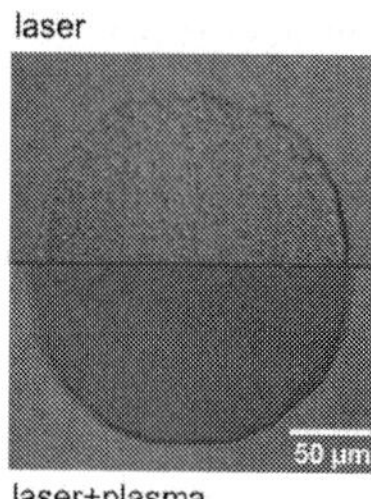

Figure 10. Microscopy image of single-pulse laser ablated spots on fused silica without (top) and with (bottom) plasma pretreatment, each ablated at the particular ablation threshold of 6 J/cm² (top) and 1.3 J/cm² (bottom), respectively.

this laser–plasma-induced effect, the absorption for each following laser pulse and consequently the depth ablation rate is constantly increased with respect to pure laser ablation as shown in Figure 11 [38].

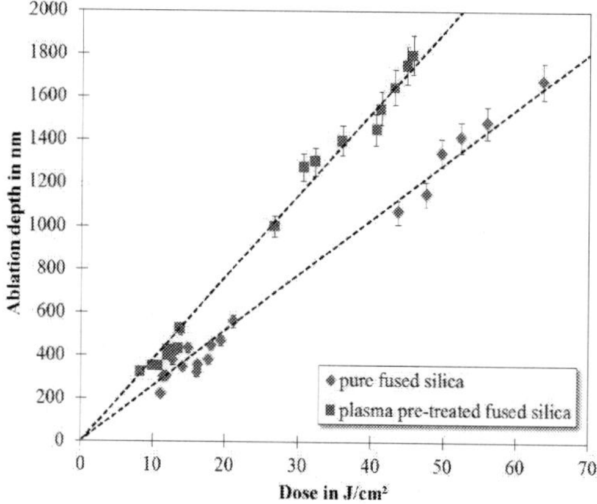

Figure 11. Dose-dependent ablation depth of fused silica ablated by excimer laser irradiation without (blue rhombi) and with plasma pretreatment (red squares).

In addition to a decrease in ablation threshold, an increase in ablation rate and an improvement of the contour accuracy the above-described plasma treatment allows the use of energy-efficient laser sources. Due to the plasma-induced increase in UV absorption, 4th-harmonic Nd:YAG lasers instead of excimer lasers can be applied for machining fused silica [39]. Even though the highest efficiency of the plasma pretreatment is found in the case of fused silica as substrate material, this approach can also be successfully applied to optical glasses [40] or technical glasses such as photovoltaic cover glasses [41]. However, due to the significantly denser network of multi-component glasses, the implantation of hydrogen and the formation of the above-described related mechanisms that come along with the presence of hydrogen are limited.

The above-described plasma pretreatment strongly influences the surface polarizability k_s and the molecule polarizability α_m of any substrate. According to the Clausius–Mosotti equation, given by

$$\frac{n^2-1}{n^2+2} = \frac{N_A \rho \alpha_m}{3\epsilon_0 M},\tag{10}$$

where N_A is the Avogadro constant, ρ is the density, ε_0 is the vacuum permittivity, and M is the molar mass, α_m is directly related to the refractive index n [42]. Since the polarizability is increased and taking Snell's law into account, this leads to a refocusing of incoming focused laser irradiation and a more localized input of laser energy, respectively, coming along with an increased fluence close to the substrate surface. This effect further contributes to the above-described mechanisms in the case of sequential laser–plasma processing.

4. Summary

The combination of both atmospheric pressure plasmas and laser irradiation allows increasing the total process efficiency and machining quality of a number of laser-based processes such as cleaning, modification and material removal. For this purpose, either plasma-chemical or plasma-physical effects or mechanisms can be employed. Chemical plasma effects involve the reduction, oxidation, or excitation of bonds, which contribute to laser-induced removal of pollutants or layers. Plasma-physical effects such as collision-induced de-excitation of metastables and high electric field strengths at work piece surfaces can be used in order to improve the modification of coatings and to increase the laser ablation rate. Due to the multiplicity of laser and plasma parameters, the partial complexity of the underlying mechanisms and interactions and the wide range of different materials, plasma-enhanced laser materials processing is still at an early stage. However, the suitability of this approach for a number of different applications was shown over the last years as presented in this chapter. In order to implement this technique in an industrial scale where focused laser beams are usually scanned for large-scale treatment, appropriate and adapted plasma sources should be developed. Due to the principle of DBD plasma sources, such upscaling can be realized quite easily. To summarize, it can be stated that the coupling of laser irradiation and atmospheric pressure plasmas facilitates the valorization of a variety of laser processes based on cost- and energy-efficient nanosecond laser sources.

5. Nomenclature

A: area

D: laser dose

d_p: optical penetration depth

d_th: thermal penetration depth

E: laser pulse energy

$E_\mathrm{g}(T)$: temperature-dependent bad gap

f: high-voltage pulse frequency

f_rep: laser pulse repetition rate

K: thermal conductivity

$k(\lambda)$: wavelength-dependent extinction coefficient

k_s: surface polarizability

M: molar mass

N_A: Avogadro constant

$n(\lambda)$: wavelength-dependent refractive index

P_av: average laser power

U: high-voltage

w_0: beam waist radius

z_R: Rayleigh length

α_m: molecule polarizability

$\alpha(\lambda)$: wavelength-dependent absorption coefficient

ε_0: vacuum permittivity

λ: wavelength

ϱ: density

τ: laser pulse duration

Φ: laser fluence

Acknowledgements

The authors would like to thank all their coworkers who contributed to the findings and results presented in this chapter, namely, S. Brückner, M. Dammann, A. Gredner, J. Heine, J. Hoffmeister, J. Ihlemann, M. Kretschmer, V. Le Meur, A. Luca, N. Mainusch, F. Peters, S. Rösner, S. Roux, A. Schmiedel, C. Schmiedel, K. Schmidt, D. Tasche, and T. Weihs. The works leading to the presented results were supported by the European Regional Development Funds and the Workgroup Innovative Projects of Lower Saxony in the frame of the "Lower Saxony Innovation Network for Plasma Technology" and the research project "Plasma Cleaning of Wall Painting and Architecture Surfaces." The authors further thank the German Federal Ministry of Economics and Technology for funding in the frame of the research project "PROKLAMO" as well as the German Federal Ministry of Education and Research for the sponsorship in the frame of the research project "SiPlaH." The support by the Volkswagen Foundation and the Federal State Lower Saxony by the sponsorship of the research professorship for laser–plasma hybrid technology at the University of Applied Sciences and Arts in the frame of the state funding program Niedersächsisches Vorab is additionally gratefully acknowledged.

Author details

Christoph Gerhard[1*], Wolfgang Viöl[1,2] and Stephan Wieneke[1,2]

*Address all correspondence to: christoph.gerhard@ist.fraunhofer.de

1 Application Center for Plasma and Photonics, Fraunhofer Institute for Surface Engineering and Thin Films, Göttingen, Germany

2 Laboratory of Laser and Plasma Technologies, University of Applied Sciences and Arts, Göttingen, Germany

References

[1] Marozau I., Shkabko A., Dinescu G., Döbeli M., Lippert T. Logvinovich D. et al.. RF-plasma assisted pulsed laser deposition of nitrogen-doped SrTiO3 thin films. Applied Physics A. 2008;93:721–727. DOI: 10.1007/s00339-008-4702-0

[2] Pak S. W., Suh J., Lee D. U., Kim E. K.. Growth of ZnTe:O thin films by oxygen–plasma-assisted pulsed laser deposition. Japanese Journal of Applied Physics. 2011;51(1s): 01AD04. DOI: 10.1143/JJAP.51.01AD04

[3] Kajita S., Ohno N., Takamura S., Sakaguchi W., Nishijima D.. Plasma-assisted laser ablation of tungsten: reduction in ablation power threshold due to bursting of holes/ bubbles. Applied Physics Letters. 2007;91:261501. DOI: 10.1063/1.2824873

[4] Arata Y., Abe N., Oda T.. Fundamental phenomena in high power CO_2 laser welding. Transactions of JWRI. 1985;14(1):5–11.

[5] Bagger C., Olsen F. O.. Review of laser hybrid welding. Journal of Laser Applications. 2005;17(1):2–14. DOI: 10.2351/1.1848532

[6]   Mahrle A., Schnick M., Rose S., Demuth C., Beyer E., Füssel U.. Process characteristics of fibre-laser-assisted plasma arc welding. Journal of Physics D. 2011;44:345502. DOI: 10.1088/0022-3727/44/34/345502

[7]   Mahrle A., Rose S., Schnick M., Beyer E., Füssel U.. Laser-assisted plasma arc welding of stainless steel. Journal of Laser Applications. 2013;25(3):032006. DOI: 10.2351/1.4798338

[8]   Zhang J., Sugioka K., Midorikawa K.. Laser-induced plasma-assisted ablation of fused quartz using the fourth harmonic of a Nd+:YAG laser. Applied Physics A. 1998;67(5):545–549. DOI: 10.1007/s003390050819

[9]   Hanada Y., Sugioka K., Obata K., Garnov S. V., Miyamoto I, Midorikawa K.. Transient electron excitation in laser-induced plasma-assisted ablation of transparent materials. Journal of Applied Physics. 2006;99:043301. DOI: 10.1063/1.2171769

[10]   Norton J. F., McMullen J. G.. Laser-formed apertures for electron beam instruments. Journal of Applied Physics. 1963;34:3640–3641. DOI: 10.1063/1.1729284

[11]   Matthias E., Reichling M., Siegel J., Käding O. W., Petzoldt S., Skurk H. et al.. The influence of thermal diffusion on laser ablation of metal films. Applied Physics A. 1994;58(2):129–136. DOI: 10.1007/BF00332169

[12]   Siegel J., Matthias E., Ettrich K., Welsch E.. UV-laser ablation of ductile and brittle metal films. Applied Physics A. 1997;64(2):213–218. DOI: 10.1007/s003390050468

[13]   Bellmann M., Gerhard C., Haese C., Wieneke S., Viöl, W.. DBD plasma improved spot repair of automotive polymer surfaces. Surface Engineering. 2012;28(10):754–758. DOI: 10.1179/1743294412Y.0000000059

[14]   Heine J., Damm R., Gerhard C., Wieneke S., Viöl W.. Surface activation of plane and curved automotive polymer surfaces by using a fittable multi-pin DBD plasma source. Plasma Science and Technology. 2014;16(6):593–597. DOI: 10.1088/1009-0630/16/6/10

[15]   Cheruthazhekatt S., Černák M., Slavíček P., Havel J.. Gas plasmas and plasma modified materials in medicine. Journal of Applied Biomedicine. 2010;8(2):55–66. DOI: 10.2478/v10136-009-0013-9

[16]   Shun'ko E. V., Belkin V. S.. Cleaning properties of atomic oxygen excited to metastable state 2s22p4(1S0). Journal of Applied Physics. 2007;102(8):083304. DOI: 10.1063/1.2794857

[17]   Hansen R. W. C., Bissen M., Wallace D., Wolske J., Miller T.. Ultraviolet/ozone cleaning of carbon-contaminated optics. Applied Optics. 1993;32(22):4114–4116. DOI: 10.1364/AO.32.004114

[18] Jeong J. Y., Babayan S. E., Tu V. J., Park J., Henins I., Hicks R. F. et al.. Etching materials with an atmospheric-pressure plasma jet. Plasma Sources Science and Technology. 1998;7(3):282–285. DOI: 10.1088/0963-0252/7/3/005

[19] Pflugfelder C., Mainusch N., Ihlemann J., Viöl W.. Removal of unwanted material from surfaces of artistic value by means of Nd:YAG laser in combination with cold atmospheric-pressure plasma. In: Castillejo M., Moreno P., Oujja M., Radvan R., Ruiz J., editors. Lasers in the Conservation of Artworks; 17–21 September 2007; Madrid, Spain. Boca Raton: CRC Press Taylor Francis Group; 2008. p. 55–58.

[20] Pflugfelder C., Mainusch N., Hammer I., Viöl W.. Cleaning of wall paintings and architectural surfaces by plasma. Plasma Processes and Polymers. 2007;4(S1):S516-S521. DOI: 10.1002/ppap.200731218

[21] Schmiedel C., Schmiedel A., Viöl W.. Combined plasma laser removal of parylene coatings. In: Proceedings of the 19th International Symposium on Plasma Chemistry (IPCS19); 27–31 July 2009; Bochum, Germany. International Plasma Chemistry Society; 2009. p. 239–243.

[22] Mainusch N., Pflugfelder C., Ihlemann J., Viöl W.. Plasma jet coupled with Nd:YAG laser: new approach to surface cleaning. Plasma Processes and Polymers. 2007;4(S1):S33-S38. DOI: 10.1002/ppap.200730302

[23] Sameshima T., Usui S., Sekiya M.. XeCl excimer laser annealing used in the fabrication of poly-Si TFTs. IEEE Electron Device Letters. 1986;7(5):276–278. DOI: 10.1109/EDL.1986.26372

[24] Varshni Y.P.. Temperature dependence of the energy gap in semiconductors. Physica. 1967;34:149–154. DOI: 10.1016/0031-8914(67)90062-6

[25] Gredner A., Gerhard C., Wieneke S., Schmidt K., Viöl W.. Increase in generation of poly-crystalline silicon by atmospheric pressure plasma-assisted excimer laser annealing. Journal of Materials Science and Engineering B. 2013;3(6):346–351.

[26] Le Meur V., Loewenthal L., Gerhard C., Viöl W.. On the debris formation during atmospheric pressure plasma-assisted laser engraving of stainless steel. In: Gerhard C., Wieneke S., Viöl W., editors. Laser Ablation: Fundamentals, Methods and Applications. 1st ed. New York: Nova Science Publishers; 2015. p. 165–177.

[27] Gerhard C., Roux S., Brückner S., Wieneke S., Viöl W.. Low-temperature atmospheric pressure argon plasma treatment and hybrid laser–plasma ablation of barite crown and heavy flint glass. Applied Optics. 2012;51(17):3847–3852. DOI: 10.1364/AO.51.003847

[28] Gerhard C., Roux S., Brückner S., Wieneke S., Viöl W.. Atmospheric pressure argon plasma-assisted enhancement of laser ablation of aluminum. Applied Physics A. 2012;108(1):107–112. DOI: 10.1007/s00339-012-6942-2

[29] Gerhard C., Roux S., Peters F., Brückner S., Wieneke S., Viöl W.. Hybrid laser ablation of Al2O3 applying simultaneous argon plasma treatment at atmospheric pres-

sure. Journal of Ceramic Science and Technology. 2013;4(1):19–24. DOI: 10.4416/JCST2012-00034a

[30] Bogaerts A., Gijbels R.. Comparison of argon and neon as discharge gases in a direct-current glow discharge a mathematical simulation. Spectrochimica Acta Part B. 1997;52(5):553–565. DOI: 10.1016/S0584-8547(96)01658-8

[31] Hoffmeister J., Brückner S., Gerhard C., Wieneke S., Viöl W.. Impact of the thermal lens effect in atmospheric pressure DBD-plasma columns on coaxially guided laser beams. Plasma Sources Science and Technology. 14;23:064008. DOI: 10.1088/0963-0252/23/6/064008

[32] Gerhard C., Tasche D., Brückner S., Wieneke S., Viöl W.. Near-surface modification of optical properties of fused silica by low-temperature hydrogenous atmospheric pressure plasma. Optics Letters. 2012;37(4):566–568. DOI: 10.1364/OL.37.000566

[33] Skuja L.. Optically active oxygen-deficiency-related centers in amorphous silicon dioxide. Journal of Non-Crystalline Solids. 1998;239(1–3):16–48. DOI: 10.1016/S0022-3093(98)00720-0

[34] Gerhard C., Weihs T., Tasche D., Brückner S., Wieneke S. Viöl W.. Atmospheric pressure plasma treatment of fused silica, related surface and near-surface effects and applications. Plasma Chemistry and Plasma Processing. 2013;33(5):895–905. DOI: 10.1007/s11090-013-9471-7

[35] Hoffmeister J., Gerhard C., Brückner S., Ihlemann J., Wieneke S., Viöl W.. Laser micro-structuring of fused silica subsequent to plasma-induced silicon suboxide generation and hydrogen implantation. In: Schmidt M., Vollertsen F., Geiger M., editors. Laser Assisted Net Shape Engineering 7 (LANE 2012); 12–15 November 2012; Fürth, Germany. Elsevier; 2012. p. 613–620. DOI: 10.1016/j.phpro.2012.10.080

[36] Brückner S., Hoffmeister J., Ihlemann J., Gerhard C., Wieneke S., Viöl W.. Hybrid laser–plasma micro-structuring of fused silica based on surface reduction by a low-temperature atmospheric pressure plasma. Journal of Laser Micro/Nanoengineering. 2012;7(1):73–76. DOI: 10.2961/jlmn.2012.01.0014

[37] Gerhard C., Brückner S., Wieneke S., Viöl W.. Atmospheric pressure plasma-enhanced laser ablation of glasses. In: Gerhard C., Wieneke S., Viöl W., editors. Laser Ablation: Fundamentals, Methods and Applications. 1st ed. New York: Nova Science Publisher; 2015. p. 151–163.

[38] Tasche D., Gerhard C., Ihlemann J., Wieneke S., Viöl W.. The impact of O/Si ratio and hydrogen content on ArF excimer laser ablation of fused silica. Journal of the European Optical Society—Rapid Publications. 2014;9:14026. DOI: 10.2971/jeos.2014.14026

[39] Gerhard C., Kretschmer M., Viöl W.. Plasma meets glass—Plasma-based modification and ablation of optical glasses. Optik & Photonik. 2012;7(4):35–38. DOI: 10.1002/opph.201290098

[40] Gerhard C., Heine J., Brückner S., Wieneke S., Viöl W.. A hybrid laser–plasma abla‐
tion method for improved nanosecond laser machining of heavy flint glass. Lasers in
Engineering. 2013;24(5–6):391–403.

[41] Gerhard C., Dammann M., Wieneke S., Viöl W.. Sequential atmospheric pressure
plasma-assisted laser ablation of photovoltaic cover glass for improved contour accu‐
racy. Micromachines. 2014;5(3):408–419. DOI: 10.3390/mi5030408

[42] Bach H., Krause D.. Analysis of the composition and structure of glass and glass ce‐
ramics. 1st ed. Berlin Heidelberg: Springer Verlag; 1999. 528 p. DOI:
10.1007/978-3-662-03746-1

Chapter 2

Cold Plasma Produced Catalytic Materials

Jacek Tyczkowski

Additional information is available at the end of the chapter

http://dx.doi.org/10.5772/61832

Abstract

The cold plasma techniques are widely applied to create new materials possessing unique properties, which cannot be prepared by any other methods. Among the many interesting substances produced with the participation of cold plasma, a special place is occupied by materials with catalytic properties. The chapter gives a brief review of various cold plasma methods used for the preparation of catalytic materials – from the plasma modification of conventional catalysts via plasma-enhanced classical synthesis of catalysts to the advanced thin catalytic films fabricated by plasma sputtering processes but primarily by plasma deposition from metalorganic precursors (PECVD). Recently, the catalytic films have attracted considerable attention due to the possibility of depositing them as very thin coatings on virtually all supports without any change in their geometry. Such coatings open the way for new reactor designs, so-called structured reactors, designated for various chemical processes. They can also be used as catalytic deposit on the surface of electrodes for fuel cells and photoelectrodes for water splitting processes. Recent developments in this field and further prospects for thin catalytic films are discussed, all the more so because it is one of the main areas of research in our department.

Keywords: Catalysts, plasma treatment, PECVD, structured reactors, fuel cells, water splitting

1. Introduction

For a long time, plasma techniques have been used in the creation of new materials possessing unique properties, which cannot be fabricated by other methods. A particularly important technique, giving plenty of possibilities, is the plasma deposition of entirely new, advanced materials in the form of solid coatings having unusual molecular structure and sophisticated nanomorphology. The plasma processes can also be used to modify conventional materials through treating them during or after "classical" synthesis, which generally leads to new

substances with disparate properties, often more suitable than those of the unmodified materials.

Among the many interesting products fabricated by plasma techniques, materials with catalytic properties occupy a special place. There exist a lot of catalytic reactions, for example, combustion of organic volatile compounds, CO_2 methanation, Fischer–Tropsch synthesis, water photo-splitting, fuel cell electrode processes, and many others, which challenge chemists all over the world to seek better catalysts or more effective catalyst preparation methods. It seems that the plasma techniques pave the way for novel solutions in this field.

In general, plasma techniques used in the preparation of catalysts can be categorized into one of two groups depending on the plasma type: thermal (equilibrium) plasma and cold (none-quilibrium) plasma [1−3]. The thermal plasma is employed in the preparation process mainly through plasma spraying of catalytically active compounds [4−6] as well as washcoats [7] on various carriers and usually in the form of thicker coatings (>> 1 μm). However, it can also be utilized for the synthesis of ultrafine catalysts whose particle diameters are in the range from a few to a few tens of nanometers and specific area is highly developed [2, 8]. Lately, graphene nanoflakes for catalytic applications have been produced by thermal plasma [9, 10]. Also, this plasma technique is used to recover and regenerate the spent conventional catalysts [11, 12]. Very recently, a more sophisticated method of the catalyst preparation employing thermal plasma has been discovered. For example, very small Ni particles (<100 nm) attached to MgO nano-rods of 10–20 nm in diameter were fabricated by an RF thermal plasma flame, into which a solid precursor consisting of Ni and MgO powders (~5 μm and ~100 nm, respectively) was injected [13]. In turn, Nehe et al. [14] proposed a new promising technique of depositing nanostructured films, namely the solution precursor plasma spray (SPPS) technique. This method was employed to the deposition of CuO–ZnO–Al_2O_3 layers that were later successfully used to catalyze the methanol reforming for the production of hydrogen (H_2) gas.

On the other hand, the cold plasma, in which all processes proceed at much lower temperatures (up to a few hundred degrees centigrade) than those in the thermal plasma (much higher than 1273 K), is a very promising tool for the preparation of catalytic materials [1-3, 15]. Recently, these materials have attracted considerable attention due to the possibility of depositing them as very thin coatings (<< 1 μm) on virtually all supports without any change in their geometry. Such coatings open the way for new reactor designs, so-called structured reactors, for various chemical catalytic processes [16]. They can also be used as catalytic deposits on the surface of fuel cell electrodes without practically any changes in their electrical conductivity and gas permeability. Further advancement is also expected with regard to new sophisticated thin films for photocatalytic splitting of water molecules and efficient production of hydrogen [17]. Apart from the thin catalytic coating deposition, the cold plasma is also utilized in preparations of conventional catalysts, through both plasma enhanced "classical" synthesis and plasma modification of as-prepared "conventional" catalysts [2, 15].

The cold plasma preparation of catalytic films, carried out by plasma polymerization processes from organometallic complexes as precursors, and study of their properties are the main research fields in our laboratory. However, for the sake of completeness, cold plasma involvements in the preparation of "conventional" catalysts as well as sputtering processes are also discussed in this chapter.

2. Cold plasma for "conventional" catalysts

There are two ways of application of plasma to the conventional production of catalysts: plasma treatment during formation of catalyst and plasma modification of catalysts that have been already conventionally formed for commercial application. The first way can be used to replace the thermal calcination or reduction of a catalyst, to modify its structure, for example, its dispersion and particle sizes, as well as to immobilize catalytic species on supports. Due to chemical treatment with active plasma, catalysts prepared in this way are usually very different from those synthesized only conventionally. Evident changes in the properties of a conventional catalyst can also be achieved by plasma treatment of its final form. Very often, the activity and selectivity of such a catalyst can be significantly improved this way [3, 15 and references therein].

2.1. Plasma-enhanced preparation of "conventional" catalysts

The results published by Chen et al. in 2004 can serve as the classic example of a comparison between a conventional catalyst and such a catalyst prepared by applying the plasma technique [18]. The catalyst was composed of Pd and α-Al$_2$O$_3$ and was tested in the selective hydrogenation of acetylene to ethylene. Both procedures of preparation are shown in Figure 1. In the plasma procedure, conventional thermal calcination and reduction processes are replaced with plasma treatment. Apart from the fact that plasma processing is much quicker, cleaner, and easier to control, it was found that the catalyst prepared by the plasma procedure exhibited a significant increase in activity and selectivity, compared with the conventional samples, especially at the lower reaction temperature. Typical results are given in Figure 2, where a drastic difference between the plasma and the conventionally fabricated Pd/α-Al$_2$O$_3$ catalysts is evident.

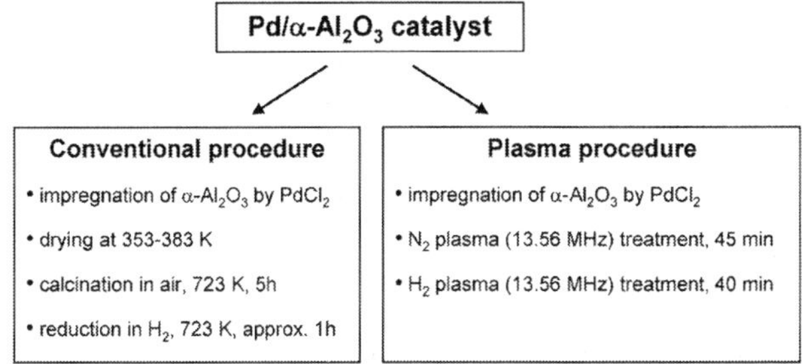

Figure 1. Two ways of preparing the Pd/α-Al$_2$O$_3$ catalyst (on the basis of Ref. [18]).

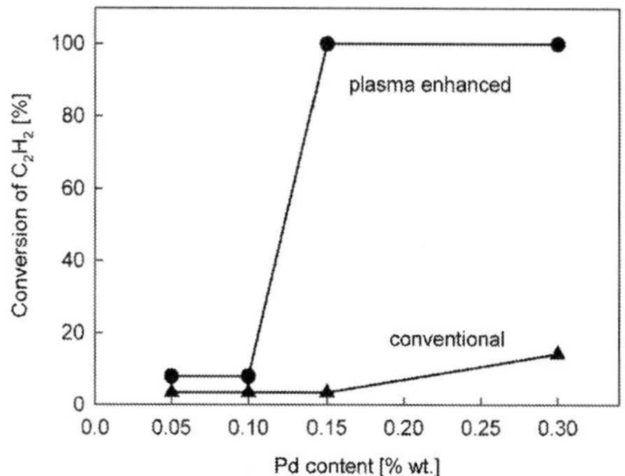

Figure 2. Conversion of C_2H_2 to C_2H_4 versus Pd content for the conventional and plasma-enhanced Pd/α-Al$_2$O$_3$ catalysts (on the basis of Ref. [18]).

The investigations simultaneously performed by Liu et al. [19] confirmed that plasma treatment during the fabrication of conventional catalysts is critical to the formation of their structure. They have shown that Ar plasma treatment between the preparation steps of drying after impregnation (PdCl$_2$ solution) and the thermal calcination evidently modifies the structure of palladium catalysts supported by a zeolite (Pd/HZSM-5). An enhanced dispersion of PdO and increase in Brönsted and Lewis acidities, which leads to a remarkable improvement in the catalyst stability, have been observed. It has also been found that combustion of methane (to carbon dioxide and water) over the plasma-treated catalyst is close to 100% efficient at 723 K, but it is only approximately 50% efficient at the same temperature over the catalyst not subjected to plasma treatment. In turn, the investigations performed by Zhu et al. [20] on a Mo-Fe/ HZSM-5 catalyst, tested for the non-oxidative aromatization of methane, have shown that Ar plasma treatment benefits the formation of carbonaceous species associated with the active species of MoC$_X$ and disfavors the coke formation, which in turn leads to deactivation of the catalyst. The crucial role of plasma treatment in the catalyst structure creation was also shown in 2004 by Legard et al. [21]. They used remote hydrogen microwave plasma (2.45 GHz) for the preparation of gold-based metallic catalysts. Unfortunately, gold tends to sinter easily and the conventional reduction of gold performed in hydrogen at high temperatures very often leads to large particles that are not catalytically active. The particles obtained by using plasma are less than 5 nm in size, and are stable during thermal treatment.

In the following years, we could observe a significant growth of interest in employing cold plasma in the conventional catalyst preparation at various stages of this procedure [22–34]. These methods involve plasma initial decomposition of precursors, plasma replacement of calcination and reduction processes, plasma pretreatment of supports, as well as plasma breaking of thin films to form specific nano-sized catalysts. In general, all these plasma processes, in comparison with the entirely conventional method, lead to unusual catalyst structures, built of smaller particles, better dispersion, stronger interaction between the catalyst and the support, and consequently, to much higher activity, enhanced selectivity, and better stability during a given catalyzed reaction. The anti-carbon deposition and anti-sintering performance are also improved by the plasma treatment. Various types of cold plasma have been tested for this purpose, such as low-pressure glow discharges, atmospheric cold plasma jets, dielectric-barrier discharges, and corona discharges. The plasmas are generated by microwaves (MW), radio frequency (RF, mainly 13.56 MHz), audio frequency (AF, in the range of kHz), alternating current (AC, in the range of Hz), direct current (DC), and even a hot filament, using non-polymerized gases, e.g., Ar, H_2, O_2, N_2, CO_2, NH_3.

Recently, much attention has been paid to the use of dielectric-barrier discharge (DBD) for plasma decomposition of inorganic precursors, such as carbonate mixtures of $CuCO_3$ + $ZnCO_3$ (for CuO–ZnO catalyst) [35] and $NiCO_3$ + $MgCO_3$ (for Ni/MgO catalyst) [36], as well as $Ni(NO_3)_2$ (for Ni catalyst) [37,38] and $Co(NO_3)_2$ (for Co_3O_4 catalyst). This interest stems from the fact that the DBD plasma has turned out to be more reactive than a typical glow discharge. For example, in contrast with the DBD plasma [37], the glow discharge does not lead to the full decomposition of the $Ni(NO_3)_2$ precursor. A specific hydrate is formed during the glow discharge treatment and thermal decomposition has to be carried out before further reactions [39]. The DBD plasma technique also enables simultaneous preparation, in the same process, a support and catalytically active component, as it was done by Hua et al. [36], who treated a powdered mixture of $NiCO_3$ and $MgCO_3$ with the DBD hydrogen plasma to prepare a Ni/MgO catalyst. For comparison, the Ni/MgO catalyst was also prepared conventionally (MgO powder was impregnated with an aqueous solution of $Ni(NO_3)_2$, then calcined at 973 K in air, and finally reduced at 1123 K in hydrogen). An evident difference in the Ni particle sizes between the plasma-fabricated and the conventional catalysts is clearly visible (Figure 3). It has also been found that the plasma-prepared Ni/MgO catalyst exhibits much higher activity and stability in CO_2 reforming of methane.

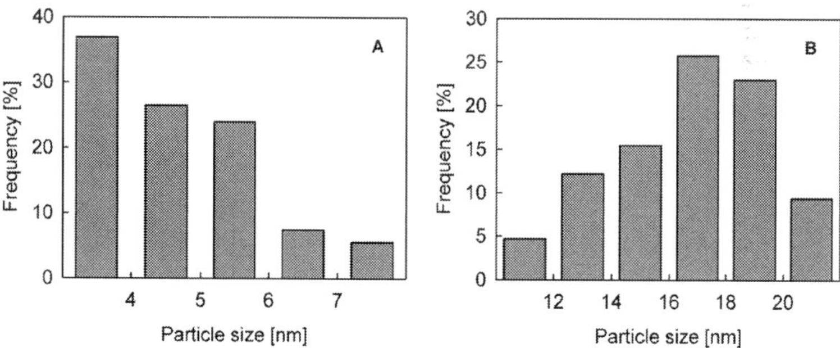

Figure 3. Ni particle size distributions in plasma-treated (A) and conventional (B) Ni/MgO catalysts [36].

It should be emphasized that plasma treatment can generate specific changes in the molecular structure formed during the preparation of a catalyst, which cannot be other-wise obtained. Molecular mechanisms of reactions running with the participation of cold plasma, constituting a complex state containing highly active species, such as ions, electrons, radicals, and excited molecules, are generally entirely different from "classic" chemical processes that occur during the conventional preparation. For example, Chen et al. [40] showed that plasma treatment (N_2 plasma, 13.56 MHz) favored significant enrichment of Co_3O_4 (cobalt spinel structure) on the surface of a cobalt catalyst prepared by the sol–gel method for combustion of methane, in comparison with only thermal calcination. It is suggested that the bombardment of highly active species on the surface during plasma treatment, inter alia, leads to the breakage of –Si–O–Co– bonds formed in the sol–gel process and creates the Co_3O_4 structure. Such a structure considerably enhances the catalytic performance. Figure 4 presents a comparison between catalysts produced without and with the plasma step. The CH_4 conversion is approximately two times higher in the case of plasma treatment.

The plasma treatment can also be used in more sophisticated processes, where the electronic structure of a material, which is crucial for its photocatalytic activity, will be modified accordingly. Recent studies performed on the so-called black TiO_2 demonstrate that an enhanced solar absorption and excellent photocatalytic activity have been achieved through the introduction of disorder surface layers on the crystalline TiO_2 nanoparticles by plasma hydrogenation. Hydrogen ion bombardment of the particle surface produces large amounts of oxygen vacancies and Ti–H bonds, thereby forming the disorder layer and, consequently, a different electronic structure of the whole nanoparticle. In the surface, tails of localized states occur, which narrows the band gap (to about 2.8 eV from 3.3 eV), but on the other hand, a smaller crystalline core causes widening of its band gap (to about 3.5 eV) [41].

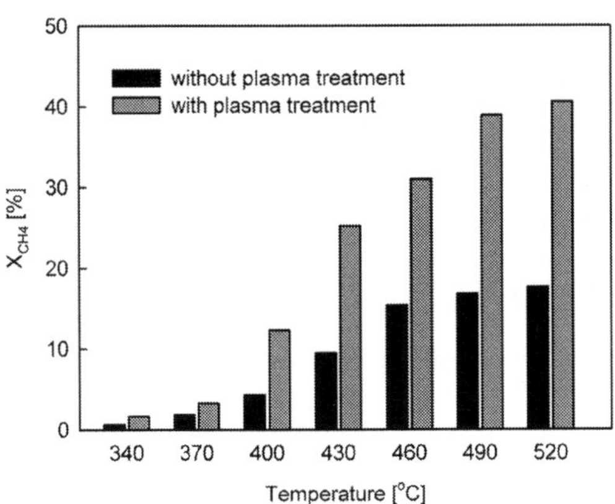

Figure 4. Influence of plasma treatment on catalytic performance [40].

In plasma treatment, a particularly important role is played by electrons that are captured by the treated surface and thus create negative potential with reference to the plasma bulk. Under these conditions, positively charged species possessing high energy bombard the surface causing a considerable transformation in its molecular structure practically at room temperature. Besides, the trapped electrons, by their mutual repulsion, elongate or distort bonds, which facilitate dispersion processes [42].

As an example of the plasma decomposition of an inorganic precursor, the process of NiO/Al_2O_3 preparation from nickel nitrate impregnated to Al_2O_3 grains can be given [43]:

$$Ni(NO_3)_2 + M^* \rightarrow NiO + 2\,NO + {}^3/_2 O_2 + M, \tag{1}$$

where M^* is an energetic species.

Another example is the plasma formation of a Pd/HZSM-5 catalyst from palladium chloride [42]:

$$PdCl_2 + M^* \rightarrow Pd^{2+} + 2\,Cl^- + M \tag{2}$$

The catalyst can be further reduced by plasma to form pure metal Pd:

$$Pd^{2+} + 2\,e \rightarrow Pd^0 \tag{3}$$

As can be seen, an electron mechanism (involving free electrons) is proposed to explain the plasma reduction. Such a mechanism gives much broader capabilities than typical chemical reduction processes, which usually require high temperatures, resulting in disadvantageous aggregation of metal particles, and are not environmentally friendly. The plasma reduction can be performed at low temperatures in various types of plasma, even in oxygen plasma [44]. However, since the electron mechanism of plasma reduction is governed by electrochemistry and the standard potential of M^{n+}/M^0 pairs determines whether the metal salt can be reduced or not, there are some cases where the plasma reduction process cannot be performed. It has been suggested that the reduction can be observed only when the standard potential is positive. Thus, $PdCl_2$ (Pd^{2+}/Pd^0, E = +0.92 V) can be reduced by plasma treatment (Eq. (3)). On the other hand, $Ni(NO_3)_2$ decomposed into metal oxide (Eq. (1)) cannot be reduced this way (Ni^{2+}/Ni^0, E = −0.25 V) [42].

The plasma treatment is also engaged in appropriate adaptation of catalytic supports through developing their surfaces and increasing the catalyst dispersion. For example, the study focused on Al_2O_3-supported Ni catalysts suggests that the catalysts with Al_2O_3 subjected to plasma treatment before impregnation are relatively easier to reduce and exhibit higher activities under mild reduction conditions [45].

Often, the plasma treatment of supports is utilized to functionalize their surfaces, which enables efficient attachment of the catalysts. Médard et al. [46] used CO_2 plasma to form carboxylic groups on the polyethylene surface, which were able to form pure covalent bonds with metallocene catalysts ($Indene_2MCl_2$, where M = Zr, Ti), then tested them successfully in styrene polymerization. Lopez et al. [47,48] applied similar treatment to poly(vinylidene fluoride) membranes. The membranes were treated with Ar and then with NH_3 plasma in order to obtain a surface rich in amino groups, which are suitable anchor sites for the immobilization of tungsten-based catalysts. In particular, tungstate ions (WO_2^{4-}), catalyzing the oxidation of secondary amines to nitrones, as well as decatungstate ($W_{10}O_{32}^{4-}$) and phosphotungstate ($W_{12}PO_{40}^{3-}$) ions, which can both be used as catalysts in the degradation of organic pollutants such as phenol, were investigated. The plasma technique was also used to modify the interface between the catalyst nanoparticles and the support. Gold nanoparticles (for hydrogenation of acetylene) were anchored on a SiO_2 support through (3-aminopropyl)triethoxysilane (APTES) molecules that should subsequently be removed. It turned out that O_2 plasma destroyed these molecules giving much higher dispersion of Au nanoparticles (~3 nm) than thermal treatment at 773 K [49].

Among the materials that can be used as catalyst supports in fuel cells, carbon nanotubes (CNTs) and nanofibers (CNFs) have captured increasing attention owing to their high electrical conductivity, large surface-to-volume ratio, and corrosion resistance. However, the non-reactive and hydrophobic nature of the nanocarbons make deposition of catalytically active nanoparticles technically difficult. To overcome this problem, their modification by changing chemical composition of the surface has proven to be efficient. The plasma treatment allows the introduction of appropriate functional groups capable of chemical anchoring metal nanoparticles onto the surface, without affecting the bulk structure and morphology of nanocarbon supports. In this manner, carboxylic and phenolic groups, with control of their ratio, were formed by air or O_2 MW plasma on CNFs [50] and CNTs [51] to attach Pd and Ru nanoparticles, respectively. Very recently, oxygen-plasma-functionalized CNTs have been presented as supports for Pt–Ru catalysts applied in direct methanol fuel cells for electrochemical methanol oxidation. It has been shown that O_2 plasma treatment leads to the formation of carbonyl (–CO) and carboxylic (–COO) groups on the CNT surface. Pt–Ru nanoparticles dispersed for an optimum plasma treatment time exhibit high catalytic activity toward oxidation of methanol [52]. Iron(II) phthalocyanine, used as an electrocatalyst for the oxygen reduction reaction (ORR) in fuel cells, was also deposited on carbon nanoparticles (Vulcan XC-72) with the aid of plasma pretreatment of the support (N_2, Ar, Ar+O_2, and NH_3 RF plasmas were tested). In some cases, a fivefold increase in the electrocatalytic activity was observed [53].

Considering the issue of carbon nanotubes, we should also mention the use of cold plasma in the preparation of catalysts for the synthesis of CNTs. Gao et al. [54, 55] employed N_2 microwave plasma for this purpose. A very thin Fe film was deposited on a Si substrate by the pulse laser technique and then it was broken by plasma treatment. The nano-sized Fe islands having the density of 1.9×10^{15} m^{-2} and diameters of about 15 nm were obtained. CNTs synthesized on this catalyst were well-aligned, vertically arranged, and had almost the same diameters and density as the nano-islands. Similarly, the nickel catalyst was obtained by converting a Ni thin film into nanoparticles by treatment with microwave H_2 or H_2/N_2 plasmas [56, 57]. Recently, the direct current (DC) plasma as a source of effective ion bombardment, which is crucial for the creation of nano-islands, has been used in the CNT synthesis. It has turned out that gold nanoparticles produced in this way from a 0.5 nm thick Au film act as efficient catalysts promoting the growth of single-walled CNTs [58]. A small addition of C_2H_2 to H_2 during the DC plasma treatment of Fe thin films causes the simultaneous formation of Fe nanoparticles and coating them with a very thin carbon layer, which reduces sintering of the nanoparticles during the growth of CNTs. This results in a very dense CNT forest, reaching $2.2–2.4 \times 10^{16}$ m^{-2} [59].

2.2. Plasma modification of as-prepared "conventional" catalysts

Many catalysts that are produced for practical use or are already used commercially still need further improvement of their efficiency and selectivity. There have been a lot of research works on the subject and it appears that also in this case, good results can be achieved by plasma treatment. For example, Ar plasma treatment (a corona discharge) of the Ni catalyst used in partial oxidation of methane to syngas causes an increase in the catalytic activity and Ni dispersion. Moreover, the plasma treatment improves the catalytic stability by preventing the deposition of carbon on the Ni catalyst [60]. Similar changes in catalytic properties have been observed for Ni and Pt catalysts treated with He plasma in the dielectric-barrier discharge (DBD) under atmospheric pressure [61]. The DBD technique was also used in order to improve the activity of MnOx catalysts in the catalytic oxidation of nitrogen oxide (NO). The effect of plasma treatment is shown in Figure 5, where increased conversion of NO to NO_2 by O_2 at low temperature (323–523 K) is recorded [62, 63].

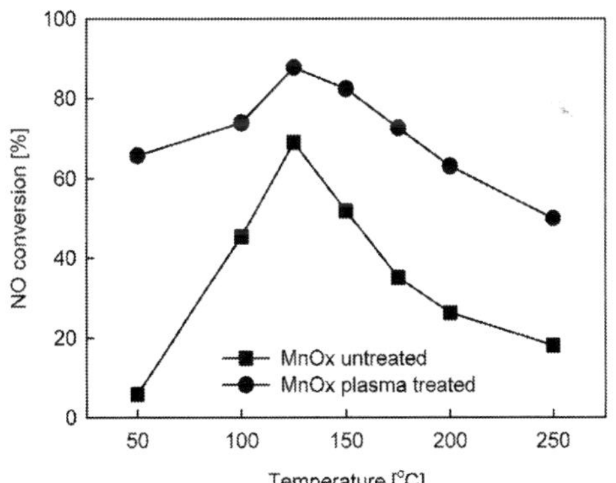

Figure 5. Oxidation activity of MnO$_X$ with and without plasma treatment [63].

An evident improvement in the catalytic properties after the plasma modification of the as-prepared catalysts has recently been reported more and more often. Such catalysts as Pd/TiO$_2$ for the selective hydrogenation of acetylene [64, 65], Fe–Cu on active carbon for the hydrolysis of carbon disulfide [66], CuO nanowires for oxidation of carbon monoxide [67], or CuO/TiO$_2$ employed in the reduction of nitrogen oxides [68] have shown higher efficiencies, only when the plasma treatment process was added after the catalyst preparation. Of course, in any case, it is necessary to choose an appropriate type of plasma and select optimized parameters of the treatment process. An example is given in Figure 6, where one can observe the effect of plasma exposure time (Ar MW plasma) on the conversion and selectivity for the above-mentioned CuO/TiO_2 catalyst tested in the reaction [68]:

$$NO + CO \rightarrow \tfrac{1}{2} N_2 + CO_2 \tag{4}$$

This result indicates that a remarkable increase in activity and selectivity is achieved depending on the plasma treatment time. However, this clear improvement is attributed not only to a greater dispersion of the catalyst but also to changes in its chemical structure. It has been suggested that highly active oxygen species (O^*) are formed on the surface during plasma treatment, which leads to positive changes in the reaction path.

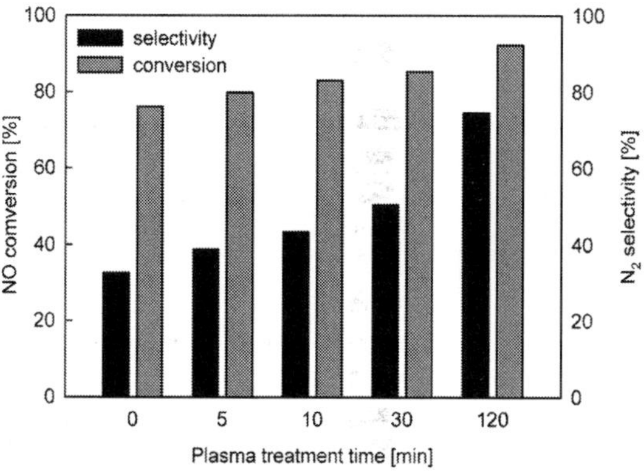

Figure 6. Catalytic performances for reaction (4) over CuO/TiO_2 catalysts following different microwave plasma treatment times [68].

The plasma treatment has also turned out to be very useful for the regeneration of spent catalysts. It has been found that the rate of reduction is several times higher for the plasma-treated deactivated catalysts than for the untreated ones. Furthermore, activity of the regenerated catalyst is usually higher than that of the fresh catalyst [1]. Application of the plasma method is especially important in the case of nanocatalysts, whose catalytic activity depends strongly on their size, so only low-temperature regeneration methods are acceptable. The plasma regeneration (O_2 DBD plasma) of TiO_2-supported gold nanoparticle catalysts (Au/TiO_2) has been successfully used to a great enhancement of catalytic activity for CO oxidation over the completely deactivated Au/TiO_2 [69]. Recently, a more sophisticated plasma treatment has been used to regenerate tungsten carbide (WC), which is considered a promising replacement for precious metal-based electrocatalysts for fuel cells. Mild treatment of WC foil with atomic oxygen generated in a plasma source operating in the atom mode (with an ion trap) allows a controlled removal of graphitic carbon from the WC surfaces without causing oxidation of WC [70].

3. Plasma sputtering of catalytic nanoparticles

Ultrafine particles are a particularly interesting form of catalysts due to their large specific surface and less-perfect crystal lattice with a large number of vacancies, which induces high catalytic activity. One of the most efficient methods of producing such particles of sizes from several to tens of nanometers is the cold plasma sputtering [71]. In this technique, positive ions that are produced in plasma generated in an inert gas, for example Ar, bombard the target surface and cause sputtering of its material. The sputtered material condenses on the substrate that is located outside of the plasma region. If some reactive gases are used (e.g., O_2, N_2, CH_4), the target material takes part in chemical processes during sputtering and finally a new converted material is deposited. This is the so-called reactive plasma sputtering. Simultaneously, two or more different targets can be specified, and the process can also be enhanced by magnetic field (magnetron plasma sputtering). So, as one can see, the plasma sputtering has great technological potential for the production of nanocatalysts. By choosing the appropriate process parameters, the structure, size, and quantity of nanoparticles can be controlled. As an example, the dependence of diameter of Pd nanoparticles on the sputtering time (Ar plasma, 100 MHz) is shown in Figure 7.

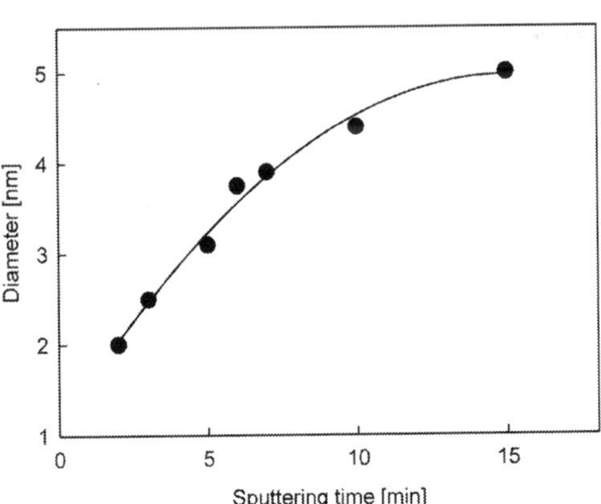

Figure 7. Pd nanoparticles diameter on the time of plasma sputtering (on the basis of Ref. [72]).

The metal and metal oxide nanoparticles are used primarily as catalytic materials for fuel cell electrodes, and mainly in this respect, the plasma sputtering has been intensively investigated over the past 15 years [17]. It has been shown that Pt, Pd, Rh, Pt_xRh_Y, Pt_xRu_Y, Pt-RuO_X, Ni_xZr_Y, PtNiZr, NiZrPtRu, CoO_X, NbO_X, NbO_XN_Y, Pd_xAu_Y, and others can be plasma-sputtered and deposited in the electrocatalytic active form [73–85]. In many cases, they exhibit higher activity than conventionally prepared electrodes, but the possibility to significantly reduce the amount of catalytic material used is the most important. It has been shown that the catalytic electrodes could be prepared with a platinum loading down to 0.005 mgPt/cm², which is drastically lower than that for conventional Pt electrodes (0.5–1.0 mgPt/cm²), with no detrimental effect on the fuel cell performance. Figure 8 presents the specific power on the current density for a PEM fuel cell with a Pt-sputtered anode, for various loadings of the catalyst. The advantageous effect of reducing the amount of catalyst is clearly visible [86].

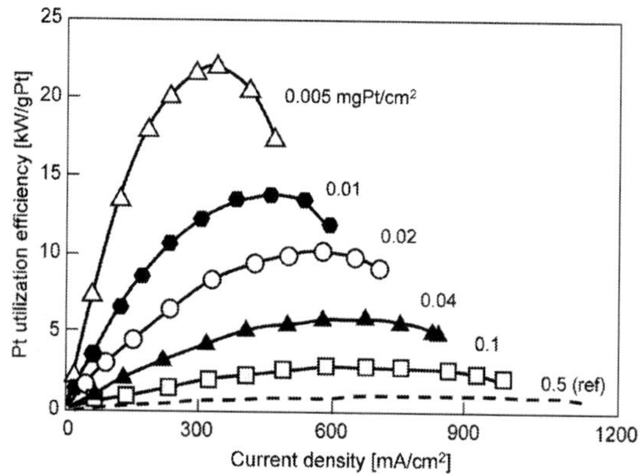

Figure 8. Platinum utilization efficiency versus current density for plasma-prepared cathodes with different Pt loading (mgPt/cm²). For comparison, the results for "classic" Pt cathode with Pt loading of 0.5 mgPt/cm² are presented [86].

The method of plasma spraying has proven to be very useful in the preparation of 3D-catalysts [17]. Already in the early 2000s, Brault et al. [87, 88] applied this method to introduce Pt clusters into a porous carbon material forming 3D-electrodes for PEM fuel cells. They obtained Pt nano-clusters of 3.5 nm, which penetrated the porous carbon electrode up to 2 μm deep. It should also be emphasized that although the electrode work was comparable to commercially available electrodes, it had the platinum density 4.5 times lower and hence was significantly more effective. Recently, a new pathway in the design of 3D-electrodes has been proposed, namely a combination of the methods of plasma polymerization (PECVD) and plasma sputtering. This dual-plasma process, i.e., the synthesis of catalytic thin films made of Pt nanoclusters (3–7 nm) embedded in a porous hydrocarbon matrix, was carried out by simultaneous plasma-polymerization of ethylene and sputtering of a platinum target. The metal content in the films could be controlled over a wide range of atomic percentages (5– 80%) [89]. Great possibilities of this method have encouraged attempts to produce non-noble metal catalysts for PEM fuel cells. Metals such as copper, cobalt, and iron have been sputtered and embedded into the matrix formed via the PECVD of pyrrole [90–93]. Plasma polymerized pyrrole (pp-pyrrole) exhibits high electrical conductivity and gas permeability as well as good chemical stability, so that together with the sputtered metal clusters forms an excellent nanocomposite, which can act as a 3D-catalyst.

4. Plasma-deposited (PECVD) thin films with catalytic properties

One of the most promising methods of producing new catalytic structures is the plasma deposition of thin films from organic and metalorganic precursors most frequently supplied to the plasma reactor in the gas phase. This method of thin-film deposition, well known as the plasma polymerization or plasma-enhanced chemical vapor deposition (PECVD), has already been used to fabricate a lot of thin-film materials for various practical applications [94]. In the late 1980s, a possibility of using PECVD to produce thin films having potential catalytic properties, such as Pd, Rh, Pt, was already mentioned [95–98]; however, only recently the involvement of this method in the field of catalysis has become a reality [3, 15, 17].

4.1. Background

Metalorganic compounds (their molecules are composed of a metal atom surrounded by organic or organic-like (e.g., CO) ligands) are key precursors for the PECVD of thin films displaying catalytic properties, in which the catalytic activity is related to the presence of metal or metal oxide clusters in amorphous or nanocrystalline forms. However, thin films possessing potential catalytic properties plasma-polymerized (pp-) from typical organic (without metal) precursors, wherein organic functional groups act as active centers, have also attracted attention. It was found quite a long time ago that the films containing organic moieties of specific electronic structure, for instance, the quinone-type groups formed in pp-4-vinylpyri‐ dene [99] or the protonated amine groups created in pp-allylamine [100], exhibit electrocata‐ lytic activity. Recently, this idea has returned – thin films have been deposited onto silica– alumina powders by plasma polymerization of 1,2-diaminocyclohexane to prepare amine‐ immobilized solid base catalysts. The existence of amine moieties has been confirmed by solid-state ^{13}C NMR, FTIR, and XPS analyses. The silica–alumina powders with these films have shown strong base catalytic activity, e.g., in Knoevenagel condensation reaction between benzaldehyde and ethyl cyanoacetate [101].

As already mentioned, the metalorganic precursors are introduced into plasma reactors mainly in the gaseous form. Sometimes, however, this poses a problem because of the low vapor pressure of such compounds. To get around this difficulty, an innovative method of production of catalytic films by the PECVD has been proposed lately. In this method, compounds are injected into a plasma reactor by means of, for example, an atomizer, in the form of aerosols created from liquid solutions of precursors. Cobalt oxide thin films for catalytic applications were deposited this way using the aerosol from a solution of cobalt carbonyl ($Co_2(CO)_8$) and hexane [102, 103]. Likewise, a series of hybrid silica-based catalysts containing various metals (Ti, V, Zr, Sn, Mn, Fe, Co) were produced utilizing nano-sized droplets (15–50 nm) sprayed from a solution of a selected metalorganic precursor, tetraethoxysilane ($Si(OC_2H_5)_4$), and an organic solvent, and then injected into a DBD plasma reactor [104]. Precursors can also be introduced directly to the plasma chamber as a solid phase mixture with a powdered support. During plasma operation, the precursor is transferred into the gas phase and then involved in the deposition processes [105–107].

The plasma deposition (PECVD) of thin films from metalorganic precursors should also include a fairly sophisticated method called the plasma-assisted atomic layer deposition (ALD). This is true nanotechnology, allowing ultra-thin films of a few nanometers to be deposited in a precisely controlled way via subsequent cycles that generate layer-by-layer growth [108]. Very recently, the ALD has emerged as an interesting tool for the atomically precise design and synthesis of catalytic materials with a controlled distribution of size, composition, and active site [109].

Occasionally, an incipient wet impregnation of the support by a solution of metalorganic precursor, followed by plasma treatment (e.g., with O_2 plasma) is applied to preparation of potential catalysts, such as CoO_X on zirconia [110], or CrO_X on silica [111]. However, this procedure is not a typical PECVD process in the gas phase.

Examples of metalorganic precursors used for the plasma deposition of thin films having potential catalytic activity are summarized in Table 1. Apparently, the PECVD method can be successfully used to produce a wide variety of films, comprising both metal and metal oxide catalytically active phases.

Metalorganic precursor	Denotation	References
platinum(II) acetylacetonate	$Pt(acac)_2$	[107,112,113]
palladium(II) acetylacetonate	$Pd(acac)_2$	[114]
palladium(II) hexafluoroacetylacetonate	$Pd(hfac)_2$	[115,116]
rutenium(II) bis(ethylcyclopentadienyl)	$Ru(EtCp)_2$	[117,118]
cobalt(III) acetylacetonate	$Co(acac)_3$	[119]
cobalt(II) bis(2,2,6,6-tetramethyl-3,5-heptanedione)	$Co(TMHD)_2$ $Co(dpm)_2$	[120]
cobalt(II) bis(cyclopentadienyl)	$CoCp_2$	[121]
cobalt(I) cyclopentadienyldicarbonyl	$CpCo(CO)_2$	[122-124]
dicobalt octacarbonyl	$Co_2(CO)_8$	[102,103]
titanium(IV) tetraisopropoxide	TTIP	[125-131]
titanium(IV) butoxide	$TNBT; Ti(OBu)_4$	[132]
titanium(IV) ethoxide	$Ti(OEt)_4$	[133]
titanium(IV) diisopropoxidebis(2,2,6,6-tetramethyl-3,5-heptanedionate)	$Ti(O-i-Pr)_2(thd)_2$	[134]
zirconium(IV) acetylacetonate	$Zr(acac)_4$	[113]

Metalorganic precursor	Denotation	References
zirconium(IV) tetra(*tert*-butoxide)	ZTB	[135]
zirconium-n-propoxide	ZNP	[104]
iron(III) acetylacetonate	Fe(acac)$_3$	[136]
iron(II) bis(hexafluoroacetylacetonate) (N,N,N',N'-teramethylenediamine)	Fe(hfa)$_2$TMEDA	[137]
tert-butylferrocene	TBF	[138]
iron(0) pentacarbonyl	Fe(CO)$_5$	[139,140]
copper(II) acetylacetonate	Cu(acac)$_2$	[124]
chromium(III) acetylacetonate	Cr(acac)$_3$	[105,106]

Table 1. Examples of metalorganic precursors for PECVD of catalytic films.

Generally, there are two possible ways to start the decomposition of metalorganic molecules in the PECVD process. The ligands of the precursor can be decomposed in the gas phase, and the products of the process then participate in the formation of the film on the substrate surface, or the precursor molecules are first adsorbed on the surface without decomposition. Then, their decomposition occurs during the interaction of plasma with the substrate. Investigations of the preparation process of CrO$_X$ deposited from chromium(III) acetyloacetonate (Cr(acac)$_3$) in microwave plasma on a Zr-based support have evidently shown that during the PECVD process adsorption of the precursor molecules on the support probably takes place by cleavage of one ligand. Further plasma operation causes the gradual destruction of ligands and formation of a CrO$_X$ structure. This chemical process is schematically presented in Figure 9 [106]. On the other hand, the study on plasma-polymerized ZrO$_2$ films from zirconium(IV) tetra-tert-butoxide in a mixture with oxygen has shown that the gas-phase reactions have a direct impact on the deposition process. Mass spectra of the plasma have revealed a number of species, including CH$_3$$^+$, C$_2H_4$$^+$, CHO$^+$, C$_2H_5$$^+$, CH$_3CO^+$, C$_4H_6O^+$, Zr$^+$, ZrO$^+$, ZrO$_2H^+$, ZrO$_3H_3$$^+$, ZrO$_4H_5$$^+$, ZrO$_3C_3H_7$$^+$, which initiate processes of the film formation [135].

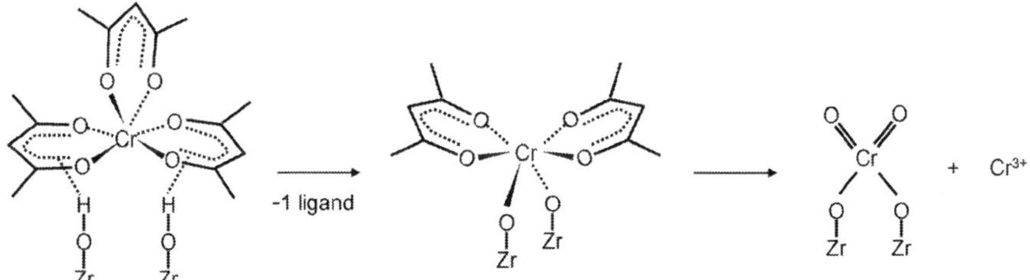

Figure 9. Schematic representation of possibly adsorption of Cr(acac)$_3$ on Zr-based support and stepwise plasma destruction of the precursor molecule, which leads to CrO$_X$ with Cr^{3+} and Cr^{6+} species [106].

By carefully controlling the process parameters, the ligands can be completely removed, leaving only "pure" metal or its oxides. Very often, however, some amount of carbon is present in deposits. To eliminate the carbon contamination, treatment with oxygen plasma, heated substrates during the deposition, and heat treatment after the deposition have been practiced [3, 15].

An appropriate choice of the PECVD process conditions also permits control of the film structure. Amorphous films such as TiO_2 films of 7–120 nm thickness (deposited from titanium(IV) butoxide) can be obtained and after thermal treatment at 713 K transformed into nanocrystalline anatase films having a grain size of about 15 nm [132]. Similar results concerning the conversion of TiO_2 from amorphous in crystalline phase of anatase structure as a result of thermal treatment were shown by Cho et al. [129] for films deposited from titanium tetraisopropoxide. However, nanocrystalline structure can also be produced directly in the PECVD process, as is the case with TiO_2 films deposited on the heated substrate, where the crystalline domains of 40–90 nm have been obtained [130]. The structure of the films can be controlled by the type of support as well. When TiO_2 films were produced form $Ti(O\text{-}i\text{-}Pr)_2(thd)_2$, the use of quartz substrates resulted in the formation of films showing predominantly the anatase crystallographic structure. On the other hand, only the rutile phase was found in the case of silicon wafers and nearly amorphous phase for metallic Ni [134].

Finally, the molecular structure of deposited films can be controlled using mixtures of metalorganic precursors. For example, interesting films of CoO_X doped with Cu were prepared by the PECVD from a gas mixture of $CpCo(CO)_2$ and $Cu(acac)_2$. It was found that pure cobalt oxide films were mainly composed of Co_3O_4 in the form of nanoclusters, whereas the Cu doped films were much more complex: CoO_X (also Co_3O_4), mixed Co–Cu oxides, and CuO_X nanoclusters were detected. Preliminary catalytic tests showed that the films of CoO_X doped with Cu initiated the catalytic combustion of n-hexane at a lower temperature, compared with the pure cobalt oxide films [124].

Among the biggest advantages of films produced by the PECVD method, beyond their structural diversity that cannot be achieved by any other method, is their very thin form (in the order of nm). This is of particular importance, on the one hand, to save materials, and on the other hand, such thin films are ideal for new constructions of catalytic structured reactors and electrodes for fuel cells. Currently, the most extensively studied plasma-deposited materials with possible applications as catalysts include some noble metal-based films (e.g., Pt, Pd, Ru) and non-noble transition metal-based films, which exist mainly in the form of oxides (e.g., TiO_2, CoO_X, FeO_X).

4.2. Noble metal–based films

Platinum-based materials (Pt or Pt alloys) are the best catalysts for many reactions, especially for hydrogen and methanol oxidation as well as oxygen reduction, which are at the core of fuel cell technologies. Unfortunately, Pt is a precious and very expensive metal. Besides, the stability of Pt and Pt alloys becomes a serious problem. Hence, extensive research has been underway to overcome these difficulties and new methods to ensure consumption of smaller amounts of platinum and at the same time provide more stable and effective catalysts are being sought.

In addition to plasma sputtering, which is a very promising technique (see: Sec. 3), the plasma polymerization method (PECVD) has also been tested as a tool for the preparation of platinum thin-film catalysts. A platinum metalorganic complex, ($Pt(acac)_2$), was used by Dhar et al. [112] as a precursor for the PECVD carried out in an RF discharge. The plasma-polymerized film was then calcined to drive off organic material, leaving behind a catalyst-loaded substrate. The same procedure was used to prepare a composite consisting of a ZrO_2 support and Pt catalyst. The support and catalyst were deposited on a metallic substrate by the PECVD as alternate layers from $Zr(acac)_4$ and $Pt(acac)_2$, respectively. It was found that Pt agglomerates were embedded in the zirconia support [113]. Recently, the mechanism of formation of Pt nanoparticles on carbon black powder used as a support, during plasma deposition from $Pt(acac)_2$ has been more closely investigated. It has been proposed that some oxygenated surface groups or structural defects are formed at the surface of carbon under the influence of plasma. These surface defects act as anchoring sites for the nucleation of Pt nanoparticles. Thus, by controlling the processes of precursor decomposition and formation of surface defects, through an appropriate choice of plasma parameters, the size and concentration of nanoparticles can be controlled [107].

Another important noble metal, which has been widely used as a catalyst in various syntheses, is palladium. In case of palladium, some attention has been paid to the plasma assisted ALD method. Pd thin films were deposited on substrates at low temperatures (≤ 373 K) by means of a remote inductively coupled hydrogen plasma and frequent use of $Pd(hfac)_2$ as a precursor. This way, Ten Eyck et al. [115] obtained films of the thickness from 1.66 to 3.87 nm after 150 cycles of pulsed ALD, depending on the type of substrate. An additional O_2 plasma step in each cycle yielded virtually 100% pure palladium thin films composed of nanometric crystalline grains, the size of which could be controlled by the number of deposition cycles [116]. Investigations performed on Pd films prepared by the typical PECVD process (from $Pd(acac)_2$) showed that the as-deposited films were amorphous and only after thermal treatment at 623 K, Pd nanocrystalline clusters of 5–10 nm size were formed [114].

The group of important noble metals used in catalysis includes ruthenium as well. As in the case of Pd plasma-deposited films, particular attention has been paid to the plasma-assisted ALD method. By using a $Ru(EtCp)_2$ precursor and $N_2/H_2/Ar$ or NH_3/Ar plasmas as reactants for the appropriate steps in the deposition cycles, very thin films of ruthenium (with the growth rate of about 0.17 nm/cycle on SiO_2 substrates), containing very low amounts of oxygen and carbon, were produced [117,118]. The same precursor ($Ru(EtCp)_2$) was used in our laboratory to perform typical PECVD process for preparing ruthenium films. XPS analyses indicated that the as-deposited films contained only metallic ruthenium (Ru^0); however, after calcination at 773 K in air, a fraction of RuO_2 (Ru^{4+}) appeared. Spectra of Ru 3d5/2 obtained for the as-deposited and calcined films are shown in Figure 10. The films were tested in the catalytic reactions, such as methanation of CO_2 [141] and water splitting [142].

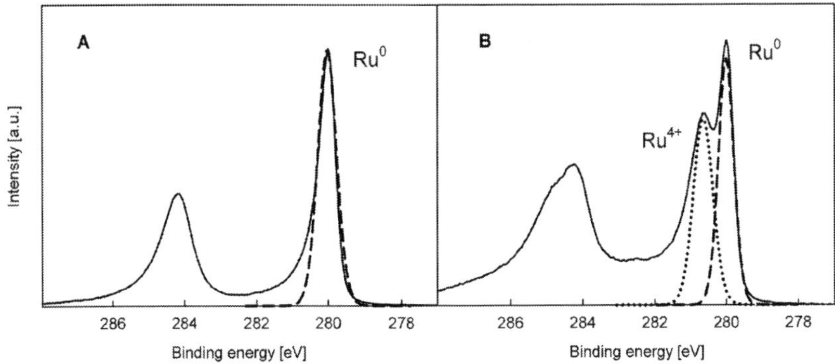

Figure 10. XPS spectra of Ru 3d5/2 for as-deposited (A) and calcined (B) films obtained by PECVD from Ru(EtCp)$_2$.

4.3. Non-noble transition metal–based films

Transition metals, mainly in the form of oxides, are well known as catalysts in many chemical and photochemical reactions. Among these oxides, for a long time TiO$_2$ has belonged to the most intensively studied compounds, mainly because of its unique photocatalytic activity [143]. The plasma deposition of TiO$_2$ thin films is prominent among the methods of producing this material and many results have already been published on this subject; some of them were mentioned above [129, 130, 132, 134], others can be found in the reviews (for example, see [15, 144]). In most reported works, either TiCl$_4$ or Ti alkoxides (mainly titanium tetraisopropoxide) are used as precursors of the PECVD process, resulting in amorphous or crystalline films having the nonstoichiometric (TiO$_X$) or stoichiometric (TiO$_2$) structure. Physicochemical and catalytic properties of all of these films are strongly dependent on their structure, which can be effectively controlled by the deposition conditions [125–128, 131, 133].

Other metal oxides important from the standpoint of catalysis are cobalt oxides (CoO$_X$) and iron oxides (FeO$_X$). Their thin films produced through the PECVD have recently been of particular interest.

4.3.1. CoO$_X$-based films

The first attempts to produce films from cobalt metalorganic precursors by the PECVD method were undertaken in the early 1980s [145,146], but they have been more closely studied only recently [102, 103, 119–122, 124, 136]. This is related to wide possibilities of application of CoOx films in new structural reactors for catalytic synthesis as well as in electro- and photo-catalyses carried out in fuel cells and water splitting processes. Various precursors have been used (see, Table 1), but one of the most convenient is CpCo(CO)$_2$, because it is liquid at room temperature and highly stable, which facilitates control of the plasma process.

The nanostructure of plasma-deposited CoO$_X$ films, which is crucial for catalytic properties, can be controlled by the type of precursor, plasma process conditions, and nature of supports.

For example, small particles of CoO_x 2–10 nm in diameter were found in the plasma-deposited films on TiO_2 supports [119]. When CoO_x films were produced on a glass substrate at elevated temperatures (423–673 K), columnar grains of the average diameter between 35 and 60 nm were formed on the film surface [147]. Highly pure and strongly oriented Co_3O_4 thin films, with features dependent on the used substrate and adopted growth temperature, were obtained by the PECVD on single crystals, such as MgO(100) and $MgAl_2O_4$(100) [120]. In turn, the plasma deposition from liquid solutions of Co compounds sprayed into the plasma reactor gave thin films containing nanocrystals 15 or 40 nm in size, depending on the type of precursor and the PECVD process conditions [102]. If such films are post-treated with Ar and O_2 plasma, the crystal nano-catkins, which can increase the contact area of the catalyst, are formed containing particles of Co_3O_4 3–12 nm in size [103].

The research conducted in our laboratory on CoO_x films produced by the PECVD from $CpCo(CO)_2$ has confirmed that the structure of these films can be controlled by the conditions of the deposition [3, 122, 123]. As deposited, thin films (25–750 nm) composed of a hydrocarbon matrix and amorphous CoO_x were obtained. However, only moderate thermal treatment was enough to transform the amorphous films into films possessing the nanocrystalline structure of cobalt spinel (Co_3O_4). The formation of such nanocrystals was confirmed not only by electron diffraction analysis but also by Raman spectroscopy measurements, an example of which is shown in Figure 11.

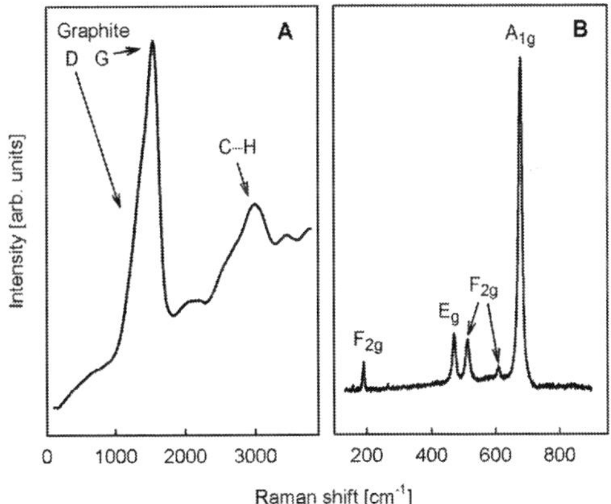

Figure 11. Raman spectra for: (A) thin film of hydrocarbon matrix with amorphous CoO_x, (B) thin film of nanocrystalline Co_3O_4 [123].

It has also been found that the size of nanocrystallites is controllable by the thermal treatment. To investigate this effect precisely, an energetic laser beam was used as the heat source. The dependence of the crystallite size on the time of laser treatment is illustrated in Figure 12. As can be seen, the average size of the Co_3O_4 nanocrystallites increases with the treatment time reaching a constant value. On the other hand, the concentration of nanocrystallites is almost constant throughout the treatment. The maximum size and concentration of the nanocrystallites can in turn be controlled by the parameters of the plasma process. It has already been shown [123] that with an increase in the precursor flow rate, the maximum size of nanocrystallites, after the thermal treatment, increases.

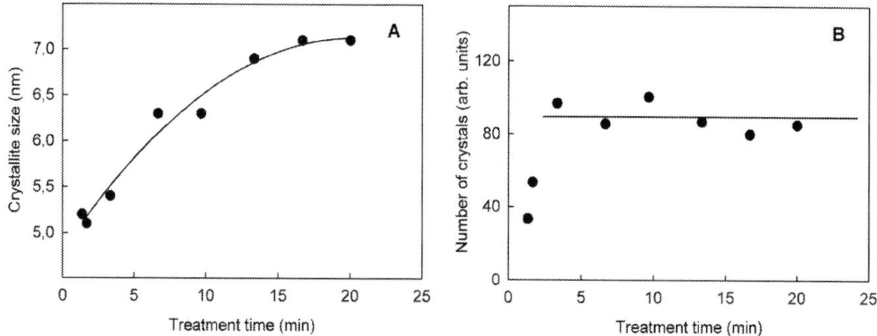

Figure 12. The size (A) and concentration (B) of Co_3O_4 nanocrystallites in a film plasma deposited from $CpCo(CO)_2$, in dependence on the laser treatment time.

The above results represent a further step toward the molecular engineering of catalysts, where the desired nanostructure of material and thus its catalytic activity can be designed.

4.3.2. FeO_X -based films

The recent interest in the plasma-deposited FeO_X thin films has been prompted primarily by the possibility of using these films as photocatalysts in the water splitting process [137, 139, 140, 148, 149]. Various types of FeO_X films can be prepared, depending on the precursor used (see, Table 1) and deposition process conditions. Already twenty years ago, Fujii et al. [136] showed that using the same precursor and constant RF plasma power, a wide range of FeO_X structures was obtained by controlling the O_2 flow rate and substrate temperature (T_S) during the deposition process. These results are summarized in Figure 13. As can be seen, at low values of the O_2 flow rate, amorphous films are formed independent of T_S. When the flow rate increases, thin films of spinel-type iron oxide (Fe_3O_4), which are composed of both Fe^{2+} and Fe^{3+}, are created. Usually, however, it is not a pure fraction, but rather Fe_3O_4–γ-F_2O_3 intermediate phase. At the O_2 flow rate higher than 30 cm^3/min, thin films composed of Fe^{3+} are formed, namely α-Fe_2O_3 (at T_S = 600–750 K) and β-Fe_2O_3 (at T_S = 450–650 K), although the films of single phase β-Fe_2O_3 cannot be obtained and this structure occurs only in coexistence with the spinel-type iron oxide. However, by appropriate selection of the PECVD conditions, it is possible to obtain films containing pure β-Fe_2O_3. Such films with columnar arrays of β-Fe_2O_3, characterized by a preferred (100) growth direction and highly porous structure, have been successfully plasma deposited from a fluorinated precursor ($Fe(hfa)_2TMEDA$) in a mixture with Ar and O_2 [137].

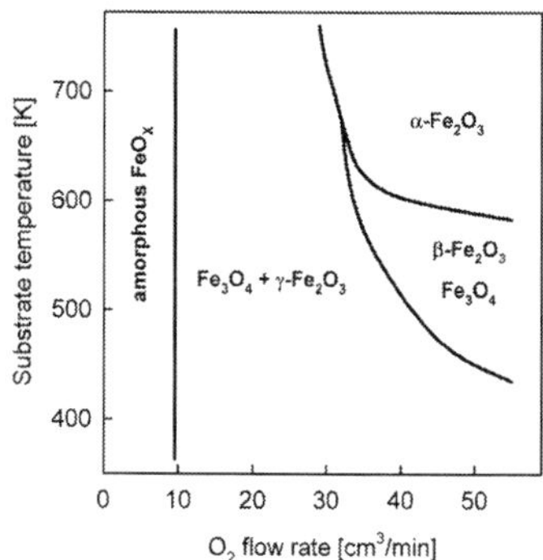

Figure 13. Formation diagram of iron oxide films deposited from Fe(acac)$_3$ at a constant RF plasma power (400 W) (on the basis of Ref. [136]).

The structure of deposited iron oxide films can also be modified by post-treatment. In Figure 14, examples of the XPS spectra of Fe 3p, concerning films deposited from Fe(CO)$_5$ in our laboratory, are shown. A clear change in the structure as a result of thermal treatment (723 K, in air) is observed. The as-deposited film contains Fe^{2+} and Fe^{3+}, which in combination with the results of X-ray diffraction and Raman spectroscopy indicates a spinel-like structure. After thermal treatment, only the completely oxidized form Fe^{3+} is present on the film surface. It has been found that in these films Fe^{3+} is associated with γ-Fe$_2$O$_3$ nanoparticles [150], which are already known as an effective catalyst [151]. An opposite process, i.e. the change of the film structure containing only Fe^{3+} to that containing a mixture of Fe^{3+} and Fe^{2+}, can be carried out using hydrogen plasma treatment. This way, hematite (α-Fe$_2$O$_3$) films were transformed to Fe$_3$O$_4$:α-Fe$_2$O$_3$, with precise control of the Fe^{3+} to Fe^{2+} ratio. This treatment substantially improved optical absorption, enhanced electrical conductivity, and improved transport properties due to a valence dynamic in the film among different iron oxidation states [140].

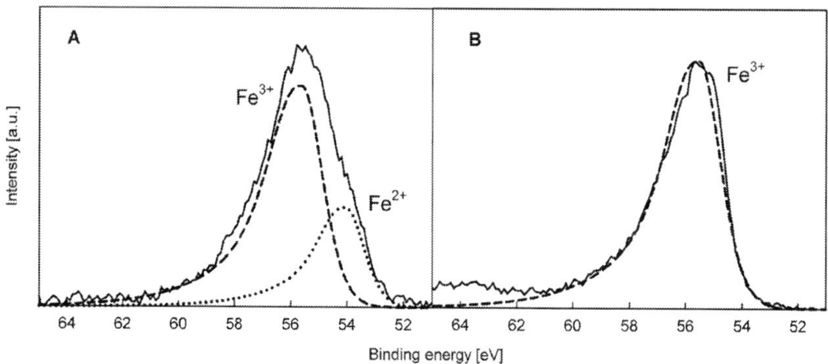

Figure 14. XPS spectra of Fe 3p for as-deposited (A) and calcined (B) films obtained by PECVD from Fe(CO)$_5$.

A wide range of possibilities provided by the PECVD method allows us to design and produce more sophisticated structures at the molecular level. It is also true for iron oxide films, where, for example, Barreca et al. [137] replaced a part of oxygen atoms with fluorine atoms in the film structure. They showed that a noticeable switch from n- to p-type conductivity occurred upon increasing the fluorine amount in the films. This transformation in the electronic structure resulted in the extended charge carrier lifetime, making F-doped β-Fe_2O_3 films an efficient water oxidation catalyst. Recently, more complex nanosystems based on iron oxide films have also been prepared. Such nanosystems, for instance, including Fe_2O_3–CuO [148] and Fe_2O_3–Co_3O_4 [149], are synthesized by using a two-step plasma-assisted strategy. First, the iron oxide nanostructured thin film (as a host) is deposited by the PECVD and then CuO or Co_3O_4 nanoparticles (as guests) are over-deposited on the host matrix by means of plasma sputtering of Cu or Co and annealing in the air. The obtained structures based on the combination of p-type (CuO and Co_3O_4) and n-type (Fe_2O_3) semiconducting oxides, which create the p-n junctions responsible for the improved separation of electron–hole pairs, are highly promising as catalysts for the photochemical splitting of water.

In addition, the advanced structures based on plasma-deposited iron oxides include Fe_2O_3 thin films that very recently have been produced by plasma-enhanced atomic layer deposition (ALD) from tertiary butyl ferrocene and O_2 plasma. The films deposited below 523 K are amorphous but can be converted into hematite (α-Fe_2O_3) with (104) preferential orientation and average crystallite size of 35 nm by annealing in He [138].

4.4. Potential applications of the plasma-deposited (PECVD) catalytic films

The three main features of the PECVD method are the source of its attractiveness and broad prospects in the production of new catalytic species, namely – the ability of preparing very thin ($\ll$ 1 μm) films, the possibility of precise control over molecular structure and nanostructure of the films by selecting appropriate precursors and conditions of the manufacturing process, and finally, the possibility of obtaining 3D structures through the use of copolymerization of two or more precursors in the PECVD process or combining the PECVD and plasma sputtering processes together.

Current research on the potential possibilities of the application of catalysts produced by the PECVD is focused on the structured reactors, where the deposition of very thin catalytic coatings is the key problem, and on the PEM fuel cells and photoelectrochemical cells for water splitting, which need cheap and efficient electro- and photocatalysts.

4.4.1. Structured reactors

Development of the new and improvement of the existing heterogeneous catalytic processes are actually tasks of great practical importance in such areas as chemical and petrochemical industries, as well as in environment protection, particularly in catalytic combustion of volatile organic compounds (VOCs) and CO_2 conversion into value-added products. Conventionally used reactors for catalytic combustion and CO_2 conversion are packed bed (filled with a bed of catalytic grains) and monolithic reactors. The first type can be regarded as a slightly old-fashioned solution showing high flow resistance and low effectiveness of the catalyst grains, which in turn leads to high spending on catalyst material. Monolithic reactors are currently the standard equipment used for catalytic combustion since they provide better performance compared with packed bed reactors. However, monolithic reactors often fail in this process. This is because the so-called fully developed laminar flow occurs throughout most of the monolithic channels. Therefore, mass transfer at the developed laminar flow is rather low and weakly affected by the fluid velocity. Taking into account the high dilution of VOCs in large gas streams, the process is mainly controlled by the mass transfer, which imposes the use of long reactors to secure high conversion of VOCs. This in turn gives rise to an undesired high pressure drop and poor catalyst exploitation. An ideal reactor for combustion of VOCs should thus allow both high mass transfer coefficients and reasonably low flow resistance. The same requirements apply to many other heterogeneous catalytic reactions [16,152].

One of the possible paths for progress in this area is the development and application of new geometrical forms of catalysts. Structural reactors equipped with metallic fillers made of wire gauzes displaying highly enhanced transport properties proved to be an interesting option for all types of catalytic processes. Although such a solution based on noble metal catalysts has been known for a long time, for example, reactors filled with stacked platinum woven wire gauzes used for ammonia oxidation, the extension of this concept to other non-noble metal-based catalysts faces serious problems. The crucial difficulty is the deposition of appropriate thin, uniform, well-adhered, and catalytically effective films on wire supports that form fine microstructural meshes, such as the one shown in Figure 15. The most important is the deposition process should not change the elaborate geometry of the meshes.

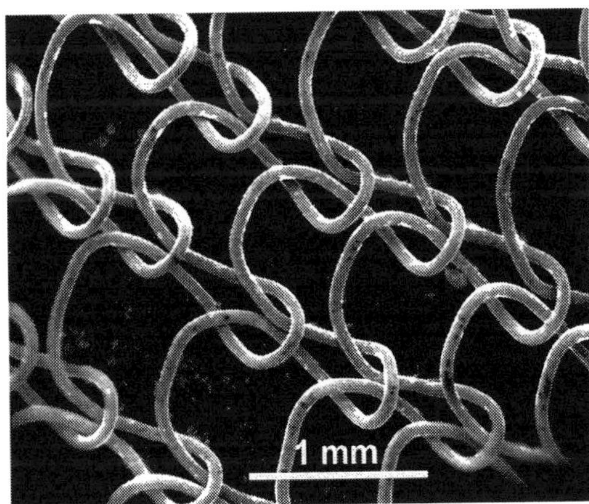

Figure 15. SEM image of a wire gauze made of kanthal with thin CoO_x film, used in our experiments with structured reactors.

The PECVD is one of the most promising methods of producing thin catalytic films on metal gauzes. Research on this subject was started in our laboratory in 2007. Catalytic combustion of n-hexane, as a representative of VOCs, and CoO_X as well as CoO_X–CuO_X based thin films produced by the PECVD, as catalysts of the combustion process, have been intensively investigated providing excellent results [122, 153–156]. It has been found that the plasma-deposited cobalt oxide catalyst showing a dispersed spinel (Co_3O_4) structure (see: Sec. 4.3.1) proves active in the n-hexane combustion as compared with commercial Pt catalysts. Even better results have been obtained for Cu-doped CoO_X films that have the lowest reaction initiation temperature (493 K). It has also been concluded that the gauze carrier enhances the mass transport in the reactors preventing the diffusional limitation of the reaction rate. Compared with the standard monolithic converter, it allows a reduction in the reactor length by around 50 times with a 20% increase in the pressure drop. These results encouraged undertaking the research on a larger scale with a prospect of possible future industrial applications. Experiments with n-hexane combustion on the plasma-deposited Co_3O_4 catalysts were successfully carried out in a large structured reactor (gas stream up to 10 m³/h STP) [16].

More recently, the research on CO_2 methanation process using thin films of catalysts plasma deposited on wire gauze carriers has been initiated by our team. The CoO_X (see: Sec. 4.3.1) and Ru-based (see: Sec. 4.2) films have been tested. Sample results are shown in Figure 16. Although these are only preliminary results, they confirm the catalytic activity of the films in the process of CO_2 methanation and encourage further research in this area.

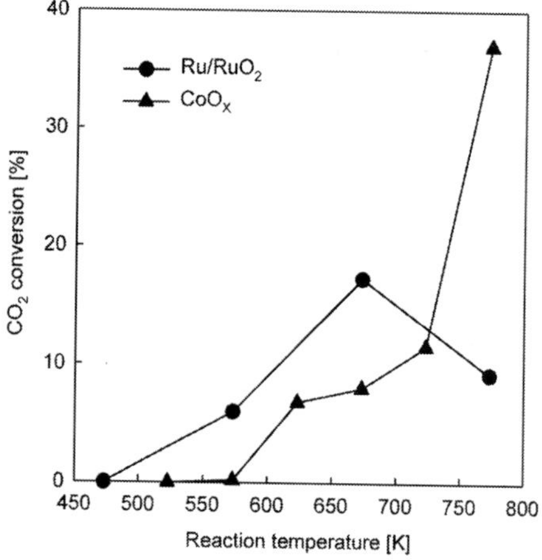

Figure 16. Conversion of CO_2 to CH_4 as a function of the process temperature for two different plasma-deposited catalytic films [141].

4.4.2. Fuel cells

Current studies on thin films produced by the PECVD, which can be used as catalytic systems in proton exchange membrane fuel cells (PEMFC), are primarily focused on finding new solutions that could compete with platinum electrodes – would be cheaper and exhibit at least comparable catalytic properties. Taking into account the promising electrocatalytic activity in the oxygen reduction reaction (ORR) demonstrated by nanoparticles of cobalt oxides [157], an attempt to produce such a material for the PEMFC electrodes by the PECVD method has been undertaken in our laboratory. The CoO_X-based films obtained from the $CpCo(CO)_2$ precursor (see: Sec. 4.3.1) were deposited on carbon paper substrate, and they were tested as the cathode in a hydrogen PEMFC (the anode was made of platinum). The preliminary results admittedly indicated the electrocatalytic activity of the films in the oxygen reduction reaction; however, it was far from the activity of Pt cathodes [17, 123]. Optimization of the conditions of plasma deposition and an enhanced procedure for the preparation of membrane electrode assemblies resulted in a significant improvement of the CoO_X electrocatalytic activity. Figure 17 presents current–voltage characteristics of the PEMFC with the cathode made of CoO_X deposited at various times (with various amounts of CoO_X) and Pt anode. The characteristic of the cell with both Pt electrodes is also shown. As can be seen, the cell with the Pt anode and CoO_X cathode is characterized by the open circuit voltage of 635 mV, which is already close to 963 mV obtained for the "pure" Pt cell [158]. Further progress in this area is expected, for example, by using doped CoO_X films and increasing the dispersion of Co_3O_4 nanoclusters.

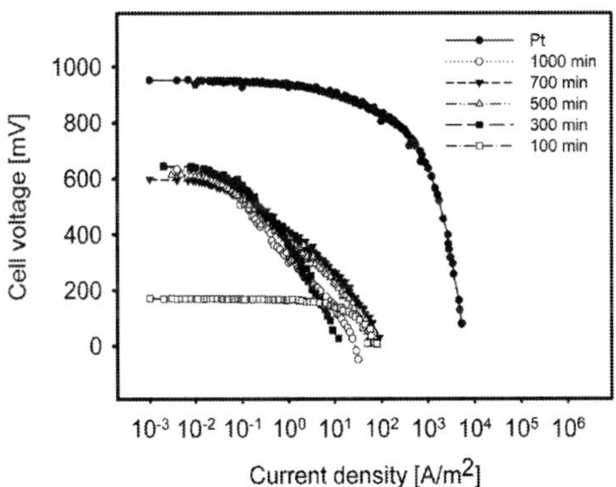

Figure 17. Current–voltage characteristics of PEMFC with the cathode of varying amounts of CoO_X deposit. The characteristic for the reference cell with both Pt electrodes is shown for comparison [158].

Very recently, we have obtained interesting results concerning the PEMFC, in which plasma-deposited FeO_X-based films were used as the cathode. The films have been deposited from $Fe(CO)_5$ (see: Sec. 4.3.2). An example of the preliminary results recorded for the fuel cell with such a cathode (and Pt anode) is shown in Figure 18. Although the results are far from expected, they point the way to further research in this area [150].

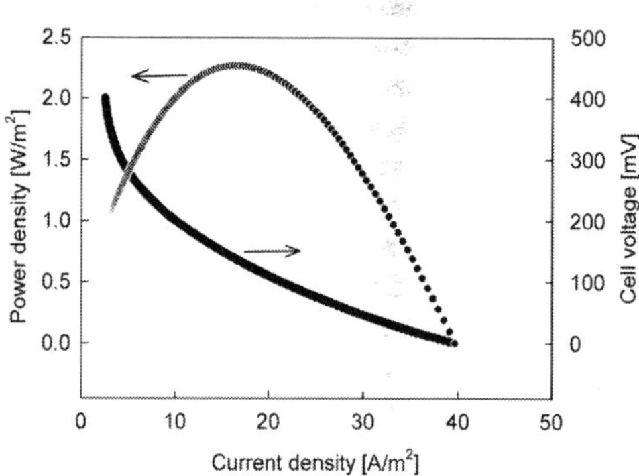

Figure 18. Power and potential versus current density for PEMFC with FeO_X cathode [150].

4.4.3. Water splitting systems

The most desirable method of hydrogen production, which represents a sustainable fuel of the future, is photoelectrochemical (PEC) splitting of water by visible light. A very important issue in the PEC hydrogen generation is the development of a high-performance photoelectrode that exhibits high efficiency in the conversion of solar energy into chemical energy, resistance to corrosion in aqueous environments, and low processing costs. However, after four decades of intensive research, since the first report on water photo-splitting [159], no material has been found to simultaneously satisfy all the criteria required for widespread PEC application. No wonder that a quest for new materials for photoelectrodes is still ongoing. The PECVD technique is also involved in this search [17].

Particular attention has recently been paid to FeO_x-based films (see: Sec. 4.3.2). Among iron oxides, hematite (α-Fe_2O_3) is considered as an attractive photoanode material, largely due to its abundance, chemical stability in aqueous environments, and light absorption in the visible range of the solar spectrum ($E_g = 2.1$–2.2 eV). However, the reported water-splitting efficiency of α-Fe_2O_3 is much lower (< 3.0 %) than the maximum theoretical value (~13 %), mainly due to the low absorption coefficient and slow reaction kinetics, resulting in a high carrier recombination rate and short hole diffusion length. Singh et al. [139] achieved the photocurrent density of about 1.1 mA/cm² at 0.9 V/SCE (with onset potential 0.6 V/SCE) for α-Fe_2O_3 films applied as a photoanode in 1M NaOH electrolyte (deposited from $Fe(CO)_5$). Naturally, this value depends on the used light intensity and development of the substrate surface, thus it is difficult to compare this particular result with those obtained in other laboratories. Nevertheless, it can be concluded that the films deposited in our laboratory from the same precursor under conditions that allow formation of γ-Fe_2O_3 nanoparticles reveal similar, if not better, photoactivity. For example, the onset potential is about 0.45 V/SCE.

A considerable improvement was observed in the photoactivity of α-Fe_2O_3 films treated by hydrogen plasma, which led to the formation of a complex structure of Fe_3O_4:α-Fe_2O_3 type. In this case, enhanced photocurrent densities (3.5 mA/cm^2 at 1.8 V/RHE) and reduced photocurrent onset potentials (from 1.68V/RHEfor non-treated films to 1.28 V/RHE) were measured in 1M NaOH electrolyte [140].

In search for further opportunities to improve photocatalytic properties of FeO$_X$ thin films, tuning of the system nano-organization and controlled tailoring of the chemical composition have been proposed. In particular, oriented columnar nanostructures offer the possibility of absorbing a significant fraction of light while providing short carrier transport distances to the electrolyte, thus minimizing recombination losses. Recently, columnar arrays of β-Fe$_2$O$_3$ showing a preferred (100) growth direction and highly porous structure have been successfully obtained in Ar/O$_2$ plasma by Barreca et al. [137]. The decoration of such columnar arrays with other transition metal oxides, such as cobalt oxide [149] and copper oxide [148], in the form of nanoparticles, might profitably suppress charge carrier recombination and improve Fe$_2$O$_3$ catalytic activity in water splitting. For example, the Fe$_2$O$_3$–Co$_3$O$_4$ heterostructure nanosystem reveals a photocurrent increase of 40% with respect to bare Fe$_2$O$_3$, which clearly proves that nanosystems produced with the involvement of PECVD are very promising [149].

Some interesting results have been obtained very recently in our laboratory for Ru^0/RuO$_2$ films produced by the PECVD (see: Sec. 4.2). In Figure 19, the photocurrent density (with subtracted dark current) measured under illumination from a 150 W Xe arc lamp equipped with an AM 1.5 filter, as a function of the applied voltage, is presented. As shown in the figure, the achieved value of the photocurrent is 9.1 mA/cm^2 at 0.6 V/SCE, which, compared with the results obtained in the same experimental system for Fe$_2$O$_3$ films (0.65 mA/cm^2 at 0.6 V/SCE) and Co$_3$O$_4$ films (0.48 mA/cm^2 at 0.6 V/SCE), is a very promising finding [142].

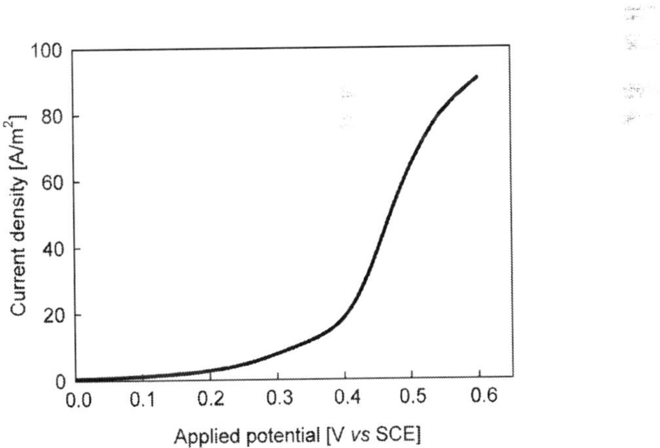

Figure 19. Photocurrent–voltage characteristic for Ru^0/RuO$_2$ photoanode prepared by PECVD (measurements in 1M NaOH electrolyte) [142].

5. Conclusions and outlook

I hope that after reading this chapter, the reader will no longer have any doubt that the enormous potential is hidden in the cold plasma technology. This technology gives us almost unlimited possibilities of modifying various materials and producing entirely new structures. This is particularly true with respect to the field of catalysis.

On one hand, the high-temperature processes such as decomposition of precursors, pyrolysis, calcination, reduction, and regeneration can be replaced by low-temperature plasma treatment, which minimizes problems caused by high temperature, such as aggregation, crystallite size growth, and sublimation, often leading to the creation of a completely different structure of the resulting catalyst. Generally, the conventional catalysts prepared with the involvement of cold plasma methods exhibit much higher activity, enhanced selectivity, and better stability. On the other hand, the cold plasma techniques, such as plasma sputtering and especially plasma polymerization (PECVD), have paved the way for the design at the molecular level and production of new advanced catalysts. Very thin films of precisely controlled molecular structure, 3D nanostructured systems, nanocomposites, all these put the cold plasma at the forefront of the twenty-first century methods of catalyst synthesis.

It is hard to imagine further progress in the construction of structured reactors without very thin catalytic films easily and cheaply deposited on precisely designed, sophisticated fillers designated for these reactors. The current development of hydrogen technologies also pins great hopes on new catalyst systems produced using the cold plasma. Although we are only at the beginning of the road, the results already obtained in the field of plasma-prepared catalytic electrodes for fuel cells and photocatalytic electrodes for PEC cells are very promising.

The ability to design new plasma nanomaterials possessing the desired catalytic properties and their preparation through controlled processes carried out in the cold plasma still pose a serious scientific challenge. As of now, the relation between cold plasma processes and the resulting structure and catalytic properties of deposited films is not entirely clear either. Further research in this field will undoubtedly lead to significant progress in science and provide attractive catalytic systems for technology.

Acknowledgements

I would like to thank the members of my team and my PhD students who are engaged in research on plasma-deposited catalytic films: Prof. P. Kazimierski, Dr. J. Balcerzak, Dr. J. Sielski, R. Kapica, Ł. Jóźwiak, M. Marchwicka, and W. Redzynia, for their help and assistance. I thank Prof. A. Kołodziej from the Institute of Chemical Engineering of Polish Academy of Science in Gliwice and Prof. J. Łojewska from Jagiellonian University in Kraków for their excellent scientific collaboration. I am also grateful to Dr. A. Kubiczek for his assistance in the preparation of this chapter.

This work was supported by the Polish National Science Center, on the basis of decision DEC-2012/07/B/ST8/03670. The financial support is gratefully acknowledged.

Author details

Jacek Tyczkowski[*]

Address all correspondence to: jacek.tyczkowski@p.lodz.pl

Department of Molecular Engineering, Faculty of Process and Environmental Engineering, Lodz University of Technology, Lodz, Poland

References

[1] Kizling MB, Järås SG. A review of the use of plasma techniques in catalyst preparation and catalytic reactions. Appl Catal A-Gen 1996;147:1–21.

[2] Liu CJ, Vissokov GP, Jang BWL. Catalyst preparation using plasma technologies. Catal Today 2002;72:173–84.

[3] Tyczkowski J, Kapica R. Cold plasma in the nanotechnology of catalysts. Pol J Chem Technol 2007;9:36–42.

[4] Kapoor A, Goyal SK, Bakhshi NN. Fischer–tropsch studies on a plasma-sprayed iron catalyst in a tube-wall reactor. Can J Chem Eng 1986;64:792–802.

[5] Dalai Ak, Bakhshi NN, Esmail MN. Conversion of syngas to hydrocarbons in a tube-wall reactor using Co–Fe plasma-sprayed catalyst: experimental and modeling studies. Fuel Process Technol 1997;51:219–38.

[6] Blanchard J, Abatzoglou N, Eslahpazir-Esfandabadi R, Gitzhofer F. Fischer-Tropsch synthesis in a slurry reactor using a nanoiron carbide catalyst produced by a plasma spray technique. Ind Eng Chem Res 2010;49:6948–55.

[7] Ismagilov ZR, Podyacheva OY, Solonenko OP, Pushkarev VV, Kuz'min VI, Ushakov VA, Rudina NA. Application of plasma spraying in the preparation of metal-supported catalysts. Catal Today 1999;51:411–7.

[8] Hinokuma S, Murakami K, Uemura K, Matsuda M, Ikeue K, Tsukahara N, Machida M. Arc plasma processing of Pt and Pd catalysts supported on γ-Al_2O_3 powders. Top Catal 2009;52:2108–11.

[9] Pristavita R, Meunier JL, Berk D. Carbon nano-flakes produced by an inductively coupled thermal plasma for catalyst applications. Plasma Chem. Plasma Process 2011;31:393–403.

[10] Maunier JL, Mendoza-Gonzalez NY, Prostavita R, Binny D, Berk D. Two-dimensional geometry control of graphene nanoflakes produced by thermal plasma for catalyst applications. Plasma Chem Plasma Process 2014;34:505–21.

[11] Wong FF, Lin CM, Chang CP, Huang JR, Yeh MY, Huang JJ. Recovery and reduction of spent nickel oxide catalyst via plasma sintering technique. Plasma Chem Plasma Process 2006;26:585–95.

[12] Wong FF, Chiang KC, Chen CY, Chen KL, Huang JJ, Yeh MY. Recovery and reduction of spent alumina-supported cobalt-molybdenum oxide catalyst via plasma sintering technique. Plasma Chem Plasma Process 2008;28:353–63.

[13] Seo JH, Lee MY, Kim JS. Radio-frequency thermal plasma preparation of nano-sized Ni catalysts supported on MgO nano-rods for partial oxidation of methane. Surf Coat Tech 2013;228:S91–6.

[14] Nehe P, Sivakumar G, Kumar S. Solution precursor plasma spray (SPPS) technique of catalyst coating for hydrogen production in a single channel with cavities plate type methanol based microreformer. Chem Eng J 2015;277:168–75.

[15] Witvrouwen T, Paulussen S, Selt B. The use of non-equilibrium plasmas for the synthesis of heterogeneous catalysts. Plasma Process Polym 2012;9:750–60.

[16] Kołodziej A, Łojewska J, Tyczkowski J, Jodłowski P, Redzynia W, Iwaniszyn M, Zapotoczny S, Kuśtrowski P. Coupled engineering and chemical approach to the design of a catalytic structured reactor for combustion of VOCs: Cobalt oxide catalyst on knitted wire gauzes. Chem Eng J 2012;200–202:329–37.

[17] Tyczkowski J. Cold plasma – a promising tool for the development of electrochemical cells. In: Shao Y, editor. Electrochemical Cells – New Advances in Fundamental Researches and Applications. Rijeka, Croatia: InTech; 2012. pp. 105–138.

[18] Chen MH, Chu W, Dai XY, Zhang XW. New palladium catalysts prepared by glow discharge plasma for the selective hydrogenation of acetylene. Catal Today 2004;89:201–4.

[19] Liu CJ. Yu K, Zhang YP, Zhu X, He F, Eliasson B. Characterization of plasma treated Pd/HZSM-5 catalyst for methane combustion. Appl Catal B-Environ 2004;47:95–100.

[20] Zhu X, Yu K, Li J, Zhang YP, Xia Q, Liu CJ. Thermogravimetric analysis of coke formation on plasma-treated Mo-Fe/HZSM-5 catalyst during non-oxidative aromatization of methane. React Kinet Catal Lett 2006;87:93–9.

[21] Legrand JC, Diamy AM, Riahi G, Randriamanantenasoa Z, Polisset-Thfoin M, Frais‐
      sard J. Application of a dihydrogen afterglow to the preparation of zeolite-supported
      metallic nanoparticles. Catal Today 2004;89:177–82.

[22] Liu CJ, Cheng DG, Zhang YP, Yu KL, Xia Q, Wang JG, Zhu XL. Remarkable enhance‐
      ment in the dispersion and low-temperature activity of catalysts prepared via novel
      plasma reduction-calcination method. Catal Surv Asia 2004;8:111–8.

[23] Zhang YP, Ma PS, Zhu X, Liu CJ, Shen Y. A novel plasma-treated Pt/NaZSM-5 cata‐
      lyst for NO reduction by methane. Catal Commun 2004;5:35–9.

[24] Zhu X, Huo PP, Zhang YP, Liu CJ. Characterization of argon glow discharge plasma
      reduced Pt/Al$_2$O$_3$ catalyst. Ind Eng Chem Res 2006;45:8604–9.

[25] Brüser V, Savastenko N, Schmuhl A, Junke H, Herrmann I, Bogdanoff P, Schröder K.
      Plasma modification of catalysts for cathode reduction of hydrogen peroxide in fuel
      cells. Plasma Process Polym 2007;4:594–8.

[26] Cheng DG, Okumura K, Xie Y, Liu CJ. Stability test and EXAFS characterization of
      plasma prepared Pd/HZSM-5 catalyst for methane combustion. Appl Surf Sci
      2007;254:1506–10.

[27] Lihong H, Wei C, Junqiang X, Jingping H, Min Y. Effect of glow discharge plasma on
      rhodium-based catalyst for oxygenates synthesis. Front Chem Eng China 2007;1:16–9.

[28] Zhu X, Huo P, Zhang YP, Cheng DG, Liu CJ. Structure and reactivity of plasma treat‐
      ed Ni/Al$_2$O$_3$ catalyst for CO$_2$ reforming of methane. Appl Catal B-Environ
      2008;81:132–40.

[29] Chu W, Wang LN, Chernavskii PA, Khodakov AY. Glow-discharge plasma-assisted
      design of cobalt catalysts for Fischer-Tropsch synthesis. Angew Chem Int Edit
      2008;47:5052–55.

[30] Liu G, Chu W, Shi X, Dai X, Yin Y. Plasma-assisted preparation of Ni/SiO$_2$ catalyst
      using atmospheric high frequency cold plasma jet. Catal Commun 2008;9:1087–91.

[31] Xu H, Chu W, Shi L, Deng S, Zhang H. Effects of glow discharge plasma on Cu–Co–
      Al-based supported catalysts for higher alcohol synthesis. React Kinet Catal Lett
      2009;97:243–7.

[32] Li Y, Jang BWL. Investigation of calcination and O$_2$ plasma treatment effects on TiO$_2$-
      supported palladium catalysts. Ind Eng Chem Res 2010;49:8433–8.

[33] Huang C, Bai S, Lv J, Li Z. Characterization of silica-supported cobalt catalysts pre‐
      pared by decomposition of nitrates using dielectric-barrier discharge plasma. Catal
      Lett 2011;141:1391–8.

[34] Zhang MB, Zhu XL, Liang X, Wang Z. Preparation of highly efficient Au/C catalysts
      for glucose oxidation via novel plasma reduction. Catal Commun 2012;25:92–5.

[35] Kuai PY, Liu CJ, Huo PP. Characterization of CuO-ZnO catalust prepared by decomposition of carbonates using dielectric-barrier discharge plasma. Catal Lett 2009;129:493–8.

[36] Hua W, Jin L, He X, Liu J, Hu H. Preparation of Ni/MgO catalyst for CO_2 reforming of methane by dielectric-barrier discharge plasma. Catal Commun 2010;11:968–72.

[37] Yan X, Liu Y, Zhao B, Wang Z, Wang Y, Liu CJ. Methanation over Ni/SiO_2: effect of the catalyst preparation methodologies. Int J Hydrogen Energ 2013;38:2283–91.

[38] Li Y, Wei Z, Wang Y. Ni/MgO catalyst prepared via dielectric-barrier discharge plasma with improved catalytic performance for carbon dioxide reforming of methane. Front Chem Sci Eng 2014;8:133–40.

[39] Shi P, Liu CJ. Characterization of silica supported nickel catalyst for methanation with improved activity by room temperature plasma treatment. Catal Lett 2009;133:112–8.

[40] Chen MH, Chu W, Zhu JJ, Dong L. Plasma assisted preparation of cobalt catalysts by sol–gel method for methane combustion. J Sol-Gel Sci Technol 2008;47:354–9.

[41] Teng F, Li M, Gao C, Zhang G, Zhang P, Wang Y, Chen L, Xie E. Preparation of block TiO_2 by hydrogen plasma assisted chemical vapor deposition and its photocatalytic activity. Appl Catal B-Environ 2014;148/149:339–43.

[42] Cheng DG. Plasma decomposition and reduction in supported metal catalyst preparation. Catal Surv Asia 2008;12:145–51.

[43] Liu C, Zou J, Yu K, Cheng DG, Han Y, Zhan J, Ratanatawanate C, Jang BWL. Plasma application for more environmentally friendly catalyst preparation. Pure Appl Chem 2006;78:1227–38.

[44] Cheng DG, Zhu X. Reduction of Pd/HZSM-5 using oxygen glow discharge plasma for a highly durable catalyst preparation. Catal Lett 2007;118:260–3.

[45] Jung B, Helleson M, Shi C, Rondinone A, Schwartz V, Liang C, Overbury S. Characterization of Al_2O_3 supported nickel catalysts derived from RF non-thermal plasma technology. Top Catal 2008;49:145–52.

[46] Médard N, Soutif JC, Lado I, Esteyries C, Poncin-Epaillard F. Synthesis and characterization of new supported metallocene catalysts using a cold plasma treatment: application to molecular mechanics and dynamics computational modeling. Macromol Chem Phys 2001;202:3606–16.

[47] Lopez LC, Buonomenna MG, Fontananova E, Iacoviello G, Drioli E, d'Agostino R, Favia P. A new generation of catalytic poly(vinylidene fluoride) membranes: Coupling plasma treatment with chemical immobilization of tungsten-based catalysts. Adv Funct Mater 2006;16:1417–24.

[48] Lopez LC, Buonomenna MG, Fontananova E, Drioli E, Favia P, d'Agostino R. Immobilization of tungsten catalysts on plasma modified membranes. Plasma Process Polym 2007;4:326–33.

[49] Liu X, Mou CY, Lee S, Li Y, Secrest J, Jang BWL. Room temperature O_2 plasma treatment SiO_2 supported Au catalysts for selective hydrogenation of acetylene in the presence of large excess of ethylene. J Catal 2012;285:152–9.

[50] Korovchenko P, Renken A, Kiwi-Misker L. Microwave plasma assisted preparation of Pd-nanoparticles with controlled dispersion on woven activated carbon fibres. Catal Today 2005;102/103:133–41.

[51] Chen J, Zhu ZH, Ma Q, Li L, Rudolph V, Lu GQ. Effects of pre-treatment in air microwave plasma on the structure of CNTs and the activity of Ru/CNTs catalysts for ammonia decomposition. Catal Today 2009;148:97–102.

[52] Chetty R, Maniam KK, Schuhmann W, Muhler M. Oxygen-plasma-functionalized carbon nanotubes as supports for platinum-ruthenium catalysts applied in electrochemical methanol oxidation. Chem Plus Chem 2015;80:130–5.

[53] Wirth S, Harnisch F, Quade A, Brüser M, Brüser V, Schröder U, Savastenko NA. Enhanced activity of non-noble metal electrocatalysts for the oxygen reduction reaction using low temperature plasma treatment. Plasma Process Polym 2011;8:914–22.

[54] Gao JS, Umeda K, Uchino K, Nakashima H, Muroaka K. Plasma breaking of thin films into nano-sized catalysts for carbon nanotube synthesis. Mater Sci Eng 2003;A352:308–13.

[55] Gao JS, Umeda K, Uchino K, Nakashima H, Muroaka K.. Control of sizes and densities of nano catalysts for nanotube synthesis by plasma breaking method. Mater Sci Eng 2004;B107:113–8.

[56] Chang SC, Lin TC, Li TS, Huang SH. Carbon nanotubes grown from nickiel catalyst prepared with H_2/N_2 plasma. Microelectron J 2008;39:1572–5.

[57] Jian SR. Effects of H_2 plasma pretreated Ni catalysts on the growth of carbon nanotubes. Mater Chem Phys 2009;115:740–3.

[58] Lee SH, Kwak EH, Kim HS, Lee SW, Jeong GH. Evolution of gold thin films to nanoparticles using plasma ion bombardment and their use as a catalyst for carbon nanotube growth. Thin Solid Films 2013;547:188–92.

[59] Zhang C, Xie R, Chen B, Yang J, Zhong G, Robertson J. High density carbon nanotube growth using a plasma pretreated catalyst. Carbon 2013;53:339–45.

[60] Li ZH, Tian SX, Wang HT, Tian HB. Plasma treatment of Ni catalyst via a corona discharge. J Mol Catal A-Chem 2004;211:149–53.

[61] Zhu YR, Li ZH, Zhou YH, Lv J, Wang HT. Plasma treatment of Ni and Pt catalysts for partial oxidation of methane. React Kinet Catal Lett 2006;87:33–41.

[62] Li K, Tang X, Yi H, Ning P, Xiang Y, Wang J, Wang C, Peng X. Research on manganese oxide catalysts surface pretreated with non-thermal plasma for NO catalytic oxidation capacity enhancement. Appl Surf Sci 2013;264:557–62.

[63] Tang X, Li K, Yi H, Ning P, Xiang Y, Wang J, Wang C. MnOx catalysts modified by nonthermal plasma for NO catalytic oxidation. J Phys Chem C 2012;116:10017–28.

[64] Shi C, Hoisington R, Jang BWL. Promotion effects of air and H_2 nonthermal plasmas on TiO_2 supported Pd and Pd–Ag catalysts for selective hydrogenation of acetylene. Ind Eng Chem Res 2007;46:4390–5.

[65] Li Y, Jang BWL. Non-thermal RF plasma effects on surface properties of Pd/TiO_2 catalysts for selective hydrogenation of acetylene. Appl Catal A-Gen 2011;392:173–9.

[66] Yi H, Zhao S, Tang X, Song C, Gao F, Zhang B, Wang Z. Low-temperature hydrolysis of carbon disulfide using the Fe–Cu/AC catalyst modified by non-thermal plasma. Fuel 2014;128:268–73.

[67] Feng Y, Zheng X. Plasma-enhanced catalytic CuO nanowires for CO oxidation. Nano Lett 2010;10:4762–6.

[68] Gao F, Liu B, Sun W, Wu Y, Dong L. The influence of microwave plasma pretreated CuO/TiO_2 catalysts in NO + CO reaction. Catal Today 2011;175:34–9.

[69] Kim HH, Tsubota S, Daté M, Ogata A, Futumura S. Catalyst regeneration and activity enhancement of Au/TiO_2 by atmospheric pressure nonthermal plasma. Appl Catal A-Gen 2007;329:93–8.

[70] Yang X, Kimmel YC, Fu J, Koel BE, Chen JG. Activation of tungsten carbide catalysts by use of an oxygen plasma pretreatment. ACS Catal 2012;2:765–9.

[71] Rossnagel SM, Cuomo JJ, Westwood WD, editors. Handbook of Plasma Processing Technology; Fundamentals, Etching, Deposition, and Surface Interactions. Park Ridge, New Jersey, USA: Noyes Publ; 1990. 523 p.

[72] Thomann AL, Rozenbaum JP, Brault P, Andreazza-Vignolle C, Andreazza P. Pd nanoclusters grown by plasma sputtering deposition on amorphous substrates. Appl Surf Sci 2000;158:172–83.

[73] Estrada W, Fantini MCA, de Castro SC, Polo da Fonseca CN, Gorenstein A. Radio frequency sputtered cobalt oxide coating: Structural, optical and electrochemical characterization. J Appl Phys 1993;74:5835–41.

[74] Witham CK, Chun W, Valdez TI, Narayanan SR. Performance of direct methanol fuel cells with sputter-deposited anode catalyst layers. Electrochem Solid-State Lett 2000;3:497–500.

[75] Brault P, Thomann AL, Rozenbaum JP, Cormier JM, Lefaucheux P, Andreazza C, Andreazza P. The use of plasma in catalysis: catalyst preparation and hydrogen production. Ann Chim Sci Mat 2001;26:69–77.

[76] Thomann AL, Rozenbaum JP, Brault P, Andreazza C, Andreazza P, Rousseau B, Estrade-Szwarckopf H, Berthet A, Bertolini JC, Cadera Santos Aires FJ, Monnet F, Mirodatos C, Charles C, Boswell R. Plasma synthesis of catalytic thin films. Pure Appl Chem 2002;74:471–4.

[77] Park KW, Choi JH, Ahn KS, Sung YE. PtRu alloy and PtRu–WO$_3$ nanocomposite electrodes for methanol electrooxidation fabricated by a sputtering deposition method. J Phys Chem B 2004;108:5989–94.

[78] Whitacre JF, Valdez T, Narayanan SR. Investigation of direct methanol fuel cell electrocatalysts using a robust combinatorial technique. J Electrochem Soc 2005;152:A1780–9.

[79] Alvisi M, Galtieri G, Giorgi L, Giorgi R, Serra E, Signore MA. Sputter deposition of Pt nanoclusters and thin films on PEM fuel cell electrodes. Surf Coat Tech 2005;200:1325–9.

[80] Saha MS, Gullá AF, Allen RJ, Mukerjee S. High performance polymer electrolyte fuel cells with ultra-low Pt loading electrodes prepared by dual ion-beam assisted deposition. Electrochim Acta 2006;51:4680–92.

[81] Whitacre JF, Valdez TI, Narayanan SR. A high-throughput study of PtNiZr catalysts for application in PEM fuel cells. Electrochim Acta 2008;53:3680–9.

[82] Xinyao Y, Zhongqing J, Yuedong M. Preparation of anodes for DMFC by co-sputtering of platinum and ruthenium. Plasma Sci Technol 2010;12:224–9.

[83] Ohnishi R, Katayama M, Kakanabe K, Kubota J, Domen K. Niobium-based catalysts prepared by reactive radio-frequency magnetron sputtering and arc plasma methods as non-noble metal cathode catalysts for polymer electrolyte fuel cells. Electrochim Acta 2010;55:5393–400.

[84] Mougenot M, Caillard A, Simoes M, Baranton S, Coutanceau C, Brault P. PdAu/C catalysts prepared by plasma sputtering for the electro-oxidation of glycerol. Appl Catal B-Environ 2011;107:372–9.

[85] Fofana D, Natarajan SK, Benard P, Hamelin J. High performance PEM fuel cell with low platinum loading at the cathode using magnetron sputter deposition. ISRN Electrochem 2013;2013:ID 174834, 6 pages.

[86] Caillard A, Charles C, Ramdutt D, Boswell R, Brault P. Effecy of Nafion and platinum content in a catalyst layer processed in a radio frequency helicon plasma system. J Phys D Appl Phys 2009;42:ID 045207, 9 pages

[87] Brault P, Caillard A, Thomann AL, Mathias J, Charles C, Boswell RW, Escribano S, Durand J, Sauvage T. Plasma sputtering deposition of platinum into porous fuel cell electrodes. J Phys D Appl Phys 2004;37:3419–23.

[88] Caillard A, Brault P, Mathias J, Charles C, Boswell RW, Sauvage T. Deposition and diffusion of platinum nanoparticles in porous carbon assisted by plasma sputtering. Surf Coat Tech 2005;200:391–4.

[89] Dilonardo E, Milella A, Cosma P, d'Agostino R, Palumbo F. Plasma deposited electrocatalytic films with controlled content of Pt nanoclusters. Plasma Process Polym 2011;8:452–8.

[90] Walter C, Brüser V, Quade A, Weltmann KD. Structural investigations of composites produced from copper and polypyrrole with a dual PVD/PE-CVD process. Plasma Process Polym 2009;6:803–12.

[91] Walter C, Kummer K, Vyalikh D, Brüser V, Quada A, Weltmann KD. Using a dual plasma process to produce cobalt-polypyrrole catalysts for the oxygen reduction reaction in fuel cells. I. Characterization of the catalytic activity and surface structure. J Electrochem Soc 2012;159:F494–500.

[92] Walter C, Kummer K, Vyalikh D, Brüser V, Weltmann KD. Tuning the process gas in a dual plasma process to enhanced the performance of cobalt–polypyrrole catalysts for the oxygen reduction reaction in fuel cells. J Electrochem Soc 2013;160:F1088–95.

[93] Walter C, Kummer K, Vyalikh D, Brüser V, Weltmann KD. Iron–polypyrrole catalyst for the oxygen reduction in proton exchange membrane fuel cells produced with a dual plasma process using varying magnetron powers and process gases. Plasma Chem Plasma Process 2014;34:785–99.

[94] Tyczkowski J. Electrical and optical properties of plasma polymers. In: Biederman H, editor. Plasma Polymer Films. London: Imperial College Press; 2004. pp. 143–216.

[95] Suhr H, Etspüler A, Feurer E, Oehr C. Plasma-deposited metal-containing polymer films. Plasma Chem Plasma Process 1988;8:9–17.

[96] Feurer E, Suhr H. Thin palladium films prepared by metal-organic plasma-enhanced chemical vapour deposition. Thin Solid Films 1988;157:81–6.

[97] Etspüler A, Suhr H. Deposition of thin rhodium films by plasma-enhanced chemical vapor deposition. Appl Phys A 1989;48:373–5.

[98] Feurer E, Kraus S, Suhr H. Plasma chemical vapor deposition of thin platinum films. J Vac Sci Technol A 1989;7:2799–802.

[99] Doblhofer K. Electrodes covered with thin, permeable polymer films. Electrochim Acta 1980;25:871–8.

[100] Doblhofer K, Eiselt I. Electrochemical performance of fixed-charge polymer films prepared on electrodes by plasma polymerization. Thin Solid Films 1984;118:181−5.

[101] Song SJ, Jung KW, Cho MD, Yang EJ, Sim HJ, Kim MY, Jang HG, Seo G, Cho DL. Preparation of amine-immobilized solid base catalysts by plasma polymerization of 1,2-diaminocyclohexane. Appl Catal A-Gen 2012;429/430:85−91.

[102] Guyon C, Barkallah A, Rousseau F, Giffard K, Morvan D, Tatoulian M. Deposition of cobalt oxide thin films by plasma-enhanced chemical vapour deposition (PECVD) for catalytic applications. Surf Coat Tech 2011;206:1673−9.

[103] Chen GL, Guyon C, Zhang ZX, Da Silva B, Da Costa P, Ognier S, Bonn D, Tatoulian M. Catkin liked nano-Co_3O_4 catalyst built-in organic microreactor by PEMOCVD method for trace CO oxidation at room temperature. Microfluid Nanofluid 2014;16:141−8.

[104] Witvrouwen T, Dijkmans J, Paulussen S, Selt B. Optimization of the synthesis conditions of a DBD plasma reactor for the synthesis of hybrid silica-based catalysts. Plasma Process Polym 2014;11:464−71.

[105] Dittmar A, Hoang DL, Martin A. TPR and XPS characterization of chromia–lanthana–zirconia catalyst prepared by impregnation and microwave plasma enhanced chemical vapour deposition methods. Thermochim Acta 2008;470:40−6.

[106] Dittmar A, Herein D. Microwave plasma assisted preparation of disperse chromium oxide supported catalysts: adsorption and decomposition of chromium acetylacetonate. Surf Coat Tech 2009;203:992−7.

[107] Laurent-Brocq M, Job N, Eskenazi D, Pireaux JJ. Pt/C catalyst for PEM fuel cells: Control of Pt nanoparticles characteristics through a novel plasma deposition method. Appl Catal B-Environ 2014;147:453−63.

[108] George SM. Atomic layer deposition: an overview. Chem Rev 2010;110:111−31.

[109] O'Neill BJ, Jackson DHK, Lee J, Canlas C, Stair PC, Marshall CL, Elam JW, Kuech TF, Dumesic JA, Huber GW. Catalyst design with atomic layer deposition. ACS Catal 2015;5:1804−25.

[110] Koyano G, Watanabe H, Okuhara T, Misono M. Structure and catalysis of cobalt oxide overlayers prepared on zirconia by low-temperature-plasma oxidation. J Chem Soc Faraday T 1996;92:3425−30.

[111] Ruddick VJ, Badyal PS. Nonequilibrium plasma activation of supported Cr(III) Phillips catalyst precursors. J Phys Chem B 1997;101:9240−3.

[112] Dhar R, Petrow PD, Liddell KNC, Ming Q, Moeller TM, Osman MA. Plasma-enhanced metal-organic chemical vapor deposition (PEMOCVD) of catalytic coatings for fuel cell reformers. IEEE T Plasma Sci 2005;33:138−46.

[113] Dhar R, Petrow PD, Liddell KNC, Ming Q, Moeller TM, Osman MA. Synthesis of Pt/ZrO$_2$ catalyst on Fecralloy substrates using composite plasma-polymerized films. IEEE T Plasma Sci 2005;33:2035−45.

[114] Kapica R, Redzynia W, Tyczkowski J. Characterization of palladium-based thin films prepared by plasma-enhanced metalorganic chemical vapor deposition. Mater Sci-Medzg 2012;18:128−31.

[115] Ten Eyck GA, Senkevich JJ, Tang F, Liu D, Pimanpang S, Karaback T, Wang GC, Lu TM, Jezewski C, Lanford WA. Plasma-assisted atomic layer deposition of palladium. Chem Vapor Depos 2005;11:60−6.

[116] Weber MJ, Mackus AJM, Verheijen MA, Longo V, Bol AA, Kessels WMM. Atomic layer deposition of highly-purity palladium films from Pd(hfac)$_2$ and H$_2$ and O$_2$ plasmas. J Phys Chem C 2014;118:8702−11.

[117] Wojcik H, Junige M, Bartha W, Albert M, Neumann V, Merkel U, Peeva A, Gluch J, Menzel S, Munnik F, Liske R, Utess D, Richter I, Klein C, Engelmann HJ, Ho P, Hossbach C, Wenzel C. Physical characterization of PECVD and PEALD Ru(-C) films and comparison with PVD ruthenium film properties. J Electrochem Soc 2012;159:H166−76.

[118] Park T, Choi D, Choi H, Jeon H. The properties of Ru films deposited by remote plasma atomic layer deposition on Ar plasma-treated SiO$_2$. Phys Status Solidi A 2012;209:302−5.

[119] Dittmar A, Kosslick H, Müller JP, Pohl MM. Characterization of cobalt oxide supported on titania prepared by microwave plasma enhanced chemical vapor deposition. Surf Coat Tech 2004;182:35−42.

[120] Barreca D, Devi A, Fischer RA, Bekermann D, Gasparotto A, Gavagnin M, Maccato C, Tondello E, Bontempi E, Depero LE, Sada C. Strongly oriented Co$_3$O$_4$ thin films on MgO(100) and MgAl$_2$O$_4$(100) substrates by PE-CVD. Cryst Eng Comm 2011;13:3670−3.

[121] Donders ME, Knops HCM, Sanden MCM, van deKessels WMM, Notten PHL. Remote plasma atomic layer deposition of Co$_3$O$_4$ thin films. J Electrochem Soc 2011;158:G92−6.

[122] Tyczkowski J, Kapica R, Łojewska J. Thin cobalt oxide films for catalysis deposited by plasma-enhanced metal-organic chemical vapor deposition. Thin Solid Films 2007;515:6590−5.

[123] Tyczkowski J. New materials for innovative energy systems produced by cold plasma technique. Funct Mater Lett 2011;4:341−4.

[124] Kapica R, Redzynia W, Kozanecki M, Chehimi MM, Sielski J, Kuberski SM, Tyczkowski J. Plasma deposition and characterization of copper-doped cobalt oxide nanocatalysts. Mater Sci-Medzg 2013;19:270—6.

[125] Battiston GA, Gerbasi R, Gregori A, Porchia M, Cattarin S, Rizzi GA. PECVD of amorphous TiO_2 thin films: Effect of growth temperature and plasma gas composition. Thin Solid Films 2000;371:126—31.

[126] da Cruz NC, Rangel EC, Wang J, Trasferetti BC, Davanzo CU, Castro SGC, de Moraes AB. Properties of titanium oxide films obtained by PECVD. Surf Coat Tech 2000;126:123—30.

[127] Maeda M, Watanabe T. Evaluation of photocatalytic properties of titanium oxide films prepared by plasma-enhanced chemical vapor deposition. Thin Solid Films 2005;489:320—4.

[128] Randeniya LK, Bendavid A, Martin PJ, Preston EW. Photoelectrochemical and structural properties of TiO_2 and N-doped TiO_2 thin films synthesized using pulsed direct current plasma-activated chemical vapor deposition. J Phys Chem C 2007;111:18334—40.

[129] Cho DL, Min H, Kim JH, Cha GS, Kim GS, Kim BH, Ohk SH. Photocatalytic characteristics of TiO_2 thin films deposited by PECVD. J Ind Eng Chem 2007;13:434—7.

[130] Borras A, Sanchez-Valencia JR, Widmer R, Rico VJ, Justo Angel, Gonzalez-Elipe AR. Growth of crystalline TiO_2 by plasma enhanced chemical vapor deposition. Cryst Growth Des 2009;9:2868—76.

[131] Borras A, Sanchez-Valencia JR, Garrido-Molinero J, Barranco A, Gonzalez-Elipe AR. Porosity and microstructure of plasma deposited TiO_2 thin films. Micropor Mesopor Mat 2009;118:314—24.

[132] Karches M, Morstein M, von Rohr PR, Pozzo RL, Giombi JL, Baltanás MA. Plasma-CVD-coated glass beads as photocatalyst for water decontamination. Catal Today 2002;72:267—79.

[133] Sobczyk-Guzenda A, Owczarek S, Szymanowski H, Gazicki-Lipman M. Amorphous and crystalline TiO_2 coatings synthesized with the RF PECVD technique from metal-organic precursor. Vacuum 2015;117:104—11.

[134] Kment S, Kluson P, Zabova H, Churpita A, Chichina M, Cada M, Gregora I, Krysa J, Hubicka Z. Atmospheric pressure barrier torch discharge and its optimization for flexible deposition of TiO_2 thin coatings on various surfaces. Surf Coat Tech 2009;204:667—75.

[135] Cho BO, Wang J, Chang JP. Metalorganic precursor decomposition and oxidation mechanisms in plasma-enhanced ZrO_2 deposition. J Appl Phys 2002;92:4238—44.

[136] Fujii E, Torii H, Tomozawa A, Takayama R, Hirao T. Iron oxide films with spinel, corundum and bixbite structure prepared by plasma-enhanced metalorganic chemical vapor deposition. J Cryst Growth 1995;151:134−9.

[137] Barreca D, Carraro G, Gasparotto A, Maccato C, Sada C, Singh AP, Mathur S, Mettenbörger A, Bontempi E, Depero LE. Columnar Fe_2O_3 arrays via plasma-enhanced growth: Interplay of fluorine substitution and photoelectrochemical properties. Int J Hydrogen Energ 2013;38:14189−99.

[138] Ramachandran RK, Dendooven J, Detavernier C. Plasma enhanced atomic layer deposition of Fe_2O_3 thin films. J Mater Chem A 2014;2:10662−7.

[139] Singh AP, Mettenbörger A, Golus P, Mathur S. Photoelectrochemical properties of hematite films grown by plasma enhanced chemical vapor deposition. Int J Hydrogen Energ 2012;37:13983−8.

[140] Mettenbörger A, Singh T, Singh AP, Järvi TT, Moseler M, Valldor M, Mathur S. Plasma-chemical reduction of iron oxide photoanodes for efficient solar hydrogen production. Int J Hydrogen Energ 2014;39:4828−35.

[141] Redzynia W, Balcerzak J, Sielski J, Kierzkowska-Pawlak H, Tyczkowski J. Promising PEMOCVD cobalt and ruthenium oxide layers for CO_2 methanation process. In: Tamulevicius S, editor. Advanced Materials and Technologies 2015; 27-31 August 2015; Palanga, Lithuania. Kaunas: Kaunas Univ Technol; 2015. p. 58.

[142] Marchwicka M, Jóźwiak Ł, Tyczkowski J. Plasma deposited Ru-based thin films for photoelectrochemical water splitting. Mater Sci-Medzg: submitted to editor.

[143] Lee SY, Park SJ. TiO_2 photocatalyst for water treatment applications. J Ind Eng Chem 2013;19:1761−9.

[144] Martinu L, Zabeida O, Klemberg-Sapieha JE. Plasma-enhanced chemical vapor deposition of functional coatings. In: Martin PM, editor. Handbook of Deposition Technologies for Films and Coatings: Science, Applications and Technology.Kidlington, Oxford, UK: Elsevier; 2010. pp. 392−465.

[145] Doblhofer K, Dürr W. Polymer-metal composite thin films on electrodes. J Electrochem Soc 1980;127:1041−4.

[146] Morosoff N, Patel DL, Lugg PS, Crumbliss AL. Chemical properties of metallated plasma polymers. J Appl Polym Sci: Appl Polym Symp 1984;38:83−98.

[147] Fujii E, Torii H, Tomozawa A, Takayama R, Hirao T. Preparation of cobalt oxide films by plasma-enhanced metalorganic chemical vapour deposition. J Mater Sci 1995;30:6013−8.

[148] Carraro G, Gasparotto A, Maccato C, Bontempi E, Bilo F, Peeters D, Sada, C, Barreca D. A plasma-assisted approach for the controlled dispersion of CuO aggregates into β iron(III) oxide matrices. Cryst Eng Comm 2014;16:8710−6.

[149] Carraro G, Maccato C, Gasparotto A, Kaunisto K, Sada C, Barreca D. Plasma-assisted fabrication of Fe_2O_3–Co_3O_4 nanomaterials as anodes for photoelectrochemical water splitting. Plasma Process Polym 2015;DOI: 10.1002/ppap.201500106.

[150] Jóźwiak Ł, Marchwicka M, Kazimierski P, Tyczkowski J. Plasma deposited thin films of iron oxide as oxygen reduction electrocatalysts in proton exchange membrane fuel cells. Mater Sci-Medzg: submitted to editor.

[151] Bartolome L, Imran M, Lee KG, Sangalang A, Ahn JK, Kim DH. Superparamagnetic γ-Fe_2O_3 nanoparticles as an easily recoverable catalyst for the chemical recycling of PET. Green Chem 2014;16:279−86.

[152] Lopatin S, Mikenin P, Pisarev D, Baranov D, Zazhigalov S, Zagoruiko A. Pressure drop and mass transfer in the structured cartridges with fiber-glass catalyst. Chem Eng J 2015;282:58–65.

[153] Łojewska J, Kołodziej A, Łojewski T, Kapica R, Tyczkowski J. Cobalt catalyst deposited on metallic microstructures for VOC combustion: Preparation by non-equilibrium plasma. Catal Commun 2008;10:142−5.

[154] Łojewska J, Kołodziej A, Kapica R, Knapik A, Tyczkowski J. In search for active non-precious metal catalyst for VOC combustion: evaluation of plasma deposited Co and Co/Cu oxide catalysts on metallic structured carriers. Catal Today 2009;147S:S94−8.

[155] Łojewska J, Kołodziej A, Łojewski T, Kapica R, Tyczkowski J. Structured cobalt oxide catalyst for VOC combustion. Part I: Catalytic and engineering correlations. Appl Catal A-Gen 2009;366:206−11.

[156] Tyczkowski J, Kapica R, Łojewska J, Kołodziej A. Method for obtaining a thin layer of catalyst material on substrates made of electrically conductive material. Patent. 2014;PL217586.

[157] Manzoli M, Boccuzzi F. Characterisation of Co-based electrocatalytic materials for O_2 reduction in fuel cells. J Power Sources 2005;145:161−8.

[158] Kazimierski P, Jóźwiak Ł, Sielski J, Tyczkowski J. Cobalt oxide-based catalysts deposited by cold plasma for proton exchange membrane fuel cells. Thin Solid Films 2015;594:5–11.

[159] Fujishima A, Honda K. Electrochemical photolysis of water at a semiconductor electrode. Nature 1972;238:37−8.

Chapter 3

Plasma Nitriding of Titanium Alloys

Afsaneh Edrisy and Khorameh Farokhzadeh

Additional information is available at the end of the chapter

http://dx.doi.org/10.5772/61937

Abstract

Titanium alloys are found in many applications where weight saving, strength, corrosion resistance, and biocompatibility are important design priorities. However, their poor tribological behavior is a major drawback, and many surface engineering processes have been developed to enhance wear in titanium alloys such as nitriding. Plasma (ion) nitriding, originally developed for ferrous alloys, has been adopted to address wear concerns in titanium alloys. Plasma nitriding improves the wear resistance of titanium alloys by the formation of a thin surface layer composed of TiN and Ti_2N titanium nitrides (e.g., compound layer). Nonetheless, plasma nitriding treatments of titanium alloys typically involve high temperatures (700–1100°C) that promote detrimental microstructural changes in titanium substrates, formation of brittle surface layers, and deterioration of mechanical properties especially fatigue strength. This chapter summarizes the previous and ongoing investigations in the field of plasma nitriding of titanium alloys, with particular emphasis on the authors' recent efforts in optimization of the process to achieve tribological improvements while maintaining mechanical properties. The development of low-temperature plasma nitriding treatments for $\alpha + \beta$ and near-β titanium alloys and further wear improvements by alteration of near-surface microstructure prior to nitriding are also briefly reviewed.

Keywords: Titanium alloys, plasma nitriding, compound layer, nitrogen diffusion zone, fatigue crack initiation, severe plastic deformation, surface failure mechanisms

1. Introduction

Titanium alloys are attractive candidates for aerospace, automotive, and biomedical industries due to their many advantages including, but not limited to, high strength-to-weight ratio, excellent corrosion/oxidation resistance, biocompatibility, and high fatigue resistance. However, their poor tribological behavior, for example, high and unstable coefficient of friction, high wear rate, and susceptibility to galling and scuffing, restricts their use for

applications where sliding is inevitable. Nitriding is a surface hardening process in which nitrogen atoms are utilized for surface modification and has been practiced for wear improvement in titanium alloys for many years. In conventional nitriding techniques such as salt bath treatments, gas nitriding, and hot isostatic pressing (HIP), nitrogen atoms are provided by liquid, gas, or solid media [1, 2]. In plasma nitriding, also known as ion nitriding and glow discharge nitriding, active nitrogen is provided by an ionized nitrogen-containing atmosphere [3–7].

The growing interest in this process is due to its high efficiency, cost-effectiveness for mass production, absence of pollution, and ease of process control [8–11]. Also, it is preferred over other nitriding technologies due to feasibility at lower temperatures, dimensional control, retention of surface finish, and versatility for achieving desired properties through modification of process parameters [8–13].

Plasma nitriding improves the tribological properties of titanium alloys through the formation of a thin surface layer mainly consisting of TiN and Ti_2N, that is, the compound layer. The nitrided microstructure also consists of a region of nitrogen-stabilized α-titanium, that is, α-case, and a nitrogen diffusion zone (typically 15–25 μm deep) underneath the compound layer [14, 15]. The plasma-nitrided microstructure is strongly correlated with the composition of titanium substrate and the nitriding process parameters such as duration, temperature, pressure, and composition of the nitriding medium, among which temperature has the most significant influence [12, 15–17].

Plasma nitriding of titanium alloys is typically carried out in the temperature range of 700–1100°C for 6–80 hours in a nitrogen-containing medium (N_2, N_2–Ar, N_2–H_2, or N_2–NH_3) [18–22]. The high temperatures involved in the process lead to grain growth, overaging, and microstructural transformations in titanium substrates that decrease the fatigue strength and ductility [10, 22, 23]. Moreover, the brittle nature of the compound layer and α-case and the high stiffness mismatch between the compound layer and the titanium substrate lead to premature failure initiation from the surface. As such, improvements in mechanical properties have been reported by practicing plasma nitriding at lower temperatures [10, 24–26]. The slow nitriding kinetics of titanium alloys at low temperatures can be enhanced by alteration of the microstructure in the surface vicinity via thermochemical (e.g., laser-assisted treatments and electron beam (EB) melting) and mechanical (e.g., mechanical attrition and shot peening) pretreatment processes [27, 28]. As a result, the depth and hardness of nitrogen diffusion will increase and thus the load bearing capacity of the plasma-nitrided surfaces will be improved [29]. Duplex treatments have also been introduced and developed in surface engineering of titanium alloys. Duplex treatments combine the positive effect of deep hardened layers achieved in the diffusion zone with low friction and high wear resistance of surface coatings leading to significant enhancements in load bearing capacity and wear resistance [30–36].

This chapter is organized into seven sections as follows. After the introduction, Section 2 gives a brief description of the fundamentals of plasma nitriding and the nitriding process parameters for titanium alloys. Section 3 presents a review of the literature on the effect of plasma nitriding on the microstructure of titanium alloys. The tribological and mechanical properties of plasma-nitrided titanium alloys are discussed in Section 4 and Section 5, respectively. Recent developments in the modification of plasma nitriding treatment via the introduction of duplex treatments and surface pretreatments are presented in Section 6, followed by a summary of the main findings in Section 7.

2. Fundamentals of plasma nitriding

In a conventional plasma nitriding treatment, an external voltage applied between the nitriding furnace (anode) and the workpiece (cathode) ionizes the nitriding gas and provides active nitrogen flux for surface modification [3, 4]. The nitriding atmosphere is generally composed of a buffer gas (e.g., Ar or H_2) and the reactant gas (e.g., N_2). A schematic illustration of voltage versus current density characteristic for glow discharge is shown in Figure 1. Plasma nitriding is conducted in the abnormal glow region where a uniform, stable glow covers the workpiece, and the current density is directly proportional to voltage drop and thus is easily controlled [13, 37]. As the applied voltage is provided by a direct current (DC) power supply, the continuous voltage and heat input may result in localized heating, overheating of thin sections, hollow cathode effect[1], arcing, and other damages to the surface. These problems can be avoided by utilizing a pulsed power supply where the heat input is controlled with duty cycles, usually in the order of 10–50% of the total cycles, without affecting the nitrogen activity or nitriding time [38]. In addition to DC and alternating current (AC) power supplies, plasma may be generated by radio frequency (r.f.) excitation [4, 39], microwaves, electromagnetic induction [4, 40], and electron emission configurations [4]. Some of the recent advances include intensifying the glow discharge using thermionically assisted triode glow discharge, plasma immersion ion implantation, electron cyclotron resonance, etc. The intensified plasma-assisted processes (IPAP) generate a higher density of energetic ions and increase the kinetics and efficiency of nitriding, making plasma nitriding feasible at low temperatures (less than 500°C), pressures, and durations. As a result, significant improvements in surface hardness, depth of nitrided layers, and tribological and fatigue properties have been reported [41–46].

Under the influence of the applied electric field, electrons accelerating from the cathode (workpiece) toward the anode collide with gas molecules of the nitriding atmosphere and generate an environment of positive and negative ions, electrons, and neutral and energetic atoms (e.g., plasma) within a few millimeters from the cathode. Subsequently, the positive ions accelerate toward the cathode over a comparatively short distance and cause a number of phenomena such as sputtering, diffusion to bulk, heating by radiation and collision, surface diffusion, plasma reaction, etc. As a result of these phenomena, surface modification takes place and the generated secondary electrons maintain the abnormal glow. This mechanism eliminates expensive precleaning operations and enhances nitriding kinetics at temperatures lower than those of the conventional nitriding techniques [47, 48].

---

1 Localized heating at narrow channels, holes, closely located workpieces, etc. where the generation of charge carriers is faster than their annihilation and the glow discharges contact each other or overlap, leading to a rise in current density and temperature.

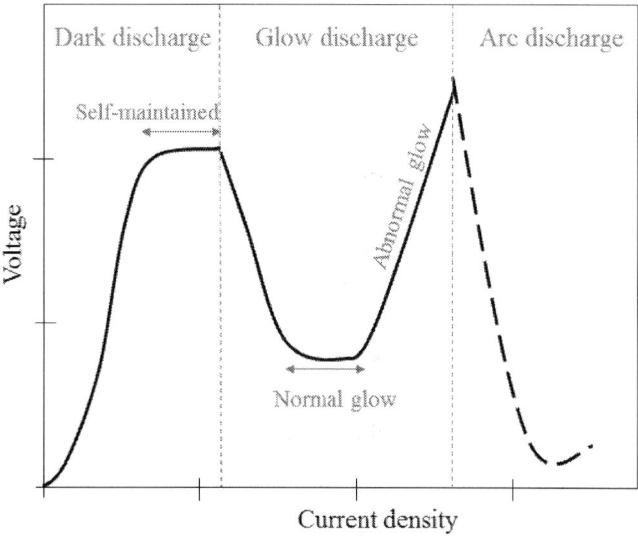

Figure 1. A schematic illustration of voltage vs. current density relationship for an electric glow discharge

Several mechanisms have been proposed to explain the mass transfer mechanism during plasma nitriding including the adsorption of atomic nitrogen model, ion adsorption model, nitrogen implantation model, Kölbel's model, and the NH^+ bombardment model. According to Kölbel's model, plasma nitriding initiates with sputter cleaning of the surfaces, followed by the reaction of sputtered atoms with nitrogen and subsequent deposition of nitride compounds on the surface. The NH^+ radical model is based on the generation of $N_mH_n^+$ molecular ions and radicals in the presence of hydrogen in the nitriding gas mixture. It is proposed that the adsorbed NH^+ radicals on the surface of workpiece dissociate and liberate nitrogen atoms that diffuse into the structure [4, 12, 13, 18].

Plasma nitriding treatments for titanium alloys are typically carried out at 700–1100°C for 6–80 hours. The atmosphere is composed of a nitrogen containing mixture, for example, N_2, N_2–Ar, N_2–H_2, or N_2–NH_3 mixtures, with pressures that vary in the range of 0.5–1.3 kPa [18–22]. In addition to the nitriding time and temperature, the pressure and composition of the nitriding gas also have a significant effect on the microstructure of plasma-nitrided titanium alloys [15, 17]. For instance, the presence of hydrogen in the nitriding mixture enhances nitrogen diffusivity by removing the inherent oxide layer on the surface of titanium alloy that interferes with the nitriding process. In another mechanism suggested by Tamaki et al. [18], the presence of hydrogen in the nitriding medium results in the formation of H^+, NH^+, and NH^{2+} radicals that have a catalytic effect on nitriding kinetics. Conversely, the presence of Ar in the nitriding mixture reduces the depth of nitrogen diffusion and causes random formation of Ti_2N and TiN through sputtering/re-deposition reactions or homogeneous reactions in the plasma [4, 18, 49, 50]. The following sections review the previous investigations on the microstructure and properties of plasma-nitrided titanium alloys focusing especially on the authors' recent work in the field.

3. Microstructure of plasma-nitrided titanium alloys

3.1. Classification of titanium alloys

Titanium has an allotropic phase transformation from its low-temperature hexagonal close-packed (HCP) crystal structure (α-phase) to body-centered cubic (BCC) crystal structure (β-phase) at 882°C, that is, β-transus temperature. Alloying elements have a strong influence on this allotropic transformation; elements such as aluminum, boron, nitrogen, and oxygen increase the β-transus temperature and are referred to as α-stabilizers. Conversely, β-stabilizing elements decrease the β-transus temperature and stabilize the β-phase at room temperature through either eutectoid transformations (iron, hydrogen) or solid solutions (vanadium, molybdenum, and niobium). Thus, depending on the type and content of their alloying elements, titanium alloys can have a single-phase α-structure (α-Ti alloys), α phase with a small (2–5 vol.%) β-phase content (near-α Ti alloys), single-phase β-structure (β-Ti alloys), or a combination of both phases ($\alpha + \beta$ alloys). The families of α- and near-α titanium alloys are famous for their microstructural stability at high temperatures and corrosion resistance; however, their microstructures cannot be modified by heat treatments. Commercially, pure (CP) titanium, Ti–5Al–2.5Sn, Ti–3Al–2.5Sn, Ti–8Al–1Mo-1V, and Ti–6Al–2Sn–4Zr–2Mo are some of the commercial alloys of this family. In contrast, a wide range of microstructures and properties can be achieved in $\alpha + \beta$ and β-titanium alloys by heat treatments and thermomechanical routes. Ti–6Al–4V, Ti–3Al–2.5V, Ti–6Al–2Sn–4Zr–6Mo, Ti–6Al–6V–2Sn, and Ti–4Al–4Mo–4Sn–0.5Si are some of the popular $\alpha + \beta$ titanium alloys. The class of β (and near-β) titanium alloys contains a high concentration of β-stabilizing elements to retain a fully β-phase microstructure at room temperature. These alloys offer good hardenability and formability and resistance to corrosion and hydrogen embrittlement. Ti–13V–11Cr–3Al, Ti–10V–2Fe–3Al, Ti–11.5Mo–6Zr–4.5Sn (β-III), and Ti–3Al–8V–6Cr–4Zr–4Mo (β-C) are among commercially used β-Ti alloys [1-7].

The behavior of Ti–6Al–4V ($\alpha + \beta$ titanium alloy) and Ti–10V–2Fe–3Al (near-β titanium alloy) alloys during plasma nitriding were investigated by the authors. The extra-low interstitial (ELI) grade Ti–6Al–4V had a mill-annealed microstructure consisting of equiaxed α-grains with retained β-particles at α-grain boundaries and fine β-particles inside the α-grains (Figure 2a). The Ti–10V–2Fe–3Al microstructure consisted of elongated primary α-plates in various orientations and aspect ratios inside β-grains with the occasional presence of α-phase at β-grain boundaries (Figure 2b). The chemical composition, average grain size, and β-phase content of the as-received alloys are presented in Table 1.

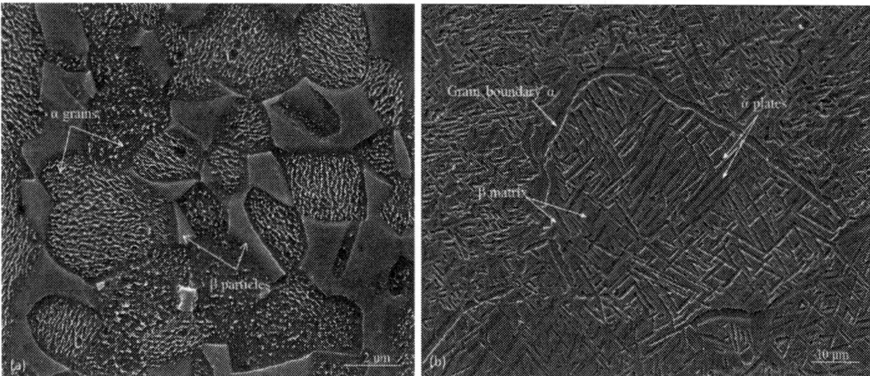

Figure 2. Scanning electron microscopy (SEM) images showing the typical microstructure of the as-received titanium alloys used in plasma nitriding investigations. (a) Ti–6Al–4V had a mill-annealed microstructure consisting of equiaxed α-grains, retained β-particles at α-grain boundaries, and fine recrystallized β-particles inside the α-grains. α-grains were delineated by the different orientation of fine recrystallized β-particles inside the α-grains (etched in Kroll's solution) [51]. (b) Ti–10V–2Fe–3Al microstructure consisted of elongated primary α-plates in various orientations and aspect ratios inside β-grains (marked by arrows) with the occasional presence of α-layers at β-grain boundaries (etched in glycerol + hydrofluoric acid (1:1) solution) [52].

	Chemical Composition (Wt. %)								Hardness (HV)	Grain size (µm)	β content** (Vol. %)
	Fe	V	Al	C*	O*	N*	H*	Ti			
Ti-6Al-4V (ELI)	0.17	4.10	5.98	0.007	0.13	0.012	0.009	balance	381.5 ± 22.8	3.8	24
Ti-10V-2Fe-3Al	1.73	10.65	2.99	-	-	-	-	balance	359.3 ± 15.4	42.4	63.5

* H, O, and N contents were determined by inert gas fusion and the C content by combustion

** The β phase content was determined following a procedure adopted from Tiley et al. [53] by overlaying a regular grid of points on the SEM images and dividing the number of points within the β phase by the total number of points.

Table 1. Properties of the as-received Ti-6Al-4V and Ti-10V-2Fe-3Al alloys used in the plasma nitriding experiments. Chemical compositions were measured using inductively coupled plasma optical emission spectrometry (ICP-OES).

3.2. Effect of plasma nitriding on microstructure of titanium alloys

Plasma nitriding of titanium alloys results in the formation of a *compound layer* on the surface that mainly consists of TiN, δ phase, with FCC crystal structure ($Fm\bar{3}m$ space group) and Ti$_2$N, ε phase, with tetragonal crystal structure (P4$_2$-mnm space group). TiN is stable over a wide range of nitrogen contents, TiN$_x$ (0.43 < × < 1.08), and has a typical hardness of 2500 HV. Ti$_2$N is stable over a small composition of ~33.3 at.% N and has a maximum hardness of 1500 HV [49, 50, 54–56]. In plasma nitriding investigations of Raveh et al. [14], the compound layer was composed of a TiN layer on top of a mixture of randomly oriented polycrystalline TiN and highly oriented Ti$_2$N. They also reported clusters of fine (50–100 A°) precipitates consisting of Al, V, Cr, and Fe elements underneath the compound layer. The compound layer has a typical thickness of 1–4 μm and may contain traces of Ti$_4$N$_{3-x}$, TiN$_{0.26}$, and TiO$_2$ in addition to TiN and Ti$_2$N [6, 14, 15, 52, 57].

Underneath the compound layer, a region of nitrogen-stabilized α-titanium, that is, *α-case* and a nitrogen *diffusion zone* form by interstitial solid solution of nitrogen atoms in the titanium structure. The formation of α-case, with typical hardness of 800–1000 HV, is not favorable due to its negative effect on the ductility and fatigue strength of titanium alloys. When α + β and β-titanium alloys are nitrided at high temperatures (typically higher than 800°C), α-case forms a continuous α-Ti layer underneath the compound layer as shown in Figure 3; however, at lower temperatures (< 800°C), coarse and columnar α-stabilized grains were reported underneath the compound layer [10, 49, 50, 54, 55, 58]. Similarly, α-case may be formed during plasma nitriding of α-titanium alloys; however, it is more difficult to distinguish unless hydride precipitation in the substrate creates contrast against the α-case [10].

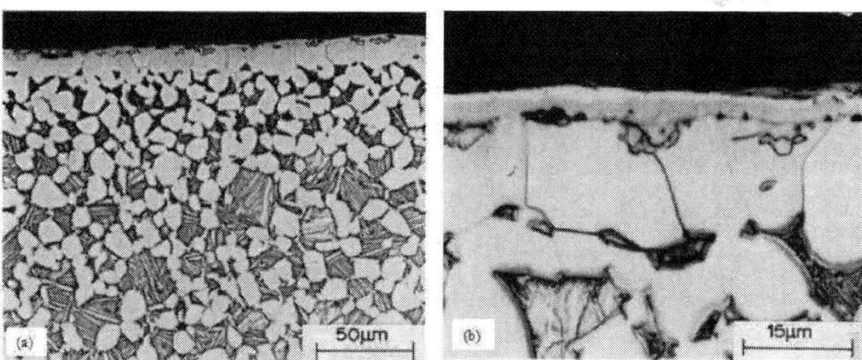

Figure 3. Optical micrographs of Ti–6Al–4V alloy after plasma nitriding in nitrogen for 3 hours at 1000°C showing the formation of an α-case underneath the compound layer at (a) low magnification and (b) high magnification [10].

Nitrogen atoms harden the diffusion zone with a profile that has its maximum at the near surface and gradually decreases toward the bulk. The depth of nitrogen diffusion is dependent on the process parameters as well as the phase composition of titanium alloys due to different solubility limits and diffusion rates of nitrogen in α- and β-titanium [5, 7, 59]. Nitrogen diffuses three times faster in β-Ti compared with α-Ti but has limited solubility in the β-phase [60, 61]. A study conducted by Il'in et al. [59] revealed that the addition of α-stabilizing elements such as Al slowed down the diffusion of nitrogen in α-titanium alloys. Conversely, β-titanium alloys had a shallow diffusion zone due to inadequate solubility of nitrogen in the β-phase. A higher depth of diffusion zone was achieved in α + β Ti alloys that possess a combination of high nitrogen solubility and diffusivity.

The effect of plasma nitriding process parameters on the microstructure and properties of titanium alloys was studied by several researchers [6, 16, 57, 62]. According to da Silva et al.

[16], the composition of plasma-nitrided Ti–6Al–4V surfaces was a very complex function of the nitriding parameters, for example, time, temperature, gas mixture composition, and pressure, among which temperature had the most significant influence. The x-ray diffraction (XRD) studies on plasma-nitrided surfaces revealed that nitriding gas pressure had a minor influence on the nitrided microstructure; however, increasing the partial pressure of nitrogen in the mixture resulted in preferential orientation of Ti_2N and promoted the formation of TiN. They interpreted that the shift observed in β-Ti XRD reflections corresponded to lattice distortions in this phase and likely residual strains induced by nitriding. In addition to TiN and Ti_2N titanium nitrides, they also reported XRD reflections of V_2O, orthorhombic TiO_2, and nitrogen-deficient ζ-Ti_4N_{3-x} nitride in the compound layers formed at certain nitriding conditions. Yildiz et al. [62] found that time and temperature of nitriding were directly correlated with the surface hardness and roughness, thickness of the compound layer, TiN content in the compound layer, and depth of the diffusion zone. They also reported that Al atoms from the Ti–6Al–4V substrate tend to segregate underneath the compound layer and form an Al-enriched layer, which impedes the inward diffusion of nitrogen during plasma nitriding. The presence of an Al-enriched layer underneath the compound layer was also confirmed by other researchers [6, 57].

The authors' investigations on optimizing the plasma nitriding process parameters were aimed at preventing the formation of α-case and microstructural changes in titanium substrates which are detrimental to some mechanical properties for example fatigue [52, 63]. Plasma nitriding treatments were performed in an industrial unit[2] consisting of a DC power supply and control console, nitriding chamber and base assembly, hydraulic lifting system for lifting the chamber from the base, temperature measurement system, evacuation system, gas supply, inlet/outlet, and pressure gauges. A schematic illustration of the nitriding set-up is given in Figure 4. No auxiliary heat source was utilized, and the heat was generated by the glow discharge.

Titanium coupons were placed at the geometric center of a cathodic cage in order to avoid common problems associated with DC plasma nitriding such as arcing damage, hollow-cathode, and edging effect. Prior to each run, the chamber was cleaned through a couple of filling/evacuation steps using Ar-H_2 (1:1) mixture. Subsequently, the coupons were sputter-cleaned at the process temperature in an Ar-H_2 (1:1) atmosphere for a few hours after a stable glow discharge was maintained. The nitriding gas mixture was purged into the chamber and the nitriding cycle was completed. Consequently, the coupons were cooled down to ~150°C inside the furnace under vacuum. Plasma nitriding was carried out at different temperatures (500–800°C), durations (4–87 hours), and pressures (67–533 Pa) in N_2/Ar and/or N_2/H_2 mixtures with nitrogen contents ranging between 1.5% and 75%. After each plasma nitriding treatment, microstructural and chemical analyses were performed and the samples were subjected to analytical microscopy examinations.

Phase constituents of the nitrided surfaces were identified using a Rigaku DMAX-1200 x-ray diffractometer and a Rigaku MiniFlex x-ray diffractometer equipped with Cu K_α tubes. The

---

2 Exactatherm Ltd., 2381 Anson Dr., Mississauga, ON, Canada

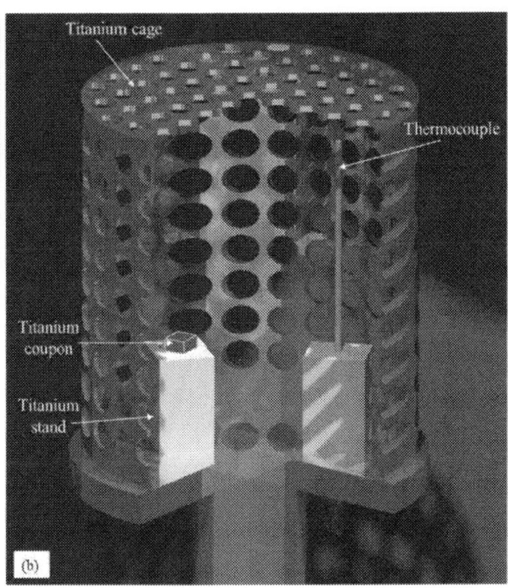

Figure 4. (a) A schematic illustration of the plasma nitriding set-up, no external heating source was used and (b) titanium coupons were placed at the geometric center of a cathodic cage (the cage was also made of Ti alloy to minimize contamination) [51].

developed residual stress level on the surface after plasma nitriding was measured with the $\sin^2\psi$ x-ray diffraction method using a laboratory non-destructive residual stress measurement LXRD system[3]. Surface morphology and cross-sectional microstructure of the nitrided alloys were examined under an FEI Quanta 200 Field Emission Gun (FEG) scanning electron microscope (SEM) equipped with an x-ray energy dispersive spectroscopy (EDS) detector. For cross-sectional study of microstructures, the nitrided surfaces were plated using an electroless nickel coating treatment prior to metallographic preparation to protect the near-surface microstructure against mechanical damage. The cross sections were then cut using a low-speed diamond saw and mounted in epoxy. The metallographic procedure began with serial wet grinding with silicon carbide abrasive papers in successive steps, followed by polishing with diamond. Ti–10V–2Fe–3Al surfaces had an additional fine polishing step using 0.06 μm colloidal silica suspension. The polished samples were cleaned in an ultrasonic bath of ethanol prior to microscopic examinations.

A scanning/transmission electron microscope (S/TEM) equipped with an energy dispersive x-ray spectrometer (Titan 80–3000) was used to investigate the microstructure of nitrided titanium alloys in the surface vicinity. Moreover, x-ray microanalysis coupled with electron diffraction patterns and electron energy-loss spectroscopy (EELS) was used to obtain structural information and chemical composition of the nitrided microstructures. TEM thin-foil speci-

---

3 Proto Manufacturing Ltd., 2175 Solar Crescent, OldCastle, ON, Canada

mens were prepared using focused ion beam (FIB) milling lift-out technique [64] (dual-beam Zeiss NVision 40[4]) with thicknesses of around 50 nm suitable for high-resolution TEM (HRTEM) imaging and EELS. Ion milling was performed using a gallium ion source at 30 kV and a thin (1–2 µm) layer of platinum was deposited on the milling location to protect the surface features from the milling damage.

The microstructure of plasma-nitrided Ti–6Al–4V consisted of a 1.9-µm compound layer on the surface followed by a 44.4-µm-deep nitrogen diffusion zone (Figure 5). The compound layer consisted of Ti_2N, TiN, and $TiN_{0.3}$ nitrides (Figures 6 and 7). Plasma-nitrided Ti–10V–2Fe–3Al had a 0.5-µm compound layer consisting of TiN ($Fm\bar{3}m$), Ti_2N ($P4_2$-mnm), V_2N ($P\bar{3}1m$), with the possibility of Fe_8N (I4-mmm) (Figures 8 and 9). Vanadium nitride, rarely reported in the literature, was observed along titanium nitride grain boundaries possibly due to faster diffusion of vanadium along grain boundaries (acting as short-circuit paths) at the nitriding temperature [18, 36-37].

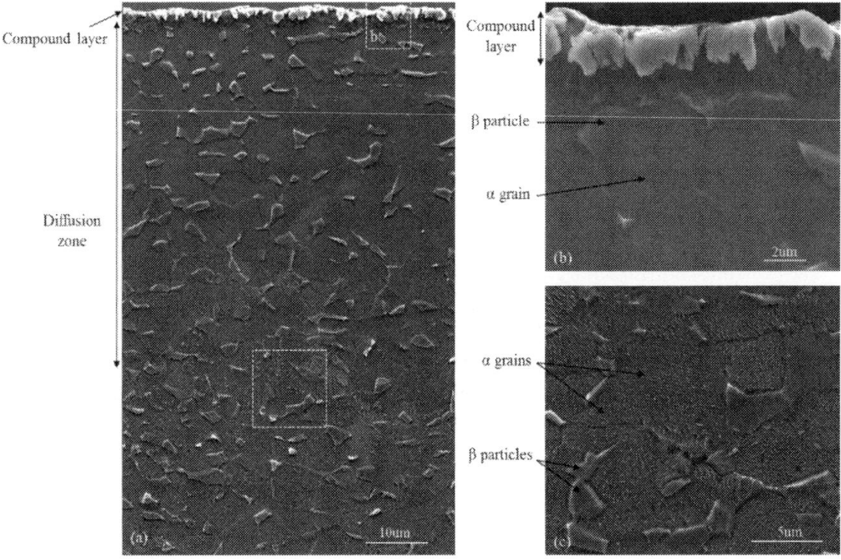

Figure 5. Typical cross-sectional SEM images of plasma-nitrided Ti–6Al–4V microstructure at 600°C consisting of a thin compound layer (1.9 ± 0.5 µm) and a deep (44.4 ± 5.6 µm) diffusion zone. Diffusion zone was delineated by the dissolution of fine recrystallized β-particles inside the α-grains (because nitrogen is an α-stabilizing element) and confirmed with the microhardness-depth profile measurements. (b) Higher magnification image of the enclosed area marked as "b" in Figure 5a confirming the absence of α-case underneath the compound layer. (c) Higher magnification image of the enclosed area marked as "c" in Figure 5a confirming that no phase transformations were observed in the titanium substrate [26].

---

4 The Canadian Centre for Electron Microscopy (CCEM), Brockhouse Institute for Materials Research, B161 A. N. Bourns Building, 1280 Main Street West, Hamilton, ON, Canada

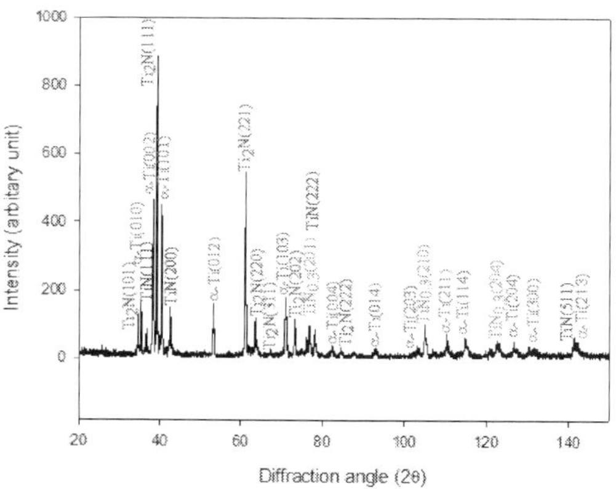

Figure 6. X-ray diffraction (XRD) analysis showing the presence of TiN, Ti$_2$N, and TiN$_{0.3}$ in the compound layer [63].

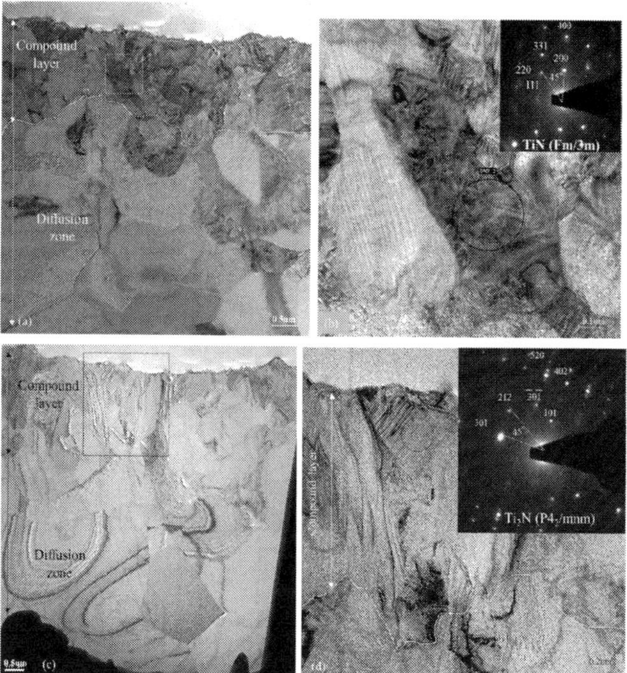

Figure 7. Bright-field TEM images and the corresponding selected area electron diffraction patterns of the plasma-nitrided Ti–6Al–4V microstructure showing the formation of TiN (space group: $Fm\bar{3}m$) and Ti$_2$N (space group: P4$_2$-mnm) nitrides in the compound layer [63].

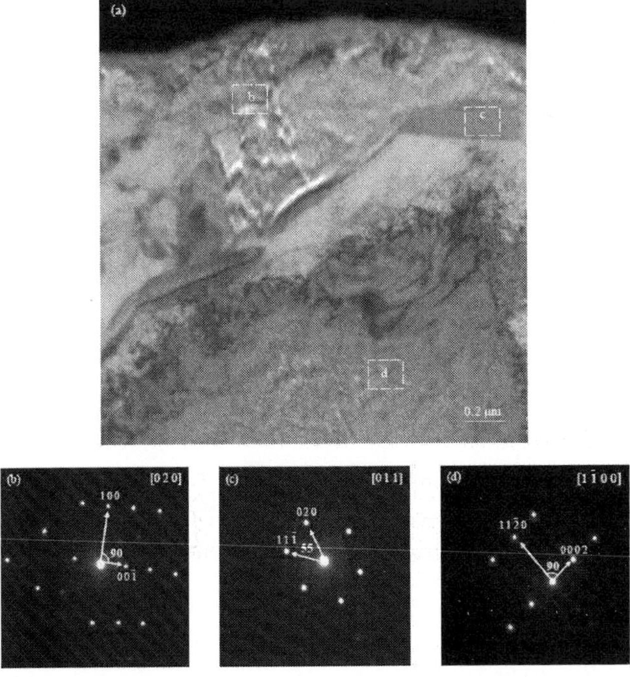

Figure 8. A bright-field TEM image of the plasma-nitrided Ti–10V–2Fe–3Al microstructure and the corresponding convergent beam electron diffraction (CBED) patterns confirming the formation of Ti$_2$N (space group: P4$_2$-mnm) and V$_2$N (space group: $P\bar{3}1m$) in the compound layer. V$_2$N nitrides were located at the grain boundaries [52].

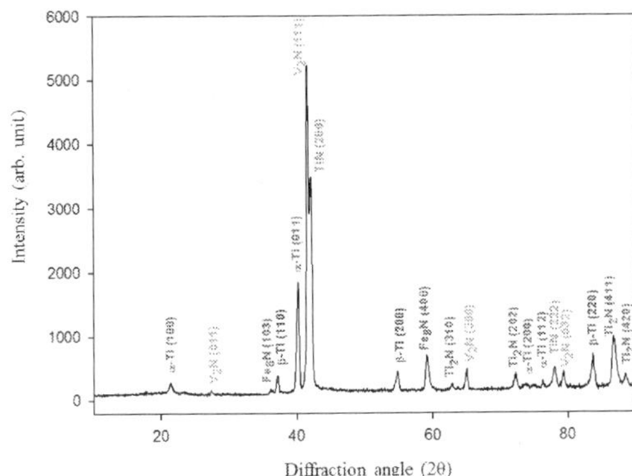

Figure 9. X-ray diffraction (XRD) pattern of the surface of plasma-nitrided Ti–10V–2Fe–3Al indicating the presence of titanium nitrides (TiN, Ti$_2$N), vanadium nitride (V$_2$N), and the possibility of iron nitride (Fe$_8$N) [52].

The depth of nitrogen diffusion was confirmed by microhardness versus depth profiles due to the solid-solution strengthening effect of interstitial nitrogen atoms in titanium structure. Microhardness profiles were obtained by microindentations at a maximum load of 245 mN (25 grf) using a diamond Vickers tip on tapered cross sections (CSM Instruments Micro-Combi Tester) and the microhardness values were determined based on load-penetration depth curves [65]. A deep nitrogen diffusion zone (35 ± 5 μm) in Ti–10V–2Fe–3Al after an 8-hour treatment (Figure 10) was attributed to the high β-phase content of the alloy with high nitrogen diffusivity [18, 22, 30, 47]. The prolonged sputtering of the surfaces prior to the nitriding process in this research also has accelerated the nitrogen diffusion by formation of several dislocations inside the α-grains. It has also been reported that the Ar+H$_2$ sputtering increased the nitriding kinetics in ferrous alloys by providing easy diffusion paths for nitrogen interstitials and eliminating inherent surface oxides [48, 49].

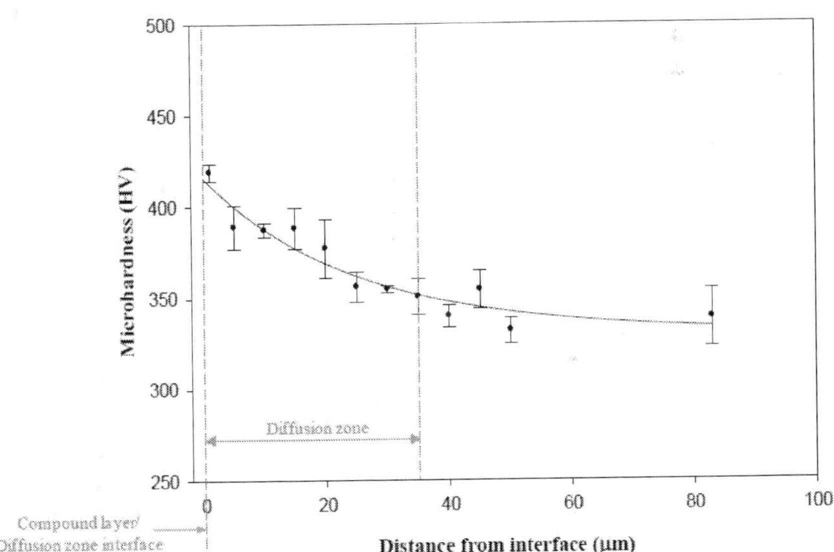

Figure 10. The microhardness-depth profile of the plasma-nitrided Ti–10V–2Fe–3Al. The diffusion zone extended to a depth of ~35 ± 5 μm underneath the compound layer [52].

4. Tribological properties of plasma-nitrided titanium alloys

Titanium alloys are renowned for their poor tribological characteristics, their strong adhesion tendency, high wear rates, susceptibility to seizure and galling, and high and unstable coefficient of friction [66–71]. Severe material transfer occurs during sliding of titanium alloys against themselves and other alloys[5], ceramics, or even polymers [72–75]. This behavior is attributed to their low resistance to plastic shearing, low work hardening, the brittle nature and low sliding resistance of their inherent surface oxides, and possibly microstructural embrittlement by oxygen dissolution [67, 76–78].

Although it is well established that plasma nitriding improves the wear resistance of titanium alloys, the tribological behavior of plasma-nitrided titanium surfaces is a function of the hardness of surface layers, surface roughness, the presence of brittle nitriding microstructural features, and the depth of nitrogen diffusion zone. Thus, the wear resistance of plasma-nitrided titanium surfaces is dependent on the process parameters, microstructure, and chemical composition of the titanium substrate and sliding conditions.

One of the earliest investigations on wear behavior of plasma-nitrided titanium alloys was performed by Bell et al. [10]. They reported that the high wear rate and high coefficient of friction of Ti–6Al–4V (COF = 0.3) were effectively improved by plasma nitriding (800°C, 12 hours, N_2 atmosphere). They attributed the low coefficient of friction (COF = 0.05), low wear rate, and anti-scuffing characteristics of the plasma-nitrided surfaces to the hardness of the compound layer.

Taktak and Akbulut [79] performed plasma nitriding at temperatures ranging from 700°C to 900°C for different durations and reported that the treatment performed at the highest temperature for the longest duration exhibited the best wear resistance and friction behavior. This was correlated with the compound layer thickness and hardness; however, the compound layer failure at high applied loads resulted in a transition to high wear rates, close to those of the untreated alloy. Higher plasma nitriding temperatures of 900°C, 1000°C, and 1100°C were studied by Shashkov [80] in different titanium alloys for different durations in N_2 atmosphere. They reported some wear resistance for plasma-nitrided titanium alloys at 900°C and reduction of wear resistance for nitrided alloys at higher temperatures and durations. They clarified that plasma nitriding of titanium alloys at elevated temperatures was accompanied by embrittlement due to the formation of brittle nitride layers and microstructural changes in the bulk such as grain coarsening.

The tribological behavior of Ti–6Al–4V alloy plasma-nitrided at temperatures ranging between 450°C and 520°C were studied by Yilbas et al. [81] against ruby counterpart in lubricated conditions. They reported that the scuffing and high friction coefficients, observed in the untreated alloy after a few sliding cycles, were avoided in the plasma-nitrided surfaces before the breakthrough of nitrided layers. This breakthrough was delayed when plasma nitriding was performed at 520°C due to a higher depth and hardness of the diffusion zone.

The wear mechanisms of plasma-nitrided Ti–6Al–4V alloy were studied by Molinari et al. [55]. The plasma nitriding treatments were carried out at different temperatures of 700°C, 800°C, and 900°C in a N_2–H_2 (4:1) gas mixture for 24 hours and tested against disks that had seen the same treatments. They found that the wear resistance was dependent on the microstructure and surface roughness of the nitrided surfaces and wear test conditions. At the lowest sliding

---

5 Except Ag–10Cu alloy and Babbitt metal (grade 2) that showed no galling against Ti–4Al–4Mn alloy

speed (0.3 m/s), nitriding inhibited oxidation-dominated wear in Ti–6Al–4V and decreased wear by providing resistance against adhesion and microfragmentation. However, compound layer failure that occurred at low-to-moderate loads for low-temperature (700°C) nitriding and at high loads for elevated temperature (800°C and 900°C) nitriding led to third-body abrasion and high wear rates. Plasma nitriding reduced wear at high sliding velocities (0.6 m/s and 0.8 m/s) by restricting the extensive plastic deformation and delamination wear of the titanium substrate. The elevated temperature plasma treatments exhibited better wear performance at high loads due to their higher thickness of the compound layer and depth of diffusion zone, but lower wear resistance at low loads possibly due to the increased residual stress level in the nitride layers and higher surface roughness.

The same authors also studied the wear mechanisms of Ti–6Al–4V alloy plasma nitrided at 800°C under lubricated rolling–sliding conditions [66]. Their findings indicated that the efficiency of lubricants was reduced by plasma nitriding, resulting in higher friction, possibly due to the better wettability and higher ionic character of the inherent titanium oxides compared with the titanium nitrides of the compound layer. However, the wear rates decreased by plasma nitriding and had an inverse relationship with the nitriding duration. The insufficient wear improvement after a short-duration process (8 hours) was related to the inadequate depth of nitriding and lack of support for the compound layer. On the other hand, an elongated nitriding treatment (24 hours) promoted a deeper diffusion zone that retained the compound layer on the surface under sliding conditions.

The positive role of a deep diffusion zone in tribological behavior of plasma-nitrided titanium alloys was also established by Nolan et al. [82], upon comparing plasma-nitrided and physical vapor deposition (PVD)-coated surfaces. Significantly lower mass loss was observed for plasma-nitrided Ti–6Al–4V with a 2-μm compound layer on top of an approximately 40-μm diffusion zone compared with the PVD-coated surface with a 2-μm TiN layer. Examination of worn surfaces and cross sections indicated that the compound layer endured the applied loads without significant subsurface plastic deformation. They proposed that the strengthening effect of the solid solution nitrogen atoms within the diffusion zone provided mechanical support for the compound layer and improved the wear resistance. Moreover, a gradual hardness profile decreased the chances of interfacial debonding due to a gradual change of elastic modulus and hardness from the surface toward the bulk.

The effect of nitriding microstructural features on the wear performance of plasma-nitrided titanium alloys was further elucidated by Cassar et al. [83]. They plasma-nitrided Ti–6Al–4V alloy using a triode set-up at low (200 V) and high (1000 V) voltages and evaluated the wear behavior against WC–Co and sapphire. The low-voltage treatment formed a thin compound layer on a deeply strengthened diffusion zone and despite the lower hardness of the compound layer, it provided better wear resistance compared with the high-voltage treatment especially at high loads. The thin compound layer could conform to the substrate deformation without experiencing spallations as the maximum shear stress was constrained within the diffusion zone. Bloyce et al. [30] explained that when the maximum applied stress during sliding exceeds the substrate yield strength, it experiences plastic deformation as evidenced by elongated grains within the subsurface and material pile-up. However, the TiN/Ti$_2$N surface layer cannot conform to the plastic deformation due to its low fracture toughness.

Recent studies by the authors investigate the failure micromechanisms of plasma-nitrided Ti–6Al–4V under dry sliding conditions using microscratch tests [63]. The tests were carried out using a diamond Rockwell tip with a 200-µm radius (CSM Instruments Micro-Combi tester) according to the ASTM C1624 standard [84]. Lateral displacement speed of 1 mm/min and loading rate of 10N/min were selected for the experiments. In addition to normal and tangential force (resolution of 0.3 mN), penetration depth (resolution of 0.3 nm) and acoustic emission signals were recorded during each test. Two types of tests were carried out: (i) progressively increasing load tests and (ii) constant load tests (loads selected according to the critical loads determined in (i)). During the progressive tests, the applied load linearly increased from 0.03 to 20 N as the tip moved at a constant displacement speed on the surface. The critical loads corresponding to the onset of coating failure events were identified using acoustic emission signals and subsequent SEM observations. In-depth site-specific cross-sectional analysis of scratch paths was carried out using a LEO (Zeiss) 1540 XB dual-beam FIB/SEM fitted with an EDS detector[6]. A thin layer of osmium (~3 nm) was deposited on the surfaces prior to the milling process to increase conductivity, prevent charging, and enhance the secondary electron yield for imaging. After locating the region of interest, a 2-µm-thick layer of platinum was deposited on the selected area to prevent surface features from damage during the milling process. The ion milling began by cutting a trench normal to the deposited platinum layer, using Ga ions from a Ga-based liquid metal ion-source (LMIS), at an accelerating voltage of 30 kV. All the FIB trenches were made parallel to the scratch direction of the scratch paths. The ion-milled trenches were imaged using a 1 kV electron beam located at a 54° angle.

The results indicated that failures initiated by the formation of microvoids within the compound layer likely atgrain boundaries. Increasing the applied load led to the formation of microcracks that propagated intergranularly within the compound layer and finally into the diffusion zone. The low resistance of nitrided surfaces to crack initiation and growth was correlated with the brittle nature and low damage tolerance of the compound layer [10]. However, it was observed that the deeply strengthened diffusion zone improved the adhesion of compound layer to the underlying substrate and prevented the spallation/delamination events under sliding contact. Moreover, subsurface crack propagation was retarded by ductile β-particles in the diffusion zone (Figure 11). The high resistance of β-phase to crack propagation in plasma-nitrided titanium alloys was confirmed by testing a plasma-nitrided near-β Ti–10V–2Fe–3Al alloy under the same conditions (Figure 12). It was proposed that the optimum microstructure consisted of a thin compound layer supported by a deep diffusion zone consisting of β-particles. Keeping that in mind, a low-temperature plasma nitriding treatment was developed for Ti–10V–2Fe–3Al alloy [52]. Evaluation of the tribological properties of the plasma-nitrided Ti–10V–2Fe–3Al surfaces using microscratch tests revealed that the coefficient of friction values were reduced by more than 72% and the nitrided surfaces exhibited an excellent load-bearing capacity. The compound layer did not experience any failures, for

---

6 Western Nanofabrication Facility, University of Western Ontario, London, ON, Canada

example, buckling or spallation, and maintained a good interfacial bonding to its substrate up to the high applied load of 17 N where partial chipping of the coating was observed at the scratch rims. Despite a low thickness, the compound layer suppressed the adhesion tendency and transfer of titanium due to high hardness and chemical inertness of TiN and Ti_2N [32, 33], and high toughness and adhesion resistance of V_2N [34, 35]. The deep diffusion zone (35 ± 5 μm) underneath the compound layer also contributed to high load-bearing capacity of the nitrided alloy by providing a good mechanical support for the compound layer [20, 39–42, 45-46]. The diffusion zone inhibited premature failures at the surface by mitigating the significant difference in the deformation behaviors of the compound layer and Ti–10V–2Fe–3Al substrate and acting like a "functionally-graded interface" [43, 44].

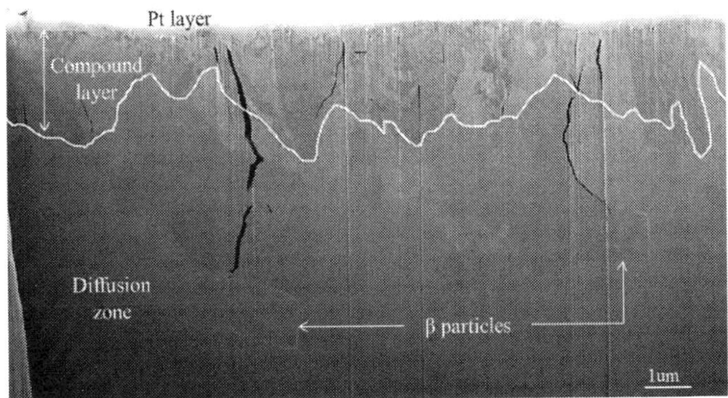

Figure 11. A typical SEM image of the FIB-milled cross section of the scratch track on the surface of plasma-nitrided Ti-6Al-4V at the highest load of 20 N showing that the microcracks initiating from the surface were stopped at β-particles in the diffusion zone [63].

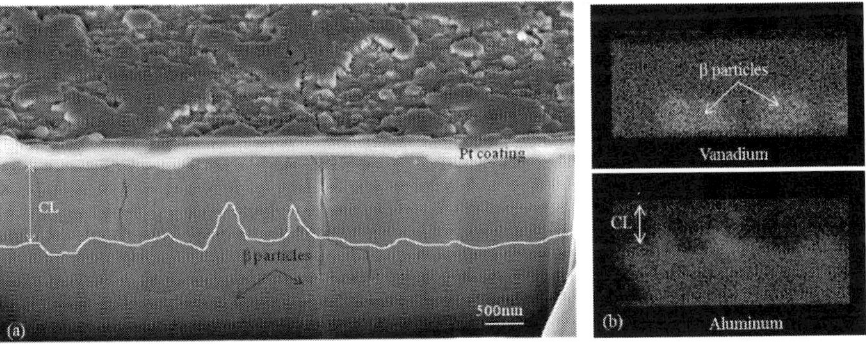

Figure 12. (a) A typical SEM image of the FIB-milled cross section of the scratch track on the surface of plasma-nitrided Ti-10V-2Fe-3Al and (b) the corresponding EDS mpas for aluminum and vanadium confirming that the β-particles inhibited further crack propagation in the subsurface region [63] (CL: compound layer, DZ: Diffusion zone).

Surface roughness of plasma-nitrided titanium alloys is another influential factor on their sliding behavior. Salehi et al. [85] investigated the effect of surface topography on the wear behavior of PVD TiN-coated and plasma-nitrided Ti–6Al–4V surfaces. They found that the compound layer thickness and surface roughness both increased with plasma nitriding-treatment temperature and proposed that sputtering is responsible for the changes in surface roughness. They reported that under self-mating conditions, oxidational wear dominated with significant mass loss for surfaces of higher roughness and thicker compound layers. Sliding wear tests against alumina counterface revealed that plasma nitriding at high temperatures (850°C and 950°C) yielded low wear rates at high loads due to thick compound layers but high wear rates at low loads due to high surface roughness. Plasma-nitrided surfaces at low temperatures (700°C and 750°C) improved the wear resistance below a critical load, above which the compound layer spalled off and the coefficient of friction increased abruptly. Similar observations were made by Yildiz et al. [62].

5. Mechanical properties of plasma-nitrided titanium alloys

Nitriding improves the fatigue strength of steels due to the precipitation of submicroscopic alloying element nitrides and compressive residual strains in the diffusion zone. Conversely, nitriding of titanium alloys does not promote precipitation strengthening mechanisms [86] and also leads to inevitable adverse effects on the toughness and fatigue properties [10, 24, 87–89]. In this section, the correlation between plasma nitriding and mechanical properties of titanium alloys is presented, and since the microstructural features of gas-nitrided Ti alloys affect fatigue properties in the same manner, gas nitriding investigations have been occasionally referred to in this section.

Due to the slow kinetics of nitrogen diffusion in titanium alloys, relatively high temperatures (>700°C) are used to produce sufficient depth of nitriding. Typical plasma nitriding treatments, performed in the temperature range of 700–1100°C for several hours, cause bulk microstructural changes such as significant grain growth and overaging that adversely affect the fatigue behavior [10, 22, 23]. Moreover, it has been reported by several researchers that the formation of brittle features in the surface vicinity and bulk microstructural changes are the main contributing factors in premature failure initiation and deterioration of mechanical properties of titanium alloys after nitriding [10, 49, 50, 54, 55, 58, 90, 91].

According to Morita et al. [90] and Tokaji et al. [92], the formation of the compound layer and grain growth were equally responsible for the fatigue deterioration of gas-nitrided titanium alloys. Morita et al. [90] proposed that the high stiffness gradient between the compound layer and Ti substrate ($E_{\text{compound layer}} = 426\,\text{GPa}$, $E_{\text{Ti}} = 100\,\text{GPa}$) led to stress concentration at the surface. The resulting stress concentration cancelled the effect of compressive residual stresses and led to the premature fracture initiation at the surface due to the low fracture toughness of the compound layer. They also reported that the fatigue strength had a Hall–Petch relationship with the grain size of nitrided titanium and the negative effect of grain growth was more pronounced for elongated treatments at high temperatures. Consequently, promising results were obtained by inhibiting grain growth through a low-temperature nitriding process.

Tokaji et al. [92] separated the effect of nitriding on the substrate microstructure by subjecting the titanium alloys to the same heating cycle under vacuum. The results indicated a slight increase in the fatigue limit compared with the annealed Titanium alloys after a 4-hour nitriding treatment, but 15 hours of nitriding significantly reduced the fatigue strength. They believed that the compound layer and diffusion zone were both responsible for impaired fatigue properties and observed improvements in fatigue life by complete removal of the compound layer and partial removal of the diffusion zone (Figure 13). Similar results were found by Nishida and Hattori [24] and Rodriguez et al. [93] after plasma nitriding of Ti–6Al–4V. They reported that the fatigue strength of plasma-nitrided Ti alloys was lower than the untreated alloy but was similar to – or even slightly better than – that of the vacuum-annealed alloy with the same heating history. The fatigue endurance limit was found to be inversely related to the plasma nitriding temperature and increased by about 30 MPa when the brittle compound layer was removed after plasma nitriding. Rodriguez et al. [93] reported that the bulk microstructural changes from mill-annealed to Widmanstätten structure were caused by a reduction in low-cycle fatigue resistance. They proposed that imposed strains during cyclic loading at low loads were not sufficient for compound layer failure, and thus the residual compressive strains retarded the fatigue crack nucleation. However, the compound layer failed at high cyclic strains due to a high surface roughness and the different stiffness between Ti–6Al–4V and titanium nitrides.

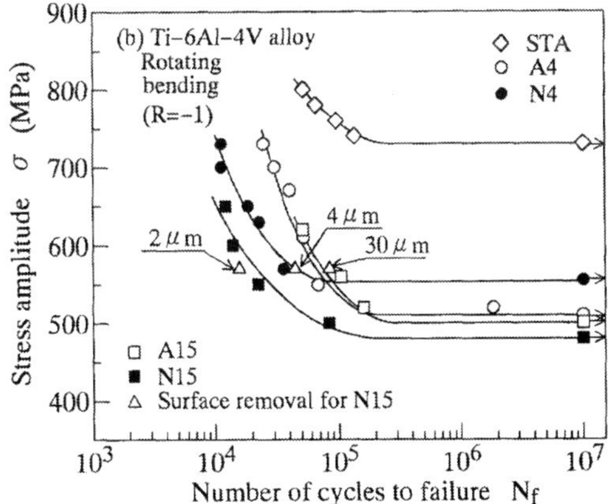

Figure 13. The S–N curve obtained by rotation bending fatigue tests on Ti–6Al–4V alloy. STA: solution treated at 950°C for 1 hour and aged at 540°C for 4 hours; A4 and N4: annealed and nitrided at 850°C for 4 hours; A15, N15: annealed and nitrided at 850°C for 15 hours [92].

Several researchers including Bell et al. [10, 25] and Guiraldenq et al. [94] correlated the impaired fatigue properties of titanium alloys after plasma nitriding to the thickness of the α-case. The α-case is a brittle layer (typical hardness 800–1000 HV) formed beneath the compound layer at high nitriding temperatures ($\geq$ 800°C), and its formation has proven to be detrimental for ductility and fatigue strength [10, 49, 50, 54, 55, 58].

Raveh et al. [50] believed that in addition to the surface hardness, fatigue crack initiation was also a function of residual strains in the nitrided layers, crystallographic orientation of different phases, and segregation of alloying elements near the surface region. The fatigue crack initiation resistance decreased with TiN content of the compound layer and the surface hardness due to its brittle nature. Moreover, incorporation of Ar in the nitriding plasma also decreased the crack initiation resistance, possibly as a result of inducing microstrains in the nitrided layers.

Conversely, according to Sobiecki and Rudnicki [95], plasma nitriding (at 800°C for 3 and 12 hours) increased the fatigue strength of Ti–1Al–1Mn alloy from 350 to 390 MPa. A plasma nitriding treatment at 500°C for 6 hours in nitrogen–hydrogen (3:1) gas mixture in the work of Rajasekaran and Raman [96] also resulted in improvements in uniaxial plain fatigue and fretting fatigue behavior of Ti–6Al–4V alloy. Lower surface roughness, generation of compressive residual stresses on the surface, and higher surface hardness were considered the main factors responsible for the fatigue improvements. The surface hardness after plasma nitriding (390 $HV_{0.2}$) was slightly higher than the untreated material (330 $HV_{0.2}$), and the surface roughness was reduced by plasma nitriding (R_a decreased from 0.80 to 0.55 µm) due to smoothing of the pretreatment grinding marks by sputtering. Furthermore, compressive residual stresses developed on the surface (in the order of 40 MPa) due to nitrogen diffusion in the titanium lattice. Fretting test results also indicated lower friction forces and shallower fretting scars for nitrided samples at all stress levels. Contradictory results were reported by Ali and Raman [88] who carried out plasma nitriding on Ti–6Al–4V alloy at 520°C for 4 and 18 hours and found that both plain and fretting fatigue lives were reduced by plasma nitriding. The samples that were nitrided in a nitrogen–hydrogen (3:1) atmosphere showed inferior results compared with those nitrided in a pure nitrogen atmosphere likely due to the higher hardness of surface layers in the former nitriding conditions.

In a recent investigation by the authors, a plasma nitriding treatment was designed for Ti–10V–2Fe–3Al β-titanium alloy to address its susceptibility to adhesion without any detrimental effects on bulk microstructure. Uniaxial tensile tests were conducted at room temperature on small-sized specimens (6 mm in diameter and 24 mm in length at the gauge section) conforming to the specifications outlined in the ASTM E8M standard [97]. The tensile tests were performed at a cross-head speed of 0.5 mm/min (MTS Criterion-43) and strain values were measured using a clip-on axial extensometer to an accuracy better than ± 0.5% of the applied strain. Subsequently, the engineering stress–strain curves were constructed and the offset yield strength, ultimate tensile strength (UTS), and ductility (tensile elongation) were determined. It was found that the modified nitriding process led to improvements in the tensile (from 881.0 to 922.5 MPa) and yield (from 821.5 to 873.0 MPa) strengths of the alloy with a minor decrease in the elongation (from 18.9% to 15.9%). The reduction of tensile elongation was possibly due to the mutual influence of precipitation of α-phase during age hardening and the solid-solution strengthening effect of nitrogen interstitials. Similar observations were made by Akahori et al. [98] after gas nitriding experiments on Ti–Nb–Ta–Zr (β-titanium) alloy. More importantly, as confirmed by TEM observations and tensile test results, the formation of brittle ω precipitates (forming in the range of 250–450°C [2, 4, 10, 12]) and α-case was avoided by optimizing the nitriding parameters.

Rotation bending fatigue tests (fully reversed cyclic loading, $R = -1$) were utilized in another investigation to evaluate the performance of plasma-nitrided Ti-6Al-4V alloy under cyclic loading conditions [26]. The tests were conducted using an R. R. Moore fatigue tester (Instron) at a frequency of 3000 rpm. The first samples were tested at ~20% of the tensile strength where failures were expected to occur at a relatively short number of cycles. The testing stress was decreased for the rest of the samples until at least three samples did not fail at 10^7 cycles (run-out). The highest stress at the run-out was taken as the fatigue endurance limit. The fatigue samples had an hour-glass configuration with a tapered length of 50 mm and a diameter of 5.6 mm in the thinnest section. The stress–life (S–N) curves were constructed and used to evaluate the resistance to fatigue crack initiation. The fracture surfaces were then analyzed under the SEM (Quanta 200 FEG-SEM) at an accelerating voltage of 10 kV to investigate the failure micromechanisms.

The fatigue tests were also performed on untreated Ti-6Al-4V alloy under the same conditions for the purpose of comparison. The gauge section of the untreated samples was mechanically polished using 1-μm diamond suspension in the last step to an average surface roughness (R_a) and root-mean-squared roughness (RMS) of 0.14 and 0.18 μm, respectively. The surface roughness measurements were obtained using an optical profilometer (Wyko, Veeco NT1100).

A low-temperature plasma nitriding (600°C) was exploited to achieve a microstructure consisting of a thin compound layer (<2 μm) supported by a deep diffusion zone (>40 μm) without any significant changes in the substrate (only a ~40% increase in the average grain size). It was found that the low-temperature plasma nitriding resulted in improvements in fatigue strength of plasma-nitrided Ti-6Al-4V compared with conventional nitriding treatments (Figure 14). Moreover, the plasma-nitrided Ti-6Al-4V alloy at 600°C illustrated a ductile type of failure under uniaxial tensile loads. Periodic transverse cracks observed on the plasma-nitrided surfaces perpendicular to the loading direction without any signs of spallation confirmed a well-bonded interface between the compound layer and the substrate. Similar observations were reported by Chen et al. [99] for TiN coatings on stainless steel substrates under tensile loads. Conversely, plasma nitriding at an elevated temperature of 900°C led to substantial deterioration of strength, toughness, and fatigue life. This was attributed to premature failure of thick and brittle nitrided surface layers – a 5.6-μm-thick compound layer and a 19.3-μm α-case – acting as stress risers and promoting a brittle type of failure in the alloy as well as substantial grain growth (370% increase in the average grain size compared with the untreated alloy) and phase transformation in the bulk microstructure from equiaxed to coarse lamellar grains (~5 times higher average grain size value) as shown in Figure 15.

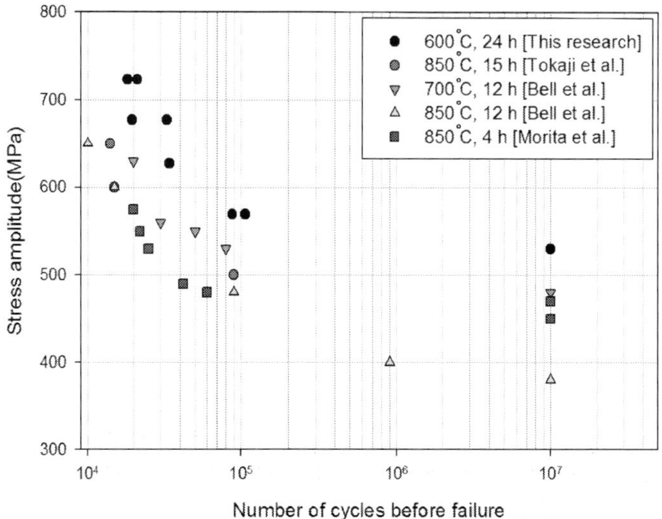

Figure 14. Comparison of the fatigue data obtained for plasma-nitrided Ti–6Al–4V with the conventional gas and plasma nitriding treatments in the literature [26].

According to fractographic investigations of the fatigue failures in low-temperature plasma-nitrided Ti-6Al-4V alloy, the fatigue failure initiated at the surface in the low cycle region (N $\leq 10^5$ cycles) and propagated in a ductile manner leading to the final rupture. No failures were observed in the high cycle region (N $> 10^5$ cycles) and the nitrided alloy endured cyclic loading until the tests were stopped at 10^7 cycles. The thin compound layer restricted the extent of premature crack initiation from the surface, whereas the deep diffusion zone with a well-bonded interface decreased the likelihood of fatigue initiation at (or below) the compound layer interface. Although the compound layer exhibited signs of plastic deformation under cyclic loading conditions, fragmentation of the compound layer and brittle failure features observed at the fatigue crack initiation sites suggested that brittle fracture of the compound layer was the precursor to the fatigue failure (Figure 16). The observation of striation-like features within the diffusion zone in the vicinity of the crack origin indicated the possibility that the crack formation was triggered by intrusions and extrusions and plastic deformation of the underlying diffusion zone. The fatigue cracks consequently propagated in the same manner as the untreated alloy. The beneficial effect of compressive residual stresses in the compound layer (−530 MPa) was overshadowed by the stress concentration in this layer due to the modulus mismatch with the substrate. Surface roughness was also an influential parameter on the fatigue strength of the nitrided alloy as by polishing the nitrided surfaces, a higher number of cycles were dedicated to the formation of fatigue cracks compared with the as-treated condition resulting in an improved fatigue life (Figure 17). Similar results were reported by Cassar et al. [91] and Novovic et al. [100] for the fatigue behavior of surfaces with average roughness values of 0.1 μm and higher.

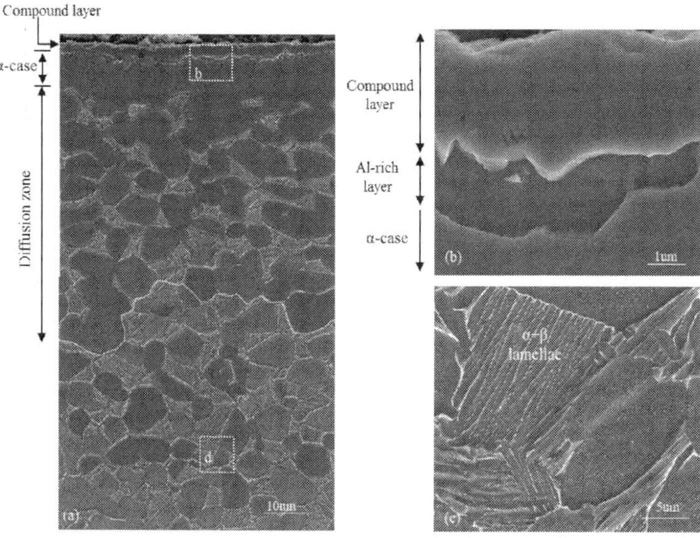

Figure 15. Typical cross-sectional SEM micrographs of the microstructure of plasma-nitrided Ti–6Al–4V treated at 900°C [26].

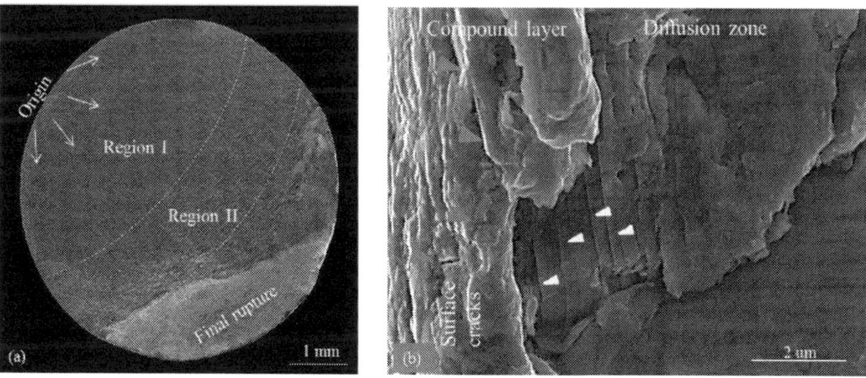

Figure 16. (a) A typical SEM image of the general view of the fatigue fracture surface of plasma-nitrided Ti–6Al–4V and (b) striation-like features and microcracks in the surface vicinity [26].

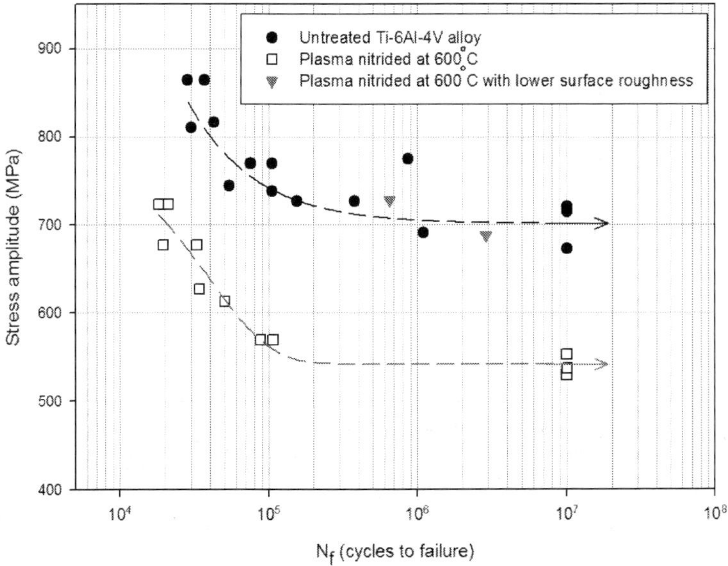

Figure 17. Stress–life (S–N) curves showing the effect of plasma nitriding on fatigue behavior of Ti–6Al–4V. It was found that decreasing the surface roughness with polishing resulted in an increase in the fatigue strength [26].

6. Recent development in plasma nitriding titanium alloys

As explained in Section 5, the main reasons for the initiation of premature failures in nitrided titanium alloys are the brittle nature of the compound layer and the α-case, the incompatibility in deformation behavior of the compound layer and the substrate and lack of sufficient adhesion at their interface. As such, the compound layers cannot conform to the extensive plastic deformation of titanium substrates and eventually experience adhesive (from the interface) or cohesive (within the compound layer) failures. Thus, the mechanical response mismatch between the nitride layers and substrates should be minimized in order to enhance the load-bearing capacity of the nitrided surfaces. The effective practical measures are based on eliminating discrete nitride layer–substrate boundaries by the design and fabrication of multilayer coatings and coatings with gradient diffusion boundaries [29, 31, 36, 82, 83, 101]. A summary of recent investigations are presented in this section.

6.1. Duplex treatments

Duplex treatments were introduced to the field of surface engineering of titanium alloys to exploit the benefits of wear-resistant coatings and the deep hardening effect of diffusion treatments [30–34, 102]. Significant improvements in wear resistance and long-term durability of titanium surfaces have been achieved by deposition of diamond-like carbon (DLC) coatings after diffusion hardening by plasma nitriding [103] or thermal oxidation [104, 105].

The beneficial role of plasma nitriding pretreatment in accommodating subsurface deformation and preventing premature failures in DLC-coated Ti–6Al–4V was confirmed in the work of Meletis et al. [103]. In their experiments, radio-frequency plasma nitriding treatment was utilized for substrate surface hardening, followed by deposition of DLC coatings with a silicon interlayer. In addition to substantial reduction of wear and coefficient of friction, it was found that the nitriding pretreatment retarded catastrophic failure of DLC coatings during sliding at high contact loads.

In another investigation by Dong [36], the combined effect of electron beam melting and plasma nitriding treatments resulted in significant improvements in wear resistance and load-bearing capacity of Ti–8.5Si titanium alloy. Electron beam melting generated a deep strengthened zone (800 HV to a depth of 600 µm) in the alloy before plasma nitriding that provided support for the nitride layer under sliding contact. As a result, the surface nitride layer endured the stresses applied during wear tests without spalling, leading to a three orders of magnitude reduction of wear rate compared with the untreated titanium alloy.

Several researchers also introduced duplex treatments that are composed of plasma nitriding and physical vapor deposition (PVD) or chemical vapor deposition (CVD) of nitride coatings. Rie et al. [106, 107] used plasma-assisted CVD (PACVD) technique to produce a thick TiN layer on top of plasma-nitrided substrates to enhance wear and corrosion resistance for biomedical applications. Their findings indicated that the duplex-treated surfaces were free of wear against ultrahigh-molecular-weight polyethylene under the tested conditions, and the corrosion resistance and biocompatibility of the untreated titanium alloy were preserved. Wear improvements were also confirmed by Ma et al. [108] through plasma nitriding and plasma-enhanced chemical vapor deposition (PCVD) of TiN coatings of Ti–6Al–4V substrates. They correlated the wear enhancement to higher surface hardness of PCVD coating and its strong adhesion to plasma-nitrided substrate. Other coatings, such as CrN, CrAlN, and WC/C, were also deposited on plasma-nitrided titanium alloys, in addition to TiN, by Cassar et al. [35, 91, 109]. Evaluation of the duplex-treated surfaces by impact tests, ball-on-disk wear tests, rotation bending fatigue tests, and microscratch tests revealed improvements in wear resistance, adhesion, and resistance to fatigue crack initiation. This was attributed to higher load support and adhesion strength offered by hard-coated surfaces and deeply strengthened substrates. It was noted that chromium aluminum nitride (CrAlN) coatings exhibited superior wear performance due to their high hardness and low elastic modulus (higher H^3/E^2 parameter).

6.2. Surface pretreatments

Alteration of the microstructure in the surface vicinity via thermochemical (e.g., laser-assisted treatments, electron beam melting, etc.) and mechanical (e.g., mechanical attrition, shot peening, etc.) pretreatment processes [27, 28] has been successfully used for increasing the nitriding efficiency of titanium alloys. Microstructural changes thus introduced, such as significant grain refinement and nonequilibrium defects, act as diffusion shortcuts and enhance the nitriding kinetics and efficiency during the posterior plasma nitriding treatment even at low temperatures [110–115].

Shot peening, extensively used for fatigue improvement, is based on introducing compressive stresses in the surface layers via impingement of high-velocity shots, under controlled conditions. Various types of shots such as cast steel shots, cut wire shots, glass beads, and zirconium shots are available, and the air blast can be delivered by either suction blast or pressure blast depending on the budget, target substrates, and the required precision. In addition to the development of a residual stress profile near the surface, shot peening results in surface roughening, strain hardening, and formation of a subsurface plastic deformation zone. Surface roughening favors early nucleation of fatigue cracks, whereas strain hardening and compressive residual stresses increase the resistance to plastic deformation and fatigue crack propagation by providing crack closure mechanism. Recently, improvements in the sliding wear behavior and tensile properties have been reported by shot peening due to its surface hardening effect [116].

Experimental studies showed that shot peening can introduce a severely deformed layer in the surface vicinity characterized by significant refinement of grains and generation of nonequilibrium defects [110–113]. The thickness of this severely deformed layer may vary in the range of 10–80 μm depending on the microstructure and properties of the substrate material; metals of a lower yield strength experience a more intense plastic deformation that extends deeper into the substrate [117]. A high-energy shot peening treatment by Han et al. [27] resulted in the formation of a nanostructured layer on the surface of Ti–4Al–2V alloy. This treatment also introduced microstructural changes such as twinning and grain refinement that extended to ~230 μm below the surface. They proposed that the grain refinement occurs through various mechanisms such as "formation of dislocation walls and tangles," "dislocation movement and twinning," and "mechanical twinning" depending on the crystal structure and stacking fault energy (SFE) of metals. Mechanical twinning was recommended to have major contributions in grain refinement of Ti–4Al–2V alloy due to its HCP structure and high stacking fault energy (SFE > 300 mJ/m^2). The high level of plastic deformation induced by shot peening resulted in activation of many slip systems and a high density of dislocations within the twin bands. Subsequently, dislocation walls were formed, which in turn generated submicron-sized grains, small misoriented subgrains and finally equiaxed fine nanograins (35 ± 5 nm). Further studies were performed by Thomas and Jackson [118] to understand the effect of temperature and alloy composition on the subsurface microstructure of shot-peened titanium alloys.

Alteration of surface microstructure by introducing high-diffusivity paths such as grain boundaries, dislocations, and atomic-level microstructural defects by shot peening can be used to accelerate the diffusion of interstitial atoms [112]. Thomas et al. [114, 115] reported that generation of a high density of twin boundaries and dislocations in the near-surface region of pure titanium and Ti–4Al–2V alloys after shot peening increased the oxygen uptake during the following thermal exposure in air. Moreover, the high stored energy associated with the higher density of these nonequilibrium defects increases chemical reactivity of surfaces and enhances the kinetics of surface reactions [113]. Several researchers have confirmed that microstructural changes brought by severe plastic deformation (SPD) in ferrous alloys enhanced the nitrogen diffusion rate during the posterior nitriding treatments and increased the efficiency of nitriding at lower temperatures [110–113]. The effect of shot peening after diffusion treatments has also been studied in the literature. In a recent study, Bansal et al. [119] designed a series of duplex treatments for Ti–6Al–4V alloy composed of diffusion treatments (nitriding and oxygen diffusion) followed by mechanical working (shot peening and planishing). The results of dry reciprocating ball-on-disk wear tests against stainless steel 440C and silicon nitride sliders indicated improvements in wear resistance by introducing a short period of low friction and lower volumetric wear.

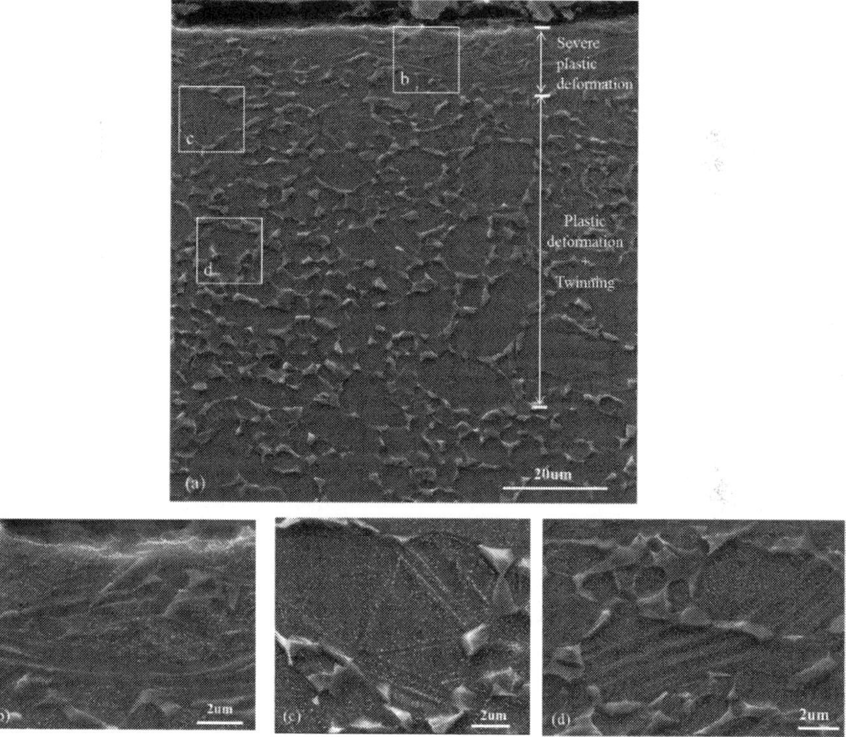

Figure 18. Typical cross-sectional SEM images showing the microstructure of Ti–6Al–4V after shot peening. A 5-μm deep severe plastic deformation (SPD) layer was formed on the surface characterized by elongation of intergranular β-particles parallel to the surface [28].

Severe plastic deformation surface layer introduced by shot peening pretreatment was found to enhance the plasma nitriding efficiency of Ti–6Al–4V alloy [28]. The high level of plastic strains induced by shot peening resulted in the formation of a 5-μm deep SPD layer on the surface characterized by twins, deformation bands, grain refinement, and a high density of

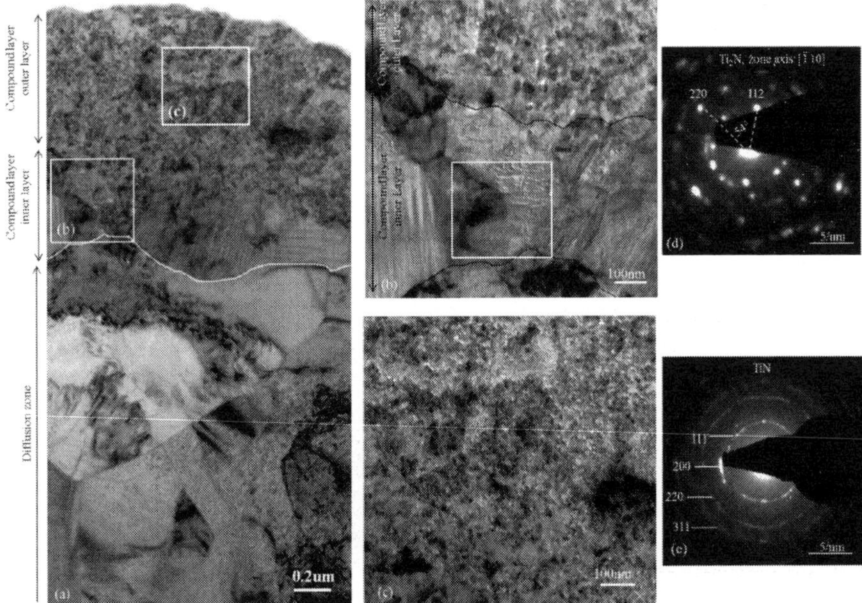

Figure 19. Typical cross-sectional bright-field TEM images of the near-surface microstructure of plasma-nitrided Ti–6Al–4V alloy with shot-peening pretreatment and the corresponding selected area electron diffraction (SAED) patterns [28].

nonequilibrium crystal defects such as subgrain boundaries and dislocation pile-ups (Figure 18 and Figure 19). This is in agreement with the 20% increase in the surface hardness value due to the work hardening of the SPD layer. The formation of the SPD layer resulted in accelerated nitriding kinetics during the plasma nitriding treatment that followed due to two main reasons: (i) grain boundaries, deformation bands, and twins provided easy diffusion paths for nitrogen interstitial atoms and (ii) high density of nonequilibrium crystal defects increased the stored energy of the surfaces and their chemical reactivity and offered additional preferential nucleation sites for titanium nitrides [111, 120]. The accelerated nitriding kinetics resulted in the formation of a 0.6-μm-thick nanocrystalline TiN layer in the compound layer (Figure 19) and increased the depth of nitrogen diffusion by almost 50% normally observed in high-temperature nitriding treatments (e.g., 900°C) [121]. As such, the plasma-nitrided surfaces that received the pretreatment step exhibited higher load-bearing capacity and better interfacial bonding compared with those without the pretreatment. The presence of nanoscale TiN grains increased the hardness and toughness of the compound layer and the compound layer endured high contact stresses for several sliding passes without chipping or spallation. In comparison, plasma-nitrided surfaces without the pretreatment experienced severe delamination and coating spallation after only two sliding passes in microscratch tests. The deep diffusion zone in the pretreated surfaces, strengthened by nitrogen interstitials, also contributed to a well-bonded interface and diminished the incompatibility in plastic deformation of the compound layer (E = 250–600 GPa [122, 123]) and substrate (E = 110 GPa [124]) by providing a more homogenized stress distribution across the interface [30, 104, 125–127]. Thus, the shot-peening pretreatment improved the nitriding efficiency of low-temperature plasma nitriding without introducing detrimental microstructural changes in the substrate that would lead to sacrificing mechanical properties.

7. Summary

Plasma nitriding is a surface hardening treatment utilized to improve the tribological properties of titanium alloys by the formation of a compound layer mainly consisting of TiN and Ti_2N titanium nitrides on the surface. However, plasma nitriding has a negative side effect on the mechanical properties of titanium alloys, especially fatigue strength, since the high temperatures involved in the treatment (700–1100°C) promote formation of brittle surface layers and detrimental microstructural changes in the bulk. Recent advancements in the field of plasma nitriding of titanium alloys are focused on the development of low-temperature (600°C or less) treatments as well as increasing the long-term durability of the nitrided surfaces through deposition of hard surface coatings such as diamond-like carbon and PVD/CVD nitride coatings. Furthermore, alteration of the surface microstructure by thermochemical and mechanical pretreatment steps is also successfully applied for increasing the nitriding efficiency during low-temperature plasma nitriding of titanium alloys.

Acknowledgements

The authors acknowledge Exactatherm Ltd., Dr. P. Lidster, Mr. G. Pigott, and Mr. Simeon Simeonov, for their help in plasma nitriding. The appreciation is extended to Natural Science and Engineering Council of Canada (NSERC) Discovery Individual Grant Program for providing financial support and the Canadian Centre of Electron Microscopy (CCEM) for TEM experimental work.

Author details

Afsaneh Edrisy* and Khorameh Farokhzadeh

*Address all correspondence to: edrisy@uwindsor.ca

Department of Mechanical, Automotive and Materials Engineering, University of Windsor, Windsor, ON, Canada

References

[1] F. Preissner and P. Minarski, "Results on nitriding titanium and Ti6Al4V with a new thermochemical treatment under high gas pressure," in Titanium '92, Science and Technology, F. H. Froes and I. Caplan, editors, Cannes, France: The Mineral, Metals and Materials Society, 1993, pp. 1979–1987.

[2] V. M. Fedirko and I. M. Pohrelyuk, "Improvement of the wear resistance of titanium alloys by thermochemical treatment in nitrogen-containing media," Materials Science, vol. 30, pp. 66–71, 1994.

[3] A. Gicouel, N. Laidani, P. Saillard, and J. Amouroux, "Plasma and nitrides: Application to the nitriding of titanium," Pure Applied Chemistry, vol. 62, pp. 1743–1750, 1990.

[4] T. Spalvins, "Advances and directions of ion nitriding/carburizing," in Proceedings of ASM's 2nd International Conference on Ion nitriding and Ion Carburizing, T. Spalvins and W. L. Kovacs, editors, Cincinnati, OH, 18-20 September 1989, ASM International, pp. 1–4.

[5] T. Bacci, G. Pradelli, B. Tesi, C. Gianoglio, and C. Badini, "Surface engineering and chemical characterization in ion-nitrided titanium and titanium alloys," Journal of Materials Science, vol. 25, pp. 4309–4314, 1990.

[6] E. Rolinski, G. Sharp, D. F. Cowgill, and D. J. Peterman, "Ion nitriding of alpha plus beta alloy for fusion reactor applications," Journal of Nuclear Materials, vol. 252, pp. 200–208, 1998.

[7] K. C. Chen and G. J. Jaung, "D.C. diod ion nitriding behavior of titanium and Ti-6Al-4V," Thin Solid Films, vol. 303, pp. 226–231, 1997.

[8] H. Akbulut, O. T. Inal, and C. A. Zimmerly, "Ion nitriding of explosively-clad titanium/steel tandems," Journal of Materials Science, vol. 34, pp. 1641–1652, 1999.

[9] C. X. Li and T. Bell, "Principles, mechanisms and applications of active screen plasma nitriding," Heat Treatment of Metals, vol. 1, pp. 1–7, 2003.

[10] T. Bell, H. W. Bergmann, J. Lanagan, P. H. Morton, and A. M. Staines, "Surface engineering of titanium with nitrogen," Surface Engineering, vol. 2, pp. 133–143, 1986.

[11] R. Gunn, "Industrial advances for plasma nitriding," in Ion Nitriding and Ion Carburizing, Proceedings of ASM's 2nd International Conference on Ion Nitriding and Ion Carburizing, T. Spalvins and W. L. Kovacs, editors, Cincinnati, OH, 18-20 September 1989, ASM International, pp. 157–163.

[12] B. Edenhofer, "Physical and metallurgical aspects of ion nitriding," Heat Treatment of Metals, vol. 1, pp. 23–28, 1974.

[13] S. Dressler, "Plasma treatments," in Surface Modification Technologies: An Engineer's Guide, T. S. Sudarshan, editor, New York: Marcel Dekker, 1989, pp. 317–419.

[14] A. Raveh, P. L. Hansen, R. Avni, and A. Grill, "Microstructure and composition of plasma-nitrided Ti-6Al-4V layers," Surface and Coatings Technology, vol. 38, pp. 339–351, 1989.

[15] S. G. Lakshmi, D. Arivuoli, and B. Ganguli, "Surface modification and characterisation of Ti–Al–V alloys," Materials Chemistry and Physics, vol. 76, pp. 187–190, 2002.

[16] S. L. R. da Silva, L. O. Kerber, L. Amaral, and C. A. d. Santos, "X-ray diffraction measurements of plasma-nitrided Ti–6Al–4V," Surface and Coatings Technology, vol. 116–119, pp. 342–346, 1999.

[17] N. N. Koval, P. M. Schanin, Y. K. Akhmadeev, I. V. Lopatin, Y. R. Kolobov, D. S. Vershinin, and M. Yu. Smolyakova, "Influence of the composition of a plasma-forming gas on nitriding in a non-self-maintained glow discharge with a large hollow cathode," Journal of Surface Investigation: X-ray, Synchrotron and Neutron Techniques, vol. 6, pp. 154–158, 2012.

[18] M. Tamaki, Y. Tomii, and N. Yamamoto, "The role of hydrogen in plasma nitriding: Hydrogen behavior in the titanium nitride layer," Plasmas and Ions, vol. 3, pp. 33–39, 2000.

[19] X. Li, D. Q. Sun, X. Y. Zheng, and Z. A. Ren, "Effect of N2/Ar gas flow ratios on the nitrided layers by direct current arc discharge," Material Letters, vol. 62, pp. 226–229, 2008.

[20] H. Yilmazer, S. Yilmaz, and M. E. Acma, "Treatment of surface properties of titanium with plasma (ion) nitriding," Defect and Diffusion Forum, vol. 283–286, pp. 401–405, 2009.

[21] T. A. Panaioti and G. V. Solov'ev, "Ion nitriding of aging ($\alpha+\beta$) titanium alloys," Metal Science and Heat Treatment, vol. 38, pp. 216–219, 1996.

[22] H.-J. Spies, "Surface engineering of aluminum and titanium alloys: An overview," Surface Engineering, vol. 26, pp. 126–134, 2010.

[23] H.-J. Spies, B. Reinhold, and K. Wilsdorf, "Gas nitriding – Process control and nitriding non-ferrous alloys," Surface Engineering, vol. 17, pp. 41–54, 2001.

[24] S. Nishida and N. Hattori, "Fatigue strength of ion-nitrided Ti-6Al-4V alloy in high cycle region," Surface Treatment: Computer Methods and Experimental Measurements, vol. 17, pp. 199–208, 1997.

[25] J. Lanagan, P. H. Morton, and T. Bell, "Surface engineering of titanium with glow discharge plasma," in Designing with Titanium, London: Institute of Metals, 1986, pp. 136–147.

[26] K. Farokhzadeh and A. Edrisy, "Fatigue improvement in low temperature plasma nitrided Ti-6Al-4V alloy" Materials Science and Engineering A, vol. 620, pp. 435–444, 2014.

[27] J. Han, G. M. Sheng, and G. X. Hu, "Nanostructured surface layer of Ti–4Al–2V by means of high energy shot peening," ISIJ International, vol. 48, pp. 218–223, 2008.

[28] K. Farokhzadeh, J. Qian, and A. Edrisy, "Effect of SPD surface layer on plasma nitriding of Ti-6Al-4V alloy," Materials Science and Engineering A, vol. 589, pp. 199–208, 2014.

[29] H. Dong, T. Bell, and A. Mynott, "Surface engineering of titanium alloys for the motorsports industry," Sports Engineering, vol. 2, pp. 213–219, 1999.

[30] A. Bloyce, H. Dong, and T. Bell, "Design and performance of a titanium duplex surface engineering system," in Surface Performance of Titanium Alloys, J. K. Gregory, H. J. Rack, and D. Eylon, editors, Warrendale, PA: The Mineral, Metals & Materials Society (TMS), 1996, pp. 23–31.

[31] C. Kwietniewski, H. Dong, T. Strohaecker, X. Y. Li, and T. Bell, "Duplex surface treatment of high strength Ti metal 550 alloy towards high load-bearing capacity," Surface and Coatings Technology, vol. 139, pp. 284–292, 2001.

[32] Y. Fu, J. Wei, B. Yan, and N. L. Loh, "Characterization and tribological evaluation of duplex treatment by depositing carbon nitride films on plasma nitrided Ti-6Al-4V," Journal of Materials Science, vol. 35, pp. 2215–2227, 2000.

[33] B. S. Yilbas, M. S. J. Hashmi, and S. Z. Shuja, "Laser treatment and PVD TiN coating of Ti-6Al-4V alloy," Surface and Coatings Technology, vol. 140, pp. 244–250, 2001.

[34] H. Dong, A. Bloyce, and T. Bell, "Oxygen thermochemical treatment combined with DLC coating for enhanced load bearing capacity of Ti–6Al–4V," Surface Engineering, vol. 14, pp. 505–512, 1998.

[35] G. Cassar, S. Banfield, J. C. Avelar-Batista Wilson, J. Housden, A. Matthews, and A. Leyland, "Impact wear resistance of plasma diffusion treated and duplex treated/PVD-coated Ti–6Al–4V alloy," Surface and Coatings Technology, vol. 206, pp. 2645–2654, 2012.

[36] H. Dong, "Current status and trends in duplex surface engineering of titanium alloys," in Heat Treating: Proceedings of the 20th Conference – Volume 1, K. Funatani and G. E. Totten, editors, Materials Park, OH: ASM International, 2000, pp. 170–177.

[37] F. Schnabaum, "Materials selection and process control for plasma-diffusion treatment of PM materials," in ASM's 2nd International Conference on Ion Nitriding and Ion Carburizing, T. Spalvins and W. L. Kovacs, editors, Cincinnati, OH, 18-20 September 1989, ASM International, pp. 81–89.

[38] U. Huchel and S. Strämke "Pulsed plasma nitriding of titanium and titanium alloys," in Proceedings of the 10th World Conference on Titanium, Chichester: Wiley-VCH, 2003, pp. 935–939.

[39] M. Raaif, F. M. El-Hossary, N. Z. Negm, S. M. Khalil, and P. Schaaf, "Surface treatment of Ti–6Al–4V alloy by rf plasma nitriding," Journal of Physics: Condensed Matter, vol. 19, pp. 1–12, 2007.

[40] A. Brokman and F. R. Tuler, "A study of the mechanisms of ion nitriding by the application of a magnetic field," Journal of Applied Physics, vol. 52, pp. 468–471, 1981.

[41] E. Meletis, "Plasma nitrided titanium and titanium alloy products," United States Patent 5443663, 1995.

[42] E. I. Meletis, "Intensified plasma-assisted processing: Science and engineering," Surface and Coatings Technology, vol. 149, pp. 95–113, 2002.

[43] A. A. Adjaottor, "A study of the effect of energetic flux bombardment on intensified plasma-assisted processing," PhD dissertation, Louisiana State University, 1997.

[44] N. Kashaev, H.-R. Stock, and P. Mayr, "Nitriding of Ti-6%Al-4%V alloy in the plasma of an intensified glow discharge," Metal Science and Heat Treatment, vol. 46, pp. 294–298, 2004.

[45] N. Kashaev, H.-R. Stock, and P. Mayr, "Assessment of the application potential of the intensified glow discharge for industrial plasma nitriding of Ti-6Al-4V," Surface and Coatings Technology, vol. 200, pp. 502–506, 2005.

[46] T. Czerwiec, H. Michel, and E. Bergmann, "Low-pressure, high-density plasma nitriding: Mechanisms, technology and results," Surface and Coatings Technology, vol. 108–109, pp. 182–190, 1998.

[47] N. L. Loh, "Plasma Nitriding," in Surface Engineering: Processes and Applications, K. N. Strafford, R. St. C. Smart, I. Sare, and C. Subramanian, editors, Pennsylvania: Technomic Publishing Company, Inc., 1995, pp. 157–178.

[48] C. Alves, O. F. de Araujo, and M. de Rômulo Ribeiro de Sousa, "Comparison of plasma-assisted nitriding techniques," in Encyclopedia of Tribology, Q. J. Wang and Y.-W. Chung, editors, New York: Springer, 2013, pp. 402–410.

[49] A. Grill, A. Raveh, and R. Avni, "Layer structures and mechanical properties of low pressure r. f. plasma nitrided Ti-6Al-4V alloy," Surface and Coatings Technology, vol. 43–44, pp. 745–755, 1990.

[50] A. Raveh, A. Bussiba, A. Bettelheim, and Y. Katz, "Plasma-nitrided α-β Ti alloy: Layer characterization and mechanical properties modification," Surface and Coatings Technology, vol. 57, pp. 19–29, 1993.

[51] K. Farokhzadeh, "Modification of Ion Nitriding of Ti-6Al-4V for Simultaneous Improvement of Wear and Fatigue Properties," PhD dissertation, Department of Me-

chanical, Automotive and Materials Engineering, University of Windsor, Windsor, ON, Canada, 2014.

[52] J. Qian, K. Farokhzadeh, and A. Edrisy, "Ion nitriding of a near-β titanium alloy: Microstructure and mechanical properties," Surface and Coatings Technology, vol. 258, pp. 134–141, 2014.

[53] J. Tiley, T. Searles, E. Lee, S. Kar, R. Banerjee, J. C. Russ, and H. L Fraser, "Quantification of microstructural features in α/β titanium alloys," Materials Science and Engineering A, vol. 372, pp. 191–198, 2004.

[54] A. Raveh, P. L. Hansen, R. Avni, and A. Grill, "Microstructure and composition of plasma-nitrided Ti-6Al-4V layers," Surface and Coatings Technology, vol. 38, pp. 339–351, 1989.

[55] A. Molinari, G. Straffelini, B. Tesi, T. Bacci, and G. Pradelli, "Effects of load and sliding speed on the tribological behaviour of Ti-6Al-4V plasma nitrided at different temperatures," Wear, vol. 203–204, pp. 447–454, 1997.

[56] H. A. Wriedt and J. L. Murray, "The N-Ti (nitrogen-titanium) system," Bulletin of Alloy Phase Diagrams, vol. 8, pp. 378–388, 1987.

[57] A. Nishimoto, T. E. Bell, and T. Bell, "Feasibility study of active screen plasma nitriding of titanium alloy," Surface Engineering, vol. 26, pp. 74–79, 2010.

[58] A. Fleszar, T. Wierzchon, S. K. Kim, and J. R. Sobiecki, "Properties of surface layers produced on the Ti-6Al-3Mo-2Cr titanium alloy under glow discharge conditions," Surface and Coatings Technology, vol. 131, pp. 62–65, 2000.

[59] A. A. Il'in, S. V. Skvortsova, L. M. Petrov, E. A. Lukina, and A. A. Chernysheva, "Effect of phase composition and structure of titanium alloys on their interaction with nitrogen during low-temperature ion nitriding," Russian Metallurgy, vol. 2006, pp. 400–405, 2006.

[60] R. J. Wasilewski and I. Kehl, "Diffusion of nitrogen and oxygen in titanium," Journal of the Institue of Metals, vol. 55, pp. 94–104, 1954.

[61] Z. Liu and G. Welsch, "Literature survey on diffusivities of oxygen, aluminum, and vanadium in alpha titanium, beta titanium, and in rutile," Metallurgical Transactions A, vol. 19A, pp. 1121–1125, 1988.

[62] F. Yildiz, A. F. Yetim, A. Alsaran, and A. Çelik, "Plasma nitriding behavior of Ti6Al4V orthopedic alloy," Surface and Coatings Technology, vol. 202, pp. 2471–2476, 2008.

[63] K. Farokhzadeh, A. Edrisy, G. Pigott, and P. Lidster, "Scratch resistance analysis of plasma-nitrided Ti–6Al–4V alloy," Wear, vol. 302, pp. 845–853, 2013.

[64] R. Wirth, "Focused Ion Beam (FIB) combined with SEM and TEM: Advanced analytical tools for studies of chemical composition, microstructure and crystal structure in geomaterials on a nanometre scale," Chemical Geology, vol. 261, pp. 217–229, 2009.

[65] W. C. Oliver and G. M. Pharr, "Measurement of hardness and elastic modulus by instrumented indentation: Advances in understanding and refinements to methodology," Journal of Materials Research, vol. 19, pp. 3–20, 2004.

[66] G. Straffelini, A. Andriani, B. Tesi, A. Molinari, and E. Galvanetto, "Lubricated rolling-sliding behaviour of ion nitrided and untreated Ti-6Al-4V," Wear, vol. 256, pp. 346–352, 2004.

[67] G. Straffelini and A. Molinari, "Dry sliding wear of Ti-6Al-4V alloy as influenced by the counterface and sliding conditions," Wear, vol. 236, pp. 328–338, 1999.

[68] H. Hong and W. O. Winer, "A fundamental tribological study of Ti/Al2O3 contact in sliding wear," Journal of Tribology, vol. 111, pp. 504–509, 1989.

[69] J. A. Ruppen, C. L. Hoffmann, V. M. Radhakrishnan, and A. J. McEvily, "The effect of environment and temperature on the fatigue behavior of titanium alloys," in Fatigue, Environment and Temperature Effects, J. J. Burke and V. Weiss, editors, New York: Springer, pp. 265-300, 1930.

[70] P. D. Nicolaou, E. B. Shell, and T. E. Matikas, "Microstructural and surface characterization of Ti-6Al-4V alloys after fretting fatigue," Materials Science and Engineering, vol. A269, pp. 98–103, 1999.

[71] J. Takeda, M. Niinomi, T. Akahori, and Gunawarman, "Effect of microstructure on fretting fatigue and sliding wear of highly workable titanium alloy, Ti-4.5Al-3V-2Mo-2Fe," International Journal of Fatigue, vol. 26, pp. 1003–1015, 2004.

[72] P. D. Miller and J. W. Holladay, "Friction and wear properties of titanium," Wear, vol. 2, pp. 133–140, 1958.

[73] H. Dong and T. Bell, "Tribological behavior of alumina sliding against Ti6Al4V in unlubricated contact," Wear, vol. 225–229, pp. 874–884, 1999.

[74] M. O. Alam and A. S. M. A. Haseeb, "Response of Ti-6Al-4V and Ti-24Al-11Nb alloys to dry sliding wear against hardened steel," Tribology International, vol. 35, pp. 357–362, 2002.

[75] J. Qu, P. J. Blau, T. R. Watkins, O. B. Cavin, and N. S. Kulkarni, "Friction and wear of titanium alloys sliding against metal, polymer, and ceramic counterfaces," Wear, vol. 258, pp. 1348–1356, 2005.

[76] S. Yerramareddy and S. Bahadur, "Effect of operational variables, microstructure and mechanical properties on the erosion of Ti-6Al-4V," Wear, vol. 142, pp. 253–263, 1992.

[77] K. G. Budinski, "Tribologial properties of titanium alloys," Wear, vol. 151, pp. 203–217, 1991.

[78] C. Feng and T. I. Khan, "The effect of quenching medium on the wear behaviour of a Ti-6Al-4V alloy," Journal of Materials Science, vol. 43 (2), pp. 788–792, 2008.

[79] S. Taktak and H. Akbulut, "Dry wear and friction behavior of plasma nitrided Ti-6Al-4V alloy after explosive shock treatment," Tribology International, vol. 40, pp. 423–432, 2007.

[80] D. P. Shashkov, "Effect of nitriding on mechanical properties and wear resistance of titanium alloys," Metal Science and Heat Treatment, vol. 43, pp. 233–237, 2001.

[81] B. S. Yilbaş, A. Z. Şahin, A. Z. Al-Garni, S. A. M. Said, Z. Ahmed, B. J. Abdulaleem and M. Sami, "Plasma nitriding of Ti-6Al-4V alloy to improve some tribological properties," Surface and Coatings Technology, vol. 80, pp. 287–292, 1996.

[82] D. Nolan, S. W. Huang, V. Leskovsek, and S. Braun, "Sliding wear of titanium nitride thin films deposited on Ti–6Al–4V alloy by PVD and plasma nitriding processes," Surface and Coatings Technology, vol. 200, pp. 5698–5705, 2006.

[83] G. Cassar, J. C. A.-B. Wilson, S. Banfield, J. Housden, A. Matthews, and A. Leyland, "A study of the reciprocating-sliding wear performance of plasma surface treated titanium alloy," Wear, vol. 269, pp. 60–70, 2010.

[84] ASTM C1624, "Standard test method for adhesion strength and mechanical failure modes of ceramic coatings by quantitative single point scratch testing," ASTM International, 2005.

[85] M. Salehi, T. Bell, and P. H. Morton, "Effect of surface topography on tribo-oxidation of titanium nitrided surfaces," Journal of Physics D: Applied Physics, vol. 25, pp. 889–895, 1992.

[86] P. A. Dearnley, T. Bell, and F. Hombeck, "Crafting the surface with glow discharge plasmas," in Surface Engineering: Processes and Applications, K. N. Strafford, R. St. C. Smart, I. Sare, and C. Subramanian, editors, Pennsylvania: Technomic Publishing Company, Inc., 1995, pp. 21–47.

[87] M. Y. P. Costa, M. L. R. Venditti, M. O. H. Cioffi, H. J. C. Voorwald, V. A. Guimarães, and R. Ruas, "Fatigue behavior of PVD coated Ti–6Al–4V alloy," International Journal of Fatigue, vol. 33, pp. 759–765, 2011.

[88] M. M. Ali and S. G. S. Raman, "Effect of plasma nitriding environment and time on plain fatigue and fretting fatigue behavior of Ti-6Al-4V," Tribology Letters, vol. 38 pp. 291–299, 2010.

[89] A. Bloyce, P. H. Morton, and T. Bell, "Surface engineering of titanium and titanium alloys," in Metals Handbook, vol. 5, Metals Park, Ohio: American Society for Metals, 1994, pp. 835–851.

[90] T. Morita, H. Takahashi, M. Shimizu, and K. Kawasaki, "Factors controlling the fatigue strength of nitrided titanium," Fatigue and Fracture of Engineering Materials, vol. 20, pp. 85–92, 1997.

[91] G. Cassar, J. C. Avelar-Batista Wilson, S. Banfield, J. Housden, M. Fenech, A. Matthews, and A. Leyland, "Evaluating the effects of plasma diffusion processing and duplex diffusion/PVD-coating on the fatigue performance of Ti–6Al–4V alloy," International Journal of Fatigue, vol. 33, pp. 1313–1323, 2011.

[92] K. Tokaji, T. Ogawa, and H. Shibata, "The effects of gas nitriding on fatigue behavior in titanium and titanium alloys," Journal of Materials Engineering and Performance, vol. 8, pp. 159–167, 1999.

[93] D. Rodriguez, J. M. Manero, F. J. Gil, and J. A. Panell, "Low cycle fatigue behavior of Ti6Al4V thermochemically nitrided for its use in hip prostheses," Journal of Materials Science, vol. 12, pp. 935–937, 2001.

[94] P. Guiraldenq, B. Haenggi, B. Cerati, and F. Gaillard, "Fretting wear and fatigue (in air and sea water) of Ti6Al4V alloy with different surface treatments," in Surface Modification Technologies VIII, T. S. Sudarshan and M. Jeandin, editors, London, UK: The Institute of Materials, 1995, pp. 17–27.

[95] T. W. J. R. Sobiecki, and J. Rudnicki, "The influence of glow discharge nitriding, oxynitriding and carbonitriding on surface modification of Ti–1Al–1Mn titanium alloy," Vacuum, vol. 64, pp. 41–46, 2002.

[96] B. Rajasekaran and S. G. S. Raman, "Plain fatigue and fretting fatigue behaviour of plasma nitrided Ti-6Al-4V," Materials Letters, vol. 62, pp. 2473–2475, 2008.

[97] ASTM E8/E8M-11, "Standard Test Methods for Tension Testing of Metallic Materials," ASTM International, 2011.

[98] T. Akahori, M. Niinomi, K. Fukunaga, and I. Inagaki, "Effects of microstructure on the short fatigue crack initiation and propagation characteristics of biomedical α/β titanium alloys," Metallurgical and Materials Transactions A, vol. 31A, pp. 1949–1958, 2000.

[99] B. F. Chen, J. Hwang, G. P. Yu, and J. H. Huang, "In situ observation of the cracking behavior of TiN coating on 304 stainless steel subjected to tensile strain," Thin Solid Films, vol. 352, pp. 173–178, 1999.

[100] D. Novovic, R. C. Dewes, D. K. Aspinwall, W. Voice, and P. Bowen, "The effect of machined topography and integrity on fatigue life," International Journal of Machine Tools and Manufacture, vol. 44, pp. 125–134, 2004.

[101] J. C. A. Batista, A. Matthews, and C. Godoy, "Micro-abrasive wear of PVD duplex and single-layered coatings," Surface and Coatings Technology, vol. 142–144, pp. 1137–1143, 2001.

[102] T. Bell, H. Dong, and Y. Sun, "Realising the potential of duplex surface engineering," Tribology International, vol. 31, pp. 127–137, 1998.

[103] E. I. Meletis, A. Erdemir, and G. R. Fenske, "Tribological characteristics of DLC films and duplex plasma nitriding/DLC coating treatments," Surface and Coatings Technology, vol. 73, pp. 39–45, 1995.

[104] H. Dong and T. Bell, "Designer surfaces for titanium components," Anti-Corrosion Methods and Materials, vol. 46, pp. 338–345, 1999.

[105] Z. X. Zhang, H. Dong, T. Bell, and B. S. Xu, "The effect of deep-case oxygen hardening on the tribological behaviour of a-C:H DLC coatings on Ti6Al4V alloy," Journal of Alloys and Compounds, vol. 464, pp. 519–525, 2008.

[106] K.-T. Rie and E. Broszeit, "Plasma diffusion treatment and duplex treatment – recent development and new applications," Surface and Coatings Technology, vol. 76–77, pp. 425–436, 1999.

[107] K.-T. Rie, T. Stucky, R. A. Silva, E. Leitao, K. Bordji, J. Y.-. Jouzeau, et al., "Plasma surface treatment and PACVD on Ti alloys for surgical implants," Surface and Coatings Technology, vol. 74–75, pp. 973–980, 1995.

[108] S. Ma, K. Xu, and W. Jie, "Wear behavior of the surface of Ti–6Al–4V alloy modified by treating with a pulsed d.c. plasma-duplex process," Surface and Coatings Technology, vol. 185, pp. 205–209, 2004.

[109] G. Cassar, S. Banfield, J. C. Avelar-Batista Wilson, J. Housden, A. Matthews, and A. Leyland, "Tribological properties of duplex plasma oxidised, nitrided and PVD coated Ti–6Al–4V," Surface and Coatings Technology, vol. 206, pp. 395–404, 2011.

[110] L. Shen, L. Wang, Y. Wang, and C. Wang, "Plasma nitriding of AISI 304 austenitic stainless steel with pre-shot peening," Surface and Coatings Technology, vol. 204, pp. 3222–3227, 2010.

[111] W. P. Tong, Z. Han, L. M. Wang, J. Lu, and K. Lu, "Low-temperature nitriding of 38CrMoAl steel with nanostructured surface layer induced by surface mechanical attrition treatment," Surface and Coatings Technology, vol. 202, pp. 4957–4963, 2008.

[112] J. F. Gu, D. H. Bei, J. S. Pan, J. Lu, and K. Lu, "Improved nitrogen transport in surface nanocrystallized low-carbon steels during gaseous nitridation," Materials Letter, vol. 55, pp. 340–343, 2002.

[113] W. P. Tong, N. R. Tao, Z. B. Wang, J. Lu, and J. T. K. Lu, "Nitriding iron at lower temperatures," Science, vol. 299, pp. 686–688, 2003.

[114] M. Thomas, T. Lindley, D. Rugg, and M. Jackson, "The effect of shot peening on the microstructure and properties of near-alpha titanium alloy following high temperature exposure," Acta Materialia, vol. 60, pp. 5040–5048, 2012.

[115] M. Thomas, T. Lindley, and M. Jackson, "The microstructural response of a peened near-a titanium alloy to thermal exposure," Scripta Materialia, vol. 60, pp. 108–111, 2009.

[116] B. K. C. Ganesh, W. Sha, N. Ramanaiah, and A. Krishnaiah, "Effect of shot peening on sliding wear and tensile behavior of titanium implant alloys," Materials and Design, vol. 56, pp. 480–486, 2014.

[117] F. Appel, "Atomic level observations of mechanical damage in shot peened TiAl," Philosophical Magazine, vol. 93, pp. 2–21, 2013.

[118] M. Thomas and M. Jackson, "The role of temperature and alloy chemistry on subsurface deformation mechanisms during shot peening of titanium alloys," Scripta Materialia, vol. 66, pp. 1065–1068, 2012.

[119] D. G. Bansal, M. Kirkham, and P. J. Blau, "Effects of combined diffusion treatments and cold working on the sliding friction and wear behavior of Ti–6Al–4V," Wear, vol. 302, pp. 837–844, 2013.

[120] S. Hogmark, S. Jacobson, and M. Larsson, "Design and evaluation of tribological coatings," wear, vol. 246, pp. 20–33, 2000.

[121] A. Czyrska-Filemonowicz, P. A. Buffat, M. Lucki, T. Moskalweicz, W. Rakowski, J. Lekki, et al., "Transmission electron microscopy and atomic force microscopy characterization of titanium-base alloys nitrided under glow discharge," Acta Materialia, vol. 53, pp. 4367–4377, 2005.

[122] E. Torok, A. J. Perry, L. Chollet, and W. D. Sproul, "Young's modulus of TiN, TiC, ZrN and HfN," Thin Solid Films, vol. 153, pp. 37-43, 1987.

[123] W. Ziaja, "Finite element modelling of the fracture behaviour of surface treated Ti-6Al-4V alloy," Computational Materials Science and Surface Engineering, vol. 1, pp. 53–60, 2009.

[124] V. Fouquet, L. Pichon, M. Drouet, and A. Straboni, "Plasma assisted nitridation of Ti-6Al-4V," Applied Surface Science, vol. 221, pp. 248–258, 2004.

[125] C. X. Li, D. Horspool, and H. Dong, "Effect of ceramic conversion surface treatment on fatigue properties of Ti6Al4V alloy," International Journal of Fatigue, vol. 29, pp. 2273–2280, 2007.

[126] H. Dong, A. Bloyce, and T. Bell, "Design and the performance of a new titanium duplex surface engineering system," presented at the Symposium on Surface Performance of Titanium, Cincinnati, OH, 7-9 October 1996.

[127] A. D. Wilson, A. Leyland, and A. Matthews, "A comparative study of the influence of plasma treatments, PVD coatings and ion implantation on the tribological performance of Ti-6Al-4V," Surface and Coatings Technology, vol. 114, pp. 70–80, 1999.

Chapter 4

Low-temperature Thermochemical Treatments of Stainless Steels – An Introduction

Rodrigo P. Cardoso, Marcio Mafra and Silvio F. Brunatto

Additional information is available at the end of the chapter

http://dx.doi.org/10.5772/61989

Abstract

Plasma technology used to perform thermochemical treatments is well established for the majority of steels, but it is not the case for the different stainless steel classes. Thus, important scientific and technological achievements can be expected in the coming years regarding plasma-assisted thermochemical treatment of such steels. The metallurgical aspects as well as the application cost-efficiency of stainless steels impose specific requirements for the thermochemical treatment, such as easy native chromium-rich oxide layer removal and surface activation at low temperature, which do not appear for other steel classes (plain, low-alloy, and tool steels). Thus, due to the highly reactive physico-chemical environment created by the plasma, plasma-assisted technology presents advantages over other "conventional" technologies like those performed in gas or liquid environments. Low temperature is needed to avoid the reduction of corrosion resistance of stainless steels, by suppressing chromium carbide/nitride precipitation, and, in this case, good surface properties are achieved by the formation of treated layers containing metastable phases. Such attributes make the low-temperature plasma thermochemical treatments of stainless steels an important R&D field in the domain of plasma technology and surface treatments, and the goal of this chapter is to introduce the reader to this important topic.

Keywords: Nitriding, Carburizing, Low-temperature thermochemical treatments, Stainless steels, Surface treatments

1. Introduction

Stainless steels are iron alloys mainly based on the Fe-Cr-Ni and Fe-Cr-C systems. The main characteristic of this class of alloys is its expressively high corrosion resistance when compared with other iron-based alloys. Thus, the main applications of these materials are related to parts exposed to corrosive operation environments, from cutlery to petroleum and nuclear indus-

tries. The corrosion resistance of these steels is related to the formation of a continuous and passive chromium-rich oxide layer on the steel surface, also termed native oxide layer. This layer is promptly formed when the part is exposed to atmospheres containing oxygen, for steels containing at least 10.5 wt.% Cr in solid solution (solved in the steel matrix), which result in the so-called stainless steels. Due to the raw materials and processing costs, stainless steels are usually more expensive than other steels and have their application restricted to cases where corrosion resistance is important. However, in most of these applications, it remains very competitive compared with other corrosion-resistant alloys.

Even for the case of high-strength stainless steels, like those of the martensitic class, the need for high-performance parts strongly justifies the use of surface-hardening treatments, aiming to enhance the wear/fatigue resistance and to extend the part service life. Keeping in mind that corrosion resistance is the dominant criterion for these steels, it is unacceptable for any applied surface treatment to cause significant corrosion resistance decrease of a treated part.

Thermochemical treatments like nitriding and carburizing are among the most known surface-hardening treatments. As an example, very good results have been achieved when (plasma) nitriding is applied to plain, low-alloy, and tool steel parts. In such cases, treatment temperatures in the order of 500°C–580°C, and for times not much longer than 36 h, are usually applied. Differently, inappropriate results are observed if similar treatment temperatures and times are used to nitride stainless steels, since their treated surfaces will present strong corrosion resistance reduction, due to chromium nitride (CrN/Cr_2N) precipitation. Thus, for stainless steels, v, making it possible to produce very hard treated surfaces with the same, or even higher, corrosion resistance than the untreated stainless steel surface. Such achievements can be obtained if the treatment temperature is set to a value high enough to activate interstitial (N and C) atom diffusion, and low enough to avoid significant Cr (substitutional) diffusion, thus keeping unaltered the chromium content in solid solution.

Another important aspect to be considered when applying a thermochemical treatment in a stainless steel is that the native oxide layer, important for corrosion resistance, also acts as a diffusion barrier for atoms from the treatment atmosphere (N and C) diffusing into the steel surface. Thus, in such thermochemical treatments, it is necessary, first of all, to remove or reduce this oxide layer to allow the diffusion of N/C atoms, aiming to alter the composition of the surface under treatment. Therefore, in the case of plasma-assisted thermochemical treatments, plasma has two main tasks:

i.   To reduce and/or remove by sputtering the native oxide layer; and

ii.   To create an environment for which the N/C chemical potential is higher than its chemical potential in the original steel surface, enabling the N/C surface alloying, and thus the surface hardening.

Considering the aforementioned main points, to present the plasma-assisted low temperature thermochemical treatments of stainless steels in a didactical way, the next two sections will discuss some fundamentals on stainless steels and plasma-assisted thermochemical treatments. After presenting the fundamentals, a section is devoted to present the typical results of low-temperature thermochemical treatments of stainless steels, and then, the chapter ends with some conclusions and final remarks.

2. Fundamentals of stainless steels and low-temperature thermochemical treatments

Depending on the alloy composition, mainly on the Cr, Ni, and C contents, stainless steels can present different microstructures at room temperature, being classified by this criterion. The three main stainless steel classes are the austenitic, martensitic, and ferritic, with duplex, and precipitation-hardening classes also present. Since Cr is the most important component regarding stainless steel corrosion resistance, it plays a major role in low-temperature thermochemical treatments, independently of the considered stainless steel class.

At low temperature, Cr has higher affinity to N/C than to Fe. Thus, under the thermodynamics basis, instead of an interstitial solid solution, chromium nitrides/carbides would be expected to be formed during the stainless steel surface alloying with such elements, leading to a consequent reduction of Cr content in solid solution. Nevertheless, the nucleation of a new phase presents some energetic barriers that need to be overcome, implying the existence of a critical nucleus size to start the precipitation of a new phase. Since the nucleation process is strongly related to atomic diffusion, and it is especially restrictive in solid state, the main parameters to be controlled to prevent nitrides/carbides nucleation process are temperature and time. If low enough temperature and time are used, no nucleation process is started and no chromium nitride/carbide is formed. In such conditions, metastable supersaturated solid solutions known as N/C-expanded phases tend to be formed. The above-mentioned aspects comprise the basis of the low-temperature thermochemical treatments used to stainless steels surface hardening.

The formation of these important metastable phases are possible since the atoms being introduced in the treated surface are interstitials (N and C), and Cr is a substitutional atom. As it is well-known from diffusion processes, interstitial atoms tend to present high diffusion coefficients and low diffusion activation energies when compared to substitutional atoms. Thus, it is possible to heat stainless steel components up to a temperature that is

i.   high enough to attain a significant diffusion of interstitial (N and C) atoms; and

ii.   low enough to avoid significant diffusion of substitutional (Cr) atoms.

At these conditions, the kinetics of the treated layer growth is acceptable and the energetic barrier for nucleation of chromium nitrides/carbides is not overcome. In stainless steels, the N/C surface alloying is accompanied by an expansion of the lattice parameters, which explains the use of the term "expanded" to the obtained phases. For austenitic stainless steels, the obtained phases can be termed as nitrogen- or carbon-expanded austenite, also known as S-phase, for the nitriding and carburizing treatment, respectively. The analog phases for martensitic stainless steels are nitrogen- or carbon-expanded martensite, respectively.

At this point, it is important to inform the reader that the precipitation of chromium nitrides/ carbides will consume Cr atoms from the steel matrix. Therefore, in the surrounding of the precipitates, the original steel matrix Cr content in solid solution will be depleted and the corrosion resistance will be locally reduced. In several cases, the Cr content in such areas can be reduced to values lower than 10.5 wt.%, which is the minimum acceptable value for stainless steels, leading the steel to be sensitized. This phenomenon is referred to as sensitization and must be strongly avoided since, as aforementioned, the high corrosion resistance of stainless steels is the main property justifying its use instead of other lower cost steels.

It is worth mentioning that the corrosion resistance in stainless steels is attributed to the formation of a passive and very stable native oxide layer that hinders the oxidation process continuity. The very negative Gibbs free energy variation for the Cr-rich native oxide layer formation is directly related to the high Cr-O atoms affinity, and to its strong stability. Thus, this native oxide layer is hard to reduce/remove at low temperatures by conventional (gas/ liquid) processes, and if it is present in the treating surface, the thermochemical treatment cannot be successfully carried out since the oxide layer prevents the diffusion of O atoms as well as the diffusion of N/C atoms. At this point, it is possible to assert that the reduction/ removal of the stable oxide layer is easily attained by plasma processes, but as the affinity of Cr with O is even higher than its affinity with N and C atoms, special attention must be spent for two technological aspects of the low-temperature thermochemical processes, to know

i.   the process gas mixture purity; and

ii.   the vacuum chamber leakage and/or degassing.

Even very low oxygen partial pressures can inhibit the reduction/removal of the native oxide layer of the stainless steel surface, practically stopping the diffusion of interstitial atoms (N and C) from the gas phase into the steel. On the other hand, if the oxygen content in the atmosphere is low enough, the action of the very reductive plasma species like atomic hydrogen (H), allied to sputtering process (usually using Ar atoms in the plasma gas mixture for this purpose), will make the native oxide layer reduction/removal easier, enabling N and C atoms to diffuse into the steel.

The last aspect that merits to be considered in this section is only important for martensitic stainless steels, which can present a metastable phase matrix (the martensite phase), for different metastability degrees, depending on the tempering stage for which the as-quenched martensite was subjected. Consequently, besides considering the possibility of chromium nitride/carbide formation instead of expanded phases during thermochemical treatment, it is also important to take into account the possibility of variations in the martensite properties. It will occur if the martensite tempering process takes place simultaneously with the thermo-chemical treatment, it can be undesirable or a strategy to reduce the treatment steps. The temperatures applied in low-temperature thermochemical treatments, usually in the 300°C–450°C range, is similar to the temperatures used for the tempering of steels. Thus, aiming to keep practically unaltered the treated material bulk properties, thermochemical treatments must be carried out at a temperature lower than that applied in the tempering treatment (as a reference, the reader can specify 50°C lower than the tempering temperature). However, if the thermochemical treatment is carried out at a higher temperature than the tempering was performed, the bulk material properties will be compatible to those obtained for a part tempered using the thermochemical treatment temperature and time.

After this brief presentation on the most important metallurgical aspects of stainless steels for understanding the main aspects related to its low-temperature thermochemical treatments, in the next section, the reader will be introduced to the fundamentals of the plasma low-temperature thermochemical treatments technology. For readers who want to go deeper into the metallurgical aspects of this important class of steels, the authors suggest the following books or review papers [1–3].

3. Fundamentals of plasma for low-temperature thermochemical treatments

This section aims to present the fundamentals of low-temperature plasma thermochemical treatments in a comprehensive way, focusing on the most important physicochemical inter-action between the plasma and the stainless steel surface. Most of the fundamentals presented are valid for all stainless steel classes and thermochemical treatments.

As indirectly mentioned in the previous section, the term low-temperature thermochemical treatment, when applied to stainless steels, refers to a treatment condition (temperature and time) where the Cr diffusion and new Cr-rich phase formation can be neglected. In such a condition, Cr is kept in solid solution, and no important reduction of the surface corrosion resistance is expected. At low-temperature treatment, even if the formation of the undesirable Cr-nitride/carbide phases is expected by thermodynamics, it does not occur since the trans-formation kinetics is limited by the atomic diffusion process, which is negligible in present case. Consequently, both the treatment and the formed phases are not in thermodynamic equilibrium.

Therefore, in the treated surface, metastable phases will be formed, the so-called expanded phases, as previously seen. These phases are supersaturated solid solutions containing N/C interstitial atoms (N for nitriding, C for carburizing, and both N and C for nitrocarburizing) in a metastable condition. The temperature limit to produce expanded phases, in detriment of precipitates, will be dependent on the steel structure, which exerts important influence on the diffusion process, being typically lower than 450°C and 400°C for austenitic and martensitic stainless steels nitriding, respectively. As a diffusion-controlled process, time is also important, varying typically from 4 to 48 h depending on the steel class, treatment temperature, and desired treated layer thickness. In fact, these values are only guidelines since it can be de-pendent on several other parameters. Thus, other aspects influencing the diffusion rate will also influence the limit temperature as, for example, the steel composition and its previous heat treatment [4, 5]. Considering the different types of stainless steels, as a general rule, limit temperature and time are higher for austenitic than for martensitic stainless steels. Taking into account the different types of thermochemical treatments, as a general rule, limit temperature and time are higher for carburizing than for nitriding, which is related to the different N-Cr and C-Cr affinities.

For low-temperature treatments, the necessary heating effect to attain the treatment temper-ature can be easily reached using only plasma heating. However, the use of hot wall reactors (with auxiliary heating systems) can be desirable for some applications, which enable the control of plasma reactivity and treatment temperature independently. Nevertheless, plasma also plays other important roles on low-temperature thermochemical treatments, which can be listed as follows:

i. It creates a very reactive environment with high N and/or C chemical potential, enabling N/C atoms to be added to the surface under treatment;

ii. It creates a very reductive environment helping the removal of the native oxide layer and preventing its re-formation (related to hydrogen species present on the gas mixture);

iii. It promotes sputtering of the native oxide layer from the stainless steel surface (mainly related to heavy species on the gas mixture, like Ar); and

iv. It heats the parts to the desired temperature by momentum transfer as energetic species bombard the part's surface (for hot wall plasma reactors, the treatment temperature is controlled by the auxiliary resistive heating system, and for other reactors, it is controlled by adjusting the plasma parameters).

All the above-mentioned "plasma roles" are related to the plasma physicochemical environment. Points iii and iv are related to plasma physical interactions with the part surface, and points i and ii are related to the nonequilibrium plasma state and so very reactive chemical environment. This very active physicochemical environment is created principally by inelastic collisions of energetic electrons with neutral atoms/molecules from the gas mixture, promoting, for example, ionization, excitation, and dissociation. These reactions occurring in the plasma environment are important for thermochemical treatments since, for example, by using typical plasma nitriding/carburizing parameters in a hot wall reactor, the atmospheres are not active if plasma is off. In other words, nitriding/carburizing will only take place if the plasma glows, promoting the formation of nitriding/carburizing species like atomic nitrogen from molecular nitrogen and/or atomic carbon from hydrocarbides (C_xH_y) molecules present in the plasma gas mixtures. The most important physical and chemical plasma–surface interactions for these treatments are schematically presented in Fig. 1.

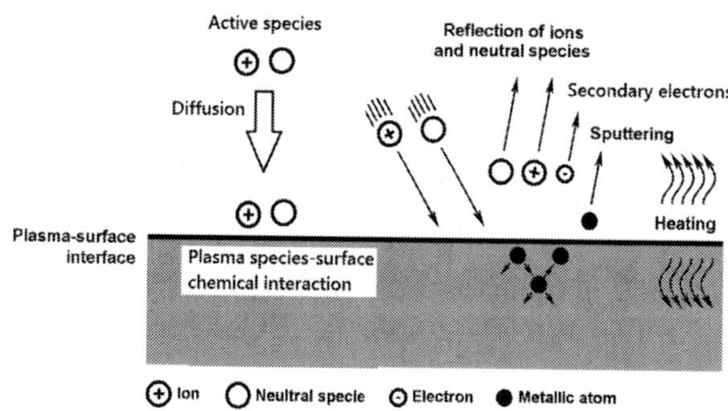

Figure 1. Main plasma–surface physicochemical interactions on low-temperature thermochemical treatments of stainless steels.

The aforementioned interactions are due to the high-energy plasma species (electrons, ions, and neutrals). The mechanism by which these species gain energy will depend on the applied plasma excitation technology, and virtually all of them could be applied for the purpose of thermochemical treatments. However, direct current (DC) plasma is the most applied in the industrial applications of low-temperature thermochemical treatments[1] due to its relative low cost and high versatility if compared to other plasma technologies. Thus, DC plasma and related physicochemical phenomena (Fig. 1) is briefly discussed, aiming to present more details on the important physicochemical aspects related to low-temperature plasma thermochemical treatments. It is worth mentioning that the plasma fundamental aspects, related to its very active physicochemical environment, are very similar for DC and other plasma technologies, being the main difference the relative importance of each phenomenon on each treatment.

The simplest way to generate DC plasma is by applying a difference of potential between two electrodes placed into a low pressures chamber with the desired gas mixture. For thermochemical treatments, these parameters are in the range of 300–1000 V and 1–10 Torr, respectively. After applying a sufficient difference of potential between two electrodes, gas breakdown will occur. If the parameters were correctly set, after a short time (on the order of some microseconds), a steady-state abnormal glow discharge will be established, covering the whole cathode surface, which is essential for the homogeneity of the treatment. In this condition, the interaction of fast[2] electrons and neutrals will create charged species by ionizing collision/reaction, and the net charge distribution in the interelectrode gap will be modified, thus modifying the potential distribution in the interelectrode gap. In an abnormal glow discharge, it will be similar to that presented in Fig. 2.

Due to this potential distribution (Fig. 2), the electric field is restricted to the cathode and anode sheath, being that the cathode fall is considerably more intense than the anode fall. Thus, the most energetic species of DC plasma will be found close to the cathode sheath boundaries. In the cathode sheath region, high-energy ions and fast neutrals (created by charge exchange collisions) are found, being accelerated to the cathode surface by the referred potential fall. However, electrons, pulled away from the cathode (part) surface (also termed "secondary electrons"; for more details, the reader is invited to see Chapman [6]) due to the impingent species bombarding the surface (see Fig. 1), enter the glow region with very high energy. Positive ions are mainly created by electron-neutral collisions in the glow region and are accelerated in the cathode sheath toward the cathode surface. The charge exchange collision occurs just when an ion travelling along the sheath collides with a neutral species. In such a case, an electron from the neutral slow specie is transferred to the fast ion (presenting kinetic energy proportional to the work performed by the electrical field on the charged particle), neutralizing it. At this moment, the original ion becomes a fast neutral, and the slow neutral

---

1 To be precise, the most industrially applied plasma excitation technology is the "pulsed-DC plasma excitation." In this case, the plasma behavior is very similar to continuous DC excitation. Pulsed DC technology is used to reduce the arcing problems, which are more frequent in industrial reactors. This reduction is achieved by introducing a period of some tens of microseconds where plasma is switched off.

2 The terms "fast" and "slow" are related to species with high and low kinetic energy, respectively. As a reference, slow species have kinetic energy near to that related to thermal agitation level.

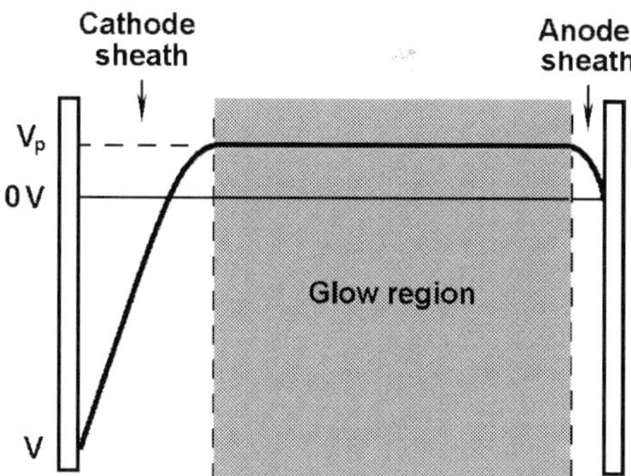

Figure 2. Typical potential distribution on a DC glow discharge, also known as DC plasma. V_p is the plasma potential, and V is the potential imposed to the cathode by the powder supply.

(presenting kinetic energy related to the environment thermal agitation level) becomes a slow ion. This new-formed ion will be also accelerated by the remainder potential fall towards the cathode. For a typical operating condition, this phenomenon is important, and as a consequence, the mean energy of ions striking the cathode is lowered and an important number of fast neutrals will also strike it. Such ions and fast neutrals are the origins of the most important physical interactions presented in Fig. 1. As previously commented, the plasma species bombardment will also promote the ejection of secondary electrons from the cathode surface, which are also represented in Fig. 1. Despite secondary electrons not being directly related to thermochemical treatments, in DC plasma, they have a fundamental role in the discharge maintenance. Since secondary electrons travel along the cathode sheath and due to their high mean free path achieve high energy, in a steady-state discharge, they are the origins of most ionization (ion creation), excitation, and dissociation collisions/reactions in the glow region. As a matter of fact, the electron collision processes are the origin of the very reactive plasma environment creation.

In most DC plasma thermochemical treatments, the part subjected to the treatment acts as the cathode of the discharge, where most of the energetic species bombardment tends to occur. In addition, since the species excitation close to the cathode is more intense (related to the high-energy electrons accelerated by the higher potential fall), generally, the cathode is subjected to a more reactive environment than that close to the anode. Thus, for parts acting as cathodes, this means that those parts are surrounded by the most active physicochemical environment of the discharge.

The aforementioned phenomena are valid for all thermochemical treatments. However, to perform different thermochemical treatments, different sources/precursors of interstitial atoms can be considered. The most common gases applied for this purpose are N_2 and CH_4, for nitriding and carburizing, respectively, which are the most applied stainless steels low-temperature thermochemical treatments. Besides the interstitial atoms precursor, other gases are also present in the discharge gas mixture to accomplish particular roles. To ensure a very reductive plasma environment, H_2 has been strongly applied. In this case, the dissociation of H_2 molecules will produce H atoms, and it will substantially increase the capability of the environment to reduce oxides. Eventually, Ar is also added to the gas mixture to enhance heating and/or sputtering effects without participating directly in the surface chemical reactions.[3] Thus, the most frequently used gas mixtures are N_2 + H_2 (+ Ar eventually) and CH_4 + H_2 + (Ar eventually) for nitriding and carburizing, respectively. The volumetric content of each gas varies for each specific application. For example, for carburizing, the maximum CH_4 content is limited by the undesired soot formation, which is potentially detrimental for the DC glow discharge stability, and also acts as a physical barrier for the diffusion of C atoms from the plasma environment onto the treating surface. In this case, the maximum reported CH_4 content is on the order of 5 vol.%. It is important to keep in mind that the maximum value is also a function of the reactor design, operating parameters and the material to be treated. Thus, in some cases, 1 vol.% CH_4 is enough to enable soot formation.

In short, the two main plasma tasks in low-temperature thermochemical treatments, as previously presented, are easily attained. The inelastic collisions/reactions in the glow region are responsible by creating active/reactive species in enough amounts to increase the N/C chemical potential on the stainless steel surface, with consequent diffusion into the material. Additionally, the reduction/sputtering of the stainless steel native oxide layer, necessary to effectively allow N/C surface alloying, is a direct consequence of the plasma–surface physicochemical interactions. To accomplish this task, plasma acts by two ways, as follows:

i.   by means of sputtering, which is a mechanical/physical interaction, being efficiently promoted by heavy species bombardment like argon; and

ii.   by means of oxide reduction, which is due to the high reductive environment created by hydrogen-containing gas mixtures that can considerably reduce the temperature for oxides reduction [7, 8]. By considering that the native oxide layer is composed by Cr-oxide only (without Fe atoms), thermodynamic data have shown that, for 1 atm pressure, a hydrogen gas purity of 99.999% is not enough to reduce Cr-oxide at temperatures up to 700°C (such statement is taken from the reading of Ellingham–Richardson diagrams for oxide formation). Thus, one can conclude that plasma is essential to remove/reduce the native oxide layer at low temperatures. It is worth noting that this point is not yet completely understood by researchers, but it is a consensus that plasma strongly contributes to remove the oxide layer by sputtering and/or to reduce it by means of reductive species, which is the case of atomic hydrogen. Despite the absence of any comments here about the role of C in oxidation–reduction reactions, C also tends to promote reduction of oxides. Thus, C atoms possibly would be important when aiming to reduce the stainless steel native oxide layer, at least for the case of low-temperature plasma carburizing treatments.

---

3 Other heavy noble gases could also be used, but Ar is the most abundant, and of lower cost, very important aspects to be considered for an industrial process.

Finally, from a technological point of view, gas purity and leakage in the plasma reactor used for low-temperature thermochemical treatments are to be also emphasized. If these "parameters" are adequate, no special attention must be paid to ensure the native oxide layer reduction, being one important advantage of plasma over other possible low-temperature gas/liquid environment thermochemical treatments of stainless steels.

On the other hand, it is important to mention that the two very probable main drawbacks of plasma process are

i.   its relatively high cost of installation (despite the low-temperature DC plasma technology and know-how that can be easily transferred to companies specialized in "conventional" plasma thermochemical treatments, with the possibility of using existing equipment); and

ii. the existence of some important geometric restrictions, which can be related to the component as well as to the discharge design. In this case, hollow cathode formation, edge effects, and difficulties to treat internal surfaces of too low diameter holes or narrow gaps are examples of the related geometric restrictions.

For readers who want to go deeper into plasma physics for materials treatment, the authors suggest to consult some books such as Chapman [6] and Lieberman and Lichtenberg [9], or the other chapters of this book.

4. Typical results obtained for low-temperature thermochemical treatments of stainless steels

Ichii et al. 1986 cited by [10] is one of the oldest references found using low-temperature thermochemical treatments and obtaining the related expanded phases. Its title is "Structure of the ion-nitrided layer of 18–8 stainless steel." Thus, apparently, one of the earliest successful low-temperature thermochemical treatments was plasma nitriding, which at first was known as ion nitriding.

Low-temperature thermochemical treatments of austenitic stainless steels are by far the most studied subject in the present area, with the majority of papers dealing with low-temperature nitriding, followed by low-temperature carburizing and then, in minor amounts, other treatments like nitrocarburizing or sequential (nitriding/carburizing) treatments. Probably this fact is related to the great number of industrial applications of this steel type. Regarding the other stainless steel classes, the application of low-temperature thermochemical treatments for martensitic stainless steel is the second more studied group. In this case, again, the majority of papers deal with the nitriding process, and only few papers have been devoted to carburizing and other thermochemical treatments. The other stainless steel classes are considerably less studied. To illustrate this point, the authors present the results of a recently performed bibliographic research on the subject in Table 1.

Studied thermochemical treatment	Low-temperature treated stainless steel class				
	Austenitic	Martensitic	Ferritic	Duplex	Precipitation Hardening
Nitriding	[10] [11] [12] [13] [17] [18] [19] [20] [21] [22] [23] [36] [40] [41] [42] [44] [46] [48] [49] [50] [51] [55] [60] [61] [62] [66] [76] [77] [79]	[4] [5] [24] [28] [30] [31] [38] [39] [52] [54] [68] [81]	[55] [69]	[43] [53] [55] [67] [70]	[31] [58]
Carburizing	[11] [14] [15] [18] [35] [36] [37] [45] [47] [48] [59] [72] [73] [74] [75] [77] [78]	[25] [26] [27] [29] [71]			[58]
Nitrocarburizing	[32] [34] [48] [56] [57]	[33] [63]			[64] [65] [80]
Sequential nitriding/ carburizing	[34] [35]				[58]

Table 1. List of works devoted to the study of low-temperature thermochemical treatments of stainless steels ranked by type of treatment and stainless steel class

As the main objective of thermochemical treatments is to promote surface hardening and the main required characteristic of stainless steels is its corrosion resistance, several works are devoted to determining both the wear and corrosion resistances of low-temperature treated surfaces. Another group of papers has been devoted to the study of more fundamental aspects, like the process kinetics, the metastable phase formation, the metastable phase structure, and the chromium nitrides/carbides precipitation phenomena.

In the next subsections, based on works referred to in Table 1, the typical results obtained by plasma low-temperature thermochemical treatments for the different classes of stainless steels are presented concerning scientific as well as technological aspects. Considering that more information is available for austenitic and martensitic stainless steels nitriding and carburizing treatments, these topics are treated in more details. For other stainless steels and treatments, only some guidelines are presented.

4.1. Austenitic stainless steels

Despite the large number of works dedicated to low-temperature thermochemical treatments of austenitic stainless steels, the large majority of research is carried out for AISI 304, 304L, 316, and 316L steels. Since nitriding and carburizing processes are very similar, typical results for both treatments are discussed together.

For austenitic stainless steels, the low-temperature treated surface is constituted of N-expanded and C-expanded austenite for nitriding and carburizing, respectively. These phases consist of supersaturated interstitial solid solutions, for which the crystalline structure lattice parameters are expanded by the presence of the interstitial atoms solved in it, as previously commented. The enlargement of the austenite lattice parameter can be easily detected by the shift of the diffraction peaks to lower 2θ angles in X-ray diffraction (XRD) measurements. In the microstructural analysis performed by means of cross-sectional micrographs of treated samples, the expanded austenite tends to present a white aspect, when the polished cross-section for analysis is etched by using Vilella or Marble etchants. The white aspect of the expanded phase layer is directly related to its higher resistance to the etchant if compared with that of the bulk material (chemically unaltered). To illustrate the aspect referred to, Figs. 3 and 4 present typical cross-sectional micrographs of nitrided and carburized AISI 316 austenitic stainless steel, respectively. By confronting Figs. 3 and 4, it is possible to notice that the growth kinetics of the carburized layer is higher than that of the nitride layer. This fact is common for most steels treated at low-temperatures, and it is attributed to the different diffusivities of N and C in the steel solid solution matrix.

When the treatment temperature is too high or the treatment time is too long, precipitation of chromium nitrides/carbides can be observed along the original austenite grain boundaries, being strongly indicative that the corrosion resistance of the treated surface was decreased. According to Egawa et al. [11], the limit temperature for low-temperature nitriding and carburizing, related to chromium nitrides/carbides precipitation in the grain boundaries, is practically independent of the steel composition. For treatments of 4 h, Egawa et al. [11] reported a limit temperature in the range of 430°C–440°C and 540°C–560°C for the low-temperature nitriding and the low-temperature carburizing of austenitic stainless steels, respectively, this difference being attributed to the different N-Cr and C-Cr chemical affinities.

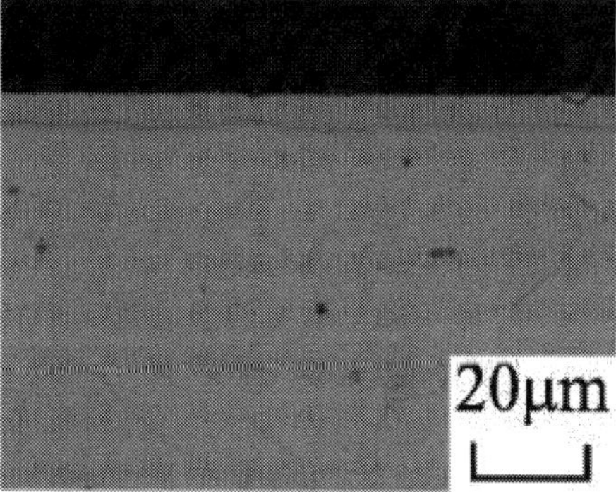

Figure 3. Cross-sectional micrograph of a low-temperature nitrided AISI 316 STEEL, treated at 400°C for 4 h (adapted from Egawa et al. [11]).

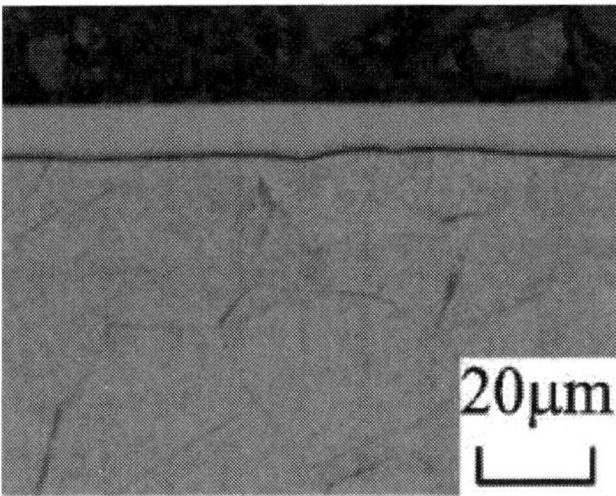

Figure 4. Cross-sectional micrograph of a low-temperature carburized AISI 316 STEEL, treated at 400°C for 4 h (adapted from Egawa et al. [11]).

For the nitriding process, the N-expanded austenite microstructure confers a surface hardness that usually attains 1200 HV, which is associated to a dissolved nitrogen content ranging between 20 and 45 at.% [10-13]. For carburizing, the typical surface hardness is around 1000 HV, which is associated to a dissolved carbon content of about 10 at.% [14–16].

Concerning the depth hardness profiles, in the low-temperature nitrided layers, a sharp hardness decrease is usually observed in the treated layer/substrate bulk interface, whereas for low-temperature carburized layers, a smooth transition tends to be observed. From a mechanical point of view, the referred smooth transition all over the treated layer leads to a smooth properties transition, being, in this case, preferable with respect to the sharp one.

The treated layers performance, regarding their potential industrial applications, is closely linked to the treated layer microstructure, and it has been the main subject of several works. The treated surface characteristics enable considerable gains to the surface wear resistance, with reported wear reductions by a factor of up to 20. Such gains are accomplished with the maintenance or even enhancement of the surface corrosion resistance. Additionally, due to the compressive stresses introduced in the treated layer, fatigue life is also increased, and compressive residual stresses between 2 and 3 GPa are commonly reported (information compiled from refs. [10, 11, 13, 17–23].

For the treated surface, corrosion resistance results found in the literature are less conclusive, but generally, if the low-temperature treatment condition is respected, there is a tendency to obtain increased corrosion resistance for the treated surfaces. The most important gains in corrosion resistance are reported for low-temperature carburizing treatments.

4.2. Martensitic stainless steels

In martensitic stainless steels, due to the high crystalline defects density of the as-quenched martensite tetragonal structure, the diffusion of atoms tends to be faster (higher diffusion coefficients for both the interstitial and substitutional atoms) than those verified for the austenite face-centered cubic structure. On the one hand, it enables higher growth kinetics to the treated layer, but, on the other hand, it also reduces the limit treatment temperature to avoid chromium nitrides/carbides precipitation. Thus, thermochemical treatments of martensitic stainless steels are usually performed at lower temperatures and for shorter times than those used for austenitic ones. Among the martensitic stainless steels, AISI 420 is the main studied steel of the present group, and it is taken as an example to illustrate this section.

In contrast to austenitic stainless steels, the low-temperature treated surface of martensitic stainless steels is usually composed of expanded phases, such as the nitrogen-expanded martensite for nitriding and carbon-expanded martensite for carburizing, as well as nitrides and carbides (ε-$M_{2-3}N$ and M_3C type, respectively).[4] Depending on the composition of the martensitic stainless steel, for which Ni is also present (it is the case of the ASTM CA-6NM grade martensitic stainless steel), and on the used plasma processing parameters, expanded austenite can be also formed together with the expanded martensite phase in the treated layer (referred to Allenstein et al. [24]).

The expanded martensite is also a supersaturated interstitial solid solution, and similar to the expanded austenite, it is detected by the shift of diffraction peaks to lower 2θ angles in XRD measurements. In the microstructural analysis of cross-sectional micrographs of the treated samples, distinct characteristics of the treated surface are observed for nitrided and carburized samples. Fig. 5 depicts the cross-sectional micrograph of a low-temperature nitrided AISI 420 martensitic stainless steel sample, showing the white aspect of a treated layer composed by N-expanded martensite and nitrides. Once again, the white aspect is related to the higher resistance of the treated layer to the etchant when compared with the substrate bulk (chemically unaltered). By confronting the results in Figs. 3 and 5, it is possible to notice that the growth kinetics of the nitrided layer is higher for the martensitic steel than that verified to the austenitic one, even at a lower treatment temperature. Such a fact is also related to the aforementioned difference of the diffusion coefficients for the different crystalline structures. Fig. 6 presents the cross-sectional micrograph of a low-temperature carburized martensitic stainless steel. In this case, a thin white-aspect layer containing cementite (termed as "outer layer"), probably in addition to C-expanded martensite phase, can be observed. The occurrence of a diffusion layer up to 40 μm in depth [25], composed of C-expanded martensite only, is also present, but it is only detected by means of microhardness depth profile measurements, as presented in the right side of Fig. 6 (for more details, the reader can check ref. [26]).

For martensitic stainless steels, the limit temperature for nitrides/carbides precipitation in the grain boundaries is around 350°C–400°C [4] and 450°C [27] for 4 h low-temperature nitriding and carburizing, respectively. As this class of steels tends to present a metastable matrix, the heat treatment performed prior to the surface treatment (in this case, the steel tempering treatment) has significant influence on low-temperature nitriding as presented by Cardoso et

---

4 ε-$M_{2-3}N$ and M_3C are closely related to ε-$Fe_{2-3}N$ and Fe_3C (cementite) phases, but as the treatment is performed at low temperature, Cr is supposedly also solved in ε-iron nitride and cementite. Hence, the most precise way to describe such phases would be by substituting Fe by M, where M would represent the main substitutional (metallic) alloying elements present on the steel. Thus, the possible more precise description for such compounds would be (Fe, Cr)$_{2-3}$N and (Fe, Cr)$_3$C, respectively.

al. [4]. Due to the similarities between nitriding and carburizing, it is expected that the same effect can be important on low-temperature carburizing.

Figure 5. Cross-sectional micrograph of a low-temperature nitrided AISI 420, treated at 350°C for 4 h.

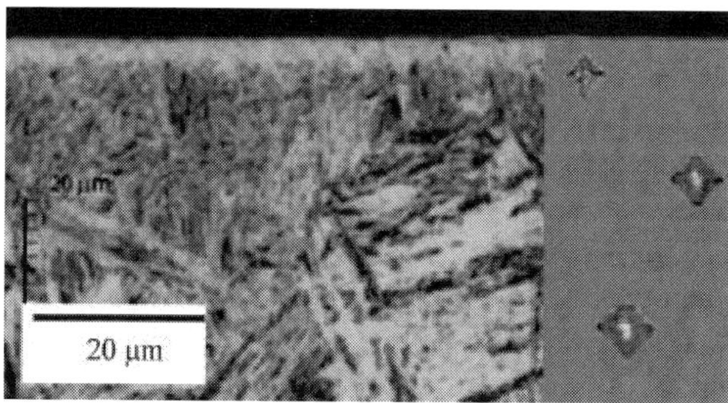

Figure 6. Cross-sectional micrograph of a low-temperature carburized AISI 420, treated at 450°C for 4 h (adapted from Scheuer [25]).

For the nitriding process, the obtained treated layer microstructure confers a surface hardness that usually attains 1470 HV, associated to a dissolved nitrogen content of 11 at.% [28]. For carburizing, the typical surface hardness is around 900 HV [27]. Up to this moment, no information has been found on the carbon content of the C-expanded martensite. Unreported measurements from the authors of this chapter indicate carbon contents on average at the white-aspect layer (outer layer) in the order of 25 at.% and around 4 at.% in the diffusion layer, but additional efforts must be conducted to confirm such results.

Concerning the depth hardness profiles, similar to austenitic stainless steels, low-temperature nitriding presents a sharp hardness decrease in the treated layer/substrate bulk interface, whereas for low-temperature carburizing a smooth hardness decrease is present.

Similar to the case of austenitic stainless steels, the treated surface characteristics enable considerable gains in the material wear resistance (a wear reduction of 97% is reported in ref. [29]), accompanied by the maintenance or even an enhancement of the surface corrosion resistance. Again, due to compressive stress in the treated layer, the fatigue life is also increased (information compiled from refs. [28–31]).

For the treated surface corrosion resistance results, few papers have been found, but it seems that the general trend is similar to that of the austenitic steels.

4.3. Ferritic, duplex, and precipitation hardening stainless steels

Works on low-temperature thermochemical treatments of ferritic stainless steels are rare in the literature and, apparently, it is more difficult to form expanded phases in such steels. The low-temperature treatments of Duplex stainless steels are roughly similar to those performed in austenitic as well as in ferritic stainless steels at the same time. However, the diffusion of interstitials from one phase to the other one, perpendicular to the surface under treatment, especially for substrates of small grain size, can be important making the treatment kinetics closer than expected for grains of different phases. However, unpublished results of the authors of this chapter, working on low-temperature Duplex steel plasma nitriding, have shown that the layer growth kinetics next to austenite grains is expressively higher than that verified next to ferrite grains. As in the other steels, the treatments promote a considerable hardening and as a consequence increase in the wear resistance.

Precipitation hardening steels tend to present similar behaviors verified for the martensitic steels. As it is the case for all stainless steels, after low-temperature thermochemical treatments, significant surface hardening is obtained, leading the wear resistance of the treated surface to become also improved. Concerning the corrosion resistance, the results presented in the literature are somehow in disagreement. Some papers present an enhancement of the surface corrosion resistance after treatment and others present a decrease of this important property.

Taking into account the results on low-temperature thermochemical treatments of stainless steels, it seems that the corrosion resistance is very sensitive to the treatment parameters, and generally, there is a set of parameters that makes it possible to enhance the corrosion resistance of the treated surface with respect to the chemically unaltered substrate bulk.

4.4. Nitrocarburizing and sequential treatments

The main idea in applying low-temperature nitrocarburizing and sequential low-temperature carburizing and nitriding treatments is related to the possibility of obtaining a treated layer with a smoother hardness transition, when compared to the nitride layers, and higher surface hardness than that obtained in the carburized surfaces. Another motivation would be verifying if the C introduction into the surface would enable the obtainment of precipitation- free treated surfaces at higher treatment temperatures than that used for the nitriding process.

According to Cheng et al. [32], for AISI 316 (austenitic) steel, the plasma nitrocarburizing possess has the advantages of both nitriding (high hardness) and carburizing (deep case depth and smooth hardness profile) treatments. However, it was observed that the addition of CH_4 to the gas mixture did not increase the limit temperature to obtain precipitation-free layers. Treating the AISI 420 (martensitic) steel, Anjos et al. [33] obtained a comportment that is closer to the nitriding process, so the advantages pointed out for austenitic stainless steels was not clearly found for the studied martensitic steel.

Studies on the sequential treatments, especially in austenitic stainless steels, also present the possibility of obtaining high hardness (typically attained from nitriding) and smooth hardness profiles (which is typically attained from carburizing) [34,35].

Considering the performance of both treatments in terms of wear, when the primary objective of high hardness and smooth hardness profiles are achieved, it tends to be better than nitriding and carburizing. Concerning corrosion resistance, the treated surface also tends to present higher corrosion resistance than the untreated surface.

5. Conclusion and final remarks

The studies on low-temperature plasma thermochemical treatments of stainless steels are relatively new, but they are already appliedand accessible industrially. Most of the present knowledge on this subject comes from works developed for austenitic stainless steels, which occur in profusion in the literature. Considering the treatments of martensitic stainless steels, the number of works is considerably lower, and for the other stainless steels, the respective works are considerably rare. The application of such treatments has proved to be very efficient in producing surfaces with high hardness, wear, and corrosion resistances and compressive residual stresses. As a consequence, the treated parts tend to present longer life when subject to severe wear, corrosive environments, and also cyclic loadings. Thus, from a technological point of view, such plasma-assisted treatments can be broadly applied. On the other hand, from the scientific point of view, several opened questions remain to be answered. As these treatments are carried out at low temperatures, the thermodynamic equilibrium cannot be attained in the solid state, and metastable phases tend to be formed. Thus, the mechanisms of formation of such phases are not yet completely understood. Considering the application potentiality, the mechanism on the origin of the corrosion resistance of surfaces composed or containing such phases is not yet completely explained. These are some examples of very basic opened questions, so it is clear that low-temperature plasma thermochemical treatments nowadays comprise an important R&D field. The development of research will very probably extend the current known range of application potentialities, enabling new industrial applications, while also being expected to bring important growth to this plasma technology in the coming years.

Author details

Rodrigo P. Cardoso[1*], Marcio Mafra[2] and Silvio F. Brunatto[1]

*Address all correspondence to: rodrigo.perito@ufpr.br

1 Federal University of Paraná (UFPR), Curitiba, Brazil

2 Federal University of Technology–Paraná (UTFPR), Curitiba, Brazil

References

[1] Folkhard E, editor. Welding Metallurgy of Stainless Steels. 1st ed. Vienna: Springer; 1988. 279 p.

[2] Lippold CJ, Damian KJ, editors. Welding Metallurgy and Weldability of Stainless Steels. 1st ed. Hoboken: John Wiley & Sons, Inc; 2005. 376 p.

[3] Lo K H, Shek C H, Lai JKL. Recent developments in stainless steels. Materials Science and Engineering. 2009;65(4–6):39–104. DOI: 10.1016/j.mser.2009.03.001

[4] Cardoso R P, Brunatto SF, Scheuer CJ. Low-temperature nitriding kinetics of stainless steel: effects of prior heat treatment. In: Colás R., Totten G.E., editors. Encyclopedia of Iron, Steel, and Their Alloys. 1st ed. New York: Taylor & Francis Group; 2015 (in press). DOI: 10.1081/E-EISA-120051669

[5] Ferreira L M, Brunatto SF, Cardoso RP. Martensitic stainless steels low-temperature nitriding: dependence of substrate composition. Materials Research. 2015;18(3):622–627. DOI: 10.1590/1516-1439.015215

[6] Chapman B. Glow Discharge Processes: Sputtering and Plasma Etching. 1st ed. New York: John Wiley & Sons; 1980. 406 p.

[7] Belmonte T, Thiébaut J M, Michel H, Cardoso R P, Maliska A. Reduction of metallic oxides by late Ar-H_2-N_2 postdischarges. I. Application to copper oxides. Journal of Vacuum Science and Technology A. 2002;20(4):1347–1352. DOI: 10.1116/1.1484096

[8] Belmonte T, Thiébaut J M, Michel H, Cardoso R P, Maliska A. Reduction of metallic oxides by late Ar–H_2–N_2 postdischarges. II. Applications to iron oxides. Journal of Vacuum Science and Technology A. 2002;20(4):1353–1357. DOI: 10.1116/1.1484097

[9] Lieberman M A, Lichtenberg A J. Principles of Plasma Discharges and Materials. 2nd ed. John Wiley & Sons, Inc. New York: John Wiley & Sons, Inc; 2005. 572 p.

[10] Menthe E, Rie K T. Further investigation of the structure and properties of austenitic stainless steel after plasma nitriding. Surface and Coatings Technology. 1999;116–119 :199–204. DOI: 10.1016/S0257-8972(99)00085-7

[11] Egawa M,Ueda N, Nakataa K, Tsujikawa M, Tanaka M. Effect of additive alloying element on plasma nitriding and carburizing behavior for austenitic stainless steels. Surface and Coatings Technology. 2010;205(s1):S246–S251. DOI: 10.1016/j.surfcoat.2010.07.093

[12] Renevier N, Collingnon P, Michel H, Czerwiec T. Low temperature nitriding of AISI 316L stainless steel and titanium in a low pressure arc discharge. Surface and Coatings Technology. 1999;111:128–133.DOI:10.1016/S0257-8972(98)00722-1

[13] Wang J, Xiang J, Peng Q, Fan H, Wang Y, Li G, Shen B. Effects of DC plasma nitriding parameters on microstructure and properties of 304L stainless steel. Materials Characterization. 2009;60:197–203. DOI: 10.1016/j.matchar.2008.08.011

[14] Souza R M, Ignat M, Pinedo C E, Tschiptschin A P. Structure and properties of low temperature plasma carburized austenitic stainless steels. Surface and Coatings Technology. 2009;204:1102–1105. DOI: 10.1016/j.surfcoat.2009.04.033

[15] Sun Y. Kinetics of low temperature plasma carburizing of austenitic stainless steels. Journal of Materials Processing Technology. 2005;168:189–194. DOI: 10.1016/j.jmatprotec.2004.10.005

[16] Tsujikawa M, Yamauchi N, Ueda N, Sone T, Hirose Y. Behavior of carbon in low temperature plasma nitriding layer of austenitic stainless steel. Surface and Coatings Technology. 2005;193:309–313. DOI: 10.1016/j.surfcoat.2004.08.179

[17] Menthe E, Bulak A, Olfe J, Zimmermann A, Rie K-T. Improvement of the mechanical properties of austenitic stainless steel after plasma nitriding. Surface and Coatings Technology. 2000;133–134:259–263.DOI:10.1016/S0257-8972(00)00930-0

[18] Liang W, Xiaolei X, Bin X, Zhiwei Y, Zukun H. Low temperature nitriding and carburizing of AISI304 stainless steel by a low pressure plasma arc source. Surface and Coatings Technology. 2000;131:563–567.DOI:10.1016/S0257-8972(00)00836-7

[19] Wang L, Xu X, Yu Z, Hei Z. Low pressure plasma arc source ion nitriding of austenitic stainless steel. Surface and Coatings Technology. 2000;124:93–96.DOI:10.1016/S0257-8972(99)00637-4

[20] Yetim A F, Yazici M. Wear resistance and non-magnetic layer formation on 316l implant material with plasma nitriding. Journal of Bionic Engineering. 2014;11:620–629. DOI: 10.1016/S1672-6529(14)60073-1

[21] Stinville J C, Cormier J, Templier C, Villechaise P. Monotonic mechanical properties of plasma nitrided 316L polycrystalline austenitic stainless steel: mechanical behaviour of the nitrided layer and impact of nitriding residual stresses. Materials Science and Engineering A. 2014;605:51–58. DOI: 10.1016/j.msea.2014.03.039

[22] Yazici M, Çomakli O,Yetim T, Yetim A F, Çelik A. The effect of plasma nitriding temperature on the electrochemical and semiconducting properties of thin passive films

formed on 316 L stainless steel implant material in SBF solution. Surface and Coatings Technology. 2015;261:181–188. DOI: 10.1016/j.surfcoat.2014.11.037

[23] Reis R F, Villanova R L, Costa K D, Durante G C. Nitrogen surface enrichment of austenitic stainless steel ISO 5832-1: SHTPN vs low-temperature plasma nitriding. Materials Research. 2015;18(3):575–580. DOI: 10.1590/1516-1439.352014

[24] Allenstein A N, Cardoso RP,Machado KD, Weber S, Pereira KMP, dos Santos C.A.L., Panossian C A L, Panossian Z, Buschinelli A JA, Brunatto SF. Strong evidences of tempered martensite-to-nitrogen-expanded austenite transformation in CA-6NM steel. Materials Science and Engineering A. 2012;552:569–572. DOI: 10.1016/j.msea. 2012.05.075

[25] Scheuer C J, Cardoso R P, Pereira R, Mafra M, Brunatto S F. Low temperature plasma carburizing of martensitic stainless steel. Materials Science and Engineering A. 2012;539:369–372. DOI: 10.1016/j.msea.2012.01.085

[26] Brunatto SF, Cardoso RP, Scheuer CJ. DC low-temperature plasma carburizing of martensitic stainless steel. In: Colás R., Totten G.E., editors. Encyclopedia of Iron, Steel, and Their Alloys. 1st ed. New York: Taylor & Francis; 2015 (in press). DOI: 10.1081/E-EISA-120051667

[27] Scheuer CJ, Cardos RP, Mafra M, Brunatto SF. AISI 420 martensitic stainless steel low-temperature plasma assisted carburizing kinetics. Surface and Coatings Technology. 2013;214:30–37. DOI: 10.1016/j.surfcoat.2012.10.060

[28] Espitia L A, Varela L, Pinedo C E, Tschiptschin A P. Cavitation erosion resistance of low temperature plasma nitrided martensitic stainless steel. Wear. 2013;301:449–456. DOI: 10.1016/j.wear.2012.12.029

[29] ScheuerCJ, Cardoso RP, das Neves JCK, Brunatto SF. Dry sliding wear behavior of low-temperature plasma carburized AISI 420 steel. In: October 4–8; Columbus OH. 2015.

[30] Xi Y T, Liu D X, Han D. Improvement of mechanical properties of martensitic stainless steel by plasma nitriding at low temperature. Acta Metallurgica Sinica (English Letters). 2008;21(1):21–29. DOI: 10.1016/S1006-7191(08)60015-0

[31] Bruhl S P, Charadia R, Simison S, Lamas D G, Cabo A. Corrosion behavior of martensitic and precipitation hardening stainless steels treated by plasma nitriding. Surface and Coatings Technology. 2010;204:3280–3286. DOI: 10.1016/j.surfcoat.2010.03.036

[32] Cheng Z, Li C X, Dong H, Bell T. Low temperature plasma nitrocarburising of AISI 316 austenitic stainless steel. Surface and Coatings Technology. 2005;191:195–200. DOI: 10.1016/j.surfcoat.2004.03.004

[33] Anjos A D, Scheuer C J, Brunatto S F, Cardoso R P. Low-temperature plasma nitrocarburizing of the AISI 420 martensitic stainless steel: microstructure and process ki-

netics. Surface and Coatings Technology. 2015;275:51–57. DOI: 10.1016/j.surfcoat. 2015.03.039

[34] Buhagiar J, Li X, Dong H. Formation and microstructural characterisation of S-phase layers in Ni-free austenitic stainless steels by low-temperature plasma surface alloying. Surface and Coatings Technology. 2009;204:330–335. DOI: 10.1016/j.surfcoat. 2009.07.030

[35] Tsujikawa M, Yoshida D, Yamauchi N, Ueda N, Sone T, Tanaka S. Surface material design of 316 stainless steel by combination of low temperature carburizing and nitriding. Surface and Coatings Technology. 2005;200:507–511. DOI: 10.1016/j.surfcoat.2005.02.051

[36] Adachi S, Ueda N. Combined plasma carburizing and nitriding of sprayed AISI 316L steel coating for improved wear resistance. Surface and Coatings Technology. 2014;259:44–49. DOI: 10.1016/j.surfcoat.2014.07.010

[37] Adachi S, Ueda N. Surface hardness improvement of plasma-sprayed AISI 316L stainless steel coating by low-temperature plasma carburizing. Advanced Powder Technology. 2013;24:818–823. DOI: 10.1016/j.apt.2012.12.011

[38] Allenstein A N, Lepienski C M, Buschinelli A J A, Brunatto S F. Improvement of the cavitation erosion resistance for low-temperature plasma nitrided Ca-6NM martensitic stainless steel. Wear. 2014;309:159–165. DOI: 10.1016/j.wear.2013.11.002

[39] Allenstein A N, Lepienski C M, Buschinelli A J A, Brunatto S F. Plasma nitriding using high H2 content gas mixtures for a cavitation erosion resistant steel. Applied Surface Science. 2013;277:15–24. DOI: 10.1016/j.apsusc.2013.03.055

[40] Alves Jr C, de Araújo F O, Ribeiro K J B, da Costa J A P, Sousa R R M, de Sousa R S. Use of cathodic cage in plasma nitriding. Surface and Coatings Technology. 2006;201:2450–2454. DOI: 10.1016/j.surfcoat.2006.04.014

[41] Balusamy T, Sankara Narayanan T S N, Ravichandran K, Park I S, Lee M H. Plasma nitriding of AISI 304 stainless steel: Role of surface mechanical attrition treatment. Materials Characterization. 2013;85:38–47. DOI: 10.1016/j.matchar.2013.08.009

[42] Bell T. Surface engineering of austenitic stainless steel. Surface Engineering. 2002;18(6):415–422. DOI: 10.1179/026708402225006268

[43] Bobadilla M, Tschiptschin A. On the nitrogen diffusion in a duplex stainless steel. Materials Research. 2015;18(2):390–394. DOI: 10.1590/1516-1439.337714

[44] Chemkhi M, Retraint D, Roos A, Garnier C, Waltz L, Demangel C, Proust G. The effect of surface mechanical attrition treatment on low temperature plasma nitriding of an austenitic stainless steel. Surface and Coatings Technology. 2013;221:191–195. DOI: 10.1016/j.surfcoat.2013.01.047

[45] Collins S R, Willians P C, Marx S V, Heuer A, Ernst F, Kahn H. Low-temperature carburization of austenitic stainless steels. In: Dossett J, Totten G E, editors. ASM Handbook, Vol 4D, Heat Treating of Irons and Steels. ASM International; 2014. p. 451–460.

[46] Czerwiec T, Renevier N, Michel H. Low-temperature plasma-assisted nitriding. Surface and Coatings Technology. 2000;131:267–277. DOI: 10.1016/S0257-8972(00)00792-1

[47] Ernst F, Cao Y, Michal G M, Heuer A H. Carbide precipitation in austenitic stainless steel carburized at low temperature. Acta Materialia. 2007;55:1895–1906. DOI: 10.1016/j.actamat.2006.09.049

[48] Fernandes F A P, Casteletti L C, Gallego J. Microstructure of nitrided and nitrocarburized layers produced on a superaustenitic stainless steel. Journal of Materials Research and Technology. 2013;2(2):158–164. DOI: 10.1016/j.jmrt.2013.01.007

[49] Fernandes F A P, Heck S C, Pereira R G, Picon C A, Nascente P A P, Casteletti L C. Ion nitriding of a superaustenitic stainless steel: Wear and corrosion characterization. Surface and Coatings Technology. 2010;204:3087–3090. DOI: 10.1016/j.surfcoat.2010.02.064

[50] Fewell M P, Mitchell D R G, Priest J M, Short K T, Collins G A. The nature of expanded austenite. Surface and Coatings Technology. 2000;131:300–306. DOI: 10.1016/S0257-8972(00)00804-5

[51] Fewell M P, Priest J M, Baldwin M J, Collins G A, Short K T. Nitriding at low temperature. Surface and Coatings Technology. 2000;131:284–290. DOI: 10.1016/S0257-8972(00)00793-3

[52] Kim S K, Yoo J S, Priest J M, Fewell M P. Characteristics of martensitic stainless steel nitrided in a low-pressure RF plasma. Surface and Coatings Technology. 2003;163–164:380–385. DOI: 10.1016/S0257-8972(02)00631-X

[53] Kliauga A M, Pohl M. Effect of plasma nitriding on wear and pitting corrosion resistance of X2 CrNiMoN 22 5 3 duplex stainless steel. Surface and Coatings Technology. 1998;98:1205–1210. DOI: 10.1016/S0257-8972(97)00240-5

[54] Kurelo B C E S, de Souza G B, da Silva S L R, Daudt N F, Alves Jr C, Torres R D, Serbena F C. Tribo-mechanical features of nitride coatings and diffusion layers produced by cathodic cage technique on martensitic and supermartensitic stainless steels. Surface and Coatings Technology. 2015;275:41–50. DOI: 10.1016/j.surfcoat.2015.03.052

[55] Larish B, Brusky U, Spies H -J. Plasma nitriding of stainless steels at low temperatures. Surface and Coatings Technology. 1999;116–119:205–211. DOI: 10.1016/S0257-8972(99)00084-5

[56] Lee I. Effect of processing temperatures on characteristics of surface layers of low temperature plasma nitrocarburized AISI 204Cu austenitic stainless steel. Trans. Nonferrous Met. Soc. China. 2012;22:S678-S682. DOI: 10.1016/S1003-6326(12)61785-3

[57] Lee I. The effect of molybdenum on the characteristics of surface layers of low temperature plasma nitrocarburized austenitic stainless steel. Current Applied Physics. 2009;9:S257-S261. DOI: 10.1016/j.cap.2009.01.030

[58] Leyland A, Lewis D B, Stevenson P B, Matthews A. Low temperature plasma diffusion treatment of stainless steels for improved wear resistance. Surface and Coatings Technology,. 1993;62:608–617. DOI: 10.1016/0257-8972(93)90307-A

[59] Li W, Li X, Dong H. Effect of tensile stress on the formation of S-phase during low-temperature plasma carburizing of 316L foil. Acta Materialia. 2011;59:5765–5774. DOI: 10.1016/j.actamat.2011.05.053

[60] Li Y, Xu H, Zhu F, Wang L. Low temperature anodic nitriding of AISI 304 austenitic stainless steel. Materials Letters. 2014;128:231–234. DOI: 10.1016/j.matlet.2014.04.121

[61] Liang W. Surface modification of AISI 304 austenitic stainless steel by plasma nitriding. Applied Surface Science. 2003;211:308–314. DOI: 10.1016/S0169-4332(03)00260-5

[62] Lin K, Li X, Sun Y, Luo X, Dong H. Active screen plasma nitriding of 316 stainless steel for the application of bipolar plates in proton exchange membrane fuel cells. International Journal of Hydrogen Energy. 2014;39:21470–21479. DOI: 10.1016/j.ijhydene.2014.04.102

[63] Lio R L, Qiao Y J, Yan M F, Fu Y D. Effects of rare earth elements on the characteristics of low temperature plasma nitrocarburized martensitic stainless steel. Journal of Materials Science and Technology. 2012;28(11):1046–1052. DOI: 10.1016/S1005-0302(12)60171-6

[64] Liu R L, Yan M F. Improvement of wear and corrosion resistances of 17-4PH stainless steel by plasma nitrocarburizing. Materials and Design. 2010;31:2355–2359. DOI: 10.1016/j.matdes.2009.11.069

[65] Liu R L, Yan M F. The microstructure and properties of 17-4PH martensitic precipitation hardening stainless steel modified by plasma nitrocarburizing. Surface and Coatings Technology. 2010;204:2251–2256. DOI: 10.1016/j.surfcoat.2009.12.016

[66] Mendes A F, Scheuer C J, Joanidis I L, Cardoso R P, Mafra M, Klein AN, Brunatto S F. Low-temperature plasma nitriding of sintered PIM 316L austenitic stainless steel. Materials Research. 2014;17(1):100–108. DOI: 10.1590/S1516-14392014005000064

[67] Nagtsuka K, Nishimoto A, Akamatsu K. Surface hardening of duplex stainless steel by low temperature active screen plasma nitriding. Surface and Coatings Technology. 2010;209:S295-S299. DOI: 10.1016/j.surfcoat.2010.08.012

[68] Nakasa K, Yamamoto A, Wang R, Sumomogi T. Effect of plasma nitriding on the strength of fine protrusions formed by sputter etching of AISI type 420 stainless steel. Surface and Coatings Technology. 2015;272:298–308. DOI: 10.1016/j.surfcoat.2015.03.048

[69] Nii H, Nishimoto A. Surface modification of ferritic stainless steel by active screen plasma nitriding. Journal of Physics: Conference Series. 2012;379(012052):1–7. DOI: 10.1088/1742-6596/379/1/012052

[70] Pinedo C E, Varela L B, Tschiptschin A P. Low-temperature plasma nitriding of AISI F51 duplex stainless steel. Surface and Coatings Technology. 2013;232:839–843. DOI: 10.1016/j.surfcoat.2013.06.109

[71] Scheuer C J, Cardoso R P, Zanetti F I, Amaral T, Brunatto S F. Low-temperature plasma carburizing of AISI 420 martensitic stainless steel: influence of gas mixture and gas flow rate. Surface and Coatings Technology. 2012;206:5085–5090. DOI: 10.1016/j.surfcoat.2012.06.022

[72] Sun Y, Haruman E. Effect of electrochemical potential on tribocorrosion behavior of low temperature plasma carburized 316L stainless steel in 1 M H 2 SO 4 solution. Surface and Coatings Technology. 2011;205:4280–4290. DOI: 10.1016/j.surfcoat.2011.03.048

[73] Sun Y, Haruman E. Tribocorrosion behaviour of low temperature plasma carburised 316L stainless steel in 0.5 M NaCl solution. Corrosion Science. 2011;53:4131–4140. DOI: 10.1016/j.corsci.2011.08.021

[74] Sun Y. Response of cast austenitic stainless steel to low temperature plasma carburizing. Materials and Design. 2009;30:1377–1380. DOI: 10.1016/j.matdes.2008.07.005

[75] Sun Y. Tribocorrosion behavior of low temperature plasma carburized stainless steel. Surface and Coatings Technology. 2013;228:S342-S348. DOI: 10.1016/j.surfcoat.2012.05.105

[76] Tsujikawa M, Egawa M, Sone T, Ueda N, Okano T, Higashi K. Modification of S phase on austenitic stainless steel using fine particle shot peening. Surface and Coatings Technology. 2013;228:S318-S322. DOI: 10.1016/j.surfcoat.2012.05.111

[77] Tsujikawa M, Egawa M, Ueda N, Okamoto A, Sone T, Nakata K. Effect of molybdenum and copper on S-phase layer thickness of low-temperature carburized austenitic stainless steel. Surface and Coatings Technology. 2008;202:5488–5492. DOI: 10.1016/j.surfcoat.2008.06.096

[78] Tsujikawa M, Noguchi S, Yamauchi N, Ueda N, Sone T. Effect of molybdenum on hardness of low-temperature plasma carburized austenitic stainless steel. Surface and Coatings Technology. 2007;201:5102–5107. DOI: 10.1016/j.surfcoat.2006.07.127

[79] Wang S, Cai W, Li J, Wei W, Hu J. A novel rapid D.C. plasma nitriding at low gas pressure for 304 austenitic stainless steel. Materials Letters. 2013;105:47–49. DOI: 10.1016/j.matlet.2013.04.031

[80] Yan M F, Liu R L, Wu D L. Improving the mechanical properties of 17-4PH stainless steel by low temperature plasma surface treatment. Materials and Design. 2010;21:2270–2273. DOI: 10.1016/j.matdes.2009.10.005

[81] Yun-tao X, Dao-xin L, Dong H. Improvement of erosion and erosion–corrosion resistance of AISI420 stainless steel by low temperature plasma nitriding. Applied Surface Science. 2008;254:5953–5958. DOI: 10.1016/j.apsusc.2008.03.189

Chapter 5

Computational Studies of the Impulse Plasma Deposition Method

Marek Rabiński and Krzysztof Zdunek

Additional information is available at the end of the chapter

http://dx.doi.org/10.5772/61985

Abstract

During the Impulse Plasma Deposition (IPD), plasma is generated in the working gas due to a high-voltage high-current discharge ignited within an inter-electrode region of a co-axial accelerator. The paper presents computational studies of working medium dynamics during the IPD discharge. The plasma has been investigated with a two-dimensional mono-fluidic snow plow model and a two-dimensional two-fluid magnetohydrodynamic code.

Computational studies of plasma dynamics have shown the function of the snow plow mechanism during the acceleration of plasma discharge, the role of the Rayleigh-Taylor instability, and the details of a sophisticated structure of a plasmoid in the region of the electrodes' front head. Simulations have significantly increased the understanding of the phenomena during the IPD; they also pointed to optimization directions. Several recent developments in the IPD surface engineering have been suggested and supported by the presented computational investigations.

Keywords: Surface Engineering, Impulse Plasma, Computational Modelling

1. Introduction

The Impulse Plasma Deposition (IPD) method of surface engineering has been developed [1, 2] in the Faculty of Materials Science of Warsaw University of Technology in the early 1980s. The principle of IPD was based on the nucleation on plasma ions, when potential energy of homogeneously generated nuclei was due to inelastic collisions with electrons. The use of pulsed plasma under reduced pressure was found to be the most effective solution to reach the conditions needed for synthesis of coatings in surface engineering.

The commonly used design of the IPD plasma accelerator consists of two coaxial metal electrodes, a cylinder and a rod, insulated from one another by a ceramic material across which the initial breakdown occurs, placed in a gas-filled vacuum chamber connected to a capacitor bank. A schematic diagram of the standard version of the accelerator is shown in Figure 1. The discharge in the standard version of the accelerator is controlled by an air spark gap. The typical ranges of the IPD process parameters are: the condenser bank capacitance $C = 100–200$ μF, the cyclically loaded voltage $U = 2–6$ kV, the external circuit inductance $L = 1.25$ μH, the electric pulse half-period $\tau_{1/2} = 40$ μs, the peak current of the order of 100 kA, the working gas pressure $p = 20–60$ Pa (nitrogen for TiN coatings or oxygen for the synthesis of Al_2O_3), and the repetition rate 0.1–1 Hz. The plasma pulses with a lifetime of about 40–120 μs, generated at a specified frequency, are ejected in the form of ion packs from the accelerator towards the substrate at a velocity of about 10^4 m/s [3, 4].

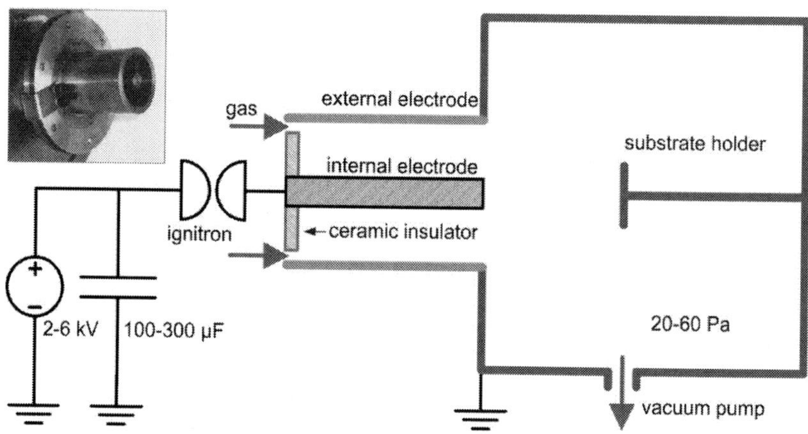

Figure 1. Scheme of the Impulse Plasma Deposition (IPD) coaxial accelerator.

During the interaction of a plasmoid with an unheated surface of the substrate, a short ($\sim10^{-3}$ s) thermal pulse acts on the surface, raising temperature to a peak value of about 2,000 K [5]. The generated heat is dissipated in the substrate material with a rate $\sim10^6$ K/s, determined by the specific thermal conductivity. In addition to the thermal effect, the interaction of plasma with the substrate surface causes subplantation of the plasma constituents, which results in the formation of a mixing zone at the interface. The features of the IPD plasma accelerator lead to a nearly complete ionization and thermodynamic non-equilibrium of the working medium. In such conditions, the probability of inelastic electron collision with the heavier plasma components, including crystallization of critical nuclei in plasma, is particularly high. Therefore, the efficiency of energy exchange between the plasma and the phase crystallizing in plasma are also high. It should be noted that the impulse plasma generated during electric discharge yields conditions significantly different from those described by the Paschen curve. Thus, the IPD technique offers a relatively easy way to increase the internal energy of the gas environment with particularly large supply of energy, difficult to accomplish by other means.

The mechanism of the IPD coating process consists of specific elementary phenomena, such as:

- the generation of a practically completely ionized non-equilibrium plasma by high-energy pulse discharge in a gas under reduced pressure;

- an acceleration of the plasma by magnetic pressure;

- a nucleation in plasma on ions;

- a dynamic interaction of the plasma with the surface of unheated substrate leading to a strong thermal activation of the surface and to sub-plantation;

- a cluster growth mechanism of a limited coalescence of nano-aggregates.

The major advantages of low-pressure impulse plasma, as used for material synthesis, especially for the synthesis of metastable high melting point phases that undergo monotropic transformation, are associated with the following features:

- the energy exchange due to inelastic electron–particle collisions is enhanced, while the nucleation barrier is lowered as a result of strongly non-isothermal character of the impulse plasma;

- short plasma lifetime causes instant immobilization of the synthesized products in the state they were created in plasma.

The IPD method of surface engineering has demonstrated its applicability in the synthesis of layers composed of materials such as diamond-like carbon, c-BN, oxides (e.g., Al_2O_3), multi-component metallic alloys of MCrAl(Y)-type, and metastable phases [4]. The IPD has been implemented at the Steel Works of Stalowa Wola, Poland, where it is used on the industrial scale for depositing TiN coatings on cutting tools. According to the best knowledge, this is the only method of plasma surface engineering of nanocrystalline wear-resistant layers on unheated and unbiased substrates. Recently, the IPD plasma generation was also combined with the magnetron sputtering technology [6, 7] – GIMS (Gas Injection Magnetron Sputtering).

2. Structure of plasma discharge region

Physical model of dynamic phenomena in the IPD accelerator with a tubular external electrode has been developed on the basis of the snow plow approximation [9]. The discharge chamber comprises two coaxial electrodes separated at one end by the insulator and opened at the other end. Working gas of uniform density and temperature fills the space between the electrodes.

At the initial stage of the discharge in the inter-electrode space, a rapidly growing electric current passes through the back wall insulator, causing ionization of atoms and formation of an axially symmetric electric current sheet on the insulator surface. The current flowing within this layer induces an azimuthal magnetic field behind. Thus, plasma starts spreading out in the form of a moving electric current sheet, accelerated by the Lorentz force, induced by the interaction of the current with its own magnetic field. The sheet propagates axially along the electrodes to the open end; it sweeps up the ionized gas, and leaves behind a vacuum region in its wake. The magnetic field in the vacuum region behind the current sheet acts much like a piston. The sheet advancing quickly along the electrodes creates a shock ahead of it; due to its strength, the shock wave pre-ionizes the gas which, affected by the current flow, converts into fully ionized plasma.

One can thus distinguish the three characteristic zones in the discharge space between the electrodes [3] (Figure 2a): an undisturbed gas in the front part of vacuum chamber, an intermediate plasma region of the discharge, and a magnetic piston next to the insulator. Within the high-vacuum region behind the current sheet, the magnetic pressure overwhelmingly predominates over the gas pressure, creating the so-called magnetic piston. The discharge region is confined within the piston front edge, which is also the back end of the preceding current sheet. In the front part of the discharge zone, the ordinary gas-dynamic shock wave separates the plasma in the intermediate region from the undisturbed gas ahead.

Since the strength of magnetic field decreases proportionally to the radius for the odd phases of the accelerator work (the first and the third half-periods of the electric discharge), the magnetic pressure at the internal electrode significantly exceeds its value at the outside electrode. The current carrying layer acquires a paraboloidal shape causing the plasma flow along the sheet towards the external electrode, just as a snow plow does. When the flow reaches the outside electrode, its radial motion is stagnated. Thus, the local pressure increases, causing the formation of a characteristic toroidal gas 'bubble' behind the sheet close to the wall. The shape of the electric current sheet results from the balance of the magnetic and fluid pressures at the contact interface between the regions of discharge and the magnetic piston. In the Lagrangian coordinate system moving with the shock wave, the pattern of the discharge area is nearly steady, although the shape of the sheet slowly changes, and the volume of gas bubble grows in time. If the cathode has the form of so-called 'squirrel-cage' composed of rods, the plasma is released outside the electrode which accelerates the current sheet motion.

The even phases of accelerator cycle (the second and the fourth half-periods) are associated with the change of electrodes polarization. Experiments [10] carried out in plasma devices with the central cathode show significant differences in the discharge pattern caused by the polarity change. The current sheet is more than twice as thick as the one formed with the internal anode; the sheet is also not so well-defined; and it is almost perpendicular to the channel walls, which causes plasma to gather on the moving interface and the motion to decelerate significantly (see Figure 2b).

Upon reaching the front end of the central electrode, the current sheet diffracts around it and pinches on the axis, forming a plasma column. At the end of each phase, due to the diminishing current, the magnetic piston disappears and reversal shocks occur. Due to the lowering amplitude of current value in the consecutive phases, the current sheet cannot approach the end of the electrode during the late half-periods.

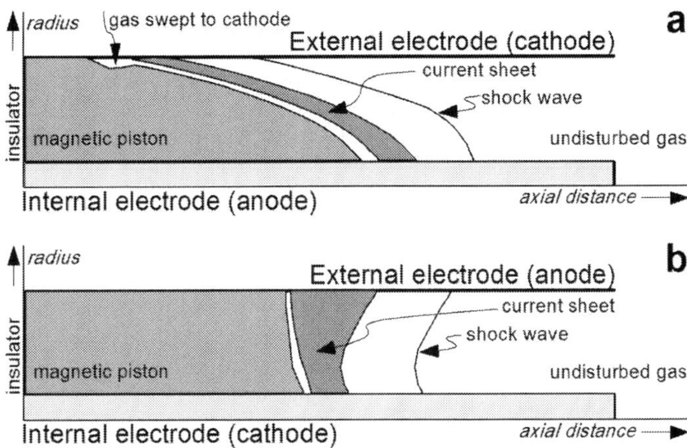

Figure 2. Schematic pattern of the discharge region for positive (a) and negative (b) polarization of electrode system.

The special feature of the energy source is that after the breakdown, the plasma spreads violently. That is where the computational modelling becomes particularly useful to predict the plasma behaviour.

3. Computational studies of IPD dynamics – snow plow approach

The first theoretical analyses used since the early 1960s for studies of plasma dynamics in the radial inverse pinches, coaxial shock tubes, plasma focus devices, plasma guns, and accelerators have been based on one-dimensional one-fluid model of the discharge. This simplified approach assumes that the current sheet has the form of an infinitely thin flat disc perpendicular to the electrode system axis and channel walls. Under the action of the Lorentz force, the current sheet is continuously accelerated, and sweeps the gas along the inter-electrode region. Equations of the model have been solved by an analogue computer.

The general snow plow approach assumes that the swept up mass is compressed into a thin layer immediately behind the shock, and the plasma discharge region is reduced into an infinitely thin current sheet. Thus, the magnetic piston edge, the current sheet behind the shock, and the shock form the same interface. The main inaccuracy of the simplified one-dimensional slab approximation follows from the systematic underestimation of the velocity values. The slab geometry causes that all the gas from the inter-electrode region is gathered on the surface of the magnetic piston, significantly decelerating its movement. Nevertheless, the plain shape of the interface is not an obligatory condition during the evaluation of theoretical model.

For the computational simulation of the flow phenomena within the IPD accelerator, the two-dimensional snow plow model has been evaluated [3, 11]. The equation set of the self-consistent model combines the description of the electric circuit with the plasma resistance and inductance, as well as the balance of magnetic and fluid pressures at the contact interface, depending on the condenser bank parameters and the plasma flow along the current sheet. The interface shape results from the dynamically established balance between magnetic and fluid pressures at both sides of the current sheet. The model is solved numerically in the Lagrangian coordinate system that moves with the plasma. In such approach, no convective term appears in the differential equation set. The snow plow approach allows a relatively simple yet accurate calculation of the current sheet dynamics – provided that the interface shape resulting from the balance of magnetic pressure behind and fluid pressure of the undisturbed gas in the front of the moving current sheet is taken into account.

The computational model has been used for several studies of the IPD discharge to assess sensitivity to various parameters of the plasma generation process [12]. As an example, Figure 3 presents the dynamics of the electric current sheet, which spreads out within both the inter-electrode space and behind the front face of the accelerator electrodes. The characteristic features observed in the diagrams are as follows:

- a delay in the current layer movement in the vicinity of the external electrode of the accelerator; it is due to the sweeping of the working gas by the moving layer, and results in the current sheet geometry as described above.

- the Rayleigh-Taylor instability forming behind the front faces of the electrodes, and resulting in two zones: one that spreads out along the system axis, and the other, above it, having the form of a torus of dense plasma. The toroidal structure is continuously supplied with the gas swept away by the current layer, which enhances the instability. The described arrangement of the plasma region behind the electrode front faces is very important from the point of view of the quality of the IPD coatings. Earlier studies of the synthesis products, and also the spectral examinations of the plasma, have suggested that each individual plasma jet consists of two fractions: one concentrated near the system axis, in which the plasma is isothermal, and the external highly non-equilibrium portion.

- for a low-energy high-pressure discharge, an additional dense plasmoid may be formed along the electrode axis at the stagnation point of the proceeding shock wave.

A distribution of the mass swept on the current sheet surface in the IPD accelerator has also been studied for different geometries of electrode system. In Figure 4, a sequence of density distributions per unit arc length is presented for different time moments of the first half-period of the electric discharge. One can observe a separation of working gas plasma (nitrogen) in the outer region of the discharge area and the titanium plasma from the evaporation zone in the front of the central electrode face while the sheet spreads out behind the front face of the electrode system.

3.1. Rayleigh-Taylor instability

Numerical studies of the IPD accelerator proved that geometry of electrodes may lead to the occurrence of the Rayleigh-Taylor (R-T) instability on the surface of the current layer [12–14]. In classical hydrodynamics, R-T instability takes place when two superposed fluids of different

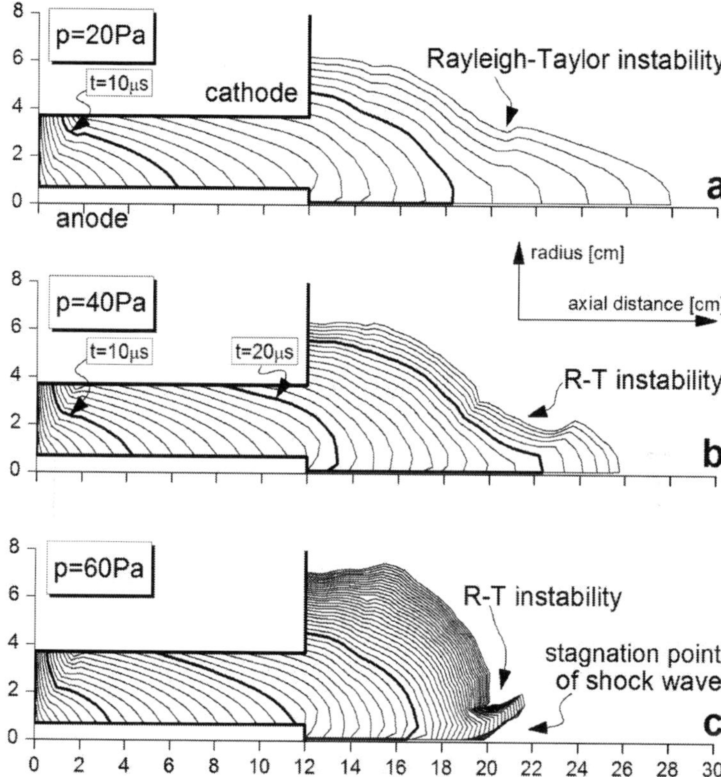

Figure 3. Time evolution of the current sheet position (plotted with 1 μs interval) for several discharge conditions (C = 100 μF, U = 6 kV, working gas – nitrogen at p = 20 / 40 / 60 Pa). Radial and axial coordinates in cm.

densities are accelerated from the heavier to the lighter medium in the direction perpendicular to their interface. This instability arises in plasma supported against acceleration by magnetic pressure. The dense plasma region in the IPD device is continuously supplied with the gas swept away by the moving snow plow structure of the current layer, which amplifies the instability. The swept gas forms the toroidal plasmoid – a coherent structure of plasma and magnetic fields. Within the plasmoid, nitrogen plasma is enriched with the erosion products, e.g. titanium in case of titanium nitride coatings. This explains why the deposited layers have non-uniform phase composition and morphology – a region on the accelerator axis is different from a region away from the axis. The plasma configuration at the front of the electrode system affects the quality of the coatings. Thus, the Rayleigh-Taylor instability formation is a critical factor in the TiN coating synthesis.

By modifying the design of the IPD accelerator, it is possible to reduce the Rayleigh-Taylor instability and to limit the erosion zone of the internal electrode. Figure 5 shows the computer

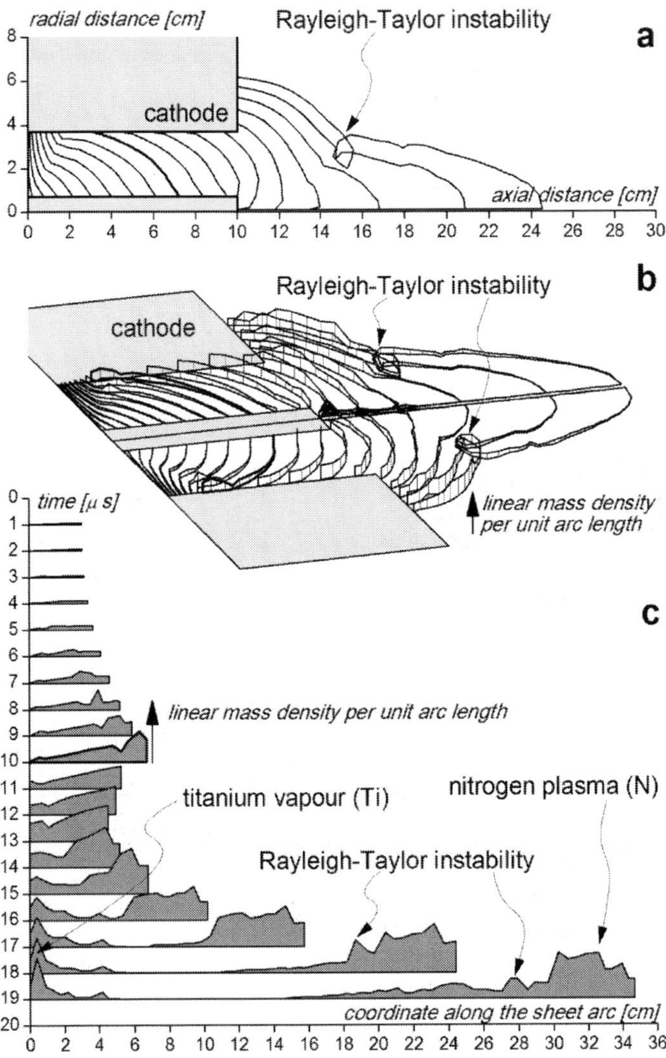

Figure 4. Time evolution of the IPD discharge (a, b) and mass density distribution (b, c) along the current sheet arc ($C =$ 100 μF, $U =$ 6 kV, working gas – nitrogen at $p =$ 40 Pa, internal electrode length – 10 cm, radius of outer electrode 3.7 cm). The sheet position and density distributions are plotted with 1 μs interval.

simulation of the plasma dynamics for this situation. The ceramic ring installed at the front face of the external electrode changes the geometry of the IPD accelerator. This modification restricts the 'climbing' of the electric current sheet upon the metallic wall of the vacuum chamber and modifies plasma flow. Comparing with the diagrams of Figure 3, the current sheet evolution shown in Figure 5 suggests that the presence of the ceramic insert reduces the tendency for the Rayleigh-Taylor instability to occur in the region behind the electrode front faces. As a result, the gas outflows along the surface of the magnetic piston towards the external electrode and reduces the plasma energy dissipation. Since there are no instabilities in the region preceding the zone of the discharge, the proceeding shock wave becomes stronger. Thus, the impulse heating is more intense at the substrate surface. This effect was experimentally confirmed with the A_2O_3 coatings: on the unheated substrate, the γ to α phase transition of deposited coating material was observed when the ceramic ring was installed at the accelerator outlet [12].

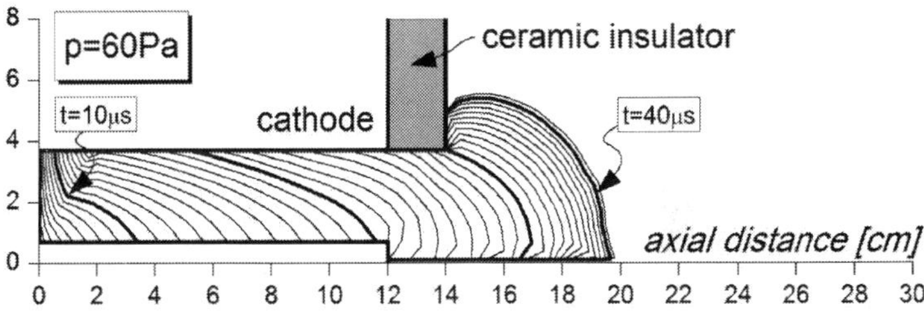

Figure 5. Time evolution of the current sheet in the IPD accelerator with the ceramic insulator at the external electrode front face. Discharge parameters are the same as in Figure 3c.

4. Experimental investigations of the deposition process

Theoretical models of the flow phenomena during a discharge in the IPD accelerator have significantly contributed to the understanding of this process. There has been also a demand for validation of computational results – in particular for experimental observations of phenomena during the plasma pinching in the front end of the electrode system. High-speed photography, ion beam, X-ray diagnostics, magnetic probes, and optical spectroscopy were used to compare observations with numerical results. Also, the morphology of deposited coatings was explored.

4.1. High-speed frame images of plasma discharge

Structure of the current sheet, as well as its dynamics, has been examined experimentally in a visible spectrum with the high-speed frame cameras (HSFC) [15]. The video signals from the CCD cameras have been digitized with a multi-channel frame grabber. In contrast to the magnetic probe measurements giving precise information on the current sheet structure in a particular point of the electrode system region, the high-speed images of the plasma show the overall pattern and the actual position of the discharge in the particular time moment.

Different geometries of the IPD accelerator electrode system have been investigated [16]. During the experimental studies of a commonly used cylindrical geometry, plasma of the

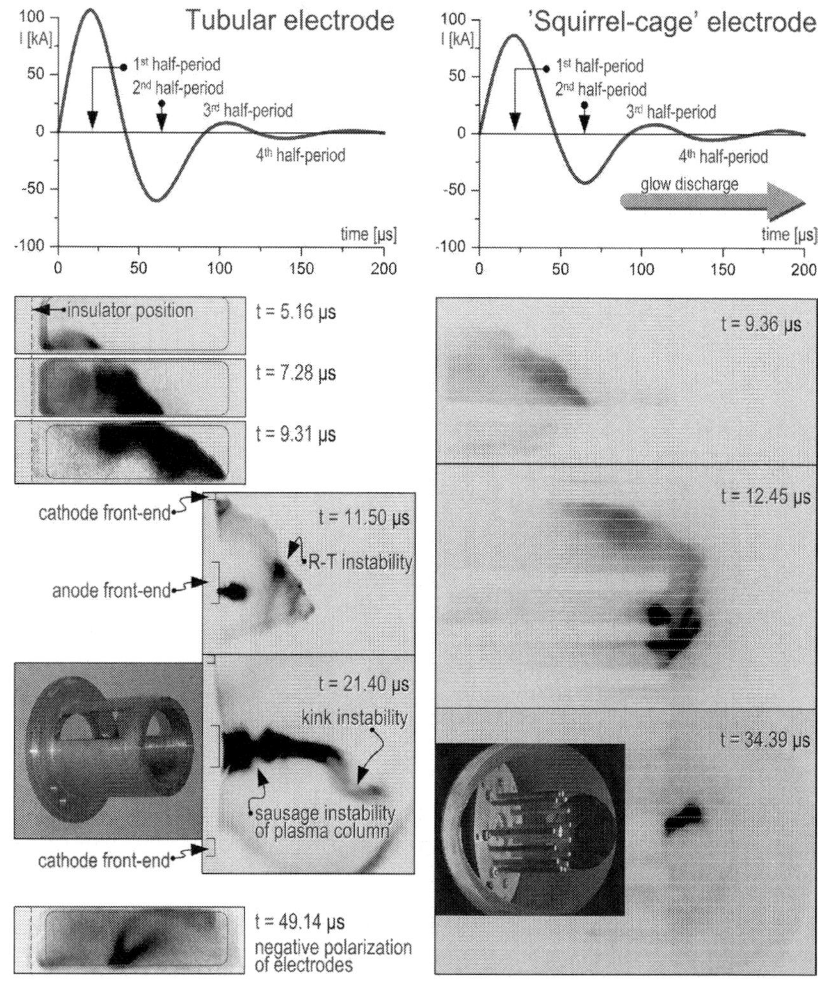

Figure 6. High-speed images in a visible spectrum for the nitrogen plasma discharge in the IPD accelerator with the tubular outer electrode (left column) and the squirrel cage type (right column). In the left column – the coaxial phase of the first half-period of current discharge, inter-electrode area seen through elongated windows cut in the outer electrode, electrode system front face, second half-period (U = 6 kV, C = 100 µF, p = 60 Pa), image width – 8.5 cm. In the right column – the IPD accelerator with squirrel cage electrodes (U = 3 kV, C = 100 µF, nitrogen at p = 20 Pa), electrodes' length – 10 cm. Internal electrode is partially hidden behind the outer electrode rods.

discharge region has been observed through the elongated windows cut in the outer electrode. In the other study, a discharge within the outer electrode system composed of stainless steel rods, arranged symmetrically around the internal titanium rod in the form of a squirrel cage, has been explored. The cage-type outer cathode has been used since the 1970s in the plasma focus devices to relieve the moving current sheet of the swept gas and to accelerate its motion and, thus, for better pinching of the plasma column and for higher neutron yield. For the IPD accelerator, however, this solution has been applied for the first time.

In Figure 6, series of photographs show discharge sequences for both tubular and rod external electrodes. It was observed that the plasma dynamics of the IPD process differs significantly, as predicted by the model of the plasma discharge. For a set of rods, the plasma is released outside the electrode, preventing mass sweeping on the tube surface. The shape of the current sheet becomes more symmetrical. A sequence of plasma discharge images proves also the most characteristic effect of the transparent external electrode: during the later phases of acceleration with the change of electrodes polarization, the current sheet is not formed on the isolator surface. The plasma pinch triggered during the first half-period of discharge transforms into a continuous glow discharge. For the tubular electrode at the end of each phase, the drop of electric current disables the magnetic piston, and the new current sheet is formed on the insulator surface. While using the cage composed of rods, the characteristic snow plow structure can be found only during the first half-period. Some evolution of the plasma column can be observed, but the main pattern of arc discharge remains constant.

4.2. Ion beams from IPD accelerator

The emission of ion beams and X-rays from the IPD accelerator have been examined experimentally [17]. To investigate the ion emissions, the Faraday Cup operating in the secondary electron emission mode was used. The 1-cm long cup consists of a metallic solid grounded cage with an aluminium collector on its lower base. In the space between grid and collector, two small magnets generate a traverse magnetic field. The cup entrance (3 mm in diameter) was screened with a metallic grid of 25% transparency posted on a ceramic insulator. During the measurements of ion emissions, the Faraday Cup detector was placed 35 or 48 cm from the electrode system front end. Signals from the ion beam diagnostic system were correlated with total discharge current time derived from the Rogowski coil and with the XET (X-ray Energy in Time) signals from photomultipliers.

The following characteristic features of ion beam emission from IPD accelerator were observed:

- 15 µs after ignition, a few to more than ten sudden oscillations of electric current were recorded. Each oscillation had a duration of about 0.1–0.5 µs, with the mean separation 2–2.5 µs between pulses; the total oscillation period reached about 20 µs. The situation has repeated itself during the second half-period after the change of electrode system polarization. The observed timing strictly correlated with stages of the IPD accelerator discharge, as predicted by the theoretical model, and observed in plasma images from the high-speed frame camera.

- The simultaneous emissions of ion beams and X-rays have been proved to correlate with the pinches of plasma column. A close correlation between number, appearance time, and duration of the ion emissions and X-ray signals was noticed, although amplitudes of both signals did not correlate.

- Ion pulses were detected at different time moments. This observation and dispersion of the peak appearance led to the conclusion that plasma streams are not generated strictly along the axis. Kinetic energy of nitrogen ions was estimated to be above 10–15 eV, and the ion velocity was about 10^4 m/s. Velocity of plasma discharge movement was estimated to be 2–4 10^3 m/s.

In summary, the observed ion beam and X-ray emissions were in an agreement with the high-speed camera images and with predictions of the plasma discharge model.

4.3. Erosion of internal electrode

The time needed for plasma to reach the electrode end and to pinch on the axis is about 15 μs, i.e one third of the current half-period. During the rest of the discharge, one can observe massive erosion of the anode front end. In the impulse plasma accelerators used earlier, practically the entire surface of the internal electrode underwent erosion. Severe deformation of the lateral sides of the electrode disturbed plasma generation and spreading. Both phenomena significantly limited the advantages of the IPD as the surface engineering method.

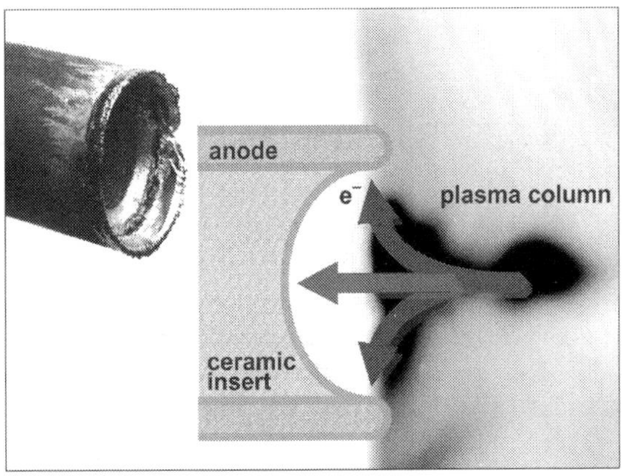

Figure 7. A physical model of electron motion in the funnel-like region of the eroded metal–ceramic electrode used in the synthesis of Al_2O_3 coatings; on the left – the final appearance of the internal electrode free end with ceramic insert (after numerous discharges), the diameter of electrode: 1.4 cm; on the right – HSFC image of plasma pinch.

The impact of eroded material (4.7 10^{-4} g/shot) from the electrodes on the dynamics of the discharge was studied through the computational methods. It has been found that when the eroded material was being added at the constant rate, it slowed down the dynamics of the discharge. When a more realistic time dependence of an eroded mass source was assumed, mainly the later phase of discharge was affected.

Observations led to formulation of the model [12] describing erosion phenomena during the IPD process. Due to the concentration of the electric field lines, an electric arc is formed towards the electrode front end. Figure 7 shows the appearance of eroded metal–ceramic internal electrode used in the IPD for aluminium oxide coatings. Initially, the front face of the ceramic insert was 'in line' with the free end of the electrode. After the prolonged use, a characteristic funnel-like erosion area in the aluminium oxide insert was found. This effect was due to acceleration of the electrons from the electric arc to the front face of the electrode. It should be emphasized that the erosion is caused by both sputtering and thermal effects. Thus, the internal electrode becomes an efficient source of the coating material components.

4.4. Tubular vs squirrel cage coaxial IPD accelerators

Using the titanium nitride as the model material for the surface engineering, the coating efficiency and the quality of the deposited layer have been examined [16] for both tubular and squirrel cage geometries of accelerator electrode system. The coatings have been analyzed with the XRD Philips PW 1140 diffractometer (Co K_α). This method deals with the commonly known standard specification data for materials available from the American Society for Testing and Materials (ATSM). The results demonstrated the substitution of titanium nitride coating with the TiN/Ti composite. The nucleation of the titanium nitride has been lowered by the insufficient excitation of nitrogen and the lack of plasma acceleration towards the substrate surface.

Our studies showed that the snow plow structure development requires prompt initiation at the insulator surface. The importance of the initial breakdown at the beginning of each current half-period and the subsequent formation of an axisymmetric current sheet structure is critical. Under optimal conditions, the consecutive packs of the plasma pulses with a lifetime of about 40–120 μs are generated at a specified frequency, and ejected from the accelerator towards the substrate at a velocity of about 10^4 m/s. From the point of view of surface engineering, the most important advantages of the low-pressure impulse plasma method are associated with the following features:

- the plasma is fully ionized and remains in the state of deep non-equilibrium,

- no external sources of force fields (electric or magnetic) are used for the activation of the gaseous phase,

- nucleation occurs on the ions within the plasma itself,

- no external substrate heating is needed, i.e. the substrate remains cold during the entire deposition process,

- internal electrode of the coaxial plasma accelerator supplies material that enters gas phase as the reactant,

- a layer is deposited through the coalescence of clusters and critical nuclei that form in the plasma; the layer thus obtained is solid and contiguous.

The application of outer electrode system composed of stainless steel rods introduces significant modifications to the IPD process. For the squirrel cage type, the smooth shape of the current sheet has been observed in the area behind the front face of the electrodes. This regularity results from the absence of the Rayleigh-Taylor instability. We believe that the observed phase composition of the coating was caused by the lack of mixing of the working gas plasma with the metal plasma. Thus, squirrel cage electrodes result in undesirable surplus of titanium in the coating material.

With squirrel cage rods, consecutive plasma sheet formations on the isolator surface were replaced with the continuous electric arc, affecting the material synthesis and deposition. The coatings obtained with the continuous arc following the snow plow discharge of the first half-period suffer from the lack of the TiN material acceleration towards the substrate yielding poor-quality coating [16]. The studies have also proved the role of the Rayleigh-Taylor instability in the titanium nitride synthesis. The lack of nitrogen plasma mixing with the titanium from the eroded electrode in the area of the instability could lead to worse quality of coatings.

These findings suggested the IPD accelerator design modifications. The first half-period was sufficient to synthesize titanium nitride for coating. The oscillatory discharge of the condenser bank was replaced by a single discharge in the electrode system with positive polarization (the rod anode and external cathode). The proposed modifications were tested, and introduced into industrial practice. The outer electrode composed of stainless steel rods (so-called 'squirrel cage electrodes') was found to be disadvantageous from the surface engineering point of view.

5. Computational studies of IPD dynamics – MHD approach

The general snow plow approach assumes that the plasma discharge region is reduced into an infinitely thin sheet. In reality, neither the current sheet nor the magnetic piston edge is infinitely thin. The detailed structure of the interface between the shock and the magnetic piston can be described by the solution of a complete magnetohydrodynamic (MHD) model. The two-dimensional MHD approach [18, 19] has been applied to investigate the sweeping of the working gas by the moving layer, as well as the details of phenomena that take place behind the discharge region. The mathematical model is based on a set of coupled transport equations for the medium composed of electrons and one kind of ions. Thus, plasma is treated as a system of two fully ionized fluids. The conservation equations for mass (the continuity equation), momentum, magnetic flux (the Faraday's law), and plasma energy densities are solved simultaneously with the Maxwell-Ohm law for the electric circuit. The cylindrical symmetry is assumed with the magnetic field restricted to the azimuthal direction.

5.1. Structure of plasma discharge

The first studies of the MHD approach focussed on the formation and evolution of the plasma bridge behind the current sheet. A relevant sequence of plasma parameter distributions for the current pulse first half-period is presented in Figure 8. One can observe the plasma structure created of the gas left in the corner between the insulator and the anode surface. The discharge structure was formed at 4 µs, in the corner close to the anode surface: the current sheet was accelerated by the Lorentz force, forming the vacuum region behind, and the remaining gas was swept to the anode. This portion of the working gas was left behind the rapidly accelerating discharge structure. In this region the acceleration is extremely rapid, because the magnetic field reaches its maximum value close to the anode surface.

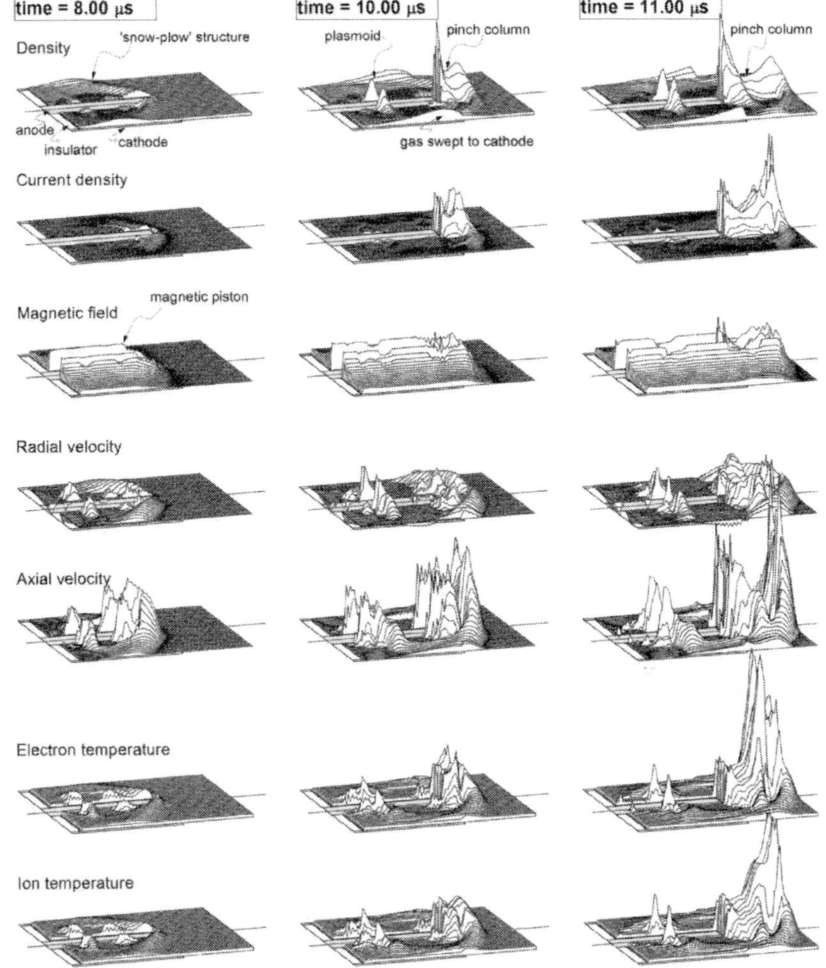

Figure 8. Sequence of plasma parameters distributions calculated for the IPD coaxial accelerator (C = 100 µF, U = 6 kV, external circuit inductance L = 1.25 µH, p = 20 Pa, electrodes' length – 10 cm, internal electrode radius – 0.7 cm, outer electrode radius – 3.7 cm).

During the next few microseconds, one can notice two streams within the working gas. The most important is the flow of the gas along the snow plow surface. The shape of the discharge region results from the balance of the magnetic and fluid pressures at the contact interface between the regions of current sheet and the magnetic piston. When the gas was decelerated on the surface of the external electrode, its pressure increased, which resulted in characteristic 'bump-on-tail' shape gas reservoir forming there. Because of significantly lower value of magnetic field strength close to the external electrode surface, the boundary of the gas reservoir is much more diffused and weakly defined.

The second process creates plasmoid-like structure close to the anode surface. The structure follows the current sheet, but with noticeably lower axial velocity. At the same time, the value of radial velocity increases. At 10 µs, the outer region of the plasmoid joints links to the boundary of the gas reservoir swept to the external electrode, forming a kind of closed connection between these structures. At the moment of the pinch ($t = 11$ µs), one can observe a fully developed structure of the plasmoid connected by a low-density bridge with gas swept to the outer electrode.

Studies presented above lead to the following conclusions:

- The significance of the snow plow mechanism has been validated within the MHD approach. This mechanism plays a key role in the complex system of discharge phenomena. This is why the previous numerical studies based on simplified snow plow code with infinitely thin current sheet described the IPD plasma dynamics quite accurately.

- The presented computational studies of the plasma dynamics have shown for the first time the details of the sophisticated structure of the spatial and time gas distributions behind the current sheet. These studies help to interpret the high-speed CCD frame camera images of earlier experimental studies [15].

The most evident proof confirming the existence of the above-described structure behind the current sheet can be found in Figure 9. One can see a strict correlation of the erosion pattern with the position of the plasmoid calculated by the presented numerical simulation. The length of smoothly eroded material quantitatively correlates with the position of the bridge between the plasmoid and the reservoir of the gas swept to the external electrode. A deep cavity close to the insulator edge corresponds to the position of the corner from which plasma is extracted.

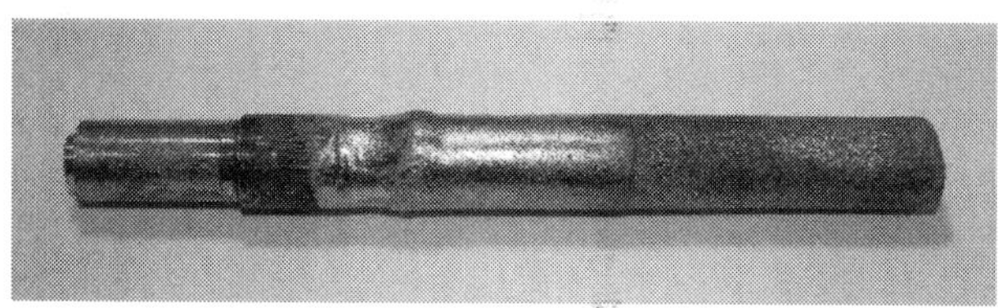

Figure 9. Erosion traces of the IPD accelerator internal electrode.

5.2. Studies of alternative discharge variants

During the numerous studies, the influence of various discharge parameters on plasma dynamics has been investigated with the positive and negative polarizations of the electrode system (an internal anode vs an internal cathode), as well as pressures outside the commonly utilized range.

As an example of the effect of electrode system polarity – for the positive polarization in the region of acceleration outlet, one can observe a separation of the plasma front. Two structures are separated because of the significantly higher velocities of the forehead portion. This phenomenon is absent when the internal electrode is a cathode.

For working gas pressure $p = 10$ Pa (i.e. lower than usually applied during IPD process), the time needed for the discharge structure to reach the internal electrode end is noticeably shorter (2–3 μs). Unfortunately, the electron temperatures at the internal electrode front end are lower, resulting in poor quality of the deposited coating. Similar simulations have also been carried out for working gas pressure of $p = 70$ Pa (higher than used in the regular conditions). Driven by the Lorentz force, the current sheet propagates axially along the electrodes to the open end. Sweeping up of the noticeably higher amount of gas leads to significant deceleration of the process dynamics. Thus, the time needed to reach the internal electrode end becomes 3–4 μs longer. The calculated distributions of plasma parameters (see Figure 10) demonstrate a partition of the discharge front into individual layers within the forehead portion of plasma. Such partition, jointly with the amounts of gas gathered on the snow plow layer, enables evaluation of Rayleigh-Taylor instability on the separated surfaces.

5.3. IPD controlled by pulse gas valve

The possibility of replacing the ignitron with a more efficient way of pulse plasma generation was studied. In particular, the application of a pulse valve controlling the plasma process by injection of the plasma forming gas was proved to be beneficial [7]. Instead of electric resistance change of the spark gap, the plasma process was initiated by injection of the working gas directly into the inter-electrode space. The gas dose was adjusted to get between-discharge pressure of the order of 10^{-4} Pa. Gas injection should be considered a useful method in optimization of the IPD coatings.

A special procedure was devised to simulate the gas valve action during computations [20]. The initial stage of the discharge was simulated under stationary pressure. After a time equal to the efficient half-width of valve pulse, the pressure of the undisturbed gas in front of the shock wave was exponentially decreased down to the value of pressure between discharges.

Computational studies proved that a longer valve action changes the dynamics of discharge only slightly, when compared with a standard accelerator working under steady pressure. On the other hand, it was also found that short gas pulse (<3 μs) leads to a very weak discharge with low plasma temperatures obtained at the end of the internal electrode. The optimum results were obtained for gas valve action reaching 4 μs and for very low values of pressure between discharges (~10^{-4} Pa).

The sequence of calculated plasma parameters for the accelerator working in the regime controlled with a pulse gas valve is presented in Figure 11. One can observe the significantly higher electron and ion temperatures of plasma in the forefront, reached after a shorter time.

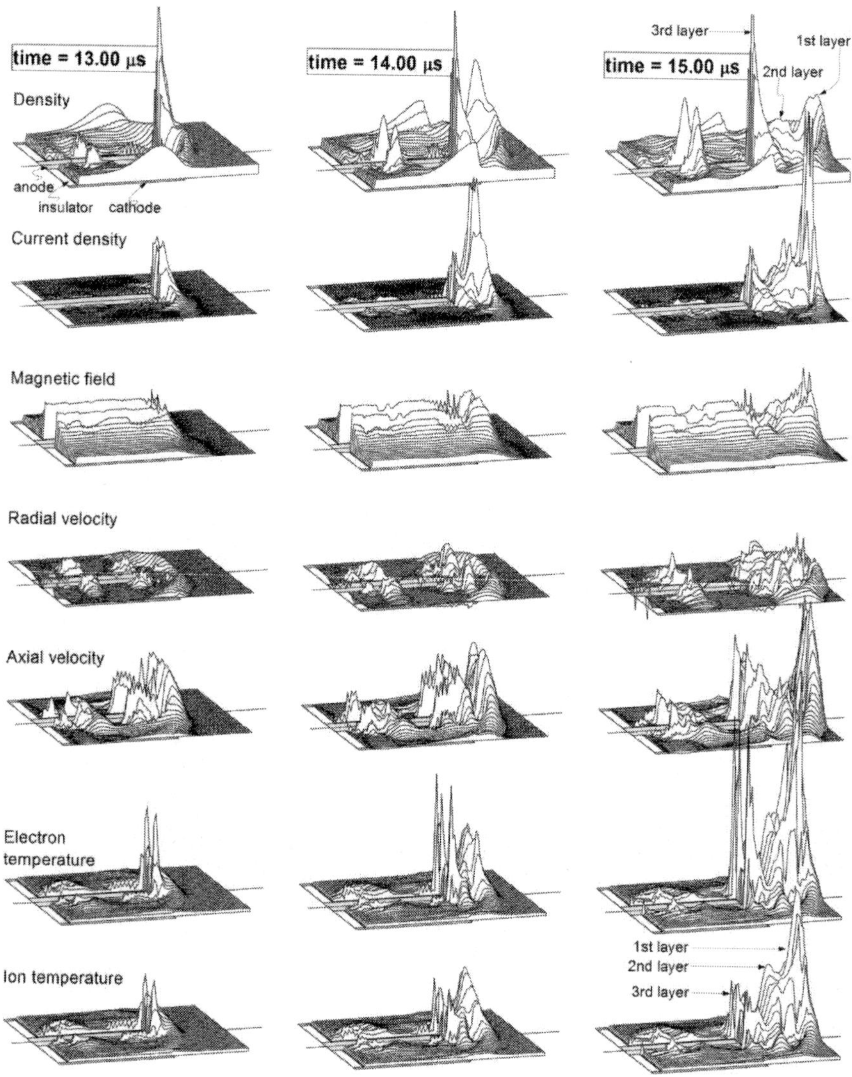

Figure 10. Sequence of calculated plasma parameters distributions for the IPD accelerator working with gas at $p = 70$ Pa – higher than in the regular conditions. ($C = 200$ µF, $U = 6$ kV, external circuit inductance $L = 1.25$ µH).

In comparison with the case of a standard IPD with the ignitron, the current layer propagates through the accelerator to the vacuum chamber under a pressure that is lower by many orders of magnitude. As a result, the internal structure of a discharge region is also notably changed; the amount of gas swept by the moving layer is lower, while the axial velocities are higher.

Significant changes in deposited coatings have been noticed [7]. The structure of titanium nitride coating was built of rectangular nanograins of a few tens of nanometres, with the lack of an amorphous phase. Such a fully nanocrystalline structure with equiaxial crystallites has been observed for the first time for the IPD method.

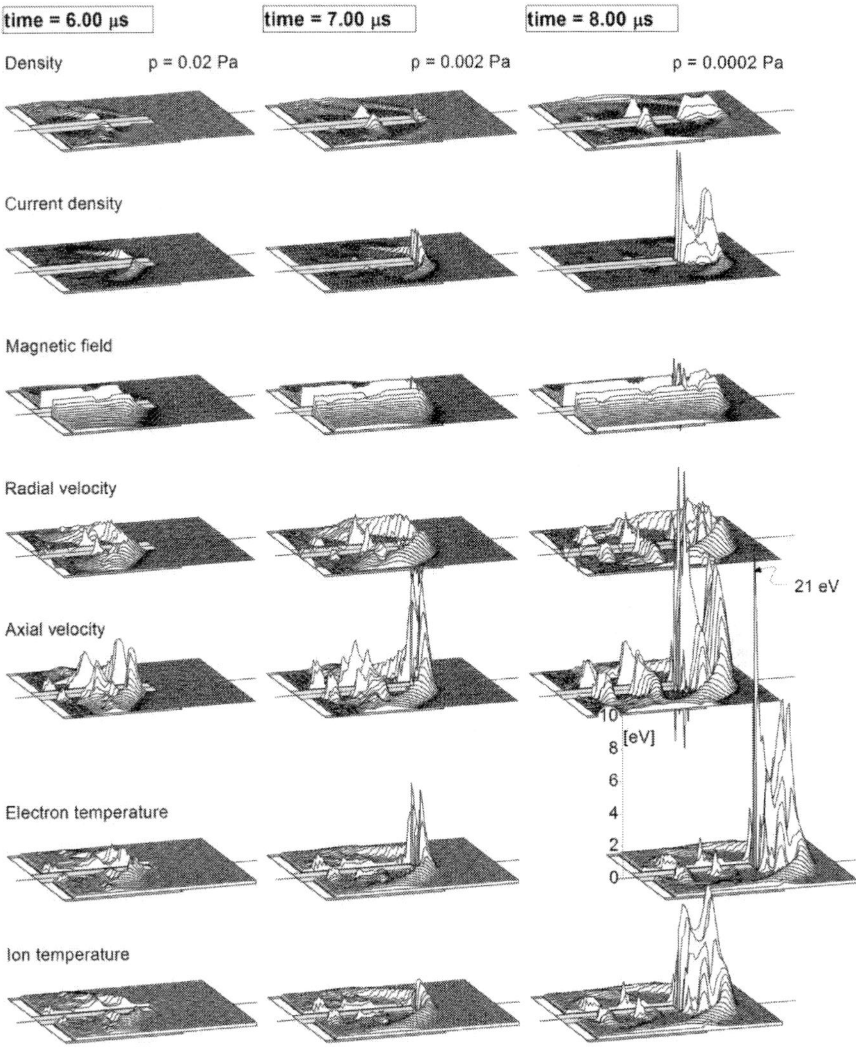

Figure 11. Sequence of calculated plasma parameters distributions for the IPD accelerator working with the pulse gas valve (switched off after 4 µs).

The TiN layers formed on cutting tools by the standard IPD method and tested under industrial conditions exhibited typical properties, i.e. the lifetime increase from several tens to several hundreds of per cent (e.g. 400%), depending on the cutting tool type. In the case of tools coated with the modified IPD accelerator with the valve control, the highly advantageous wear resistance can be associated with a qualitative change of plasma propagation conditions. It seems to be proved that an increase in the mean free path of gas molecules from 10^{-3} m, as in the standard IPD method, to a value of the order of 0.1–1 m leads to a qualitative decrease in the energy dissipated in intermolecular collisions. The TiN layers formed in this way on unheated parts of cutting tools exhibit a 16-fold increase of the tool lifetime during cutting tests [21].

6. Concluding remarks

Computational studies presented above lead to the following conclusions:

- The mathematical models allow the effective analysis of plasma dynamics. The computer simulation anticipates the evolution and space distributions of the discharge region dependent on the accelerator geometry and parameters of the plasma generation process. Numerical experiments throw also a new light on the interpretation of some phenomena within the IPD accelerator. Both qualitative and even semi-quantitative agreement between the results of the computational modelling and the experimental observations were demonstrated.

- Computational studies of the plasma dynamics have shown the significance of the snow plow mechanism during the acceleration phase of plasma discharge, the role of the Rayleigh-Taylor instability, the sensitivity to process parameters and geometry, as well as the details of sophisticated plasma structures near electrodes' front head. The latter are very important to the quality of the coatings. Thus, the computational studies have played the key role in the development of the IPD method.

Numerical simulations have significantly increased the understanding of the phenomena during the IPD method; the studies also provided guidance for the optimization and application of the Impulse Plasma Deposition technique. Several recent developments made in the IPD surface engineering have been suggested and supported with the help of computational studies presented in this paper.

Acknowledgements

The reported studies were supported by the Committee for Scientific Research (Poland) under the contracts KBN 3 P407 027 06, KBN 7T08C 050 12, KBN 7 T08C 054 17, KBN 7 T08C 032 24, KBN 819/T02/2006/31 and by the National Centre for Research and Development (Poland) within projects no. NR15-0002-10/10, 5899/B/T02/2010/38, 2013/09/B/ST/02418.

Author details

Marek Rabiński[1*] and Krzysztof Zdunek[2]

*Address all correspondence to: marek.rabinski@ncbj.gov.pl

1 National Centre for Nuclear Research, Division of Plasma Physics, Otwock, Poland

2 Warsaw University of Technology, Faculty of Materials Science, Warsaw, Poland

References

[1] Sokołowski M. Influence of the pulse plasma chemical content on the crystallization of diamond under conditions of its thermodynamic instability. J Cryst Growth 1981;54(3):519–522. DOI: 10.1016/0022-0248(81)90507-8

[2] Sokołowski M, Sokołowska A. Electric charge influence on the metastable phase nucleation. J Cryst Growth 1982;57(1):185–188. DOI: 10.1016/0022-0248(82)90265-2

[3] Rabiński M, Zdunek K. Physical model of dynamic phenomena in impulse plasma coaxial accelerator. Vacuum. 1997;48(7–9):1997. DOI: 10.1016/S0042-207X(97)00060-2

[4] Zdunek K. Concept, techniques, deposition mechanism of impulse plasma deposition - A short review. Surf Coat Technol 2007;201(9–11):4813–4816. DOI: 10.1016/j.surfcoat.2006.07.024

[5] Romanowski Z, Wronikowski M. Specific sintering by temperature impulses as a mechanism of formation of a TiN layer in the reactive pulse plasma. J Mater Sci 1992;27(10):2619–2622. DOI: 10.1007/BF00540678

[6] Zdunek K, Nowakowska-Langier K, Chodun R, Dora J, Okrasa S, Talik E. Optimization of gas injection conditions during deposition of AlN layers by novel reactive GIMS method. Mate Sci Poland 2014;32(2):171–175. DOI: 10.2478/s13536-013-0169-6

[7] Zdunek K, Nowakowska-Langier K, Dora J, Chodun R. Gas injection as a tool for plasma process control during coating deposition. Surf Coat Technol 2013;228(1):S367–S373. DOI: 10.1016/j.surfcoat.2012.05.101

[8] Zdunek K, Nowakowska-Langier K, Chodun R, Okrasa S, Rabiński M, Dora J, Domanowski P, Halarowicz J. Impulse plasma in surface engineering - a review. J Phys: Conf Ser 2014;564(1):012007. DOI: 10.1088/1742-6596/564/1/012007

[9] Fishman F, Petschek H. Flow model for large radius-ratio magnetic annular shock-tube operation. Phys Fluids 1962;5:632–633. DOI: 10.1063/1.1706671

[10] Keck C. Current distribution in a magnetic annular shock tube. Phys Fluids 1962;5:630–632. DOI: 10.1063/1.1706671

[11] Rabiński M, Zdunek K. Snow plow model of IPD discharge. Vacuum 2003;70:303–306. DOI: 10.1016/S0042-207X(02)00659-0

[12] Rabiński M, Zdunek K. Computer simulations and experimental results of plasma dynamics during the Impulse Plasma Deposition process. Surf Coat Technol 1999;116–119:679–684. DOI: 10.1016/S0257-8972(99)00128-0

[13] Rabiński M, Zdunek K. Rayleigh-Taylor instability in plasma jet from IPD accelerator. Surf Coat Technol 2003;174–175:964–967. DOI: 10.1016/S0257-8972(03)00534-6

[14] Rabiński M, Zdunek K. Modeling of flow phenomena during the Impulse Plasma Deposition process. In: IEEE, editor. EUROCON 2007: The International Conference on Computer as a Tool, vols 1–6; Sep 09–12, 2007; Warsaw, Poland. New York: IEEE; 2007. p. 2159–2165.

[15] Rabiński M, Zdunek K, Paduch K, Tomaszewski K. Experimental studies of current sheet structure in IPD coaxial accelerator. Surf Coat Technol 2001;142–144:49–51. DOI: DOI: 10.1016/S0257-8972(01)01076-3

[16] Rabiński M, Wierzbiński E, Zdunek K. Studies of squirrel cage type coaxial accelerator for IPD process. Surf Coat Technol 2005;200:788–791. DOI: 10.1016/j.surfcoat. 2005.01.061

[17] Baranowski J, Jakubowski L, Rabiński M, Zdunek K. Studies of simultaneous X-ray emission and ion beams in the IPD plasma accelerator. Czech J Phys 2002;52(Suppl. D):D188–D193.

[18] Potter D E. Numerical studies of the plasma focus. Phys Fluids 1971;14(9):1911–1924. DOI: 10.1063/1.1693700

[19] Rabiński M, Zdunek K. Computational studies of plasma dynamics in Impulse Plasma Deposition coaxial accelerator. Surf Coat Technol 2007;201:5438–5441. DOI: 10.1016/j.surfcoat.2006.07.005

[20] Rabiński M, Chodun R, Nowakowska-Langier K, Zdunek K. Computational modelling of discharges within the Impulse Plasma Deposition accelerator with a gas valve. Phys Scr 2014;T161:014049. DOI: 10.1088/0031-8949/2014/T161/014049

[21] Zdunek K, Nowakowska-Langier K, Chodun R, Kupczyk M, Siwak P. Properties of TiN coatings deposited by the modified IPD method. Vacuum 2010;85:514–517. DOI: 10.1016/j.vacuum.2010.01.024

Medical and Green Applications of Plasma

Chapter 6

Physicochemical Analysis of Argon Plasma-Treated Cell Culture Medium

Claudia Bergemann, Cornelia Hoppe, Maryna Karmazyna, Maxi Höntsch, Martin Eggert, Torsten Gerling and Barbara Nebe

Additional information is available at the end of the chapter

http://dx.doi.org/10.5772/61980

Abstract

The effects of cold plasma under atmospheric pressure are being explored for medical applications. It was found that plasma effects on cells correspond to a plasma–medium interaction; thus, plasma-treated cell culture medium alone is able to influence the cell behavior. Here, we discovered that the liquid-mediated effect of atmospheric-pressure argon plasma on mouse liver epithelial cells persists up to 21 days of storage; i.e., the liquid preserves the characteristics once induced by the argon plasma. Earlier investigations of our group revealed that temperature and pH, hydrogen peroxide production and oxygen content can be excluded as initiators of the detrimental biological changes. As we found here, the increased osmolality in the media caused by plasma treatment can also be excluded as a reason for the observed cell effects. Conversely, we found changes in the components of cell culture medium by fast protein liquid chromatography (FPLC) and decreased cell viability in plasma-treated media independent of the presence of fetal calf serum (FCS) during plasma treatment. The persistent biological effect on plasma-treated liquids observed here could open up new medical applications. Stable plasma-treated liquids could find application for dermatological, dental, or orthopedic therapy.

Keywords: Non-thermal atmospheric-pressure plasma jet, Plasma–liquid interaction, Osmolality, Size-exclusion chromatography, Cell viability

1. Introduction

Plasma medicine is an emerging field of interdisciplinary research combining physics, biology, and clinical medicine [1,2]. In the first applications, gas plasma was used in the form of hot plasma for cauterization [1,3]. Currently, the effects of cold plasma under atmospheric pressure are being explored for medical applications. Several experiments showed the achievement of

blood coagulation and bacterial decontamination, with little effect on the surrounding tissue [4–9], as well as the promotion of wound healing and tissue regeneration [10–13]. This proliferative reaction was correlated with the secretion of growth factors induced by plasma treatment [14]. Conversely, adverse effects such as anti-proliferative and anti-tumorigenic effects were reported [15–18]. Also, the detachment of cells from the substrate, an effect that accompanies the change to a rounded cell morphology, and apoptosis were reported for animal cell lines like CHO-K1 (Chinese hamster ovary epithelial cells), 3T3 (mouse fibroblasts), BAEC (bovine aorta endothelial cells), H5V (mouse endothelial cells), and RASMC (rat aorta smooth muscle cells) [1,19,20]. This is in line with the observation that the expression of cell adhesion molecules is changed upon plasma treatment [21]. Detrimental cell effects like DNA damage were shown by Kalghatgi et al. [22]. Plasma treatment of epithelial cells led to cell detachment and additional DNA damage, which was proven by the phosphorylation of the histone protein H2AX [22].

It was discovered that these effects can be mediated by the use of plasma-treated liquids alone [23]. Plasma treatment of cell culture medium resulted in DNA damage to the subsequently incubated cells [22,24]. Similar to the direct treatment, liver epithelial cells lost their substrate attachment when incubated with argon plasma-treated medium alone [25]. In addition, cell–cell contact proteins, e.g., the tight junction protein zonula occludens (ZO-1), as well as the cell membrane morphology were shown to be impaired [25,26]. Based on these observations, it could be concluded that plasma effects are not only solely based on the direct plasma–cell interaction but also on the interaction between plasma and medium, and these plasma-induced changes in the complex Dulbecco's modified Eagle's medium (DMEM) were long lasting: up to 7 days and more [26].

Further analysis revealed that neither lipid oxidation [24] nor the generation of ozone [27] was responsible for these effects. Also, changes in temperature, pH-value, or the concentration of hydrogen peroxide could be excluded as a reason [25,26]. In addition, it was shown that decreasing the oxygen concentration of the medium as a result of plasma treatment was not the reason for the respective cell defects [26]. Several publications speculated that the generation of reactive oxygen species (ROS) through plasma treatment would be an explanation for DNA damage [20,28,29]. However, the cell-damaging effect of plasma-treated medium could be detected after 1 h as well as up to 7 days after treatment [22,25,26], indicating that the component(s) remained stable for a long time. This result excludes the short-lived ROS as an inducing agent.

The present work aimed to examine the contribution of different medium components, e.g., fetal calf serum and gentamicin, to the detrimental cell effects resulting from plasma treatment. Furthermore, physicochemical parameters other than temperature, pH, oxygen concentration, or the presence of hydrogen peroxide, which had already been excluded as a reason [25, 26], were investigated. Taking into consideration that the plasma effect is mediated only by liquids [22,23,25,26,29], the cell culture medium was treated with argon plasma and cells were incubated afterward with the medium. The plasma treatment was performed using an atmospheric-pressure argon plasma jet.

2. Material and methods

2.1. Argon plasma source

The atmospheric-pressure plasma jet kINPen®09 (neoplas tools GmbH, Greifswald) was used for all experiments. This plasma source consists of a quartz capillary (inner diameter of 1.6 mm) with a needle electrode (diameter of 1 mm). A high-frequency voltage of 1.1 MHz/2–6 kV is applied at this electrode. Argon gas (purity 99.996%) is used as a feed gas with a gas flow of 1.9 slm. A gas discharge is ignited at the tip of the high-voltage needle exciting the argon gas. In this way, low-temperature plasma is generated and blown out of the capillary. The so-called plasma jet outside the nozzle has a length of 12–14 mm and is about 1 mm wide. The temperature at the visible tip of the plasma jet does not exceed 50°C. Details for the characterization of the plasma source are given by Weltmann *et al.* [12,26].

2.2. Argon plasma treatment of cell culture media

One-hundred microliters of the defined cell culture medium (see below) was supplied per well in a 96-well plate (Greiner Bio-One). The kINPen®09 was mounted vertically and the quartz capillary was positioned at the top edge of each well in the 96-well plate. The distance between the tip of the quartz capillary and the 96-well plate does not exceed 1 mm. In this way, the argon plasma had an immediate contact with the cell culture medium (Figure 1).

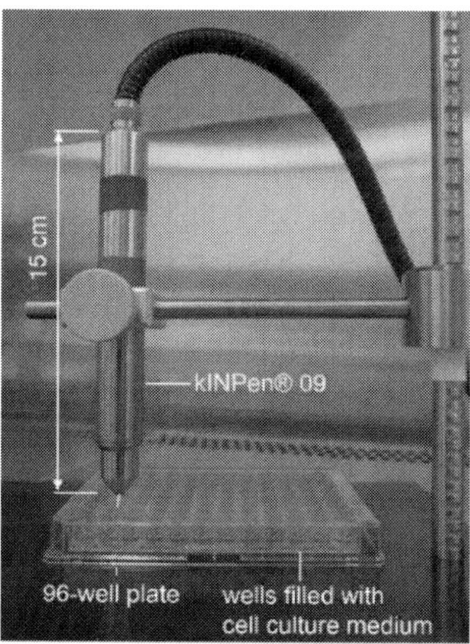

Figure 1. Experimental setup for argon plasma treatment of cell culture media. The kINPen®09 was mounted vertically. The distance between the tip of the quartz capillary and the 96-well plate does not exceed 1 mm.

Dulbecco's modified Eagle's medium (DMEM; Invitrogen) was used for all experiments. For cell culture experiments, DMEM is usually supplemented with 10% fetal calf serum (PAA Gold, PAA Laboratories) and 1% gentamicin (Ratiopharm). To explore the effect of the supplements on plasma treatment results, pure DMEM without supplements or with one supplement was treated as well. We refer to the different media as follows: pure DMEM = DMEM; DMEM with 10% FCS and 1% gentamicin = DMEM+FCS+Genta; DMEM with 10% FCS = DMEM+FCS; DMEM with 1% gentamicin = DMEM+Genta. The different media underwent argon plasma treatment for 60 or 120 s as indicated for the specific experiment. Argon gas treatment using the kINPen®09 without igniting the plasma was used as a control. Untreated medium was included as a further control. After plasma treatment, the different samples and controls were pooled in separate tubes (1.5 ml, Eppendorf) and analyzed for their physicochemical parameters. For subsequent cell experiments, the medium was stored at 37°C and 5% CO_2 for at least 24 h to ensure that the cell effects observed were neither caused by the changed oxygen concentrations nor by the persistence of hydrogen peroxide; these were shown to be balanced over 24 h after plasma treatment [26]. For investigations regarding the persistent effect of plasma-treated medium on cells, the medium was stored at 37°C and 5% CO_2 for 7 or 21 days. For all experiments, with the exception of the size-exclusion chromatography, the supplements omitted during plasma treatment were added afterward to ensure analogous conditions.

2.3. Physicochemical analysis of argon plasma-treated cell culture media

2.3.1. Osmolality

To acquire the loss of solvent during argon plasma treatment, the osmotic strength was utilized by the measurement of the freezing point depression. The respective medium (50 µl) was applied to an Osmomat 030 (Gonotec) and analyzed in triplicate. The determination of osmolality was performed in three independent experiments for DMEM+FCS+Genta, which had been argon plasma-treated for 60 or 120 s, and for the same medium after argon flow for 60, 120, or 180 s versus the untreated control.

2.3.2. Size-exclusion chromatography

Size-exclusion chromatography was performed by gel filtration on a fast performance liquid chromatography (FPLC) system equipped with a UV detector (ÄKTA purifier, GE Health-care). After argon plasma treatment, samples were pooled and the protein concentration was measured. The samples were diluted with distilled water to a concentration of 4 mg/ml protein, sterile-filtrated (0.2 µm), applied to a size-exclusion column (Superdex 200, 10/300 GL, GE Healthcare) and separated with a constant flow of 0.5 ml/min for 60 min, using a phosphate buffer (34.3 mM Na_2HPO_4, 14.5 mM NaH_2PO_4, 200 mM NaCl, pH 7.14). The samples were monitored by the UV detector (280 nm) and collected in fractions of 500 µl. Chromatograms were conducted for DMEM, DMEM+FCS, and DMEM+Genta, which were argon plasma-treated for 120 s, for the untreated control and for the same medium after argon flow for 120 s.

2.4. Epithelial cell experiments

2.4.1. Cell culture of mHepR1

The epithelial cell line mHepR1 (murine hepatocytes) [30] was used throughout the experiments. These immortalized mHepR1 cells represent a clone of the HepSV40 line derived from transgenic mice [31] and exhibit characteristic markers for epithelial cells like the tight-junction-associated protein ZO-1. The epithelial cells were cultured in DMEM supplemented with 10% FCS and 1% gentamicin at 37°C and 5% CO_2. Near confluence, cells were detached with 0.25% trypsin/0.38% EDTA (Invitrogen) for 10 min. After stopping trypsinization by the addition of cell culture medium, an aliquot of 100 µL was put into 10 mL of CASY® ton buffer solution (Roche Innovatis) and the cell number was measured in the counter CASY® Model DT (Schärfe System).

2.4.2. Immunofluorescence staining of ZO-1

Into a 24-well plate (Greiner), which was provided with a round coverslip (diameter of 12 mm, MENZEL) and 1 ml DMEM+FCS+Genta, 5×10^4 cells/well were seeded. The mHepR1 cells were incubated for 2 days at 37°C and 5% CO_2 to achieve a confluent cell layer for ZO-1 observation. Then the culture medium was replaced by plasma-treated or control DMEM+FCS+Genta, prepared as described in paragraph 2.2. and stored for 21 days at 37°C and 5% CO_2. The plate was incubated for another 24 h at 37°C and 5% CO_2 followed by immunofluorescence staining of the cells for ZO-1 protein. The cells were permeabilized with ice-cold acetone (−20°C, 200 ml, Lab-Scan) for 5 min and, after washing three times with phosphate buffer solution (PBS; PAA Laboratories), the cells were incubated with rabbit anti-ZO-1 antibody (diluted 1:100, Invitrogen) for 30 min at room temperature. After washing again with PBS, the cells were incubated with fluorescein-conjugated Alexa Fluor 488 secondary antibody goat anti-rabbit (1:100, Invitrogen) for 30 min at room temperature in the dark to avoid fading of the fluorescent dye. Finally, the cells were embedded in mounting medium (FluroShield, Sigma Aldrich) with a coverslip. The ZO-1 protein was analyzed by the inverted confocal laser scanning microscope LSM 780 (Carl Zeiss, Oberkochen, Germany), equipped with a 63× oil-immersion differential interference contrast (DIC) objective.

2.4.3. Cell viability

To monitor the viability of the cells in plasma-treated medium, the CellTiter 96®AQueous Non-Radioactive Cell Proliferation Assay (Promega) was used. For each medium sample and the control (prepared as described in paragraph 2.2.), three wells containing 100 µl of medium were prepared in a 96-well plate and 2×10^4 mHepR1 cells were added to each well. After an incubation time of 24 h (37°C and 5% CO_2), 20 µl of the MTS solution (Promega) was added to each well and incubated for 2 h (37°C and 5% CO_2) with manual shaking for every 30 min. Afterward, 60 µl of each well was transferred into a fresh 96-well plate and the absorption was determined at 492 nm (reference 620 nm) using a microplate reader (Anthos). The appropriate medium without cells served as a blank. The viability of cells was calculated corresponding to cells in the untreated control medium (100%). The determination of cell viability was performed in three independent experiments for DMEM+FCS+Genta, which had been argon plasma-treated for 60 or 120 s, for the untreated control and for the same medium after argon flow for 60, 120, or 180 s. Three additional experiments were performed for DMEM and DMEM +FCS, which were argon plasma-treated for 120 s, for the untreated control and for the same media after argon flow for 120 s. The supplements omitted during plasma treatment were added before cell culture to ensure analogous conditions.

2.4.4. Scanning Electron Microscopy (SEM)

Into a 24-well plate (Greiner), which was provided with a round coverslip and 1 ml DMEM +FCS+Genta, 5×10^4 cells/well were seeded. The mHepR1 cells were incubated for 2 days at 37°C and 5% CO_2 to achieve a confluent cell layer. Then the culture medium was replaced by plasma-treated DMEM+FCS+Genta (60 s) or control medium, prepared as described in paragraph 2.2. and stored for 7 days at 37°C and 5% CO_2. After 24 h of incubation, the mHepR1 monolayer was washed three times with PBS and subsequently fixed in glutardialdehyde (GDA, Sigma Aldrich, 0.5% in PBS) and stored for at least 24 h at 4°C. Afterward, the mHepR1 monolayer was critical point-dried. For this purpose, the GDA PBS solution was first gradually removed by drainage using acetone with increasing concentrations of 30, 50, 70, 90, and 100% for 10 s, 5, 10, and 15 min and twice for 10 min, respectively. Subsequently, the acetone was substituted by critical point drying (K 850 EMITECH). Then samples were sputter-coated with gold particles for 100°s (layer thickness approximately 20 nm), achieving a conductive dissipation for SEM. The cell surface structure was analyzed with a DSM 960 A (Carl Zeiss), operated at 10 kV at a magnification of 5,000× and a tilt angle of 60°.

3. Results and discussion

The atmospheric-pressure plasma jet kINPen09 [12] was applied throughout the experiments. The same plasma jet was used in our recent work [25], where we found that plasma effects on cells correspond to a plasma–medium interaction, and thus plasma-treated DMEM+FCS +Genta alone is able to open cell–cell contacts in a confluent cell monolayer of mHepR1 cells. Immunostaining of the zonula occludens tight junction protein (ZO-1) showed large openings between cells, which led to the complete degradation of the tight junction proteins. Normally, untreated cells represent clear, continuous ZO-1 bands between adjoining cells. In contrast, large intercellular openings were observed using plasma-treated DMEM+FCS+Genta. This effect was verified also for the medium that was stored up to 7 days after plasma treatment showing a persistent effect of plasma-treated medium on cells [26].

As we discovered here, this long-lasting effect is persistent for up to 21 days after argon plasma treatment. The tight junction protein ZO-1 of murine hepatocytes mHepR1 is shown in Figure 2. After incubation of cells with DMEM+FCS+Genta, which was plasma-treated for 120 s and stored for 21 days, the expression of ZO-1 in the cell contact zone was either reduced in different areas or disappeared due to the retraction in a centripetal direction. In consequence, the cell size increased due to the loss of cell–cell contacts in the monolayer. In normal controls, tight cell–cell contacts could be found between neighboring cells.

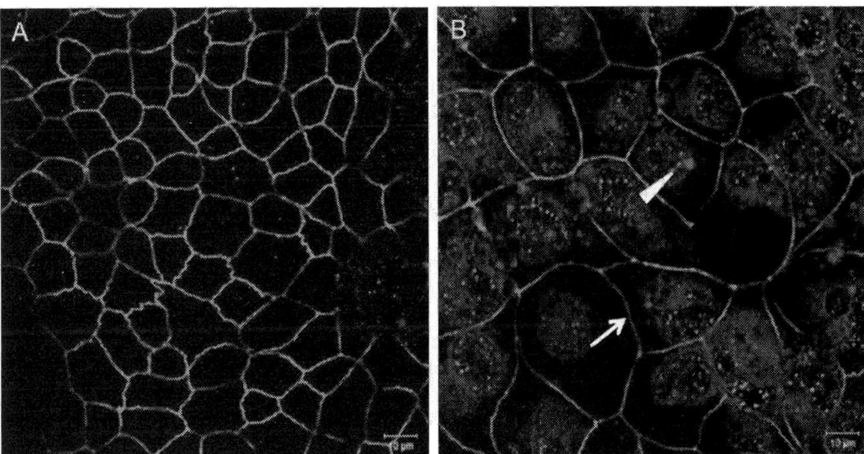

Figure 2. ZO-1 of mHepR1 cells after 21 days. (A) In control cells, the tight cell–cell contacts between neighboring cells are clearly visible. (B) After incubation of cells with plasma-treated DMEM+FCS (120 s) – which was stored for 21 days – either the expression of ZO-1 in the cell contact zone was reduced (arrow) or ZO-1 disappeared due to the retraction in a centripetal direction (arrowhead). Confocal microscopy (LSM 780, Carl Zeiss, bars 10 μm).

Accompanied with the long-lasting effect of plasma-treated medium on the distribution of ZO-1 in mHepR1 cells, we found changes in the cell surface morphology by SEM. The mHepR1 cells exhibited shortened microvilli with a lower density resulting from the plasma-treated cell culture medium [26]. These findings on microvilli shortening were revealed after a 1-day as well as a 7-day storage time of plasma-treated medium (60 s; complete DMEM). SEM images of mHepR1 cells (Figure 3), incubated for 24 h with the plasma-treated (60 s), 7-day stored DMEM+FCS+Genta illustrate this effect. Untreated and even argon-treated mHepR1 cells present elongated microvilli covering the cell surface at a high density. In contrast, the plasma-treated mHepR1 cells showed greatly shortened microvilli and the density over the entire cell surface seems to be reduced.

The formation of cell surface structures, e.g., microvilli, is essential for characteristic functions of specialized cells in tissues. Microvilli increase the cell surface and play an important role in metabolic processes: they regulate cellular functions by external signals as well as Ca^{2+} signaling [32]. Effects similar to those we observed in our work were detected by Pfister and Burstein after the treatment with ophthalmic drugs [33]. They observed the loss of surface epithelial microvilli as well as rupture of the tight junctions on corneal epithelial cells.

As a consequence, the question arises of which parameter that can persist as long as 21 days is changed in cell culture medium due to plasma treatment. As recently shown, plasma effects due to thermal damage of differential proteins in cell culture medium could be excluded for

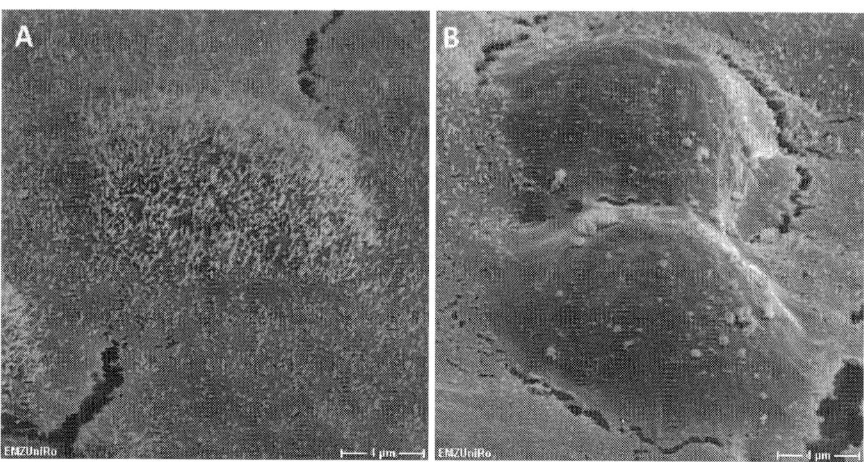

Figure 3. Microvilli of mHepR1 epithelial cells (A) in DMEM+FCS+Genta as control and (B) in the same medium, which was argon plasma-treated for 60 s and stored for 7 days (culture time 24 h). Scanning electron microscopy (SEM, DSM 910, magnification 5,000×, tilt angle 60°, and bar 4 μm).

treatment times below 180 s for the atmospheric-pressure plasma jet kINPen09 [26]. The temperature in the cell culture medium did not exceed 25°C. Plasma–liquid interactions were studied by Oehmigen *et al.* [34] and von Woedtke *et al.* [35] relating to the effective disinfection of liquids. It was shown that the inactivation of bacteria is strongly dependent on the acidification of aqueous liquids (pH decrease). In our work, we could show that the pH of the complete cell culture medium DMEM remained constant. The buffering capacity of sodium hydrogen carbonate in the medium was sufficient to maintain the pH even after a plasma treatment. Furthermore, the redox amphoteric hydrogen peroxide was found in liquids after plasma treatment in various studies [10,34]. As hydrogen peroxide also occurs in the cell metabolism, it can be degraded by various repair and protection mechanisms in the cell. Antioxidants inside the cell are able to detoxify "natural" amounts of ROS. In the presence of an overabundance of ROS, these mechanisms fail. Thus, the concentration of hydrogen peroxide and oxygen in the medium after plasma treatment was investigated earlier. Our investigations revealed that hydrogen peroxide and oxygen concentrations were balanced over the time period of up to 1 day, but the cell behavior was affected by argon plasma-treated medium stored for 1 or 7 days [26] or as described here for 21 days.

These detrimental effects on mHepR1 cells due to argon plasma-treated cell culture medium lead us to investigate other physicochemical parameters, which could change during the plasma treatment of a medium. One parameter to which cells react sensitively is the osmolality [36]. Due to evaporation processes induced by the argon gas flow, the medium volume in the well could be reduced during plasma treatment. Accordingly, the osmotic strength in the medium would increase. To examine this, we utilized the osmolality by measuring the freezing point depression for argon plasma-treated DMEM+FCS+Genta. The results in Figure 4 show an increasing osmolality for the medium after argon plasma treatment for 60 or 120 s. Interestingly, argon gas flow alone did not influence osmolality as high as argon plasma did. Thus, osmolality in the medium after 60 s of plasma treatment is higher (444.11 mOsmol/kg) than in the argon gas-treated control (384.67 mOsmol/kg). To reach the same osmolality level as after 60 s of plasma treatment, the argon gas flow needs to impact on the medium for 180 s.

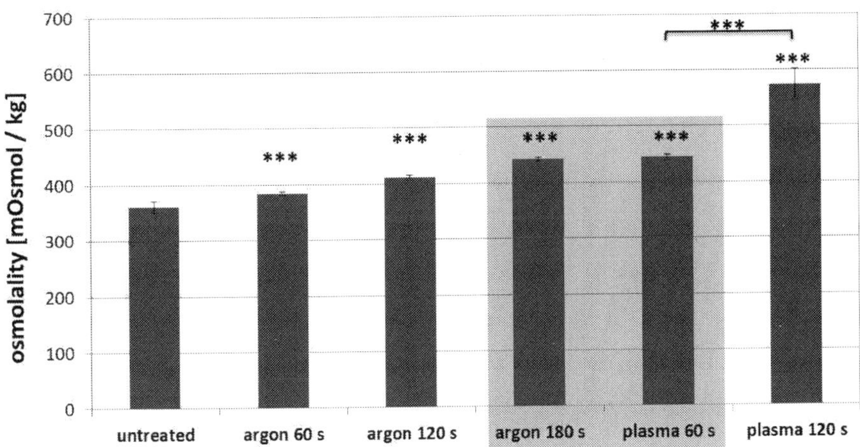

Figure 4. The osmolality of treated DMEM+FCS+Genta increased significantly with longer exposition times. Note that argon plasma treatment (plasma) over 60 s and argon gas treatment (argon) over 180 s result in similar values. (mean ± SD, *** $p < 0.001$, t-test, $n = 3$).

To determine the effect of increasing osmolality on the cell behavior, we studied the viability of mHepR1 cells cultured for 24 h in DMEM+FCS+Genta, which was treated similarly to the samples used for osmolality measurements. As can be seen in Figure 5, the viability of the cells was massively reduced due to the 60 s plasma-treated medium. Interestingly, cells in 180 s argon gas-treated medium kept a viability of around 80%, although the osmolality values of 60 s argon plasma versus 180 s argon gas were in the same range.

Thus, increasing osmolality during argon plasma treatment is not the reason for the detrimental cell effects we observed, neither the changing of the pH nor the oxygen concentration nor the persistence of hydrogen peroxide after plasma treatments, which were precluded earlier.

Besides physical parameters, which could be changed due to plasma treatment, alterations in the chemical composition of the cell culture medium were investigated. In a recent study, Kalghatgi et al. [24] demonstrated an induction of DNA damage in mammalian breast epithelial cells by plasma-treated cell culture medium. This effect was not reduced if the medium had been stored up to 1 h prior to addition to the cells. The authors hypothesized that this remaining biological effect of plasma-treated cell culture medium is caused by stable organic peroxides

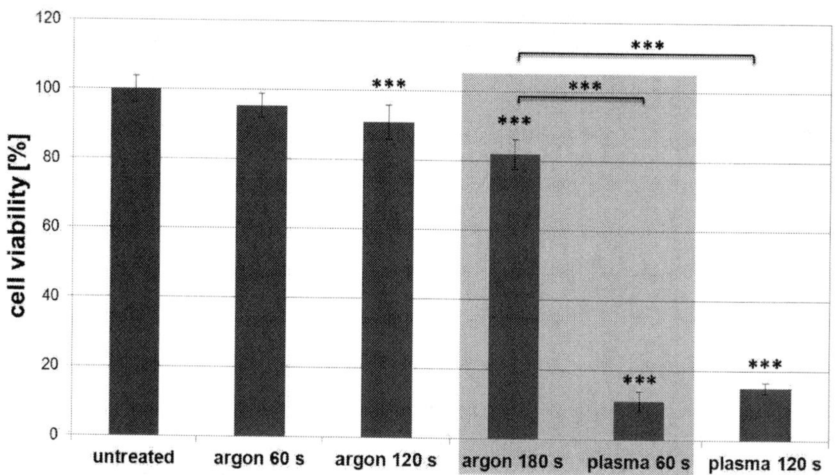

Figure 5. Difference in the viability of cells cultured in argon plasma- and argon gas-treated DMEM+FCS+Genta. Note that argon gas treatment over 180 s did not decrease cell viability as much as the argon plasma treatment did. (mean ± SD, *** $p < 0.001$, t-test, $n = 3$).

made up of medium compounds like amino acids [24]. The formation of stable peroxides from amino acids and proteins by reactive oxygen species like an OH radical is a well-known process [37]. In recent years, an active discussion has begun about the very complex ROS chemistry in plasma-treated liquids and its biological effects, including reactive species like OH radical, hyperoxide anion, and also the relatively stable hydrogen peroxide [1,38,39].

As we showed recently by size-exclusion chromatography on an FPLC analysis system, medium components are modified by argon plasma treatment [26]. We detected an additional peak upon plasma treatment of DMEM+FCS+Genta compared with the argon gas-treated control. In these experiments, we separated the components after plasma treatment by gel filtration. Interestingly, the peak height increased depending on the treatment time of the medium. In the present experiments, we examined the contribution of different medium components, FCS and gentamicin, to this additional peak and to the effects on cells. For this purpose, we investigated plasma treatment for the medium with (DMEM+FCS) and without supplements (DMEM). The different medium samples were analyzed after argon gas flow for 120 s and after argon plasma for 120 s by gel filtration with an FPLC system and compared with untreated controls (Figure 6).

All chromatograms of the analyzed media show peaks at a retention volume of about 22 ml (21.67–22.10 ml) and 26 ml (25.41–25.77 ml). The characteristic peak for albumin at about 15 ml (14.38–14.70 ml) was found in DMEM supplemented with FCS. An additional peak was found for the samples, which were treated for 120 s with argon plasma (see arrow) in basic DMEM, in DMEM+FCS (see Figure 6) and in DMEM+Genta (data not shown). No signal around the retention volume of 20 ml could be detected in untreated media and media that

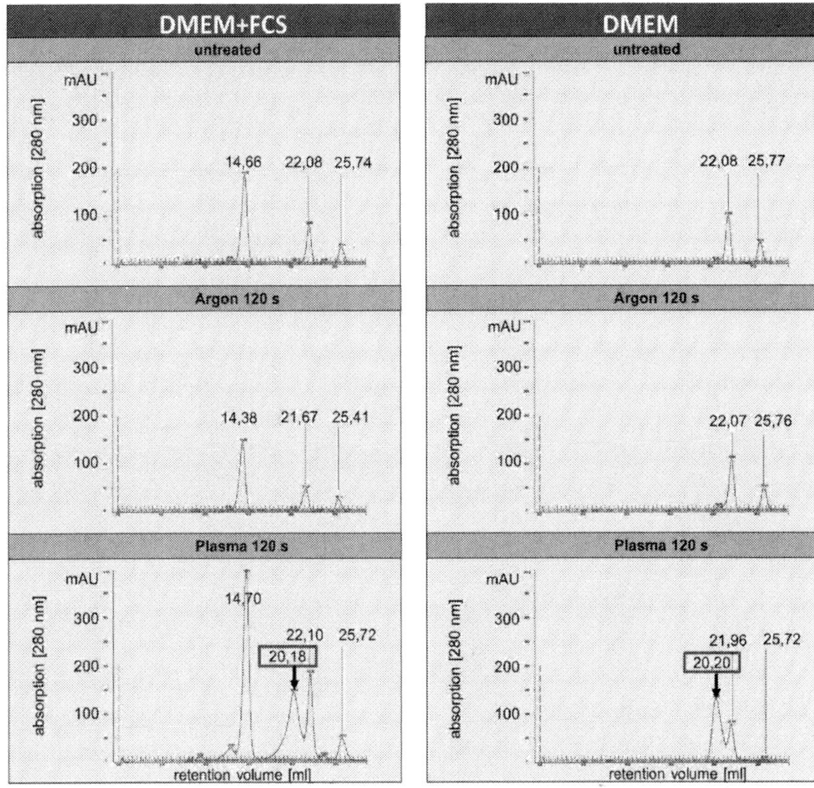

Figure 6. An additional peak at a retention volume of about 20 ml could be detected in the plasma-treated samples using size-exclusion chromatography. Note that cell culture medium DMEM without FCS also showed this signal, indicating that the source of this signal was not the interaction of argon plasma with FCS.

were only exposed to argon gas flow. This result gives rise to the idea that the visible changes in the chromatogram were caused by a change in the basic media (DMEM) independent of the supplements (FCS and gentamicin). Argon plasma seems to initiate the formation of components with higher molecular weight in the basic DMEM that appear at a lower retention time in the chromatogram.

Investigators hypothesized that ROS induces the formation of organic peroxides, which could be responsible for the damaging effects on cells in culture [21,22,37]. This conclusion is mainly based on the comparison of the incubation of cells with plasma-treated PBS, FCS, bovine serum albumin (BSA) or single amino acids [22]. DNA damage could not be shown after incubation of cells with plasma-treated PBS. However, incubation with plasma-treated BSA (in PBS) or FCS (in PBS) resulted in DNA damage of subsequently added cells. Even plasma treatment of PBS containing single amino acids resulted in adverse cell effects. Further analysis revealed a correlation between the peroxidation efficiency of single amino acids and their potential to induce DNA damage in cells [22]. These findings supported the thesis of amino acid peroxide formation and its potential contribution to DNA damage generation. The damage could be mediated either by single amino acids or by FCS, both being usual components of cell culture media.

As shown in Figure 5 and also observed in an earlier work of our group, DMEM+FCS+Genta impaired cell vitality after argon plasma treatment [25]. Based on the results deduced from size-exclusion chromatography (see Figure 6), it was deemed important to analyze the viability of mHepR1 cells dependent on the medium supplements during plasma treatment. For this purpose, DMEM and DMEM+FCS were argon plasma-treated for 60 and 120 s. Controls without treatment or only argon gas flow were also investigated. It was found that the mHepR1 cell viability decreased significantly in plasma-treated media after 24-h incubation for both approaches (Figure 7). In particular, in the presence of FCS during plasma treatment, the inhibitory effect is more obvious.

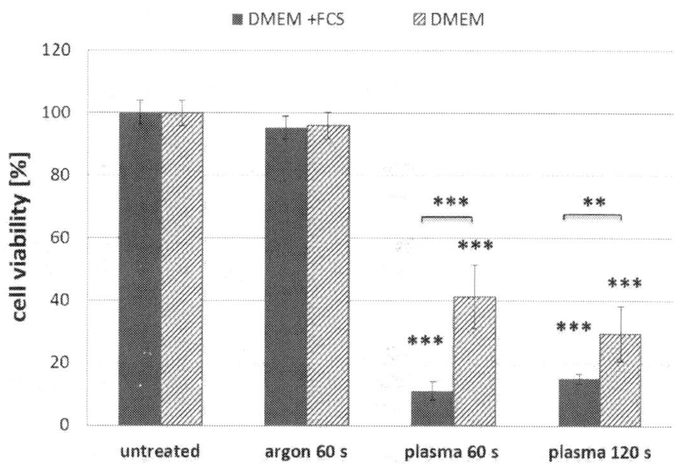

Figure 7. Cell viability after incubation of mHepR1 cells in plasma-treated DMEM and DMEM+FCS. (mean ± SD, ** $p <$ 0.005, *** $p < 0.001$, t-test, $n = 3$).

Therefore, it can be assumed that components in the basic DMEM without supplements were changed due to the impact of argon plasma and influenced cell viability.

Based on the hypothesis of Kalghatgi *et al.*, for the generation of stable organic peroxides from amino acids during plasma treatment [22] and our cell viability results in mHepR1 cells in plasma-treated DMEM versus DMEM+FCS, it can be concluded that peroxides formed from single amino acids by plasma treatment could be the reason for the detrimental cell effects we found. However, peroxidated amino acids are semi-reactive compounds and it seems possible that they generate compounds with higher molecular weight during storage, e.g., by polymerization. The identification of the newly generated substances in the cell culture medium awaits further analysis.

There is increasing evidence that plasma treatment promotes the healing process of tissue and accelerates wound healing [13,40–42]. The skin is a complex architectural multilayer cell system and processes concerning wound healing can be examined on cell monolayers as an *in vitro* model system. However, it is important to keep in sight that a cell monolayer *in vitro* is much more sensitive to agents because there are no cells in the "second row" to protect or replace cells in the apical row [43]. Here, we described effects of argon plasma-treated liquids (e.g., cell culture medium) as an indirect approach for plasma application in medicine. Transferred to *in vivo* systems, the opening of cell–cell contacts we observed by plasma-treated liquids could have a positive effect on the penetration of conventional therapeutics (antibiotics and disinfectants) on skin. Often, the application of conventional liquid antiseptics is not sufficient and sustainable as the borders and the surrounding of chronic wounds frequently consist of sclerotic skin, impeding an effectual penetration of these products [44]. With regard to disinfection, direct plasma treatment of living intact and wounded skin was found to be safe for doses even higher than required for inactivation of bacteria [45,46]. Impaired cell adhesion and reduced cell viability due to plasma-treated liquids are important starting points for further investigations concerning cancer therapy. The work presented here, focused on the basic mechanisms in the interaction of plasma-treated cell culture medium with cells and on the components in the treated medium, which are responsible for the persistent biological effect.

The vision could be the establishment of local plasma centers to prepare relatively stable plasma liquids for dermatological (chronic wounds and tumors), dental (peri-implantitis), or orthopedic (joint rinsing) applications to support or replace the conventional therapy.

4. Summary and conclusions

This study focused on the physicochemical analysis of argon plasma-treated cell culture medium DMEM with the additives FCS and gentamicin. In addition, the efficacy of plasma-treated complete cell culture medium DMEM upon storage and its impact on the cell physiology of epithelial mHepR1 cells were ascertained. We discovered that the liquid-mediated effect of atmospheric-pressure argon plasma on mouse liver epithelial cell–cell contacts and cell membrane microvilli persists up to even 21 days or 7 days of storage, respectively. Earlier investigations of our group revealed that temperature and pH (both were constant) as well as hydrogen peroxide production and oxygen content (both decreased within 1 day) can be excluded as initiators of the detrimental biological changes. As we found here, increased osmolality in the media caused by plasma treatment can also be excluded as a reason for the observed cell effects. On the other hand, we found an additional peak in size-exclusion chromatography analysis in basic DMEM after plasma treatment and significantly decreased cell viability in plasma-treated media independent of the presence of FCS during plasma treatment. High molecular compounds generated during plasma treatment of DMEM without

FCS give an impulse for further investigations on the formation, stability, and reaction of amino acid peroxides in this medium. The persistent biological effect on plasma-treated liquids observed here could open up new medical applications. Stable plasma-treated liquids could find applications for dermatological, dental, or orthopedic therapy.

Acknowledgements

The authors acknowledge the financial support of the University Medical Center Rostock, Germany (FORUN 889061, PlasmaBiomedicine) for C.H., M.K., and C.B.. M.H. is grateful for the doctoral stipendium of the state Mecklenburg-Vorpommern (Germany) and of the University of Rostock, Interdisciplinary Faculty, Department of Life, Light and Matter. The authors appreciate the BMBF Germany Pilot Program Campus PlasmaMed (13N11183). They would also like to express their thanks to the Electron Microscopic Center (EMZ) of the University Medical Center Rostock (M. Frank, G. Fulda) for qualified technical assistance.

Author details

Claudia Bergemann[1], Cornelia Hoppe[1,2], Maryna Karmazyna[1,3], Maxi Höntsch[1,4], Martin Eggert[5], Torsten Gerling[6] and Barbara Nebe[1*]

*Address all correspondence to: barbara.nebe@med.uni-rostock.de

1 University Medical Center Rostock, Department of Cell Biology, Schillingallee, Rostock, Germany

2 Technical University Dresden, DFG-Center for Regenerative Therapies, Dresden, Germany

3 Seracell Stem Cell Technology GmbH, Schillingallee, Rostock, Germany

4 Sarstedt AG and Co, Quality Management/EWZ, Nümbrecht, Germany

5 University Medical Center Rostock, Department of Internal Medicine, Center for Extracorporeal Organ Support, Schillingallee, Rostock, Germany

6 Leibniz-Institute for Plasma Science and Technology e.V., Greifswald, Germany

References

[1] von Woedtke T, Reuter S, Masur K, Weltmann KD. Plasmas for medicine. Phys Rep. 2013;530:291–320.

[2]   von Woedtke T, Metelmann HR, Weltmann KD. Clinical plasma medicine: State and perspectives of in vivo application of cold atmospheric plasma. Contrib Plasma Phys. 2014;54:104–17.

[3]   Fridman G, Friedman G, Gutsol A, Shekhter AB, Vasilets VN, Fridman A. Applied plasma medicine. Plasma Process Polym. 2008;5:503–33.

[4]   Fridman G, Peddinghaus M, Ayan H, Fridman A, Balasubramanian M, Gutsol A, et al. Blood coagulation and living tissue sterilization by floating-electrode dielectric barrier discharge in air. Plasma Chem Plasma P. 2006;26:425–42.

[5]   Kalghatgi SU, Fridman G, Cooper M, Nagaraj G, Peddinghaus M, Balasubramanian M, et al. Mechanism of blood coagulation by nonthermal atmospheric pressure dielectric barrier discharge plasma. IEEE T Plasma Sci. 2007;35:1559–66.

[6]   Daeschlein G, von Woedtke T, Kindel E, Brandenburg R, Weltmann KD, Junger M. Antibacterial activity of an atmospheric pressure plasma jet against relevant wound pathogens in vitro on a simulated wound environment. Plasma Process Polym. 2010;7:224–30.

[7]   Daeschlein G, Scholz S, Ahmed R, von Woedtke T, Haase H, Niggemeier M, et al. Skin decontamination by low-temperature atmospheric pressure plasma jet and dielectric barrier discharge plasma. J Hosp Infect. 2012;81:177–83.

[8]   Daeschlein G, Scholz S, Ahmed R, Majumdar A, von Woedtke T, Haase H, et al. Cold plasma is well-tolerated and does not disturb skin barrier or reduce skin moisture. J Dtsch Dermatol Ges. 2012;10:509–15.

[9]   Daeschlein G, Scholz S, Arnold A, von Podewils S, Haase H, Emmert S, et al. In vitro susceptibility of important skin and wound pathogens against low temperature atmospheric pressure plasma jet (APPJ) and dielectric barrier discharge plasma (DBD). Plasma Process Polym. 2012;9:380–89.

[10] Nosenko T, Shimizu T, Morfill GE. Designing plasmas for chronic wound disinfection. New J Phys. 2009;11:115013.

[11] Nastuta AV, Topala I, Grigoras C, Pohoata V, Popa G. Stimulation of wound healing by helium atmospheric pressure plasma treatment. J Phys D Appl Phys. 2011;44:105204-0.

[12] Weltmann KD, Kindel E, Brandenburg R, Meyer C, Bussiahn R, Wilke C, et al. Atmospheric pressure plasma jet for medical therapy: Plasma parameters and risk estimation. Contrib Plasma Phys. 2009;49:631–40.

[13] Heinlin J, Isbary G, Stolz W, Morfill G, Landthaler M, Shimizu T, et al. Plasma applications in medicine with a special focus on dermatology. J Eur Acad Dermatol. 2011;25:1–11.

[14] Kalghatgi S, Friedman G, Fridman A, Clyne AM. Endothelial cell proliferation is enhanced by low dose non-thermal plasma through fibroblast growth factor-2 release. Ann Biomed Eng. 2010;38:748–57.

[15] Vandamme M, Robert E, Lerondel S, Sarron V, Ries D, Dozias S, et al. ROS implication in a new antitumor strategy based on non-thermal plasma. Int J Cancer. 2012;130:2185–94.

[16] Plewa JM, Yousfi M, Frongia C, Eichwald O, Ducommun B, Merbahi N, et al. Low-temperature plasma-induced antiproliferative effects on multi-cellular tumor spheroids. New J Phys. 2014;16:043027.

[17] Kalghatgi S, Kelly C, Cerchar E, Azizkhan-Clifford J. Selectivity of non-thermal atmospheric-pressure microsecond-pulsed dielectric barrier discharge plasma induced apoptosis in tumor cells over healthy cells. Plasma Med. 2011;1:249–63.

[18] Kaushik NK, Kaushik N, Attri P, Kumar N, Kim CH, Verma AK, et al. Biomedical importance of indoles. Molecules. 2013;18:6620–62.

[19] Stoffels E, Sakiyama Y, Graves DB. Cold atmospheric plasma: Charged species and their interactions with cells and tissues. IEEE T Plasma Sci. 2008;36:1441–57.

[20] Kieft IE, Broers JLV, Caubet-Hilloutou V, Slaaf DW, Ramaekers FCS, Stoffels E. Electric discharge plasmas influence attachment of cultured CHO k1 cells. Bioelectromagnetics. 2004;25:362–68.

[21] Haertel B, Wende K, von Woedtke T, Weltmann KD, Lindequist U. Non-thermal atmospheric-pressure plasma can influence cell adhesion molecules on HaCaT-keratinocytes. Exp Dermatol. 2011;20:282–84.

[22] Kalghatgi S, Kelly CM, Cerchar E, Torabi B, Alekseev O, Fridman A, et al. Effects of non-thermal plasma on mammalian cells. Plos One. 2011;6:16270.

[23] Haertel B, Hahnel M, Blackert S, Wende K, von Woedtke T, Lindequist U. Surface molecules on HaCaT keratinocytes after interaction with non-thermal atmospheric pressure plasma. Cell Biol Int. 2012;36:1217–22.

[24] Kalghatgi S, Azizkhan-Clifford J. DNA damage in mammalian cells by atmospheric pressure microsecond-pulsed dielectric barrier discharge plasma is not mediated via lipid peroxidation. Plasma Med. 2011;1:167–77.

[25] Hoentsch M, von Woedtke T, Weltmann KD, Nebe JB. Time-dependent effects of low-temperature atmospheric-pressure argon plasma on epithelial cell attachment, viability and tight junction formation in vitro. J Phys D Appl Phys. 2012;45:025206.

[26] Hoentsch M, Bussiahn R, Rebl H, Bergemann C, Eggert M, Frank M, et al. Persistent effectivity of gas plasma-treated, long time-stored liquid on epithelial cell adhesion capacity and membrane morphology. Plos One. 2014;9:e104559.

[27] Kalghatgi S, Fridman A, Azizkhan-Clifford J, Friedman G. DNA damage in mammalian cells by non-thermal atmospheric pressure microsecond pulsed dielectric barrier discharge plasma is not mediated by ozone. Plasma Process Polym. 2012;9:726–32.

[28] Stoffels E, Kieft IE, Sladek REJ, van den Bedem LJM, van der Laan EP, Steinbuch M. Plasma needle for in vivo medical treatment: Recent developments and perspectives. Plasma Sources Sci Technol. 2006;15:S169–S80.

[29] Dobrynin D, Fridman G, Friedman G, Fridman A. Physical and biological mechanisms of direct plasma interaction with living tissue. New J Phys. 2009;11:115020.

[30] Nebe B, Rychly J, Knopp A, Bohn W. Mechanical induction of beta-1-integrin-mediated calcium signaling in a hepatocyte cell-line. Exp Cell Res. 1995;218:479–84.

[31] Paul D, Hohne M, Pinkert C, Piasecki A, Ummelmann E, Brinster RL. Immortalized differentiated hepatocyte lines derived from transgenic mice harboring Sv40 T-antigen genes. Exp Cell Res. 1988;175:354–62.

[32] Lange K. Role of microvillar cell surfaces in the regulation of glucose uptake and organization of energy metabolism. Am J Physiol-Cell Physiol. 2002;282:C1–C26.

[33] Pfister RR, Burstein N. Effects of ophthalmic drugs, vehicles, and preservatives on corneal epithelium – scanning electron-microscope study. Invest Ophth Visual. 1976;15:246–59.

[34] Oehmigen K, Winter J, Hahnel M, Wilke C, Brandenburg R, Weltmann KD, et al. Estimation of possible mechanisms of *Escherichia coli* inactivation by plasma treated sodium chloride solution. Plasma Process Polym. 2011;8:904–13.

[35] von Woedtke T, Oehmigen K, Brandenburg R, Hoder T, Wilke C, Hähnel M, et al. Plasma-liquid interactions: Chemistry and antimicrobial effects. In: Machala Z, Hensel K, Akishev Y, editors. Plasma for bio-decontamination, medicine and food security: Springer, Netherlands; 2012. pp. 67–78.

[36] Waymouth C. Osmolality of mammalian blood and of media for culture of mammalian cells. In Vitro Cell Dev Biol Plant. 1970;6:109–27.

[37] Gebicki S, Gebicki JM. Formation of peroxides in amino-acids and proteins exposed to oxygen free-radicals. Biochem J. 1993;289:743–49.

[38] van Gils CAJ, Hofmann S, Boekema BKHL, Brandenburg R, Bruggeman PJ. Mechanisms of bacterial inactivation in the liquid phase induced by a remote RF cold atmospheric pressure plasma jet. J Phys D Appl Phys. 2013;46:175203.

[39] Machala Z, Tarabova B, Hensel K, Spetlikova E, Sikurova L, Lukes P. Formation of ROS and RNS in water electro-sprayed through transient spark discharge in air and their bactericidal effects. Plasma Process Polym. 2013;10:649–59.

[40] Barekzi N, Laroussi M. Dose-dependent killing of leukemia cells by low-temperature plasma. J Phys D Appl Phys. 2012;45:422002.

[41] Ermolaeva SA, Varfolomeev AF, Chernukha MY, Yurov DS, Vasiliev MM, Kaminskaya AA, et al. Bactericidal effects of non-thermal argon plasma in vitro, in biofilms and in the animal model of infected wounds. J Med Microbiol. 2011;60:75–83.

[42] Emmert S, Brehmer F, Hänßle H, Helmke A, Mertens N, Ahmed R, et al. Atmospheric pressure plasma in dermatology: Ulcus treatment and much more. Clin Plasma Med. 2013;1:24–29.

[43] Hoene A, Prinz C, Walschus U, Lucke S, Patrzyk M, Wilhelm L, et al. In vivo evaluation of copper release and acute local tissue reactions after implantation of copper-coated titanium implants in rats. Biomed Mater. 2013;8:035009.

[44] Lademann O, Kramer A, Richter H, Patzelt A, Meinke MC, Czaika V, et al. Skin disinfection by plasma-tissue interaction: Comparison of the effectivity of tissue-tolerable plasma and a standard antiseptic. Skin Pharmacol Physiol. 2011;24:284–88.

[45] Dobrynin D, Wu A, Kalghatgi S, Park S, Shainsky N, Wasko K, et al. Live pig skin tissue and wound toxicity of cold plasma treatment. Plasma Med. 2011;1:93–108.

[46] Fluhr JW, Sassning S, Lademann O, Darvin ME, Schanzer S, Kramer A, et al. In vivo skin treatment with tissue-tolerable plasma influences skin physiology and antioxidant profile in human stratum corneum. Exp Dermatol. 2012;21:130–34.

Chapter 7

Non-thermal Plasma Technology for the Improvement of Scaffolds for Tissue Engineering and Regenerative Medicine - A Review

Pieter Cools, Rouba Ghobeira, Stijn Van Vrekhem, Nathalie De Geyterand and Rino Morent

Additional information is available at the end of the chapter

http://dx.doi.org/10.5772/62007

Abstract

Non-thermal plasma technology is one of those techniques that suffer relatively little from diffusion limits, slow kinetics, and complex geometries compared to more tradition-al liquid-based chemical surface modification techniques. Combined with a lack of sol-vents, preservation of the bulk properties, and fast treatment times; it is a well-liked technique for the treatment of materials for biomedical applications. In this book chapter, a review will be given on what the scientific community determined to be essential to ob-tain appropriate scaffolds for tissue engineering and how plasma scientists have used non-thermal plasma technology to accomplish this. A distinction will be made depending on the scaffold fabrication technique, as each technique has its own set of specific prob-lems that need to be tackled. Fabrication techniques will include traditional fabrication methods, rapid prototyping, and electrospinning. As for the different plasma techniques, both plasma activation and grafting/polymerization will be included in the review and linked to the in-vitro/in-vivo response to these treatments. The literature review itself is preceded by a more general overview on cell communication, giving useful insights on how surface modification strategies should be developed.

Keywords: Non-thermal plasma technology, surface modification, tissue engineering, scaffold fabrication, biomaterials

1. Introduction

In the 1980's, researchers started first experimenting with the idea of fabricating human tissues. But the true start of the field of tissue engineering was not until 1993, when Langer and Vacanti

wrote their legendary paper. The philosophy behind tissue engineering, as published by its two founders, is the following: *"the loss of failure of an organ or tissue is one of the most frequent, devastating and costly problems in human healthcare. A new field, tissue engineering, applies the principles of biology and engineering to the development of functional substitutes for damaged tissue."* In the two decades that followed, this new field in science expanded rapidly and continues to do so up to this day. As scientists started to unravel the pathways that drive tissue regeneration, it became clear that the fabrication of fully functioning human tissues was far more complex than initially anticipated [1, 2]. Building a tissue from the bottom up requires a profound knowledge of physics, chemistry, biology, and engineering and this from the macroscale (mm–cm) down to the molecular level (Å). Steady progress has been made over the last 25 years, resulting in some (commercial) successes such as Matrigel® and Dermagraft®. Alongside the rise of the field of tissue engineering, non-thermal plasma technology has gained importance as a cheap and efficient tool for the non-destructive modification of biomaterial surfaces. In this book chapter, an overview will be given on how plasma technology has contributed to the fabrication of better scaffolds for tissue engineering, preceded by an analysis on how to mimic the human body best, down to the molecular level.

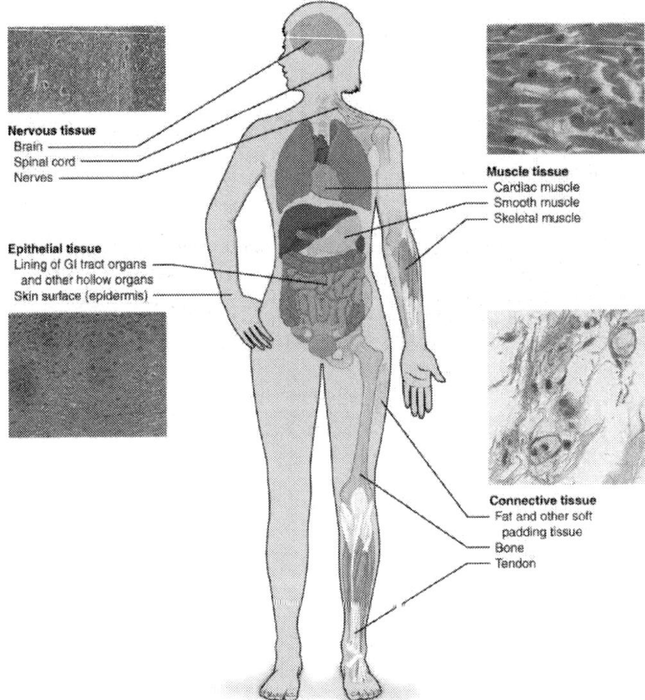

Figure 1. Different types of tissues available in the human body [132].

2. Cellular communication

2.1. Tissue deconstruction

Before understanding that how cells can be manipulated to grow into the required type of tissue, it is important to comprehend what a tissue exactly consists of. In general, a tissue is composed out of (1) a certain set of cells (osteoblast, fibroblast, epithelial cell, neuroblast, astrocyte...) from the same origin, forming the "building blocks" of a tissue, (2) an extracellular matrix (ECM), forming the "cement" of such tissue, offering a 3D structural and biochemical support, and (3) in most cases, a vascular network that allows transport of oxygen and nutrients [3–5]. Depending on the type of tissue (nervous, muscular, epithelial, or connective tissue), both the cell-types, cell ratio, and cell density, as well as the type of matrix (types of proteins and their 3D configuration) will greatly vary, as shown in Figure 1. The key to a viable tissue lies in its communication pathways: How do cells sense their environment? How do cells communicate with each other? What is triggering the different steps in a cell's life cycle? To postulate an answer to these questions, one has to go down to the (sub)cellular level.

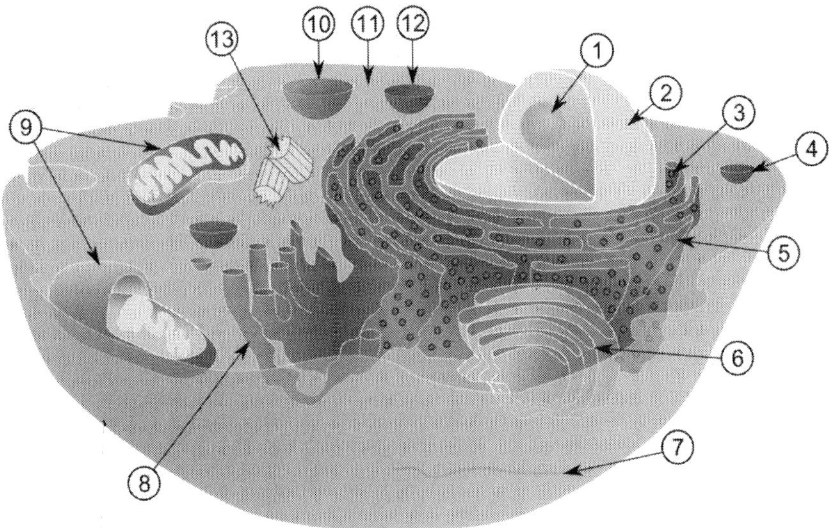

Figure 2. Schematic representation of cellular composition: (1) nucleolus, (2) nucleus, (3) ribosomes, (4) vesicle, (5) endoplasmic reticulum, (6) Golgi apparatus, (7) cytoskeleton, (8) smooth endoplasmic reticulum, (9) mitochondria, (10) peroxisome, (11) cytoplasm, (12) lysosome, (13) centrioles (open source).

2.2. Cell deconstruction

In basic biology, it is taught that a cell consists of a nucleus, containing the genetic information, a number of organelles that are responsible for the cells metabolism, and a cell membrane that separates and protects the cell interior from its environment (Figure 2) [1, 2, 4–7]. A cell constantly communicates with its environment and with other cells through signals released by these cells. Cell signaling occurs through the cell membrane via transmembrane proteins such as G-protein-coupled receptors and ion exchange channels or via diffusion of lipid soluble molecules and small molecules (Figure 3). Part of the transmembrane proteins consists out of a receptor site at the extracellular side of the membrane. When the matching signal molecule (a protein, a peptide, or an amino acid) binds with such a specific available receptor site, it will trigger a cascade of intracellular physico-chemical pathways that control the behavior of the cell to a large extent.

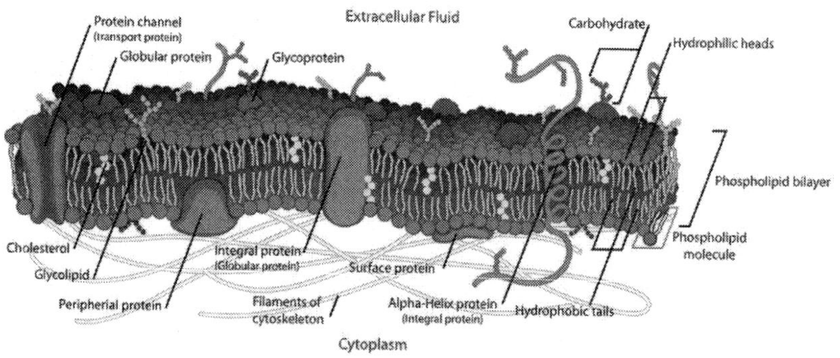

Figure 3. Schematic representation of cellular membrane, depicting different types of transmembrane proteins (open source).

2.2.1. What is ECM?

ECM is a protein-based support structure for cells, produced by cells. Initially, it was believed that it was relatively inert, but today it is known to interact intensely and specifically with cells. The main component categories of ECM are collagens (structural strength), elastin, proteoglycans (matrix resilience), and glycoproteins (cohesive bonding). These are quite general denominations for groups of proteins that show a surprisingly wide range of varieties, as it has been discovered in the past decade [8, 9]. Structure-wise, ECM is as varied as it is chemically diverse. In general, it could be described as a nanofiber matrix, mainly composed of collagen fibers, consisting of bundles of collagen fibrils that are anchored to the basement membrane, a dense sheet of collagen, lamilin, and other glycoproteins. How cells exactly interact with the ECM and how critical that interaction is, will be discussed next.

2.2.2. Cell–ECM communication

Cells sensing their environment will mostly limit themselves to contact-dependent communication pathways [8, 10, 11]. The main group of transmembrane proteins responsible for sensing and communicating with the ECM is the integrin family. Integrins are transmembrane

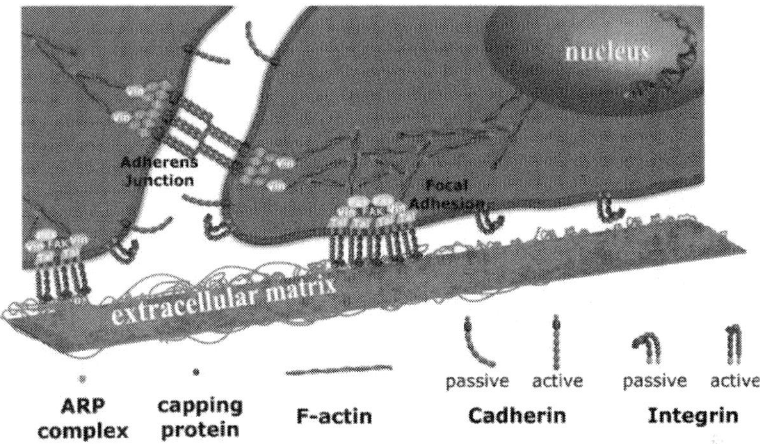

Figure 4. Schematic representation of integrin and cadherin activity [133].

glycoproteins consisting of α and β subunits with a large extracellular receptor domain that binds to certain RGD peptide sequences within the proteins (collagen, fibronectin, vitronectin...) that form up the ECM (see Figure 4). Up to this day about 20 different integrin transmembrane proteins have been discovered, each responding to different RGD peptide sequences. When an integrin binds onto the ECM, it will trigger the binding sites of the cytoplasmic tail, resulting in the formation of focal adhesion structures. These focal adhesions will bind to the cytoskeleton, triggering a cascade of intracellular signaling that will cause changes in the gene expression, which in turn affects all aspects of cellular behavior such as proliferation, differentiation, protein secretion, growth factor generation, and the generation of survival signals that prevent the cell to go into apoptosis. What makes cellular signaling (and communication) truly complex is that these transmembrane receptors rarely act alone: Most communication pathways consist out of multicomponent systems that depend on both the fixation of ligands on specific receptors as well as the transport of ions (e.g., Ca^{2+}) via ion channels [12]. Therefore, the key for designing a successful artificial matrix for the growth of a tissue is to simultaneously take into account multiple types of signaling molecules (immobilized on such artificial matrix), as well as their 3D distribution and density. In what follows next, an overview is given on how to design such 3D constructs.

2.3. Artificial ECM

The general idea behind tissue engineering is a three-step process (Figure 5): In the first step, healthy cells are extracted and isolated from the patient involved; In the second step, these cells are incubated in-vitro (in so-called bioreactors) on an artificial 3D construct that mimics the natural ECM [3–5, 13]; In the third step, after the formed tissue has reached a certain degree of maturation, the 3D construct is being implanted into the patient to replace the failing tissue or organ. The material scientist within the group of tissue engineers is mainly interested in the artificial 3D construct used during the second step and how to optimize it in such a way that the seeded cells sense no difference compared to their natural support (ECM).

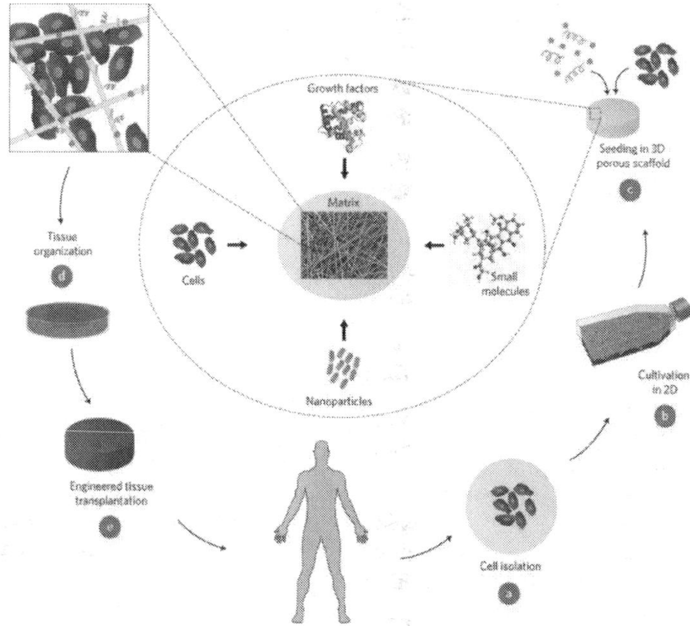

Figure 5. Schematic representation of the typical tissue engineering pathway followed [134].

2.4. Bio vs. artificial

The first step in fabricating appropriate cell support structures is the choice of material. As the support is implanted in-vivo and a second surgery has to be avoided, it is essential that the material is both biocompatible (preferably bioactive) and biodegradable [14–22]. This first condition is very generic, yet it rules out about 99 % of all materials available. Considering the remainder of the materials available, three different fabrication strategies can be followed: either work with a biopolymer that is sourced from a living organism (e.g., collagen from pigs), work with a synthetic material (e.g., polyesters or degradable metal alloys), or work with sheets of living cells as such. Each strategy has its advantages and disadvantages. Biopolymers such as collagen, gelatin, fibronectin... are all found in native tissue. Therefore, when cells are seeded onto such scaffolds, they do not lose their enzymatic activity to interact with and modify the surrounding ECM-like structure to their liking. The main downside using biopolymers is that they are extracted from foreign tissue, causing significant variability in quality as well as a constant risk for the transfer of xenographic pathogens. From a mechanical point of view, most biopolymers do not have the structural strength to be used in any load-bearing applications, limiting themselves mostly to "soft" tissue applications. Most properties of synthetic materials are complementary to biopolymers, with their advantages pointing toward reproducibility, availability and price, degradation rate control, and mechanical strength. Their most critical downside is that their surface chemistry does not resemble the natural ECM at all. The interaction between the cells and their immediate environment will therefore be very different compared to their response to native ECM, thus preventing good communication between the cells and their surroundings, resulting in sub-optimal tissue generation. Cell sheets are a scaffold-free alternative but are quite difficult to grow and handling them without damaging the delicate sheet is not without challenge [18]. In all cases, there are solutions to the previously stated problems, some of them will be discussed further along in the chapter, and others fall outside of the scope of the chapter but are discussed in other excellent review papers [18, 20, 23–26].

2.5. (Bio)scaffold fabrication

Independent of the chosen material (living or dead), it still needs to be processed into a suitable 3D structure, resembling a natural extracellular environment. In the early days of in-vitro culturing, cells were seeded onto flat sheets and their response was studied. As scientists started to unravel the communication mechanisms of cells, it was concluded that 2D surfaces were bad models for cell studies, as cell morphology and behavior are greatly influenced by what it senses on all sides (see Figure 6). Alongside other developments in tissue engineering, material scientists and engineers therefore started developing techniques that would allow them to grow cell cultures in-vitro in a 3D environment [1, 27–29]. This research partially resulted in a new field within the tissue engineering community that is now known as biofabrication. In what follows, the most commonly used scaffold fabrication techniques will be discussed in more detail, followed by how non-thermal plasma technology can add value to the performance of such fabricated structures.

2.5.1. Conventional fabrication methods

Conventional methods for manufacturing scaffolds include solvent casting and particulate leaching, gas foaming, fiber meshes and fiber bonding, phase separation, melt molding, emulsion freeze drying, solution casting, and freeze drying [30].The repetitive limitations of this family of techniques is that they are all inherently limited when it comes to pore size control, pore geometry, pore interconnectivity, and the possibility to construct internal channels for transportation of oxygen and nutrients (see Figure 7). The limited poor interconnectivity prevents good cell migration, resulting in a high density of cells near the scaffold surface, with very little cells at the center.

Traditional scaffold fabrication methods were made very popular in the early days of tissue engineering, as the use of leaching salts, porogens, and supercritical fluids were cheap fabrication methods that did not require specific and/or expensive machinery [25]. To overcome the problem of poor pore interconnectivity, researchers at that time strived to develop

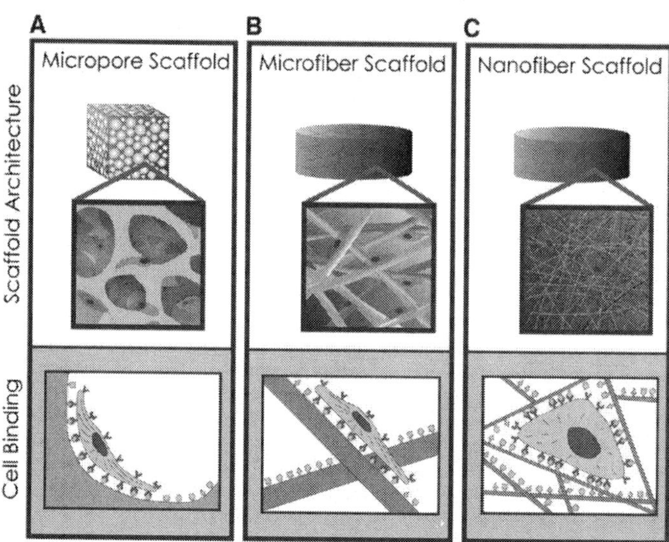

Figure 6. Differences in cell attachment efficiency between microporous and microfiber scaffolds on the one hand and nanofiber scaffolds on the other hand [135].

scaffolds with high pore ratios, up to 90 %. Although this partially resolved the pore inter-connectivity problem, it resulted in scaffolds with mediocre to poor mechanical properties [31].

A special case of a conventional scaffold fabrication method is the supercritical fluid process-ing. As it is a solvent-free technique, it is better suited than the other methods mentioned. Supercritical fluid (SCF) is created once a substance is exposed to an environment where its critical temperature and pressure are exceeded [32, 33]. A further increase in compression will therefore no longer result in liquefaction. Its physical properties are a combination of both liquid properties (density and solubility) and gas properties (diffusivity and viscosity). The most commonly used SCF is CO_2, as it is cheap, readily available, and reaches its supercritical point at near-room temperature (31°C, 7380 kPa). One of the more interesting properties of SCF's is their ability to dissolve polymers at a much lower temperature than their normal Tg. Making use of this property, Mooney et al. [6] were one of the first groups to apply this technique for the fabrication of microporous scaffolds. They dissolved a poly (lactic-co-glycolic) acid (PLGA) disc in CO_2 under SCF conditions. By rapidly decreasing the pressure, the solubility of the CO_2 decreases, causing nucleation and growth of gas bubbles within the polymer disc. The result is a sponge-like structure consisting of a high density of pores with varying size. The fact that no solvents were needed made it an excellent technique for tissue engineering applications. Several groups picked up on the technique, and it is still used to this day. Despite the lack of solvents during the production process, it suffers from the same limitations as the other conventional techniques though, inherently limiting it as a viable option for tissue engineering.

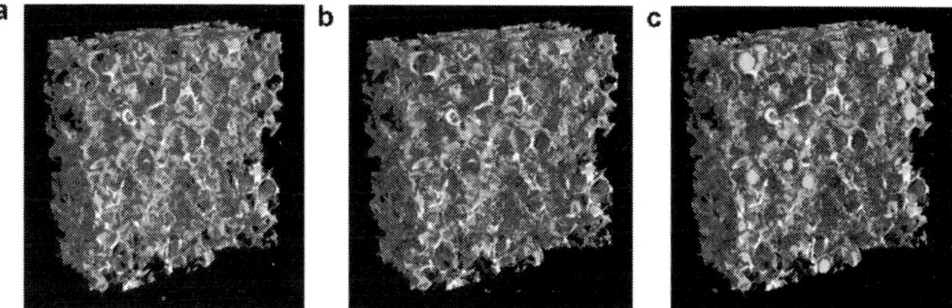

Figure 7. Micro-CT scan of traditionally fabricated scaffold via solvent casting/particulate leaching (A). B and C are showing the porous interconnectivity and pore size, respectively [136].

2.5.2. Rapid prototyping

Rapid prototyping is the collection of fabrication techniques that are able to directly translate a computer-generated design into a physical model. They are considered to be an additive process in which all parts of the construct are grown in a layer-by-layer fashion [20, 30, 34, 35]. The family of rapid prototyping techniques can be subdivided into two categories, depending on their mode of assembly: melt/dissolution deposition and particle bonding.

A melt/dissolution deposition system for tissue engineering applications can be considered as a miniature version of the traditional extrusion systems used in thermoplastic polymer processing, combined with an XYZ moving platform. The polymer, either molten or dissolved, gets forced through a small nozzle, after which it solidifies on the moving platform. Depending on nozzle size, temperature/concentration, and moving speed, a wide range of filament thicknesses can be extruded. Based on the digital design, a repetitive pattern can be printed, producing a porous structure with well-defined, well-interconnected pores in all directions, as can be seen in Figure 8. The typical pore size ranges typically from 100–1000 μm. More recently, the focus has shifted from thermoplastic biodegradable polymers to hydrogels, as several research groups have now succeeded to mix cells within the polymer feed [36, 37]. This generates hybrid scaffolds that are composed of both dead and living species.

Particle bonding systems use particles that are selectively bonded in a thin layer of powder material. These thin layers are then bonded onto each other in complex 3D structures, while being supported by unbounded material. At the end, the unreacted powder is removed and a well-defined interporous structure is obtained that exhibits both macropores and micropores (depending on the particle size of the powders). As the material is bonded via laser light or via adhesive droplets, dispersed via an inkjet type system, the amount of suitable polymers is limited.

While well-defined, geometrically complex 3D structures with excellent pore interconnectivity can be fabricated, they still exhibit a pore size that is an order of magnitude higher than what is found for natural ECM. This is not the case for that other popular scaffold fabrication technique: electrospinning.

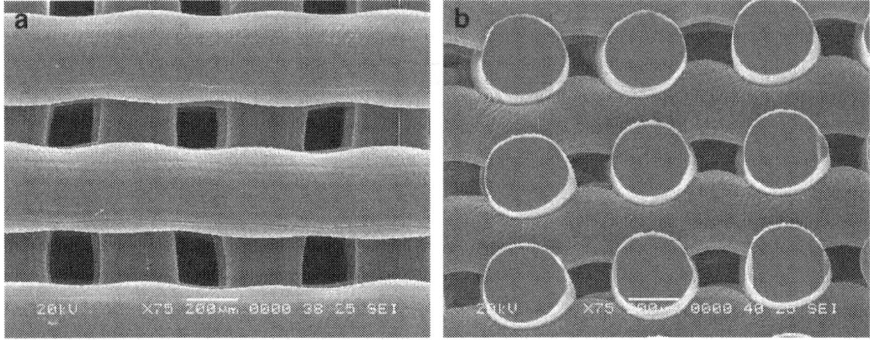

Figure 8. Top (A) and side (B) view of scaffold fabricated via microextrusion system [137].

2.5.3. Electrospinning

Electrospinning (ES) of polymers was already patented for the first time in 1939, but it was not until the early 1990 that a renewed interest in the 1D scaffold fabrication technique made it into the popular tool that it is today [38–43]. A typical ES machine consists of three major parts (see Figure 9): 1) a needle pump, feeding a polymer solution, at a certain flow, into the ES spinning chamber; 2) a high voltage source that is connected to the needle; 3) a grounded collector plate at a distance x from the needle. When applying a sufficiently high potential difference between the needle and the collector plate, the surface tension of the polymer droplet will break, resulting in the formation of a jet of polymer solution toward the collector plate. As the polymer solution travels at a high speed, the solvent will start to evaporate, resulting in the generation of nm-thick fibers that are then forming a randomly ordered 2D fiber sheet onto the collector plate (see Figure 9). The main difference between the ES and the other discussed scaffold fabrication techniques is the dimension range. As the fibers have a thickness in the order of nm, they are able to form μm- and nm-size pores, closely resembling the structures found in natural ECM. In the past 20–25 years, a wide variety of different polymers have been successfully electrospun, including most biodegradable and natural polymers. The biggest downsides of the ES technique are its lack in structural strength, thus preventing its use in load-bearing implant applications, and due to its dense packing, it makes it difficult for cells to migrate deeper into the spun matrix.

3. The added value of non-thermal plasma technology

As mentioned earlier, synthetic biodegradable polymers have a number of advantages (reproducibility, price...) and disadvantages over biopolymers and cell sheets [18]. Their main disadvantage is their lack of proper surface properties that stimulate the adhesion and proliferation of cells and differentiate them into the right type of tissue [44]. Numerous wet-

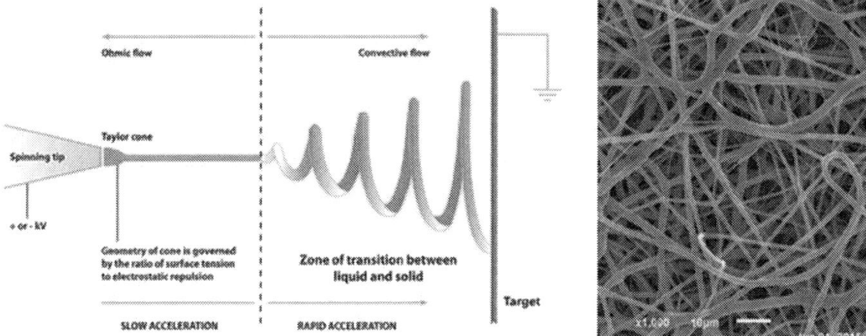

Figure 9. SEM micrograph of PCL electrospun fiber mesh (left: open source; right: original).

chemical techniques have been developed and applied, but most of them are either too aggressive, (partially) destroying the delicate structures, showing low treatment efficiency, time-consuming, or waste generating...

Non-thermal plasma technology is known as an excellent tool for surface engineering and a valid alternative for wet-chemistry-based surface engineering [14, 15, 45, 46]. It has been extensively used for surface decontamination/etching and surface treatment/activation on the one hand and the grafting/deposition of thin films with unique properties on the other hand. Its lack of solvents, fast treatment times, and preservation of bulk properties make it an ideal treatment method for delicate tissue engineered constructs [47]. Originally, research was focused on low-pressure plasma systems as they are well defined; exhibit excellent plasma stability; obtain good treatment efficiencies; and are able to deposit dense films that adhere well, are pinhole free, and show excellent stability. In the past two decades, there has been a shift toward research at (sub) atmospheric pressure as it eliminates the cost of extensive vacuum equipment and is less time-consuming, albeit not as well defined, and deposit films that are less dense.

In the following part of the chapter, an overview will be given on how both low-pressure and atmospheric pressure plasma technology have improved the performance of scaffolds for tissue engineering produced with different fabrication strategies.

3.1. Non-thermal plasma activation

Exposing a substrate to a plasma discharge fueled by an inert gas is the more basic approach for increasing the surface energy of said surface [14, 15, 48, 49]. Depending on the type of inert gas used, functional groups will be introduced directly (O_2, CO_2, air), indirectly (He, Ar) via post-oxidation processes, or a combination of both (N_2, NH_3, CF_4). At the same time, most plasma treatments will induce some degree of roughness, also contributing to the increase in wettability and surface energy, as has been extensively studied in the past on flat surfaces (see Figure 10). When applying a plasma treatment onto a scaffold structure, it is important to take the time-dependent character of the treatment into account. The gradual decrease in treatment efficiency upon storage, also known as the aging effect, is something that is often overlooked in literature. The substrate surface will try to return to a more stable energy state by lowering its surface energy again. This is either accomplished via polymer chain reorientation or the desorption of low molecular weight components. Depending on the substrate characteristics (crystallinity, molecular weight, storage atmosphere), they will lose up to 90 % in treatment efficiency over a period of only weeks. Using proper storing conditions (vacuum, low temperature) can slow down the ageing effect significantly, though it is something that has to be kept in mind when developing scaffolds for tissue engineering applications.

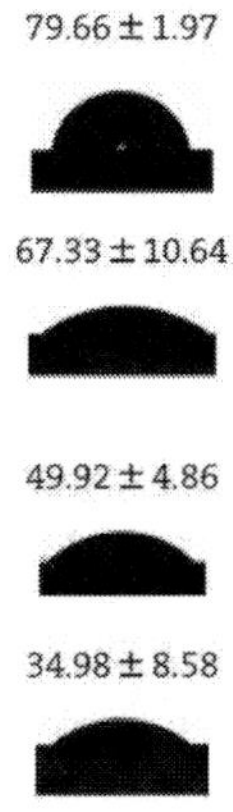

Figure 10. The change in WCA after 5 min of air plasma treatment at atmospheric pressure [129].

In the overview of the literature, it is clear that the developed non-thermal plasma surface modification strategies were in their early stages. As the scaffold fabrication methods evolved, plasma technological solutions also became more in tune with their microbiological environment. This specific chapter section therefore consists mainly out of basic surface-cell interactions, with the more advanced modification strategies following thereafter [138].

3.1.1. Traditional fabrication methods

To our best knowledge, it was not until 2002 that the first articles were published on the treatment of traditionally produced scaffolds with non-thermal plasma. Chim et al. [50] used a vibrating salt leaching method to fabricate poly(D,L) lactic acid (PDLLA) scaffolds, after which they applied an oxygen low-pressure radio frequent (RF) plasma (100 W–3 min). Static water contact angle goniometry (WCA) measurements showed a decrease in contact angle from 80° down to 60°, indicating an increase in surface polar groups, yet no X-ray photoelectron spectroscopy (XPS) or secondary ion mass spectrometry (SIMS) was performed to quantify this. Scanning electron microscopy (SEM) measurements revealed an increase in surface roughness as well as the formation of sub-μm-sized pits. The seeding of embryonic palatal mesenchyme (HEPM) cells into the scaffold resulted in a five-fold increase in cell activity compared to untreated scaffolds, while measurements of alkaline phosphatase activity (ALP) indicated an accelerated cellular differentiation into osteoblasts. Köse et al. [51] followed a similar strategy for the fabrication and treatment of poly(3-hydroxybutyrate-co-3hydroxyvalerate) (PHVB) scaffolds and their subsequent seeding with osteoblasts.

Around the same time, Claase et al. [52] used both freeze drying and salt leaching to synthesize scaffolds from different poly(ethylene oxide terephthalate)/poly(butylene terephthalate) (PEOT/PBT) mixtures, a copolymer that supposedly would enhance osseointegration. First lab tests revealed no significant increase in osseointegration, independent of composition. A CO_2 RF low-pressure plasma (49 W–30 min) was applied, but no surface characterization techniques were used to analyze its effects. Goat bone marrow cells were seeded into the scaffolds, after which a whole range of in-vitro tests revealed that the plasma treatment had a positive effect on adhesion and proliferation, but only for certain copolymer compositions (PEOT/PBT in ratio 70/30), while pore size was only of minor influence. A few years later, the experiment was repeated with rat bone narrow stromal cells, giving very similar results for the growth of cartilage rather than bone [31].

It is now considered to be general knowledge that surfaces rich in primary amines are excellent for the initial adsorption of certain proteins, thus positively stimulating cell behavior. Yang et al. [53] were, to our knowledge, the first to use an ammonia low-pressure RF plasma (50 W–5 min) on traditionally fabricated scaffolds (solvent casting + salt leaching specifically). WCA measurements on the plasma-treated poly-L-lactic acid (PLLA) and PLGA scaffolds indicated a high increase in wettability (78° → 15°), while XPS showed the incorporation of 4–6 % of nitrogen. Human foreskin fibroblasts were seeded, resulting in a significant increase in cell viability according to the performed MTT assays. Fluorescence microscopy showed an elongated morphology and a distinctively higher seeding efficiency.

Choi et al. [54] focused on the surface alteration of commercially available calcium phosphate scaffolds (200–400 µm pore size) using an atmospheric pressure plasma jet (APPJ) with either air or nitrogen as a discharge gas (15 kV, 13 mA, 5 slm, 10–20 min). WCA were reduced from 80° to <10°. XPS analysis revealed that the relatively high hydrophobicity was caused by significant carbon surface contamination, which was sharply reduced by the plasma treatment in favor of the introduction of –OH groups onto the surface. MC3T3 mouse osteoblasts were seeded on scaffolds treated with either air or nitrogen. Adhesion assays revealed a significant increase in cell adhesion compared to untreated scaffolds, even though no differences were found between treatments. Proliferation analysis revealed though that cells seeded onto nitrogen-treated surfaces proliferated at a higher rate, which can be linked back to the influence of amine rich surfaces. Morphology analysis learned that cells were still quite round, suggesting that the activation step as such is insufficient. This indicates that a grafting step or coating step might be better for initial cell-surface interactions, while still benefiting from the long-term effects of the underlying CaP substrate.

In the following 13 years, several other research groups have performed comparable plasma surface treatments on scaffolds produced via conventional fabrication methods. Their results have been summarized in Table 1. Most found significantly better cell adhesion and good cell proliferation. Ring et al. [55] also proved that a plasma treatment can lead to better neovascularization, while Shah et al. [56] showed that the results obtained in-vitro do not always translate well in-vivo, bringing a whole new set of challenges.

Author	year	scaffold type	fabrication method	reactor & gas	cell-type	effect
Safinia et al. [57]	2005	PDLLA	thermally induced phase separation	low pressure RF air &NH$_3$	/	higher wettability
Wan et al. [58]	2005	PLLA	phase separation	low pressure RF NH$_3$	M3T3 fibroblasts	better cell proliferation
Safinia et al. [59]	2007	PS/PLGA	thermally induced phase separation	APPJ	/	higher wettability
Ring et al. [55]	2010	Matriderm®	commercial	low pressure RF Ar/H$_2$	in-vivo	better neovascularizati on
Han et al. [60]	2011	PLA-PCL	solvent casting/ salt leaching	atmospheric pressure bipolar DC	L929 & MC3T3 osteoblasts	no cytotoxicity & homogeneous distribution through scaffold
Shah et al. [56]	2014	PLA	vibrating particle/ salt leaching	low pressure RF O$_2$	osteoblast & endothelial co-culture	in-vitro synergetic in-vivo location dependent
Bak et al. [61]	2014	PCL	gas foaming/ salt leaching	low pressure RF O$_2$ & N$_2$	MC3T3 osteoblasts	enhanced adhesion and proliferation
Sardella et al. [62]	2015	PCL	solvent casting/ salt leaching	low pressure RF H$_2$O/N$_2$	Saos-2 osteoblasts	50/50 H$_2$O/N$_2$ best cell adhesion and proliferation
Trizio et al. [63]	2015	PCL	solvent casting/ salt leaching	APPJ He/O$_2$	Saos-2 osteoblasts	cell-clustering, polymorphal shape slow spread throughout scaffold

Table 1. Overview of literature on plasma modification of traditionally fabricated scaffolds not discussed in the text.

3.1.2. Rapid prototyping

Rapid prototyping techniques were introduced to the tissue engineering community in the late 1990s [64–66]. Still, it took almost a decade before the first papers were published that involved plasma treatment of such structures, with the majority of the papers published in the past 5 years. The first paper, to our knowledge, was published by Wagner et al. in 2006 [67]. They developed an in-house system and printed PLGA scaffolds with 1.3 mm (!) gaps, after which they treated them with an oxygen low-pressure RF plasma (0.04 kPa, 5 min, 15 kV). No surface analysis was performed. Ovine and human osteoblasts were seeded, and results were compared to untreated scaffolds and tissue foil. Improvements in cell adhesion and proliferation were found compared to the untreated scaffolds, but results were worse compared to the tissue foil. In theory, one can still talk about a 3D scaffold in this case but with such pore sizes results will always be similar to what is found for cell-seeding on 2D films, as the gaps will be too big for cells to bridge.

In 2007, Moroni et al. [68] applied an Ar low-pressure RF plasma treatment (0.01 kPa, 30 min) on PEOT/PBT scaffolds printed on an Envisiontec® system. Both scaffolds based on generic designs as well as those based on CT scans from actual patients were used in this study. Again no surface analysis was performed. A SEM study of the scaffolds revealed a significantly higher degree of printing imperfections for the patient-based scaffolds. Yet, when in-vitro tests (MTT + GAG) were performed, the patient-based scaffolds performed remarkably better in terms of tissue formation and differentiation compared to the generic ones, both being treated with the non-thermal plasma. This indicates that biomimetic scaffold designs also play an important role for successful tissue engineered constructs.

Yildirim et al. published two papers on the modification of polycaprolactone (PCL) scaffolds with pore size 300 μm, printed using an in-house system [69, 70]. In both cases, an O_2 low-pressure RF plasma (18 W, 0–7 min) was applied. WCA measurements showed a progressive increase in wettability (58° → 22°). XPS measurements revealed an initial increase in oxygen functional group incorporation (up to 1 min treatment). Further treatment resulted in a decrease to values similar to untreated PCL. This suggests that the main effect of the oxygen plasma is etching rather than functional group incorporation. 7F2 mouse osteoblasts were seeded in both cases, giving similar results: enhanced adhesion and differentiation (higher levels of ALP and osteocalcin), and accelerated formation of mixed mineralization, all indications of increased osseointegrating properties. In 2012, Jacobs et al. [71] performed a more systematic study of different plasma discharge gasses (Ar, He and air) on PCL printed scaffolds, using a medium pressure parallel-plate system (5 kPa, 1–2 W, 30–300 s). Short treatment times resulted in the formation of a wettability gradient toward the center of the scaffold, as visualized by a standardized ink-staining method [72]. Longer treatment times resulted in a homogeneous wettability throughout the scaffolds, as confirmed by XPS (%O + 7 %). MC3T3 osteoblasts were seeded into the scaffold. Initial cell adhesion, as well as cell proliferation, greatly increased for those scaffolds treated by plasma, covering the complete interior of the scaffold (see Figure 11). Cell morphology was more elongated, and protein levels were higher.

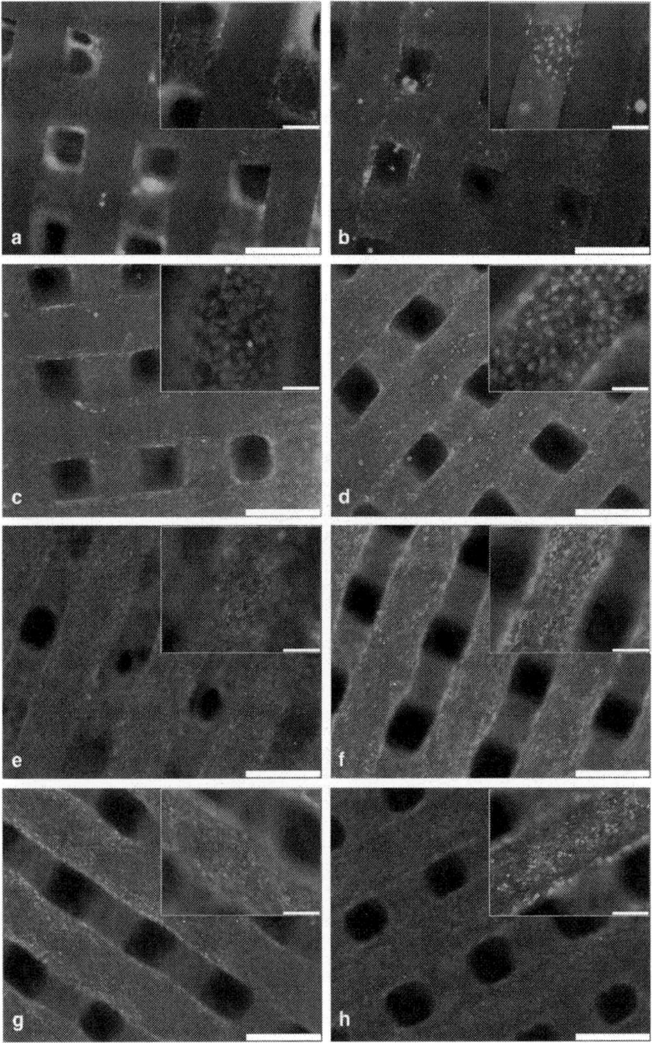

Figure 11. Fluorescent micrographs of PCL scaffold cross-sections of MC3T3 seeded osteoblasts. Left, from top to bottom (a–e): untreated scaffolds after 1–3–7–21 days of seeding. Right, from top to bottom (b–h): air plasma-treated scaffolds after 1–3–21 days of seeding [137].

3.1.3. Electrospinning

To this day, electrospun or e-spun fibers are considered to be the closest (semi)synthetic alternative to natural ECM, being one of the only techniques that is able to reproducibly fabricate scaffolds with sub-μm features on a larger, economically viable scale. As already mentioned, it is a relatively old technique (1939) that only gained renewed interest two decades ago, it was not until 2005 that the first papers were written on the plasma modification of such fibers.

Baker et al. [73] were one of the first groups searching for a non-invasive method to improve the cell attachment of finite smooth muscle cells on both randomly and aligned polystyrene (PS) fibers. An Ar RF plasma was applied (296 W, 0.01 kPa) for 5 min, and surface analysis was performed. The speed of adsorption of water was so high that it was impossible to measure correct WCA, while XPS measurements revealed a 20 % increase of oxygen, which is similar to results found for 2D films [74]. Cell attachments assays indicated a twofold increase in adhesion and actin staining images show that the cells align very well along the direction of the fibers. Despite the excellent in-vitro results, serious questions have to be raised to the use of PS as a modern implant material, as it is not biodegradable.

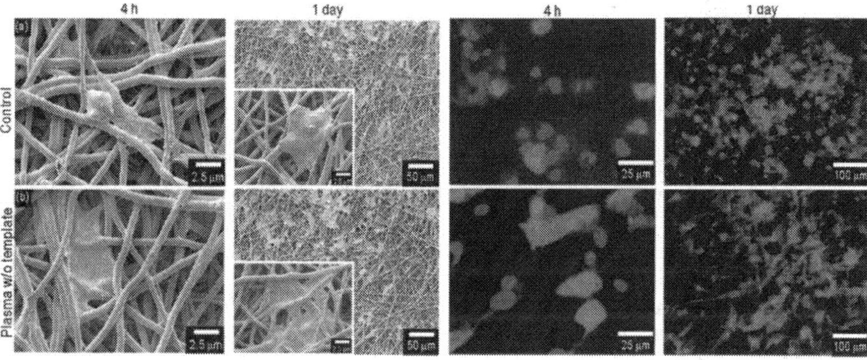

Figure 12. Improved cell infiltration into electrospun scaffolds after plasma treatment (left) and improved cell morphology (right) Jeon et al. [139].

Synthetic biodegradable polyesters such as PLA and PCL were often the material of choice for the fabrication of traditional and rapid prototyped scaffolds, and this is no different for the fabrication of electrospun materials. Several publications were published in the last 7 years where either PCL or PLA nanofibers were exposed to a plasma treatment [75–84]. Both low-pressure RF systems and atmospheric pressure plasma jets (APPJs) were used to generate plasma discharges using a variety of discharge gasses (air, NH_3, Ar, $N_2 + H_2$, O_2, or a combination of those). As was the case for the other fabrication techniques, several publications lack proper surface analysis, hence introducing a novel surface modification technique, but not characterizing what changes are taking place and to what extent. Those papers that did include a (partial) material analysis found the following results: WCA for PLGA fibers were decreased from 135° to around 45° in most cases. For PCL and PLLA, most were able to achieve full adsorption of the droplet (WCA =0°). XPS measurements showed an increase in oxygen containing functional groups between 2 and 7 % for PL(G)A, pure O_2 plasma treatment being the only one to achieve an increase of more than 4 %. The oxygen content of PCL fibers was increased up to 12 %. Those papers that used N_2 or NH_3 were able to incorporate up to 5 % in nitrogen containing groups, but no additional derivatization of alternative techniques were used to see what type of nitrogen groups were incorporated onto the surface. The changes in surface wettability came at a cost though. Tensile tests revealed a reduction between 30 and 40 % in tensile strength, indicating that plasma no longer can be viewed as a non-invasive technique when it comes to scaffolds with sub-μm dimensions. Whereas the papers in most cases lack proper surface characterization, they excel in providing good quality in-vitro and in-vivo results. Almost all studies revealed an increase in initial adhesion and cell morphology and this for practically any cell-type tested (fibroblasts, osteoblasts, smooth muscle cells, stem cells, Schwann cells...), but some studies went into even more detail. Park et al. [78] studied the adsorption of Bcl-2 proteins as a function of incorporated N-density on their e-spun PLGA scaffolds. Results showed that a mediocre hydrophilicity (WCA between 50° and 60°) resulted in the highest uptake of said protein. In parallel, they seeded fibroblasts onto the ammonia-treated scaffolds and found that the protein expression of the cells followed the same trend, with an overall reduction in the expression of stress-induced reactive oxygen species (ROS) secretion. De Valence et al. [79] tested their air plasma-activated PCL vascular grafts in-vivo, even after disappointing results in-vitro. Whereas the smooth muscle cells showed no significant increase in proliferation in-vitro, in-vivo tests revealed that cells were able to penetrate the scaffolds more efficiently and at higher densities, mainly due to the increased wettability. Cheng et al. [85] found similar results in-vivo for their Ar-/NH_3-treated PLA scaffolds – indicating that for cell infiltration, wettability is most likely the critical factor. When it comes to mineralization, Yang et al. [76] found similar results compared to Yildirim et al. [86] in the rapid prototyping section: improved mineralization within the first few hours after immersion of Ar plasma-treated scaffolds into simulated body fluid (SBF) solution. This resulted in the complete filling of the (sub)μm pores within 6 hours. After 7 days, the CaP was converted into type B carbonate apatite, the same building block found in natural apatite. Still, due the inferior properties of e-spun fibers compared to printed structures, it is to our belief that e-spun scaffolds as such will never be used for the repair of bone-related defects, unless combined with a polymeric, metallic, or ceramic support structure.

Whereas most other scaffold fabrication techniques were limited to biodegradable polyesters, electrospinning has a much broader spectrum of materials available. Biodegradable polyurethanes (PU) are a viable alternative to biodegradable polyesters, yet it is known that in some cases they trigger a more aggressive immunological response upon implantation [87]. Zandén et al. [88, 89] published two papers on the influence of O_2 RF low-pressure plasma treatment on the haemocompatibility of PU meshes and their influence on human embryonic stem cells. In both papers, they had to conclude that a plasma treatment as such is insufficient to significantly alter the haemocompatibility or to stimulate the differentiation of stem cells. Ardeshir-ylajimi et al. [90] exposed polyether sulfone scaffolds (PES) to an O_2 low-pressure microwave (MW) plasma (0.04 kPa, 10 min). In-vivo, studies with rabbits revealed that the scaffolds induced an increase in ALP levels and a significant increase in calcium content was found. The relative expressions of Runx2, Col 1, osteonectin, and osteocalcin analyzed via reverse transferase–polymerase chain reaction (RT-PCR) were all elevated within the first week.

Surprisingly, analysis of scaffolds implanted for more than 7 days showed that the expression levels dropped drastically and no more significant differences were found compared to untreated samples (which might be due to possible ageing effects of the plasma-treated scaffolds).

Electrospinning is the only scaffold fabrication technique, to our knowledge, where plasma has been successfully applied to improve the surface properties of biopolymers. In general, biopolymers are frequently used to fabricate scaffolds via other scaffolding techniques (e.g., rapid prototyping), but due to their high water content, it is less obvious to apply a non-thermal plasma, whereas electrospun materials are normally used in a dehydrated state. Baek et al. used an MW induced Ar plasma jet to modify electrospun/salt leached silk fibroin scaffolds (400 µm) intended for cartilage repair [91, 92]. Neonatal human knee articular chondrocytes were seeded onto scaffolds, resulting in a 50 % increase of initial cell attachment and a 100 % increase of proliferation compared to untreated scaffolds. Although, when the GAG content was analyzed, little to no difference was found for the treated scaffolds, indicating yet again that plasma treatment is not specific enough when it comes to the stimulation of secondary processes such as the enzymatic excretion of polysaccharides.

3.1.4. Intermediate conclusion

The overall conclusion that can be made for plasma-treated scaffolds is that it greatly increases the wettability of the scaffolds, allowing for a more efficient seeding process. This is reflected in significantly higher cell adhesion within the first 24 hours and a better distribution of the cells throughout the scaffold. After 7 days, though, the advantages of plasma treatments on cell proliferation are in most cases less substantial. This is most probably due to the non-specific nature of the introduced functional groups onto the surface of the scaffold interior. In the introduction, it was mentioned that cell signaling is highly dependent on the type and distribution density of transmembrane proteins protruding from the cell surface. Both the type of substrate functional groups and their density thus play a major role in the metabolic pathways of the cell. A more hydrophilic surface does allow cells to initially attach better, but as the correct integrin bond is missing, proper cell proliferation, and differentiation are inhibited. Nitrogen-rich surfaces perform better compared to oxygen rich surfaces, as in most cases they will contain a certain amount of primary amines, which is one of those functional groups preferred by cells for bone regeneration (or by the proteins initially excreted by the cells or present in the growth medium). However, there is often no control of functional group densities. Therefore, to better mimic the ECM environment, it would be recommended to have a higher control over the type of functional groups on the surface and their distribution, which are two parameters that can be better controlled by plasma grafting and plasma polymerization, which will be discussed next.

3.2. Plasma grafting and polymerization

Thin polymer-like films deposited via plasma polymerization are considered very different from films deposited via traditional wet-chemical techniques in the sense that the monomer gets fractionalized when exposed to plasma. This results in highly branched, highly cross-linked amorphous networks. These nm-thick films are often characterized as highly stable, dense, extremely adhesive to the substrate, and pinhole free (although not all these characteristics transfer well from low-pressure systems to atmospheric pressure systems and are influenced substantially by the set of plasma parameters used). As it is a solvent-free deposition technique, it is considered an excellent tool for tissue engineering applications.

Plasma grafting is a hybrid technique positioning itself between plasma treatment and plasma polymerization. The plasma step itself is identical to a plasma treatment, the (macro)monomer being introduced only after plasma treatment. Either the introduced reactive sites are then used directly as initiation points for free radical vinyl polymerization or afterwards by using the introduced oxygen/nitrogen functional groups to covalently bond other (macro)molecules, involving a wet-chemical step. This results in coatings with excellent retention of the functional groups, but low surface functional group density control. Both techniques are considered superior compared to plasma treatments when it comes to the variety of functional groups that can be introduced onto a scaffold surface and their stability over time.

3.2.1. Traditional fabrication methods

One of the best-known papers on plasma grafting and plasma polymerization applied on traditionally fabricated PDLLA scaffolds (supercritical CO_2) was published in 2005 by Barry et al. [93]. In this paper, an O_2 low-pressure RF plasma was used (40 Pa, 3–20 W, 3 min) to compare the penetration efficiency of allylamine when plasma grafted or plasma polymerized, respectively. XPS cross-sectional analysis revealed that for the grafting an evenly distributed nitrogen signal could be found, but at relatively low atomic concentrations (2 %) (see Figure 13). Plasma polymerized allylamine resulted in incorporation of up to 10 % nitrogen but after washing a parabolic gradient was found. They linked this gradient to limitations in diffusion of the allylamine precursor due to the limited pore interconnectivity. For the grafting step, as they allow the monomer to flow through the scaffold for up to 10 min, they claim that it eventually will penetrate all of the scaffold, but at a lower incorporation efficiency. In-vitro tests, using M3T3 fibroblasts revealed that only for the plasma polymerized scaffolds the cells were able to penetrate to the center of the scaffolds, while at the same time showing higher densities a the scaffold surface. This indicates that a certain density of polar functional groups is required to allow cells to effectively migrate throughout the scaffolds. Furthermore, it is an important indication of the diffusion limitations of traditionally fabricated scaffolds: Allylamine is a very small molecule, yet sharp gradients in coating chemical composition could be found. The diffusion limitations will increase exponentially when applying larger precursors for polymerization or trying to immobilize larger biomacromolecules such as collagen, gelatin, fibronectin...

Six years later, this research was picked up again by the same group in collaboration with the University of Bari, resulting in the publication of two papers by Intranuovo et al. [94, 95]. In the first paper, published in 2011, a very similar experiment was conducted involving low-pressure RF pulsed plasma polymerization of allylamine, with very similar results when it comes to gradient deposition. To improve the migration of the 3T3 fibroblasts used in the experiment, a hexane plasma was applied for a very short time (30 s) directly after the

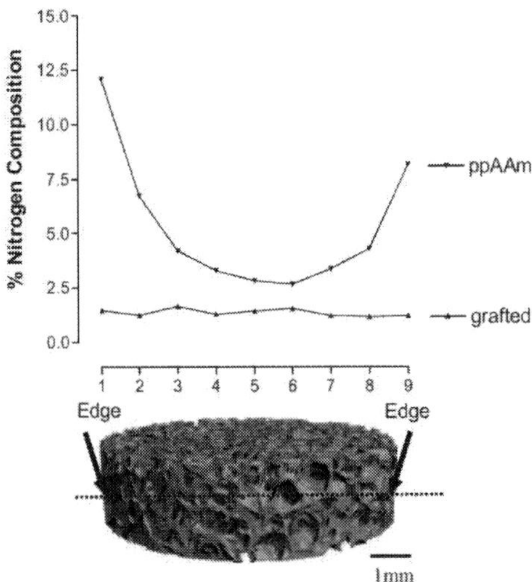

Figure 13. Distribution of nitrogen through traditionally fabricated scaffold: plasma grafting versus plasma polymerization [140].

allylamine plasma polymerization step. Hexane, being a bigger molecule, was limited even more when trying to diffuse into the scaffold. Combined with the short treatment time, this resulted in a hydrophobic outer surface of the scaffold. Upon seeding, the fibroblasts were "forced" to migrate into the scaffold. 96 hours after seeding cross-sectional imaging revealed a homogeneous distribution throughout the scaffold. In the paper published 3 years later, they again repeated the experiment, but using C_2H_4/N_2 mixture (3 min) instead, followed by a hydrogen plasma (30 s) to introduce primary amines onto the surface. In a second step, they solely used C_2H_4 plasma (90 s) to introduce reverse gradients (more nitrogen in the central part of the scaffold). Having used even smaller molecules compared to allylamine paid off, as the gradient throughout the scaffolds was far less pronounced, with up to 7 % of nitrogen functional groups in the center of the scaffolds. This also showed in the in-vitro testing, with Saos-2 osteoblasts evenly distributed for scaffolds both with and without the ethylene plasma step, thus being an example of an over-engineered solution.

In two papers published by the biomaterials research center of South Korea, O_2 low-pressure RF plasma systems (30 s, 0.1 Pa) were used to activate PLLA scaffolds, followed by an acrylic acid grafting step [96, 97]. Chondrocytes were seeded onto the scaffolds, and significant increase in adhesion and proliferation was found. Four weeks after seeding, higher levels of collagen II were measured as well as increased GAG levels, indicating the formation of cartilage. The success of the in-vitro tests was related to the acid rich surface, as well as an increase in surface roughness. Demirbilek et al. [98] focused their study on the changes in oxidative stress caused by plasma grafting. They used the freeze drying technique to fabricate their scaffolds. After plasma treatment (no details given), scaffolds were grafted with either ethylene diamine or low-molecular weight PEG. L929 osteoblasts were seeded, and oxidative stress was measured both via malondialdehyde and advanced oxidation protein products. Surprisingly, results showed that the level of oxidative stress was greatly reduced compared to cells grown on untreated scaffolds, even for PEG-grafted scaffolds, which are normally considered to be anti-fouling.

Several authors used low pressure plasma systems to activate PLGA scaffolds (O_2/CO_2, Ar and NH_3 plasma, respectively) fabricated via phase separation/ salt leaching in order to immobilize a biomacromolecule of choice [99–101]. Hu et al. [101] immobilized both polylysine (enhances electrostatic interactions) and RGD peptides, using a glutaraldehyde linker. Osteogenic precursor cells were seeded onto the scaffolds and after 14 and 28 days, alkaline phosphatase (ALP) activity and calcium assays were performed to determine the level of osteogenic differentiation. Results showed that the introduction of charge (either via NH_2 or polylysine) was not enough to stimulate differentiation, whereas the introduction of RGD peptides resulted in a significant increase in ALP activity and calcium concentration. This indicates that the RGD-peptide activity is well preserved, resulting in a stimulated osteogenic differentiation of the precursor cells. Woo et al. [100] immobilized β-glucans, a polysaccharide known for its enhanced wound healing properties, onto their scaffolds. Adipose tissue-derived stem cells were seeded onto the scaffolds, resulting in enhanced cell proliferation and improved cell morphology (more spindle like). Unfortunately, no studies were performed on differentiation. Shen et al. [99] immobilized rhBMP-2, an osteoinductive protein. Different plasma discharges were tested, and those that introduced a negatively charged surface were found to be most efficient in immobilizing the positively charged protein (O_2 and CO_2 vs. NH_3). OCT1 osteoblasts were seeded and ALP activity measurements indicated a significant increase, while SEM images showed accelerated mineralization within 22 days. Zheng et al. [102] used a different approach to obtain similar results, using a PLGA-tricalciumphosphate composite scaffold (phase separation + particle leaching). After ammonia plasma treatment, the scaffolds were dipped in a collagen solution, effectively immobilizing some of it onto the surface. Ex-vivo analysis of seeded rat dental pulp stem cells indicated greatly enhanced ALP activity (20 times higher), while von Koss staining showed increased areas of mineralization and Azan staining showed the presence of dentin proteins, all of which are indications of dentin-like bone formation.

Overall, quite a diverse group of traditional scaffold modifications has been discussed above. Compared to the plasma treatment sections, results are far more promising when it comes to cell proliferation, and especially cell differentiation. In-vitro, ex-vivo, and in-vivo experiments have shown that plasma-modified traditional scaffolds could be successfully applied to grow bone and cartilage tissue. Most papers though, avoid the topic of treatment homogeneity, while the first few papers discussed in this section, as well as papers discussed in the plasma treatment section, clearly show that this is an important issue to take into account. With this gradient problem taken into account, it is no surprise that there are very few papers available that focus on plasma polymerization of traditionally fabricated scaffolds or on the distribution homogeneity of the larger immobilized biomolecules such as rhBMP-2, β-glucans, or collagen. For both rapid prototyped structures, with their high pore interconnectivities, and electrospun fibers, with their extremely high porosity, this should not be such an issue.

3.2.2. Rapid prototyping

All papers found on plasma grafting and polymerization for rapid prototyped scaffolds were published in the last 4 years, indicating a large potential for further studies. Between 2012 and 2014, several papers were published by Declerq et al. and Berneel et al., following the same plasma grafting procedure [103–105]. PCL scaffolds (200 μm pore size) printed via a bioscaffolder (SYS + ENG) were exposed to an Ar DBD plasma for 30 s, followed by immersion in a 1M 2-aminoethylmetacrylate solution and subsequent exposure to UV radiation for 60 min. In a consecutive step, gelatin B (GelB) was immobilized on the introduced primary amines via carbodiimide wet chemistry, and finally, those samples were immersed in a fibronectin solution, resulting in an active physisorption of the protein (6 ng/mm^2). In the initial study published in 2012, both HFFs and MC3T3 osteoblasts were seeded on the scaffolds and the improvements on cell colonization for each consecutive coating step (plasma treatment → AEMA → GelB → fibronectin) were studied. Cross-sectional images 21 days after seeding show that the osteoblasts only grow on the edges of the plasma-treated samples, while for all other samples, they proliferated and migrated completely into the scaffolds (see Figure 14). For the fibronectin rich samples, the highest density of cells was found, including the formation of new natural ECM. In a second study, the performance of adipose-derived stem cells on fibronectin rich scaffolds was compared to commercially available collagen hydrogel scaffolds. Initially, the collagen scaffolds performed significantly better for seeding efficiency (×2) and protein production (+50 %) compared to the fibronectin rich scaffolds, which can most likely be contributed to the design of the scaffolds rather than the surface chemical properties. After 28 days, RT-PCR was performed to analyze the differentiation efficiency and selectivity. Results show that, although there was a delay in differentiation compared to the more compact collagen scaffolds, the fibronectin scaffolds show a better and more stimulated osteogenic differentiation compared to the commercial collagen scaffolds. Further improvement of scaffold designs, combined with higher concentrations of immobilized fibronectin should lead, in our opinion, to higher seeding efficiencies and even more pronounced differentiation. Van Bael et al. [106] repeated the exact same procedure on PCL scaffolds fabricated via selective laser sintering (SLS) without the fibronectin physisorption and found very similar results.

Sousa et al. [107] used the BioExtruder to fabricate PCL scaffolds with pore size diameters of 900 μm. An Ar low-pressure RF plasma (100 W, 30 s) was applied and exposed to ambient air to allow the formation of peroxides and hydroxyperoxides on the scaffold surface. This was followed by an immersion in an acrylic acid solution combined with exposure to UV-irradiation. In a final step, collagen was immobilized onto the surface using carbodiimide wet chemistry. M3T3 fibroblasts were seeded onto the surface, and MTT tests were performed after 7 and 14 days. Results show no significant differences toward the untreated scaffolds, but more importantly, also no significant differences were found compared to commercially available collagen scaffolds, unlike to what was found by other researchers. This is a strong indication that the type of cell used for seeding (fibroblast vs osteoblast vs. stem cell) as well as the origin of the cell (primary cell lines versus cancer cell lines) and scaffold design all play an important role on the possible in-vitro outcome of the study.

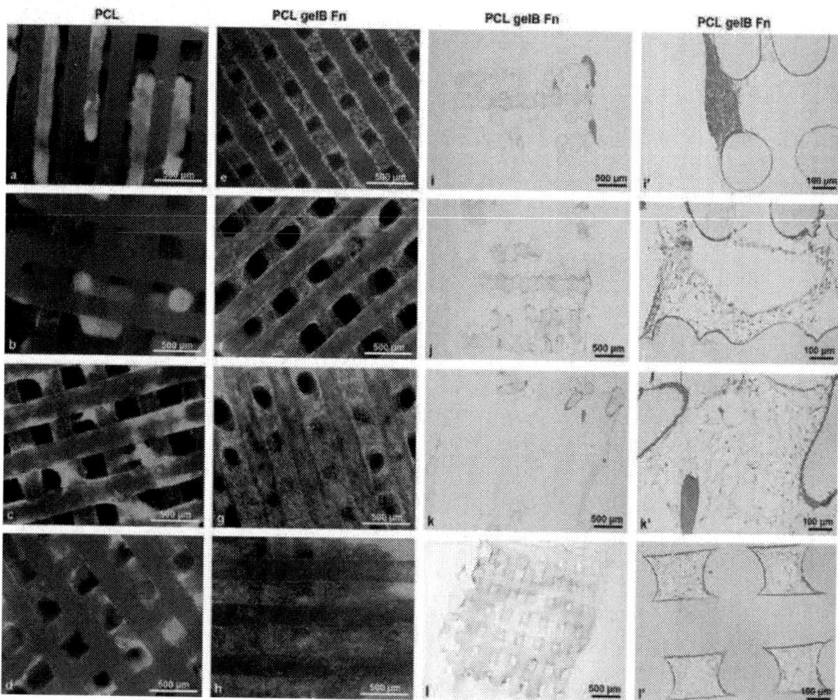

Figure 14. Viability, infiltration, and colonization on PCL and PCL GelB Fn scaffolds. Calcein AM images of HFF cells after 1, 7, 14, and 21 days on PCL scaffolds (a–d) and PCL GelB Fn scaffolds (e–h). Colonization of the 3D scaffolds by HFF cells 1, 7, 14, and 21 days post-seeding. Histological sections taken of the center of PCL GelB Fn scaffold [scale bar 500 lm (i–l)]. Details of the histological sections of the center of PCL GelB Fn scaffolds are given in i'–l' (scale bar 100 lm). Hematoxylin/eosin staining of 5-lm thick paraffin sections. The pink lining is representative for the gelatin coating. Only when cells were present, the integrity of the scaffolds could be preserved [142]

Nebe et al. [108] also used SLS to fabricate PLGA/tricalcium phosphate composite scaffolds with a 500-µm pore size. An O_2 low-pressure MW plasma was applied (500 W, 5 Pa, 0.1 slm, 10 s) followed by a plasma polymerization step using allylamine as a precursor (500 W, 50 Pa, 0.05 slm, 144 s), resulting in the incorporation of up to 16 % of nitrogen, as confirmed via XPS (no mention of cross-sectional analysis). MG-63 osteoblasts were seeded onto the scaffolds, and SEM analysis was performed in the first 30 min, showing already large differences in cell attachment to the scaffold, yet no quantitative analysis was performed.

Owen et al. [109] and Kim et al. [110] both used an acrylic acid enriched O_2 low-pressure RF system to deposit acid rich coatings on their scaffolds with 800 and 450 µm pore diameters, respectively. Owen et al. used the scaffolds as such, seeding human mesenchymal progenitors. ALP activity was measured, showing that the stiffest samples combined with the acrylic acid resulted in the best osteogenic differentiation. Kim et al. introduced an extra step, immobilizing rhBMP-2 onto their scaffolds via simple incubation. MG-63 osteoblasts were seeded onto the scaffold, and again ALP activity was measured. Results show an 80 % increase of activity compared to the coated scaffolds without immobilized protein and a 90 % increase compared with untreated scaffolds; results that are more pronounced compared to what was found by Owen et al.

Compared to the research performed on the plasma modification of traditionally fabricated scaffolds, the amount of papers on plasma-modified rapid-prototyped scaffolds is more recent and rather limited. The main reason for this is most likely the high acquisition cost of the rapid prototyping systems, ranging somewhere between 30.000 and 250.000 euro, combined with the fact that most of those systems can be found in biomaterials laboratories having limited experience with plasma technology. Those groups that did perform plasma grafting/polymerization on rapid-prototyped scaffolds have found very similar results compared to results found for modified traditional scaffolds. The main difference between the two fabrication techniques is that there is little to no mention of possible gradients in chemistry. This suggests that either: (1) it is not a problem thanks to the better pore interconnectivity or (2) no attention was paid to it. The latter seems more likely, based on the results found by Jacobs et al. [71] for a plasma-activated rapid-prototyped scaffolds. In either case, more structural studies are required to analyze this, as well as a transfer of some of the results found for traditional scaffolds to the rapid prototyped ones.

3.2.3. Electrospinning

Several of the experiments performed on other scaffolds have also been performed on electrospun scaffolds; therefore, this chapter part will mainly focus on those plasma grafting and polymerization strategies that have not been discussed for other scaffolds, while the rest will be summarized in Table 2 [141].

Author	scaffold material	reactor & gas	precursor	cell-type	effect
Shabani et al. [118]	PES	MW low pressure O_2	collagen grafting	somatic stem cells	better cell infiltration into the scaffold
Jeong et al. [119]	Silk	RF low pressure O_2	CH_4 polymerization	keratinocytes & fibroblasts	addition of O_2 into plasma increases initial adhesion
Park et al. [120]	PU	MW atmospheric plasma Ar	PLGA grafting	HUVEC	adhesion +100% proliferation +150% better endothelialization
Yao et al. [121]	PU	40 kHz low pressure plasma Ar	4-vinylpyridine + quaternization	S. aureus E. coli	Log5 and Log3 reduction in bacterial activity
Lopez-Pérez et al. [122]	PCL	RF low pressure O_2	vinyl phosphonic acid vinyl sulfonic acid	Saos-2 osteoblasts	higher vitronectin adsorption

Author	scaffold material	reactor & gas	precursor	cell-type	effect
					better proliferation for vinyl phosphonic acid
Chen et al. [123]	PLLA	DC low pressure pulsed system O_2	gelatin grafted	chondrocytes	4* increase in cell viability 4* increase in collagen production
Jia et al. [124]	PCL	RF low pressure Ar	soluble eggshell protein	human dermal fibroblasts	more elongated morphology
Park et al. [125]	PLA	RF low pressure O_2	acrylic acid polymerization	NIH 3T3	better adhesion and proliferation + complete coverage of the scaffold pores within 6 days
Paletta et al. [126]	PLLA	RF low pressure O_2	cyclic RGD grafted	mesenchymal stem cells	no enhanced differentiation due to limited RGD surface densities
Seyedjafari et al. [127]	PLLA	MW low pressure O_2	nanohydroxyapatite grafted	unrestricted somatic stem cells in-vivo	Higher ALP and mineralization + ossification and formation of trabeculi
Santos et al. [128]	Starch (100μm)	RF low pressure Ar	vitronectin & fibronectin adsorption	HUVEC	better proliferation higher concentration endothelial growth factors higher cadherin activity
Ma et al. [129]	PCL	RF low pressure air	gelatin grafted	endothelial cells	better morphology aligned along fibers
Ghaedi et al. [130]	PLA	MW low pressure O_2	collagen grafted	hepatocytes	efficient differentiation into osteoblasts and adipocytes
Ai et al. [131]	PHVB	MW low pressure O_2	collagen grafted	unrestricted somatic stem cells	enhanced proliferation

Table 2. Overview of literature on plasma polymerization/grafting on ES substrates not discussed within the text.

The flexibility of nanofibers allows them to be used in a wider variety of tissue engineering applications. An example that has been amply studied is the vascular graft. One aspect that is critical for such applications is the haemocompatibility. Wang et al. [111] and Cheng et al. [112] both used low-pressure RF systems to immobilize heparin on silk nanofibers and PLLA nanofibers, respectively. Whereas Wang et al. used an Ar plasma, Cheng et al. used an $Ar/NH_3 + H_2$ plasma. Both claim a high-grafting efficiency, however, the study of Cheng et al. shows that their treatment strategy results in a 10-times higher adsorption rate compared to an Ar plasma treatment. In-vitro tests with blood platelets showed a significant decrease in platelet attachment compared to untreated scaffolds (Cheng et al.) and a four-time longer coagulation time (Wang et al.). Finally, the seeding of bovine aortic endothelial cells showed an enhanced cell infiltration into the scaffold independent of the level of heparin immobilized, suggesting that the addition of ammonia to the plasma is unnecessary (Cheng et al.). Wang et al. tested their scaffolds in-vivo, showing excellent biocompatibility with only minor signs of inflammation. He et al. [113] used a low-pressure ICP air plasma (30 W–5 min) to immobilize collagen rather than heparin onto their PCL-PLLA electrospun scaffolds (470 ± 130 nm) for vascular graft applications. Rhodamine-stained scaffolds show an excellent distribution of collagen throughout the scaffolds. Human coronary artery endothelial cells were seeded onto the scaffolds. The collagen-coated scaffolds stimulated the spreading, attachment, and overall viability combined with a good preservation of the phenotype.

Another application that has received considerable attention is the fabrication of e-spun scaffolds for nerve-guided regeneration. Delgado-Rivera et al. [114] used a plasma-patterning technique involving a PMDS mask to generate tracks of grafted lamilin on a polyamide nanofiber mesh. RG3.6 neural precursors were seeded onto both patterned and completely treated scaffolds. Fluorescence microscopy showed that the precursor cells almost exclusively stuck to the patterns growing in a bilateral fashion. Zandér et al. [115] used an air plasma treatment (18 W–5 min) on aligned PCL nanofibers with sub-μm dimensions followed by covalent immobilization of lamilin, using carbodiimide wet-chemistry. PC12 neuron-like differentiated cells were seeded onto the e-spun scaffolds. SEM imaging revealed significantly longer neurons on scaffolds that contain higher concentrations of lamilin. The authors do admit that they have no control on how the protein is folded onto the surface, thus having no indication on the efficiency of the immobilized lamilin, which could most likely be partially addressed via gene expression analysis.

Guex et al. [116] focused their research on the repair of damaged myocardium using PCL fibers coated with a CHO-type coating (ethylene + CO_2) rich with ester bonds (O/C = 0.35). Mesenchymal stem cells were seeded onto the scaffolds and then implanted into a rat model. Four weeks after implantation, scaffolds were analyzed, showing that the plasma-coated grafts significantly stabilized the cardiac function, resulted in attenuated dilation and gave cause to lower ejection fractioning and fractional shortening all compared to untreated scaffolds.

Nuhiji et al. [117] developed a multifunctional e-spun scaffold based on the immobilization of silane groups. Oxygen low pressure RF plasma pretreatment (63 Pa, 30 W, 3 min) resulted in a 10-fold increase in 3-aminopropyl-trimethoxysilane or 3-mercaptopropyl-trimethoxysilane. The silane-enriched surfaces were then immersed in a neutravidin-enriched buffer solution overnight, resulting in the effective immobilization of the neutravidin. CD-4 antibodies were seeded onto the scaffold, resulting in specific bonding. According to the authors, changing the neutravidin for other biomacromolecules such as proteins, enzymes, and antibodies should be easy, resulting in a multitool scaffold for biosensing and tissue engineering.

A wide variety of polymerizable precursors and biomacromolecules were selected by research groups around the world to be used for plasma polymerization and plasma grafting, respectively, on electrospun scaffolds. Compared to the plasma-activated scaffolds, results were far more promising. In most cases, better cell proliferation rates were found on top of good initial cell adhesion and, when applicable, cell differentiation was drastically improved. In-vivo results were promising as well, with limited infection, good cell ingrowth, and, in a few cases, even good vascularization. Compared to other fabrication techniques, electrospinning has been studied quite extensively, resulting in a wider variety of possible treatment strategies.

4. Conclusion and outlook

A quite extensive overview has been given on the literature involving, in one way or the other, the use of non-thermal plasma technology for the modification of scaffolds for tissue engineering applications. In the past 15 years, research has evolved from "simple" plasma treatments, mostly improving cell adhesion, to quite elegant methods, often involving nothing more than a plasma step and some protein-enriched aqueous solutions. These methods resulted in scaffolds that were able to support cells that adhered better, proliferated better, and differentiated better compared to their untreated counterparts. These results show that plasma is a reliable and efficient tool that should be included in the toolbox of every tissue engineer. Comparing the three scaffold categories reviewed here, it is to our belief that the traditionally fabricated scaffolds will disappear over time, as the rapid prototyped scaffolds are superior in both pore interconnectivity and biomimetic design. Electrospinning is a technique that is complementary to the other two and should even be used in parallel with the others, combining both structural integrity and ECM-like support structures in one scaffold design. The challenges for the next few years will be to translate the technology from generic scaffolds to patient-customized scaffolds that could actually be used on patients. An increase in scaffold dimensions and complexity will be unavoidable and cause new and unexpected problems as well as increase problems still encountered today (such as gradient depositions). Secondly, more attention should be paid to the latest discoveries in cell biology, as this would accelerate the tuning process of what should be immobilized on a scaffolds' surface, thus addressing the issue that research conducted to this day is almost exclusively unidimensional, studying the effect of one component on cellular behavior, while it is known that most ques triggering the cellular metabolism are multicomponent systems. Gene expression analysis is becoming more popular, but should be systematically used to give researchers more fundamental insights into what their surface modification techniques are triggering on a molecular level. Thirdly, at this point, the level of in-vivo testing remains relatively low and should become more mainstream, as literature has shown that there are often large discrepancies between results found in-vitro and in-vivo.

Acknowledgements

This chapter has received funding from the European Research Council under the European Union's Seventh Framework Program (FP/2007–2013)/ERC Grant Agreement No. 279022.

Author details

Pieter Cools*, Rouba Ghobeira, Stijn Van Vrekhem, Nathalie De Geyterand and Rino Morent

*Address all correspondence to: Pieter.Cools@ugent.be

Research Unit Plasma Technology, Department of Applied Physics, Ghent University, Ghent, Belgium

References

[1] Howard D, Buttery LD, Shakesheff KM, and Roberts SJ: Tissue engineering: strategies, stem cells and scaffolds. *Journal of Anatomy* 2008, 213(1):66–72.

[2] Blitterswijk CAV, Moroni L, Rouwkema J, Siddappa R, and Sohier J: In: Blitterswijk CV, Thomsen P, Lindahl A, Hubbell J, Williams DF, Cancedda R, Bruijn JDD, and Sohier J (Eds): *Tissue engineering*. Burlington: Academic Press, 2008, pp. xii–xxxvi.

[3] Novosel EC, Kleinhans C, and Kluger PJ: Vascularization is the key challenge in tissue engineering. *Advanced Drug Delivery Reviews* 2011, 63(4):300–311.

[4] Helsen JA, and Missirlis Y: Tissue engineering: Regenerative medicine. *Bio materials*, Springer, Verlag Berlin Heidelberg, 2010, pp. 269–289.

[5] Lanza R, Langer R, and Vacanti JP: *Principles of tissue engineering*. Academic Press, Burlington San Diego London, 2011.

[6] Mooney DJ, Baldwin DF, Suh NP, Vacanti JP, and Langer R: Novel approach to fabricate porous sponges of poly(d,l-lactic-co-glycolic acid) without the use of organic solvents. *Biomaterials*, 1996, 17(14):1417–1422.

[7] Brown RA: *Extreme tissue engineering: concepts and strategies for tissue fabrication*. Wiley, Oxford Chichester Hoboken, 2013.

[8] Hay ED: *Cell biology of extracellular matrix*. Springer, 2013.

[9] Hynes RO: The extracellular matrix: not just pretty fibrils. *Science (New York, N.Y.)*, 2009, 326(5957):1216–1219.

[10] O'Connor C, Adams JU, and Fairman J: *Essentials of cell biology*. Cambridge: NPG Education, 2010.

[11] Alberts B, Bray D, Hopkin K, Johnson A, Lewis J, Raff M, Roberts K, and Walter P: *Essential cell biology*. Garland Science, New York, 2013.

[12] Soderling TR, Chang B, and Brickey D: Cellular signaling through multifunctional Ca^{2+}/calmodulin-dependent protein kinase II. *Journal of Biological Chemistry*, 2001, 276(6):3719–3722.

[13] Lutolf MP: Integration column: artificial ECM: expanding the cell biology toolbox in 3D. *Integrative Biology*, 2009, 1(3):235–241.

[14] Desmet T, Morent R, Geyter ND, Leys C, Schacht E, and Dubruel P: Nonthermal plasma technology as a versatile strategy for polymeric biomaterials surface modification: a review. *Biomacromolecules*, 2009, 10(9):2351–2378.

[15] Morent R, De Geyter N, Desmet T, Dubruel P, and Leys C: Plasma surface modification of biodegradable polymers: a review. *Plasma Processes and Polymers*, 2011, 8(3): 171–190.

[16] Cools P, De Geyter N, and Morent R: PLA enhanced via plasma technology: a review. *New Developments in Polylactic Acid Research*, Nova Science, New York, 2015, pp. 79–110.

[17] Cools P, Morent R, and De Geyter N: Plasma modified textiles for biomedical applications. *Advances in Bioengineering*, InTechOpen, Rijeka, 2014, pp.117–148.

[18] Elloumi-Hannachi I, Yamato M, and Okano T: Cell sheet engineering: a unique nanotechnology for scaffold-free tissue reconstruction with clinical applications in regenerative medicine. *Journal of internal medicine*, 2010, 267(1):54–70.

[19] Drury JL, and Mooney DJ: Hydrogels for tissue engineering: scaffold design variables and applications. *Biomaterials*, 2003, 24(24):4337–4351.

[20] Billiet T, Vandenhaute M, Schelfhout J, Van Vlierberghe S, and Dubruel P: A review of trends and limitations in hydrogel-rapid prototyping for tissue engineering. *Biomaterials*, 2012, 33(26):6020–6041.

[21] Lutolf M, and Hubbell J: Synthetic biomaterials as instructive extracellular microenvironments for morphogenesis in tissue engineering. *Nature Biotechnology*, 2005, 23(1):47–55.

[22] Hubbell JA: Biomaterials in tissue engineering. *Nature Biotechnology*, 1995, 13(6):565–576.

[23] Hodde J: Naturally occurring scaffolds for soft tissue repair and regeneration. *Tissue Engineering*, 2002, 8(2):295–308.

[24] Balakrishnan B, and Banerjee R: Biopolymer-based hydrogels for cartilage tissue engineering. *Chemical Reviews*, 2011, 111(8):4453–4474.

[25] Hollister SJ: Porous scaffold design for tissue engineering. *Nature Materials*, 2005, 4(7):518–524.

[26] Mironov V, Visconti RP, Kasyanov V, Forgacs G, Drake CJ, and Markwald RR: Organ printing: tissue spheroids as building blocks. *Biomaterials*, 2009, 30(12):2164–2174.

[27] Dhandayuthapani B, Yoshida Y, Maekawa T, and Kumar DS: Polymeric scaffolds in tissue engineering application: a review. *International Journal of Polymer Science*, 2011, 2011:1–19.

[28] Cheung H-Y, Lau K-T, Lu T-P, and Hui D: A critical review on polymer-based bio-engineered materials for scaffold development. *Composites Part B: Engineering*, 2007, 38(3):291–300.

[29] Tanner K: Bioactive composites for bone tissue engineering. Proceedings of the Institution of Mechanical Engineers. *Part H: Journal of Engineering in Medicine*, 2010, 224(12):1359–1372.

[30] Yeong W-Y, Chua C-K, Leong K-F, and Chandrasekaran M: Rapid prototyping in tissue engineering: challenges and potential. *Trends in Biotechnology*, 2004, 22(12):643–652.

[31] Claase MB, de Bruijn JD, Grijpma DW, and Feijen J: Ectopic bone formation in cell-seeded poly(ethylene oxide)/poly(butylene terephthalate) copolymer scaffolds of varying porosity. *Journal of Materials Science: Materials in Medicine*, 2007, 18(7):1299–1307.

[32] Quirk RA, France RM, Shakesheff KM, and Howdle SM: Supercritical fluid technologies and tissue engineering scaffolds. *Current Opinion in Solid State and Materials Science*, 2004, 8(3–4):313–321.

[33] Tai H, Mather ML, Howard D, Wang W, White LJ, Crowe JA, Morgan SP, Chandra A, Williams DJ, and Howdle SM: Control of pore size and structure of tissue engineering scaffolds produced by supercritical fluid processing. *European Cells & Materials*, 2007, 14:64–77.

[34] Bártolo PJ, Domingos M, Patrício T, Cometa S, and Mironov V: Biofabrication Strategies for Tissue Engineering. *Advances on Modeling in Tissue Engineering*, Springer, Dordrecht Heidelberg London New York, 2011, pp. 137–176.

[35] Abdelaal OA, and Darwish SM: *Characterization and Development of Biosystems and Biomaterials*. Springer, 2013, pp. 33–54.

[36] Declercq HA, Gorski TL, Tielens SP, Schacht EH, and Cornelissen MJ: Encapsulation of osteoblast seeded microcarriers into injectable, photopolymerizable three-dimensional scaffolds based on d,l-lactide and ε-caprolactone. *Biomacromolecules*, 2005, 6(3): 1608–1614.

[37] Lippens E, Swennen I, Gironès J, Declercq H, Vertenten G, Vlaminck L, Gasthuys F, Schacht E, and Cornelissen R: Cell survival and proliferation after encapsulation in a chemically modified Pluronic® F127 hydrogel. *Journal of Biomaterials Applications*, 2013, 27(7):828–839.

[38] Li D, and Xia Y: Electrospinning of nanofibers: reinventing the wheel? *Advanced Materials*, 2004, 16(14):1151–1170.

[39] Pham QP, Sharma U, and Mikos AG: Electrospinning of polymeric nanofibers for tissue engineering applications: a review. *Tissue Engineering*, 2006, 12(5):1197–1211.

[40] Cipitria A, Skelton A, Dargaville T, Dalton P, and Hutmacher D: Design, fabrication and characterization of PCL electrospun scaffolds—a review. *Journal of Materials Chemistry*, 2011, 21(26):9419–9453.

[41] Agarwal S, Wendorff JH, and Greiner A: Use of electrospinning technique for biomedical applications. *Polymer*, 2008, 49(26):5603–5621.

[42] Martins A, Reis R, and Neves N: Electrospinning: processing technique for tissue engineering scaffolding. *International Materials Reviews*, 2008, 53(5):257–274.

[43] Ma Z, Kotaki M, Inai R, and Ramakrishna S: Potential of nanofiber matrix as tissue-engineering scaffolds. *Tissue Engineering*, 2005, 11(1–2):101–109.

[44] Jacobs T, Morent R, De Geyter N, Dubruel P, and Leys C: Plasma surface modification of biomedical polymers: influence on cell-material interaction. *Plasma Chemistry and Plasma Processing*, 2012, 32(5):1039–1073.

[45] Cheruthazhekatt S, Černák M, Slavíček P, and Havel J: Gas plasmas and plasma modified materials in medicine. *Journal of Applied Biomedicine*, 2010, 8(2):55–66.

[46] Chu PK, Chen J, Wang L, and Huang N: Plasma-surface modification of biomaterials. *Materials Science and Engineering: R: Reports*, 2002, 36(5):143–206.

[47] Siow KS, Britcher L, Kumar S, Griesser HJ, Plasma methods for the generation of chemically reactive surfaces for biomolecule immobilization and cell colonization—a review. *Plasma Processes and Polymers*, 2006, 3(6-7):392–418.

[48] Noeske M, Degenhardt J, Strudthoff S, and Lommatzsch U: Plasma jet treatment of five polymers at atmospheric pressure: surface modifications and the relevance for adhesion. *International Journal of Adhesion and Adhesives*, 2004, 24(2):171–177.

[49] Hegemann D, Brunner H, and Oehr C: Plasma treatment of polymers for surface and adhesion improvement. *Nuclear Instruments and Methods in Physics Research Section B: Beam Interactions with Materials and Atoms*, 2003, 208:281–286.

[50] Chim H, Ong JL, Schantz JT, Hutmacher DW, and Agrawal C: Efficacy of glow discharge gas plasma treatment as a surface modification process for three-dimensional poly(d,l-lactide) scaffolds. *Journal of Biomedical Materials Research Part A*, 2003, 65(3): 327–335.

[51] Köse GT, Kenar H, Hasırcı N, and Hasırcı V: Macroporous poly(3-hydroxybutyrate-co-3-hydroxyvalerate) matrices for bone tissue engineering. *Biomaterials*, 2003, 24(11): 1949–1958.

[52] Claase MB, Grijpma DW, Mendes SC, de Bruijn JD, and Feijen J: Porous PEOT/PBT scaffolds for bone tissue engineering: preparation, characterization, and in vitro bone

marrow cell culturing. *Journal of Biomedical Materials Research Part A*, 2003, 64(2):291–300.

[53] Yang J, Shi G, Bei J, Wang S, Cao Y, Shang Q, Yang G, and Wang W: Fabrication and surface modification of macroporous poly(l-lactic acid) and poly(l-lactic-co-glycolic acid) (70/30) cell scaffolds for human skin fibroblast cell culture. *Journal of Biomedical Materials Research*, 2002, 62(3):438–446.

[54] Choi Y-R, Kwon J-S, Song D-H, Choi EH, Lee Y-K, Kim K-N, and Kim K-M: Surface modification of biphasic calcium phosphate scaffolds by non-thermal atmospheric pressure nitrogen and air plasma treatment for improving osteoblast attachment and proliferation. *Thin Solid Films*, 2013, 547:235–240.

[55] Ring A, Langer S, Schaffran A, Stricker I, Awakowicz P, Steinau H-U, and Hauser J: Enhanced neovascularization of dermis substitutes via low-pressure plasma-mediated surface activation. *Burns*, 2010, 36(8):1222–1227.

[56] Shah AR, Cornejo A, Guda T, Sahar DE, Stephenson SM, Chang S, Krishnegowda NK, Sharma R, and Wang HT: Differentiated adipose-derived stem cell cocultures for bone regeneration in polymer scaffolds in vivo. *Journal of Craniofacial Surgery*, 2014, 25(4):1504–1509.

[57] Safinia L, Datan N, Höhse M, Mantalaris A, and Bismarck A: Towards a methodology for the effective surface modification of porous polymer scaffolds. *Biomaterials*, 2005, 26(36):7537–7547.

[58] Wan Y, Tu C, Yang J, Bei J, and Wang S: Influences of ammonia plasma treatment on modifying depth and degradation of poly(l-lactide) scaffolds. *Biomaterials*, 2006, 27(13):2699–2704.

[59] Safinia L, Wilson K, Mantalaris A, and Bismarck A: Atmospheric plasma treatment of porous polymer constructs for tissue engineering applications. *Macromolecular Bioscience*, 2007, 7(3):315–327.

[60] Han I, Vagaska B, Lee DH, Park BJ, Shin S-W, Kim JK, Lee SJ, and Park J-C: Pre-wetting process with helium atmospheric pressure glow discharge for three-dimensional porous scaffold. *Macromolecular Research*, 2011, 19(7):711–715.

[61] Bak T-Y, Kook M-S, Jung S-C, and Kim B-H: Biological effect of gas plasma treatment on CO_2 gas foaming/salt leaching fabricated porous polycaprolactone scaffolds in bone tissue engineering. *Journal of Nanomaterials*, 2014, 2014:1–6.

[62] Sardella E, Gristina R, Senesi GS, d'Agostino R, and Favia P: Homogeneous and micro-patterned plasma-deposited PEO-like coatings for biomedical surfaces. *Plasma Processes and Polymers*, 2004, 1(1):63–72.

[63] Trizio I, Intranuovo F, Gristina R, Dilecce G, and Favia P: He/O_2 atmospheric pressure plasma jet treatments of pcl scaffolds for tissue engineering and regenerative

medicine. *Plasma Processes and Polymers*, 2015, Early View Publication; these are NOT the final page numbers, use DOI for citation DOI: 10.1002/ppap.201500104.

[64] Yang S, Leong K-F, Du Z, and Chua C-K: The design of scaffolds for use in tissue engineering. Part II. Rapid prototyping techniques. *Tissue Engineering*, 2002, 8(1):1–11.

[65] Landers R, Hübner U, Schmelzeisen R, and Mülhaupt R: Rapid prototyping of scaffolds derived from thermoreversible hydrogels and tailored for applications in tissue engineering. *Biomaterials*, 2002, 23(23):4437–4447.

[66] Hutmacher DW: Scaffolds in tissue engineering bone and cartilage. *Biomaterials*, 2000, 21(24):2529–2543.

[67] Wagner M, Kiapur N, Wiedmann-Al-Ahmad M, Hübner U, Al-Ahmad A, Schön R, Schmelzeisen R, Mülhaupt R, and Gellrich NC: Comparative in vitro study of the cell proliferation of ovine and human osteoblast-like cells on conventionally and rapid prototyping produced scaffolds tailored for application as potential bone replacement material. *Journal of Biomedical Materials Research Part A*, 2007, 83A(4):1154–1164.

[68] Moroni L, Curti M, Welti M, Korom S, Weder W, De Wijn JR, and Van Blitterswijk CA: Anatomical 3D fiber-deposited scaffolds for tissue engineering: designing a neo-trachea. *Tissue Engineering*, 2007, 13(10):2483–2493.

[69] Yildirim ED, Besunder R, Guceri S, Allen F, and Sun W: Fabrication and plasma treatment of 3D polycaprolactane tissue scaffolds for enhanced cellular function. *Virtual and Physical Prototyping*, 2008, 3(4):199–207.

[70] Yildirim ED, Pappas D, Güçeri S, and Sun W: Enhanced cellular functions on poly-caprolactone tissue scaffolds by O_2 plasma surface modification. *Plasma Processes and Polymers*, 2011, 8(3):256–267.

[71] Jacobs T, Declercq H, De Geyter N, Cornelissen R, Dubruel P, Leys C, and Morent R: Improved cell adhesion to flat and porous plasma-treated poly-ε-caprolactone samples. *Surface and Coatings Technology*, 2013, 232:447–455.

[72] Jacobs T, Morent R, De Geyter N, Desmet T, Dubruel P, and Leys C: Visualization of the penetration depth of plasma in three-dimensional porous PCL scaffolds. *IEEE Transactions on Plasma Science*, 2011, 39(11(1)):2792–2793.

[73] Baker SC, Atkin N, Gunning PA, Granville N, Wilson K, Wilson D, and Southgate J: Characterisation of electrospun polystyrene scaffolds for three-dimensional in vitro biological studies. *Biomaterials*, 2006, 27(16):3136–3146.

[74] France R, and Short R: Plasma treatment of polymers effects of energy transfer from an argon plasma on the surface chemistry of poly(styrene), low density poly(ethylene), poly(propylene) and poly(ethylene terephthalate). *Journal of the Chemical Society, Faraday Transactions*, 1997, 93(17):3173–3178.

[75] Park H, Lee KY, Lee SJ, Park KE, and Park WH: Plasma-treated poly(lactic-co-glycolic acid) nanofibers for tissue engineering. *Macromolecular Research*, 2007, 15(3):238–243.

[76] Yang F, Wolke JGC, and Jansen JA: Biomimetic calcium phosphate coating on electrospun poly(ε-caprolactone) scaffolds for bone tissue engineering. *Chemical Engineering Journal*, 2008, 137(1):154–161.

[77] Prabhakaran MP, Venugopal J, Chan CK, and Ramakrishna S: Surface modified electrospun nanofibrous scaffolds for nerve tissue engineering. *Nanotechnology*, 2008, 19(45):455102.

[78] Park H, Lee JW, Park KE, Park WH, and Lee KY: Stress response of fibroblasts adherent to the surface of plasma-treated poly(lactic-co-glycolic acid) nanofiber matrices. *Colloids and Surfaces B: Biointerfaces*, 2010, 77(1):90–95.

[79] de Valence S, Tille J-C, Chaabane C, Gurny R, Bochaton-Piallat M-L, Walpoth BH, and Möller M: Plasma treatment for improving cell biocompatibility of a biodegradable polymer scaffold for vascular graft applications. *European Journal of Pharmaceutics and Biopharmaceutics*, 2013, 85(1):78–86.

[80] Yan D, Jones J, Yuan X, Xu X, Sheng J, Lee JM, Ma G, and Yu Q: Plasma treatment of electrospun PCL random nanofiber meshes (NFMs) for biological property improvement. *Journal of Biomedical Materials Research Part A*, 2013, 101(4):963–972.

[81] Dolci LS, Quiroga SD, Gherardi M, Laurita R, Liguori A, Sanibondi P, Fiorani A, Calzà L, Colombo V, and Focarete ML: Carboxyl surface functionalization of poly(L-lactic acid) electrospun nanofibers through atmospheric non-thermal plasma affects fibroblast morphology. *Plasma Processes and Polymers*, 2014, 11(3):203–213.

[82] Sankar D, Shalumon K, Chennazhi K, Menon D, and Jayakumar R: Surface plasma treatment of poly(caprolactone) micro, nano, and multiscale fibrous scaffolds for enhanced osteoconductivity. *Tissue Engineering Part A*, 2014, 20(11–12):1689–1702.

[83] Jeon HJ, Lee H, and Kim GH: Nano-sized surface patterns on electrospun microfibers fabricated using a modified plasma process for enhancing initial cellular activities. *Plasma Processes and Polymers*, 2014, 11(2):142–148.

[84] Venugopal J, Low S, Choon AT, Kumar AB, and Ramakrishna S: Electrospun-modified nanofibrous scaffolds for the mineralization of osteoblast cells. *Journal of Biomedical Materials Research Part A*, 2008, 85(2):408–417.

[85] Cheng Q, Lee BL-P, Komvopoulos K, Yan Z, and Li S: Plasma surface chemical treatment of electrospun poly(L-lactide) microfibrous scaffolds for enhanced cell adhesion, growth, and infiltration. *Tissue Engineering Part A*, 2013, 19(9–10):1188–1198.

[86] Yildirim ED, Ayan H, Vasilets VN, Fridman A, Guceri S, and Sun W: Effect of dielectric barrier discharge plasma on the attachment and proliferation of osteoblasts cul-

tured over poly(ε-caprolactone) scaffolds. *Plasma Processes and Polymers*, 2008, 5(1): 58–66.

[87] Santerre JP, Woodhouse K, Laroche G, and Labow RS: Understanding the biodegradation of polyurethanes: from classical implants to tissue engineering materials. *Biomaterials*, 2005, 26(35):7457–7470.

[88] Zandén C, Voinova M, Gold J, Mörsdorf D, Bernhardt I, and Liu J: Surface characterisation of oxygen plasma treated electrospun polyurethane fibres and their interaction with red blood cells. *European Polymer Journal*, 2012, 48(3):472–482.

[89] Zandén C, Erkenstam NH, Padel T, Wittgenstein J, Liu J, and Kuhn HG: Stem cell responses to plasma surface modified electrospun polyurethane scaffolds. *Nanomedicine: Nanotechnology, Biology and Medicine*, 2014, 10(5):949–958.

[90] Ardeshirylajimi A, Dinarvand P, Seyedjafari E, Langroudi L, Adegani FJ, and Soleimani M: Enhanced reconstruction of rat calvarial defects achieved by plasma-treated electrospun scaffolds and induced pluripotent stem cells. *Cell and Tissue Research*, 2013, 354(3):849–860.

[91] Baek HS, Park YH, Ki CS, Park J-C, and Rah DK: Enhanced chondrogenic responses of articular chondrocytes onto porous silk fibroin scaffolds treated with microwave-induced argon plasma. *Surface and Coatings Technology*, 2008, 202(22):5794–5797.

[92] Cheon YW, Lee WJ, Baek HS, Lee YD, Park JC, Park YH, Ki CS, Chung KH, and Rah DK: Enhanced chondrogenic responses of human articular chondrocytes onto silk fibroin/wool keratose scaffolds treated with microwave-induced argon plasma. *Artificial Organs*, 2010, 34(5):384–392.

[93] Barry JJA, Silva MMCG, Shakesheff KM, Howdle SM, and Alexander MR: Using plasma deposits to promote cell population of the porous interior of three-dimensional poly(d,l-lactic acid) tissue-engineering scaffolds. *Advanced Functional Materials*, 2005, 15(7):1134–1140.

[94] Intranuovo F, Howard D, White LJ, Johal RK, Ghaemmaghami AM, Favia P, Howdle SM, Shakesheff KM, Alexander MR, Uniform cell colonization of porous 3-D scaffolds achieved using radial control of surface chemistry. *Acta Biomaterialia*, 2011, 7(9): 3336–3344.

[95] Intranuovo F, Gristina R, Brun F, Mohammadi S, Ceccone G, Sardella E, Rossi F, Tromba G, and Favia P: Plasma modification of PCL porous scaffolds fabricated by solvent-casting/particulate-leaching for tissue engineering. *Plasma Processes and Polymers*, 2014, 11(2):184–195.

[96] Ju YM, Park K, Son JS, Kim JJ, Rhie JW, and Han DK: Beneficial effect of hydrophilized porous polymer scaffolds in tissue-engineered cartilage formation. *Journal of Biomedical Materials Research Part B: Applied Biomaterials*, 2008, 85(1):252–260.

[97] Park K, Jung HJ, Kim J-J, and Han DK: Effect of surface-activated PLLA scaffold on apatite formation in simulated body fluid. *Journal of Bioactive and Compatible Polymers,* 2010, 25(1):27–39.

[98] Demirbilek ME, Demirbilek M, Karahaliloğlu Z, Erdal E, Vural T, Yalçın E, Sağlam N, and Denkbaş EB: Oxidative stress parameters of L929 cells cultured on plasma-modified PDLLA scaffolds. *Applied Biochemistry and Biotechnology,* 2011, 164(6):780–792.

[99] Shen H, Hu X, Yang F, Bei J, and Wang S: The bioactivity of rhBMP-2 immobilized poly(lactide-co-glycolide) scaffolds. *Biomaterials,* 2009, 30(18):3150–3157.

[100] Woo YI, Lee MH, Kim H-L, Park J-C, Han D-W, Kim JK, Tsubaki K, Chung K-H, Hyun SO, and Yang Y-I: Cellular responses and behaviors of adipose-derived stem cells onto β-glucan and PLGA composites surface-modified by microwave-induced argon plasma. *Macromolecular Research,* 2010, 18(1):90–93.

[101] Hu Y, Winn SR, Krajbich I, and Hollinger JO: Porous polymer scaffolds surface-modified with arginine-glycine-aspartic acid enhance bone cell attachment and differentiation in vitro. *Journal of Biomedical Materials Research Part A,* 2003, 64(3):583–590.

[102] Zheng L, Yang F, Shen H, Hu X, Mochizuki C, Sato M, Wang S, and Zhang Y: The effect of composition of calcium phosphate composite scaffolds on the formation of tooth tissue from human dental pulp stem cells. *Biomaterials,* 2011, 32(29):7053–7059.

[103] Berneel E, Desmet T, Declercq H, Dubruel P, and Cornelissen M: Double protein-coated poly-ε-caprolactone scaffolds: successful 2D to 3D transfer. *Journal of Biomedical Materials Research Part A,* 2012, 100(7):1783–1791.

[104] Declercq HA, Desmet T, Dubruel P, and Cornelissen MJ: The role of scaffold architecture and composition on the bone formation by adipose-derived stem cells. *Tissue Engineering Part A,* 2013, 20(1–2):434–444.

[105] Declercq HA, Desmet T, Berneel EE, Dubruel P, and Cornelissen MJ: Synergistic effect of surface modification and scaffold design of bioplotted 3-D poly-ε-caprolactone scaffolds in osteogenic tissue engineering. *Acta Biomaterialia,* 2013, 9(8):7699–7708.

[106] Van Bael S, Desmet T, Chai YC, Pyka G, Dubruel P, Kruth J-P, and Schrooten J: In vitro cell-biological performance and structural characterization of selective laser sintered and plasma surface functionalized polycaprolactone scaffolds for bone regeneration. *Materials Science and Engineering: C,* 2013, 33(6):3404–3412.

[107] Sousa I, Mendes A, Pereira RF, and Bártolo PJ: Collagen surface modified poly(ε-caprolactone) scaffolds with improved hydrophilicity and cell adhesion properties. *Materials Letters,* 2014, 134:263–267.

[108] Nebe B, Cornelsen M, Quade A, Weissmann V, Kunz F, Ofe S, Schroeder K, Finke B, Seitz H, and Bergemann C: Osteoblast Behavior In Vitro in Porous Calcium Phos-

phate Composite Scaffolds, Surface Activated with a Cell Adhesive Plasma Polymer Layer. Materials science forum. Trans Tech Publications, 2012, pp. 566–571.

[109] Owen R, Sherborne C, Paterson T, Green NH, Reilly GC, and Claeyssens F: Emulsion templated scaffolds with tunable mechanical properties for bone tissue engineering. *Journal of the Mechanical Behavior of Biomedical Materials*, 2015, 54:159–172.

[110] Kim BH, Myung SW, Jung SC, and Ko YM: Plasma surface modification for immobilization of bone morphogenic protein-2 on polycaprolactone scaffolds. *Japanese Journal of Applied Physics*, 2013, 52(11S):11NF01.

[111] Wang S, Zhang Y, Wang H, and Dong Z: Preparation, characterization and biocompatibility of electrospinning heparin-modified silk fibroin nanofibers. *International Journal of Biological Macromolecules*, 2011, 48(2):345–353

[112] Cheng Q, Komvopoulos K, and Li S: Plasma-assisted heparin conjugation on electrospun poly(l-lactide) fibrous scaffolds. *Journal of Biomedical Materials Research Part A*, 2014, 102(5):1408–1414.

[113] He W, Ma Z, Yong T, Teo WE, and Ramakrishna S: Fabrication of collagen-coated biodegradable polymer nanofiber mesh and its potential for endothelial cells growth. *Biomaterials*, 2005, 26(36):7606–7615.

[114] Delgado-Rivera R, Griffin J, Ricupero CL, Grumet M, Meiners S, and Uhrich KE: Microscale plasma-initiated patterning of electrospun polymer scaffolds. *Colloids and Surfaces B: Biointerfaces*, 2011, 84(2):591–596.

[115] Zander NE, Orlicki JA, Rawlett AM, Beebe Jr. TP, Quantification of protein incorporated into electrospun polycaprolactone tissue engineering scaffolds. *ACS Applied Materials & Interfaces*, 2012, 4(4):2074–2081.

[116] Guex A, Frobert A, Valentin J, Fortunato G, Hegemann D, Cook S, Carrel T, Tevaearai H, and Giraud M-N: Plasma-functionalized electrospun matrix for biograft development and cardiac function stabilization. *Acta Biomaterialia*, 2014, 10(7):2996–3006.

[117] Nuhiji E, Wong CS, Sutti A, Lin T, Kirkland M, and Wang X: Biofunctionalization of 3D nylon 6, 6 scaffolds using a two-step surface modification. *ACS Applied Materials & Interfaces*, 2012, 4(6):2912–2919.

[118] Shabani I, Haddadi-Asl V, Seyedjafari E, Babaeijandaghi F, and Soleimani M: Improved infiltration of stem cells on electrospun nanofibers. *Biochemical and Biophysical Research Communications*, 2009, 382(1):129–133.

[119] Jeong L, Yeo I-S, Kim HN, Yoon YI, Jang DH, Jung SY, Min B-M, and Park WH: Plasma-treated silk fibroin nanofibers for skin regeneration. *International Journal of Biological Macromolecules*, 2009, 44(3):222–228.

[120] Park BJ, Seo HJ, Kim J, Kim H-L, Kim JK, Choi JB, Han I, Hyun SO, Chung K-H, and Park J-C: Cellular responses of vascular endothelial cells on surface modified polyur-

ethane films grafted electrospun PLGA fiber with microwave-induced plasma at atmospheric pressure. *Surface and Coatings Technology*, 2010, 205:S222–S226.

[121] Yao C, Li X, Neoh K, Shi Z, and Kang E: Surface modification and antibacterial activity of electrospun polyurethane fibrous membranes with quaternary ammonium moieties. *Journal of Membrane Science*, 2008, 320(1):259–267.

[122] López-Pérez PM, Da Silva RM, Sousa RA, Pashkuleva I, and Reis RL: Plasma-induced polymerization as a tool for surface functionalization of polymer scaffolds for bone tissue engineering: an in vitro study. *Acta Biomaterialia*, 2010, 6(9):3704–3712.

[123] Chen J-P, and Su C-H: Surface modification of electrospun PLLA nanofibers by plasma treatment and cationized gelatin immobilization for cartilage tissue engineering. *Acta Biomaterialia*, 2011, 7(1):234–243.

[124] Jia J, Duan YY, Yu J, and Lu JW: Preparation and immobilization of soluble eggshell membrane protein on the electrospun nanofibers to enhance cell adhesion and growth. *Journal of Biomedical Materials Research Part A*, 2008, 86(2):364–373.

[125] Park K, Jung HJ, Kim J-J, Ahn K-D, Han DK, and Ju YM: Acrylic acid-grafted hydrophilic electrospun nanofibrous poly(l-lactic acid) scaffold. *Macromolecular Research*, 2006, 14(5):552–558.

[126] Paletta JRJ, Bockelmann S, Walz A, Theisen C, Wendorff JH, Greiner A, Fuchs-Winkelmann S, and Schofer MD: RGD-functionalisation of PLLA nanofibers by surface coupling using plasma treatment: influence on stem cell differentiation. *Journal of Materials Science: Materials in Medicine*, 2010, 21(4):1363–1369.

[127] Seyedjafari E, Soleimani M, Ghaemi N, and Shabani I: Nanohydroxyapatite-coated electrospun poly(l-lactide) nanofibers enhance osteogenic differentiation of stem cells and induce ectopic bone formation. *Biomacromolecules*, 2010, 11(11):3118–3125.

[128] Santos M, Pashkuleva I, Alves C, Gomes ME, Fuchs S, Unger RE, Reis R, and Kirkpatrick CJ: Surface-modified 3D starch-based scaffold for improved endothelialization for bone tissue engineering. *Journal of Materials Chemistry*, 2009, 19(24):4091–4101.

[129] Ma Z, He W, Yong T, and Ramakrishna S: Grafting of gelatin on electrospun poly(caprolactone) nanofibers to improve endothelial cell spreading and proliferation and to control cell orientation. *Tissue Engineering*, 2005, 11(7–8):1149–1158.

[130] Ghaedi M, Soleimani M, Shabani I, Duan Y, and Lotfi A: Hepatic differentiation from human mesenchymal stem cells on a novel nanofiber scaffold. *Cellular and Molecular Biology Letters*, 2012, 17(1):89–106.

[131] Ai J, Heidari KS, Ghorbani F, Ejazi F, Biazar E, Asefnejad A, Pourshamsian K, and Montazeri M: Fabrication of coated-collagen electrospun PHBV nanofiber film by plasma method and its cellular study. *Journal of Nanomaterials*, 2011, 2011:1–8.

[132] Reprinted from an Illustration from Anatomy & Physiology, Connexions Web site. http://cnx.org/content/col11496/1.6/, Jun 19, 2013 under the Creative Commons Attribution 3.0 Unported license.

[133] Reprinted from Soft Matter, 2007, Girard PP, Cavalcanti-Adam EA, Kemkemer R, and Spatz JP: Cellular chemomechanics at interfaces: sensing, integration and response, Copyright © 2007, with permission from Royal Society of Chemistry.

[134] Reprinted from Nature Nanotechnology, 2010, Dvir T, Timko BP, Kohane DS, and Langer R: Nanotechnological strategies for engineering complex tissues, Copyright © 2010, with permission from Nature Publishing Group.

[135] Reprinted from Chemical Society Reviews, 2009, Place ES, George JH, Williams CK, and Stevens MM: Synthetic polymer scaffolds for tissue engineering. Copyright © 2009, with permission from Royal Society of Chemistry.

[136] Reprinted from Plasma Processes and Polymers, 2014, Intranuovo F, Gristina R, Brun F, Mohammadi S, Ceccone G, Sardella E, Rossi F, Tromba G, and Favia P: Plasma modification of PCL porous scaffolds fabricated by solvent-casting/particulate-leaching for tissue engineering. Copyright © 2014, with permission from Wiley.

[137] Reprinted from Surface and Coatings Technology, 2013, Jacobs T, Declercq H, De Geyter N, Cornelissen R, Dubruel P, Leys C, and and Morent R: Improved cell adhesion to flat and porous plasma-treated poly-ε-caprolactone samples. Copyright © 2013, with permission from Elsevier.

[138] Reprinted from Plasma Processes and Polymers, 2007, Yildirim ED, Ayan H, Vasilets VN, Fridman A, Guceri S, and Sun W: Effect of dielectric barrier discharge plasma on the attachment and proliferation of osteoblasts cultured over poly(ε-caprolactone) scaffolds. Copyright © 2007, with permission from Wiley.

[139] Reprinted from Plasma Processes and Polymers, 2013, Jeon HJ, Lee H, and Kim GH: Nano-sized surface patterns on electrospun microfibers fabricated using a modified plasma process for enhancing initial cellular activities. Copyright © 2013, with permission from Wiley.

[140] Reprinted from Advanced Functional Materials, 2005, Barry JJA, Silva MMCG, Shakesheff KM, Howdle SM, and Alexander MR: Using plasma deposits to promote cell population of the porous interior of three-dimensional poly(d,l-lactic acid) tissue-engineering scaffolds. Copyright © 2005, with permission from Wiley.

[141] Reprinted from Journal of Biomedical Materials Research, 2012, Berneel E, Desmet T, Declercq H, Dubruel P, and Cornelissen M: Double protein-coated poly-ε-caprolactone scaffolds: successful 2D to 3D transfer. Copyright © 2012, with permission from Wiley.

Chapter 8

Plasma-Enhanced Vapor Deposition Process for the Modification of Textile Materials

Sheila Shahidi, Jakub Wiener and Mahmood Ghoranneviss

Additional information is available at the end of the chapter

http://dx.doi.org/10.5772/62832

Abstract

Nowadays many techniques are used for the surface modification of fabrics and textiles. Two fundamental techniques based on vacuum deposition are known as chemical vapor deposition (CVD) and physical vapor deposition (PVD). In this chapter, the effect of plasma-enhanced physical and chemical vapor deposition on textile surfaces is investigated and explained.

Keywords: Chemical vapor deposition, textile, fabric, modification

1. Introduction

Textile industries have been experiencing fast development with versatile products for a wide spectrum of applications being developed through technological innovations. The surfaces of textiles offer a platform for functional modifications to meet specific requirements for various applications. Surface modification of textiles gives the textile product the desired properties. Many techniques ranging from conventional wet treatments to biological methods are used for the surface modification of textiles.

In recent years, a great attention to the functionalization of textile surfaces has been attracted with the help of new technologies, such as high-energy beam processes, vapor deposition, and nanoparticle coatings.

The coupling of electromagnetic power into a process gas volume generates the plasma medium comprising a dynamic mix of ions, electrons, neutrons, photons, free radicals, metastable excited species, and molecular and polymeric fragments.

Vapor deposition (either physical or chemical) is a coating process, where the coating material is condensed from vapor phase, forming a thin film, or modifies the surface of the substrate

by creating functional groups on the surface. Different types of deposition techniques are used to meet this criterion. Most of these techniques, including vacuum deposition, are used to minimize unwanted reaction with the free space and to shape the film composition easily. Two fundamental techniques based on vacuum deposition are known as chemical vapor deposition (CVD) and physical vapor deposition (PVD).

In the CVD technique, thin film design is achieved by chemical reaction between precursors. The reaction needs hot substrate or high-temperature medium. Due to the need of gas phase for surface coating, the CVD technique is suitable for large and complex-shaped surfaces. Physical processes such as evaporation and sputtering are used in the PVD technique.

By contacting effects of the gases, numerous sputtering methods are used to form plasma. Different surfaces can be coated with deposition flux using PVD methods. To select the deposition method, some parameters are very important, such as thin film material, substrate type, uniformity, and control of the thickness.

Conventional finishing techniques applied to textiles (dyeing, stain repellence, flame retardance, antibacterial treatments, wrinkle recovery improvement, sizing, desizing, washing, and so on), generally wet chemical process, produce a lot of wastewater. Vapor phase treatment is an eco-friendly and dry technology offering an interesting replacement to add new abilities and functionalities. Some abilities such as antibacterial properties, electrical and mechanical properties, hydrophilicity, water repellence, and biocompatibility of textiles and fibers can be modified. On the other hand, bulk properties and touch of the textiles remain unaffected [1].

In this chapter, the effect of both plasma-enhanced physical and chemical vapor deposition processes on textile surfaces is investigated and explained.

2. Vapor deposition

In vapor deposition process, materials in a vapor state are condensed through condensation, chemical reaction, or conversion to form a solid material. These processes are used for coatings to alter the mechanical, electrical, thermal, optical, corrosion resistance, and wear properties of the substrates. Vapor deposition processes usually take place within a vacuum chamber [2].

Physical vapor deposition (PVD) and chemical vapor deposition (CVD) are two categories of vapor deposition processes.

Vapor deposition (either physical or chemical) is a coating process where the coating material is condensed in vacuum at the substrate from vapor phase, forming a thin film ($\leq$10 μm in the case of physical deposition and $\leq$1000 μm in the case of chemical deposition). Sometimes, the deposited material further reacts with the gaseous substances to form a final compound coating. Generally, metals and nonmetals can be deposited; a certain case depends on the applied method [3].

It is well known that thin film coating can be realized with a thickness of a few nanometers. Different types of deposition techniques are used to meet this criterion. All these techniques, including vacuum, are used to minimize unwanted reaction with the free space and to shape the film composition easily. Two main techniques based on vacuum deposition are known as CVD and PVD. In the CVD technique, thin film design is realized by chemical reaction between precursors. In the PVD technique, some processes such as sputtering, thermal evaporation, molecular beam epitaxy, and electron beam evaporation are used. In sputtering techniques, by contacting effects of the gases in the background, plasma is formed. PVD techniques are very useful and efficient for surface modifications [4].

2.1. Chemical vapor deposition (CVD)

CVD is a chemical process used to fabricate high-quality and high-performance solid materials. The process has many applications in different industries. In this process, volatile precursors can react or decompose on the substrate surface to create the proper coating. Some by-products that are produced through this process should be removed through the chamber by gas flow technique.

Some microfabrication processes, such as polycrystalline, monocrystalline, epitaxial, and amorphous, are used in CVD techniques to deposit materials in different forms. Silicon compounds, carbon nanofibers, carbon nanotubes, carbon fiber, fluorocarbons, filaments, titanium nitride, and various high-k dielectrics can be modified by this method. CVD is also used to produce synthetic diamonds [5].

CVD is practiced in a variety of formats. These processes generally differ in the means by which chemical reactions are initiated and classified by operating pressure and are listed as below:

1. APCVD (Atmospheric pressure CVD).

2. LPCVD (Low-pressure CVD) – In this method, unwanted gas-phase reactions reduced and film uniformity also improved.

3. UHVCVD (Ultrahigh vacuum CVD) – In this method, pressure is below 10^{-6} Pa (~10^{-8} torr).

Modern CVD is either LPCVD or UHVCVD, which is classified by the physical characteristics of vapor as follows:

- AACVD (Aerosol-assisted CVD) – CVD in which the precursors are transported to the substrate by means of a liquid/gas aerosol.

- DLICVD (Direct liquid injection CVD) – CVD in which the precursors are in liquid form. Liquid solutions are injected in a vaporization chamber toward injectors.

- MPCVD (Microwave plasma-assisted CVD).

- PECVD (Plasma-enhanced CVD) – PECVD technique allows deposition at lower temperatures.

- RPECVD (Remote PECVD) – Similar to PECVD except that the wafer substrate is not directly in the plasma discharge region. Removing the wafer from the plasma region allows processing temperatures down to room temperature.

- ALCVD (Atomic-layer CVD) – Deposits successive layers of different substances to produce layered crystalline films.

- CCVD (Combustion CVD) – Combustion CVD or flame pyrolysis is an open-atmosphere, flame-based technique for depositing high-quality thin films and nanomaterials.

- HFCVD (Hot filament CVD) – This process uses a hot filament to chemically decompose the source gases.

- HPCVD (Hybrid physical–chemical vapor deposition) – This process involves both chemical decomposition of precursor gas and vaporization of a solid source.

- MOCVD (Metal organic CVD) – This CVD process is based on metal organic precursors.

- RTCVD (Rapid thermal CVD) – This CVD process uses heating lamps or other methods to rapidly heat the wafer substrate.

- VPE (Vapor-phase epitaxy).

- PICVD (Photo-initiated CVD) – This process uses UV light to stimulate chemical reactions [6].

2.1.1. CVD uses

CVD is commonly used to deposit conformal films and augment substrate surfaces in ways that more traditional surface modification techniques are not capable of. CVD is useful in the process of atomic layer deposition. A variety of applications for such films exist. Also, polymerization by CVD, perhaps the most versatile of all applications, allows for super-thin coatings that possess some very desirable qualities, such as lubricity, hydrophobicity, and weather resistance, to name a few [6].

For many years, wet chemistry methods have been the main method used for coating the surfaces. Liquid processes have many disadvantages, such as the need to dispose of organic solvents, incomplete wetting of high aspect ratio structures, diffusion-limited transport of reactants, and poor control of reactant supply, compared with gas-phase processes. The CVD method is often called nanoparticle vapor deposition (NVD). In NVD, two binary chemicals are injected simultaneously or when the other chemical is present. Currently, CVD has mostly been applied in coating industries, but the use of CVD in other areas, such as in the production of powders, fibers, monoliths, and composites, is growing rapidly [7–9].

2.1.2. PECVD and its applications

PECVD is a process used to deposit thin films from a gas state (vapor) to a solid state on a substrate. In this process, thin film can be deposited on the substrate at low temperature. Reacting gases created plasma and chemical reactions are also involved. By DC discharge or RF) (AC) frequency, plasma is created between two electrodes, and the reacting gases are filled between the electrodes [6].

PECVD method is one of the most universal approaches to treat fibers, yarns, and fabrics. Compared to other treatments, such as the traditional chemical methods, plasma treatment is less material consuming and easier to operate, results in less effluent, and is much more environment friendly. It is also a good way to deposit nonpolymerizable substance at normal conditions on the surface of a substrate. In addition, the PECVD process would only affect a few topmost atomic layers of the substrate surface, ensuring that the bulk of the substrate is not affected, thus allowing a decoupling of the surface properties from the bulk properties of the material and giving the product designer a considerable new degree of freedom. When a chemical is carried into the reactor, it gains energy through inelastic collision with high-energy species especially with the electrons. The bonds of the chemicals are broken up and a new substance would be deposited on the surface of the substrates. The substance and the substrate would be constantly subjected to the plasma until the process is finished. Since high-performance fiber is somewhat inert to normal chemical reactions, the PECVD treatment would be an ideal choice to make chemical or physical changes to these fibers [10].

Low-pressure plasma processes, such as PECVD volatilize organic compounds, and graft functional groups (such as -NH$_2$ and -OH) onto hydrophobic polymers. These techniques are used for improving biomedical textiles and materials. Materials and devices' surfaces are functionalized, without any influence on their bulk properties. By plasma deposition treatment, spreading, adhesion, and cells proliferation of biomedical materials' surfaces can be optimized. Density and distribution of the ions and radicals that are produced in the discharge can be tuned by different external plasma parameters such as pressure, power input, gas flow rate, and so on [11].

The lower deposition temperatures are critical in many applications where CVD temperatures could damage the devices being fabricated [12].

Plasma PECVD offers a wide range of cost-effective material processes to manufacturers in the electronic, industrial, and medical sectors.

For separating conductive layers from each other, some dielectrics such as SiN, SiO$_2$, SiO, and SiON are used. These dielectrics provide different electrical insulating properties. For military devices, usually diamond-like-carbon (DLC) is used as a capping agent. It has radiation-hardening characteristics. Also A-Si and poly-Si have different conductive properties depending on their doping quantities [13].

The films typically deposited using PECVD are silicon compounds.

Silicon dioxide and silicon nitride are dielectric (insulating) materials commonly used in the fabrication of electronic devices. These films are also used for encapsulation to protect devices from corrosion by atmospheric elements, such as moisture and oxygen [14].

Silicon dioxide can be deposited using a combination of silicon precursor gases, such as dichlorosilane or silane, and oxygen precursors, such as oxygen and nitrous oxide, and has an important influence on infrared and ultraviolet (UV) absorption, stability, mechanical stress, and electrical conductivity.

Silicon dioxide can also be deposited from a tetraethoxysilane (TEOS) silicon precursor in an oxygen or oxygen–argon plasma. High-density plasma deposition of silicon dioxide from silane and oxygen/argon has been widely used to create a nearly hydrogen-free film with good conformality over complex surfaces [15].

Jin et al. [16] reported the relationship between the PECVD parameters and film properties. The design and synthesis of various functional hybrid film systems for SiO_x film coatings with good mechanical properties at high deposition rates have been reported [16].

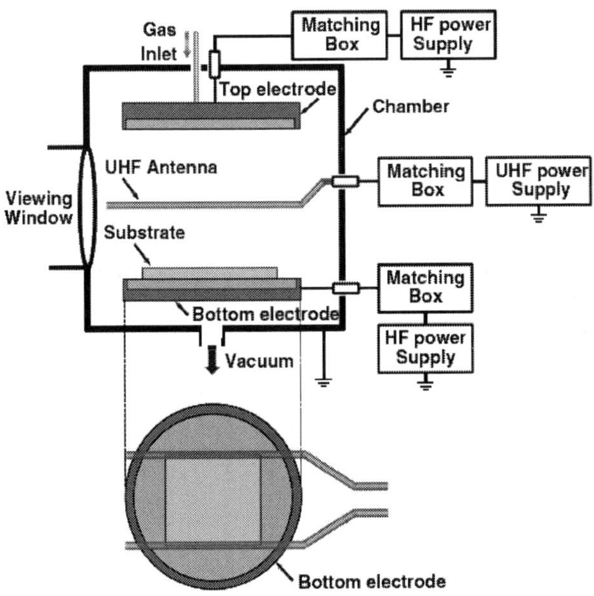

Figure 1. A schematic diagram of the PECVD with the dual frequencies UHF and HF apparatus used to deposit the SiO_x film [16].

Figure 1 shows a schematic diagram of the plasma configuration used for deposition of the SiO_x film. Two circular electrodes covered with a ceramic plate were placed within a cylindrical enclosure containing both electrodes. Two line ultra–high-frequency (UHF) electrodes covered with a ceramic textile tube were placed in the chamber. Table 1 shows the coating parameters.

During the experiment, the base pressure was approximately 3×10^{-2} Pa. A polycarbonate (PC) and Si wafer (100) substrate was placed on the bottom electrode. During the plasma process, a gas composed of octamethylcyclotetrasiloxane (OMCTS) and oxygen gas was allowed to flow to the top electrode. The OMCTS precursor has been used in this study. The SiO_x film thickness was 1 μm.

SiO_x films were deposited on polycarbonate (PC) substrates at room temperature using a hybrid plasma CVD process. SiO_x films were deposited by applying high-frequency (HF) bias with UHF power.

Parameter	Condition
Base pressure	$<3 \times 10^{-2}$ Pa
Deposition pressure	1.7×10^{-1} Pa–2.8×10^{-1} Pa
Top electrode (HF, 13.56 MHz)	120 W
UHF power (320 MHz)	0, 80, 150, 220 W
Bottom electrode (HF, 13.56 MHz)	0, 70, 90, 120 W
Film thickness	1 μm
Temperature	Room temperature
Substrates	PC and Si wafer (100)

Table 1. Deposition parameters [16]

The resulting film showed a Si–O network structure and Si–O–Si intensity [16].

Plasma-enhanced PECVD dielectric films are ubiquitous in the microelectronics industry as well as in other advanced technologies.

Hughey and Cook [17] used silicon nitride for study as a typical PECVD film.

Technologically important materials such as silicon nitride are used as a passivation layer in microelectronic devices, as an etch stop layer in interconnect stacks, and as an active element in optical waveguides. The goals of their study are then to identify specifically the deposition conditions suitable to maintain the mechanical integrity of a silicon nitride film in optical devices.

Results shown in this work provide further evidence for the generality of phenomena observed in different PECVD dielectric materials; primarily, stress increases irreversibly on thermal cycling and is caused by a reduction in the amount of bonded hydrogen [17].

In another point of view, DLC is an amorphous carbon film and has good properties like diamond, such as high electrical resistance, high chemical inertness, high hardness, high wear resistance, and low friction. DLCs are used in biotechnology and material science. DLC thin film deposition techniques and conditions strongly influence DLC quality and its thermal and optical properties as well as the adherence degree to the surfaces and its wear resistance. The properties of α-C:H thin films fabricated by PECVD are sensitive to deposition parameters [18].

Dias et al. [18] described that the photoacoustic spectroscopy and the open photoacoustic cell method are employed in the thermo-optical characterization of hydrogenated amorphous carbon thin films. The samples were fabricated in a modified PECVD system with varying deposition times.

The spectroscopic results showed that the best absorption and smallest transmission around 330–500 nm occurred in samples with 30 s to 10 min of deposition. DLC-coated glass with a 2 min deposition time showed UV transparency. Furthermore, the thermal diffusivity increased with decreasing film thickness.

Therefore, considering the heat dissipation and the absorption intensity, α-C:H-coated glass at 10 min can be suggested for coatings of aerospace carcasses. On the other hand, the α-C:H-coated glass at 2 min could be used as an anti-reflective coating on crystalline silicon solar cells as a device for generating clean energy.

A very important conclusion was that the thickness of the films (deposition time) is a relevant parameter. The changes in deposition time limit the UV transparency and dehydrogenation of thin films [18].

The formation of nanocomposite of metal oxide/carbon nanotubes has been extensively studied because of their potential industrial applications in optics, catalysis, microelectronics, and many other areas. A corona discharge enhanced CVD method has been shown to synthesize NiO-loaded carbon nanotubes (CNTs) by Yu et al. [19]. In a synchronous system, formation of CNTs and embedment with NiO nanoparticles are carried out. Preparation time is reduced significantly and the reaction temperature is much lower compared with previous methods. The diameter of these NiO nanoparticles is about 5 nm, which is characterized by TEM. Surface of NiO nanoparticles is reduced into metallic Ni and they are all encapsulated into the CNTs walls [19].

2.1.3. Application of PECVD in the textile industry

In view of the wide employment of PECVD in the surface functionalization of various materials, researchers studied different methods [20, 21].

Here are further descriptions of some of these examples.

According to Mossotti et al. [22], hexamethyldisiloxane (HMDSO) plasma polymerization of thin films using low-pressure plasma equipment was successfully developed for several applications on different substrates.

A Si:Ox:Cy:Hz film was deposited on scoured knitted wool fabrics by PECVD at different applied powers. HMDSO was used as the monomer, and argon and oxygen was used as feed gases. The polymerization was preceded by a stage of activation of substrates to make the wool fabrics more reactive to the deposition [22].

The polymerization was performed in a mixture of gases at constant flow rate: argon (20 sccm), oxygen (20 sccm), and HMDSO (3 sccm) for 8 min at a constant pressure of 2×10^{-2} mbar. Applied power varied from 30 to 50 W.

The experimental arrangement used for the plasma process was the lab-scale low-pressure equipment. A schematic diagram of experimental setup is shown in Figure 2. Two capacitively coupled cylindrical electrodes are used in the reaction chamber. Sample was placed on a roller below the electrodes. At different speeds, this roller could rotate; 4 rpm was chosen for working speed. Because of homogeneous deposition on different substrate surfaces, this configuration was excellent for plasma polymerization.

The reactor was made of steel. The power electrode was connected to an RF (13.56 MHz) generator tuned with the plasma system by an impedance matching network.

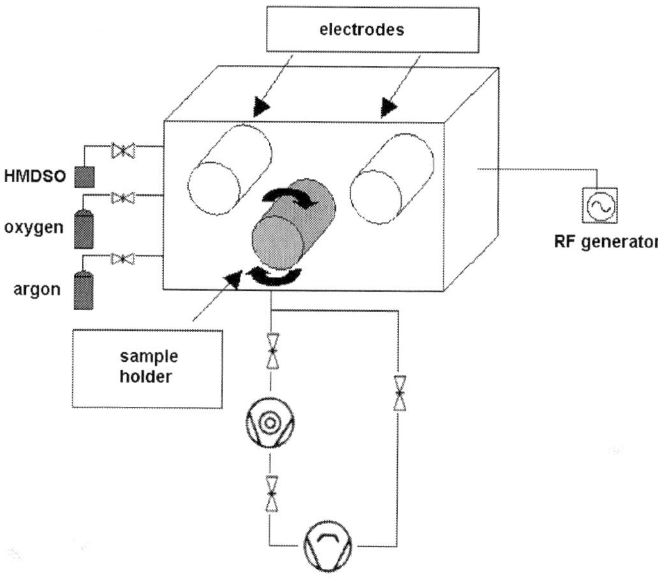

Figure 2. A schematic diagram of experimental setup [22].

Thin films (Si:Ox:Cy:Hz) were deposited on knitted wool fabrics. The achieved results of their research revealed the formation of films with a prevalent inorganic character. The results showed that this kind of treatment could represent an efficient technique to reduce pill formation on knitted wool fabrics. This plasma treatment did not affect the whiteness index of the fabric, while an increase in bursting resistance was pointed out.

Aramid fibers are one of the most popular choices for making soft body armor because of its high modulus and high strength as well as its good flexibility. Generally speaking, the ballistic performance of soft body armor is affected by parameters from two aspects, that is, the parameters associated with the projectile, including the mass, the geometry, the velocity, and the material, and those with the ballistic panel, including the fiber property, the inter-yarn friction, and the structural characteristics of the ballistic panel. To date, friction between yarns has been found to be one of the important parameters in the soft body armor design. The research of Chu et al. [10] focused on coating the aramid yarns with the PCVD technology at the atmospheric pressure (APPCVD); the schematic of the setup is shown in Figure 3. An organosilicon compound $(CH_3)_2Cl_2Si$ was selected as the precursor substance to be deposited on the surfaces of yarns [10].

The APPCVD treatment included two steps: (i) the surface activation of the Twaron yarns by nitrogen gas plasma and (ii) precursor disassociation and deposition on the activated Twaron yarns using chemical $(CH_3)_2Cl_2Si$. The chamber in the APPCVD reactor was designed to be rectangular with a size of 10 cm × 5 cm. The Twaron yarns to be processed were positioned in

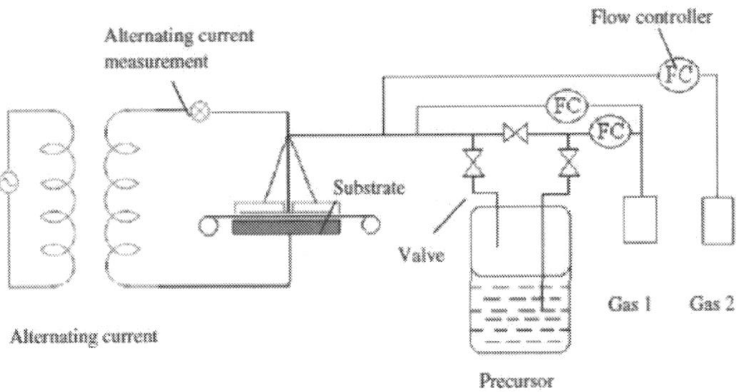

Figure 3. Schematic diagram for the setup of atmospheric pressure plasma-enhanced vapor deposition used [10].

the sample holder statically after being wrapped on a thin and smooth paper tape. The precursor chemical $(CH_3)_2Cl_2Si$ was thermostabilized outside the APPCVD reactor and was then carried into the reactor by the carrier gas, which was nitrogen. Figures 3 and 4 depict the APPCVD reactor used in the experiment for the production of plasma, which was established in Salford University. The plasma source is alternate current (AC) with the specified input voltage of 21.9 V and the frequency of 3.25 kHz. The flow rate of nitrogen gas was set to be 5 L/min and the chemical flow rate was set to be 0.2 L/min. The gap between two electrodes was set to be 3 mm in the APPCVD reactor.

Figure 4. Photograph of the atmospheric pressure plasma-enhanced vapor deposition reactor [10].

The research of Chu et al. was concerned with the surface modification of Twaron yarns to increase the inter-yarn friction for the efficient improvement of ballistic fabrics.

The chemical particles were deposited onto the surface of fibers with treatment time being 21 s, 2 min, and 4 min, and SEM photography provided evidence for the surface modification. FTIR spectra supported the existence of Si–O–Si vibration, which can be attributed to the chemical deposition. EDX analysis confirmed the deposition of the chemical through the obvious increase of Si element on the surface of the fiber. The surface modification has led to obvious increase in both the frictional coefficients (static and kinetic) and frictional force between the yarns. The yarn–yarn static frictional coefficient showed an increase from 0.1617 to 0.2969 and the kinetic frictional coefficient showed an increase from 0.1554 to 0.2436. It was also found that the tensile strength, Young's modulus, and breaking strain of the Twaron yarns are not negatively affected by the APPCVD treatment but show a slight increase.

Their research has led to the establishment of a feasible and valid method to increase the inter-yarn friction within ballistic fabrics to achieve enhanced ballistic energy absorption and to provide better protection against high-velocity projectile impact [10].

Surface modification (physical or chemical treatments) can be used for creating fire-retardant properties onto the polymer surface (where the flammability occurs) and would thus allow preserving the bulk properties of the material. In the work of Jimenez et al. [23], a new route to improve the fire-retardant properties of a polymer by deposing a thin film on its surface using plasma-assisted polymerization technique has been investigated. The plasma-assisted polymerization of organosilicon compounds is an interesting technique to create thin polysiloxane-based films on the surface of various substrates. Polysiloxane-based films are known to show good thermal stability, interesting flame properties, and an interesting hydrophobicity.

The experimental setup used to make the polysiloxane-based deposit is shown in Figure 5.

The monomer (TMDSO), premixed with oxygen, is injected into the cold remote nitrogen plasma through a coaxial injector. First, the samples were treated by the cold remote nitrogen plasma to increase the adhesion quality of the organosilicon film on the polymer. Then, the deposition step is performed without air re-exposure. Moreover, they reported that this solution gives even better performances in terms of Limited Oxygen Index (LOI) values, rate of heat release (RHR), and ignition time (IT) [23].

In the other research work, to improve textile fabric abrasion resistance, a $SiO_xC_yH_z$ thin film was realized by low-pressure plasma CVD (PCVD) at room temperature using HMDSO as the precursor compound. Organosilicon compound hexamethyldisiloxane (HMDSO) as a best reagent precursor was chosen because of its availability, liquid state, volatility at room temperature, low flammability, low toxicity, and low cost. HMDSO plasma can produce coatings with different surface properties (e.g., hydrophobic or hydrophilic behavior, friction-reducing or wear-resistant coatings, and barrier coatings); it also shows good adherence to substrates and high transparency to visible radiation. To study the film properties, different plasma exposure times on several fabrics were investigated in the range of 4–12 min by Rosace et al. [24].

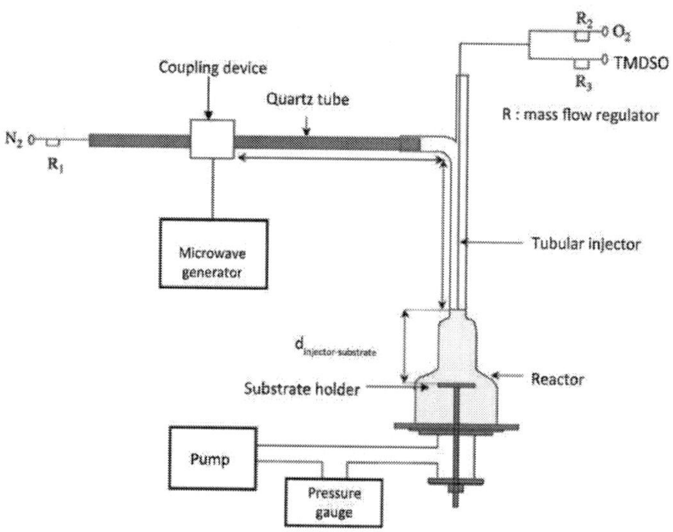

Figure 5. Experimental setup of PECVD. TMDSO: 1,1,3,3-tetramethyldisiloxane [23].

Four different textile fabrics have been used in their study to investigate the effect of HMDSO plasma deposition on different morphological kinds of fabric. The process consists of a first stage of activation of the textile fiber surface using O_2 plasma followed by a second stage based on the deposition of thin films onto previously activated textile fiber surfaces.

From the results obtained by FTIR-ATR, SEM, and EDX elemental analysis measurements, Rosace et al. concluded that organosilicon thin film on the textile fabric surface was successfully performed. Organosilicon film is an anti-wear coating against the abrasion of textile fabrics. In fact, after the application of thin layer, all tested textile fabrics showed a remarkable improvement of abrasion resistance, a lower weight loss, and a higher endpoint of Martindale test than untreated samples. But the adhesion of the coating to the substrate is not optimal. In fact, after five washing cycles, the deposited film seems to be removed from the treated fabrics. Poor washing fastness is in the relevance to the fabric's end use. Functional coatings do not involve any of the solvents and do not require high temperatures. They also have high long term and wash stability. This environment-friendly finishing can replace some wet chemical with no emissions of volatile organic compounds (VOCs) in a closed vacuum system [24].

In the other research work, plasma surface activation was made on the Polyethylene terphthalate (PET) nonwoven in a Europlasma CD 400M/PC laboratory system, as illustrated in Figure 6, by Wei et al. [25]. The material was subjected to oxygen plasma treatment at 100 W power with a gas flow rate of 1.67 cm³/s for 60 s. Plasma-induced polymerization was performed on the same PET needle-punched nonwoven material using grafting chemical (acrylic acid). First, polyester nonwoven sample was treated for 60 s and a power of 200 W with argon plasma. Then, to facilitate polymer grafting and formation of peroxides on the fiber surface, sample was exposed to atmospheric air for 15 min. Thereafter, sample was treated with 20% (w/v) aqueous acrylic acid at 50°C for 5 h. After the grafting reaction, the material was rinsed in boiling deionized water to remove any acrylic acid homopolymer and monomer and then was dried at 40°C for 24 h [25].

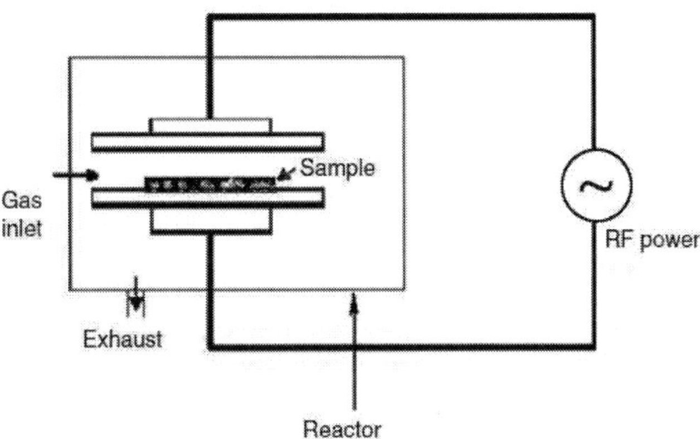

Figure 6. Plasma treatment system [25].

The PCVD deposition was performed on the PET nonwoven material using tetraethoxysilane (TEOS) and oxygen as the precursors. The RF source used was 13.56 MHz. The deposition of SiO_2 was made at the pressure of 150 Pa with a power of 100 W. The deposition lasted for 5 min.

They concluded that textile materials can greatly be modified by plasma treatments. Different effects on the textile surfaces can observe with different gas plasma treatments. Plasma-based modifications have many advantages. Plasma treatment is applicable in various forms on different substrates and materials. Biological, optical, electrical, and chemical properties and also morphology of the treated samples can be modified. It is environment-friendly and a green technique. Therefore, the great potential for significant improvements in the properties of textiles by plasma-enhanced modification is highly promising [25].

In the other example, the influence of flow rate and proportion of reactive gas and pressure in the reactive chamber on the growth of diamond films fabricated by PECVD has been studied by Deng and Zhu [26]. The samples were fabricated by PECVD. n-Si(100) wafers were used as substrates with a resistance of 90 Ω. The wafers were cleaned by standard method. The power of RF source was 500 W, and the frequency was 13.56 MHz. The temperature of substrates during the deposition process was 950°C. Before the reactive gases were introduced, the n-Si substrate was pretreated in H plasma under 200°C for 20 min. They found that these parameters (flow rate and proportion of reactive gas and pressure in the reactive chamber) influence the growth process as a whole rather than individually. It can be deduced that diamond films with different ratios of SP2 bonds will be fabricated by controlling GH_2 in the range of 16–40 ms to meet different applications [27]. The low-temperature synthesis of carbon nanofibers by microwave PECVD using a $CO/Ar/O_2$ system and their characterizations were performed by Mori and Suzuki [27].

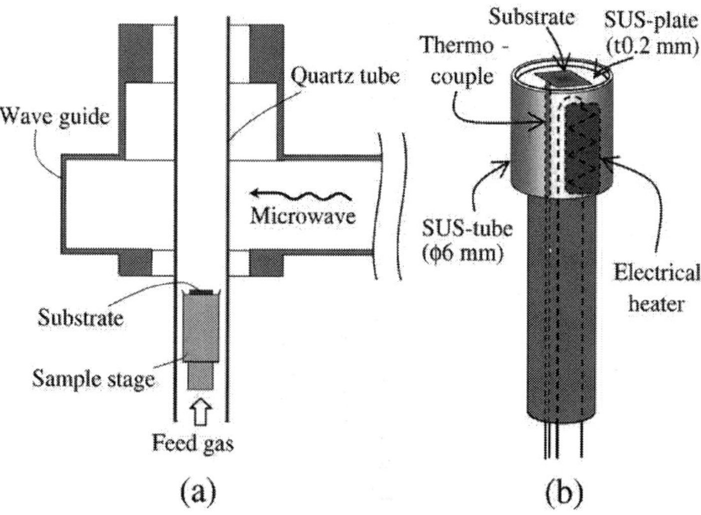

Figure 7. A schematic diagram of experimental apparatus: (a) plasma reactor and (b) sample stage [27].

Figure 6 shows a schematic diagram of the microwave plasma CVD system of Mori and Suzuki [27].

Borosilicate glass and CaF_2 plates with a thin Fe catalyst layer were used as substrates. For the CNF deposition, $CO/Ar/O_2$ microwave discharge plasma was used. In this process, CO flow rate was 10 sccm, Ar flow rate was 30 sccm, O_2 flow rate was 0–1.0 sccm, total pressure was 400 Pa, microwave power was 80 W, and deposition time was 10 min.

The diameter of bulk CNFs was about 50–100 nm and the surface of CNFs was covered by branching fibers and their nuclei with a diameter of about 5–20 nm. The growth rate of CNFs is about 4–6 nm/s [27].

A method most often used to synthesize titanium dioxide thin films for photocatalytic applications is a sol–gel technique. Other ways to deposit these films comprise reactive magnetron sputtering, ion beam sputtering, spray pyrolysis, hydrothermal technique, CVD, and metal organic CVD method. Thin TiO_2 films, deposited using the above-listed techniques, exhibit both photocatalytic properties and receptiveness to the modification of their hydrophilicity under the effect of UV light. A synthesis method that is useful for the deposition of titanium oxide films is PECVD technique.

Many researches and studies have been done about low-temperature plasma and enhancement of chemical reactions in recent years. To deposit semiconducting, insulating, and conducting thin films, RF glow discharges have used. It is also used for electrical and optical devices like optical filter systems. With PECVD technique, titanium dioxide films were stimulated and used as optical filter application.

Szymanowski et al. [28] reported an application of the RF PECVD method to the deposition of titanium oxide films, exhibiting such photo-induced properties as change of water contact angle under illumination with UV light and photocatalytic activity with respect to chemical and biological contamination of water. They have achieved good results for the RF PECVD-treated material [28].

A low-temperature process, PECVD, has been used for the synthesis of TiO_2 for optical applications. By RF PECVD technique, thin films of titanium oxide were deposited and designed for photocatalytic applications. Homogeneous films were obtained and optical properties like stoichiometric TiO_2 method were obtained. Their bactericidal activity in a combination with n UV irradiation was tested.

From the result, it can be concluded that PECVD produced titanium oxide coatings with a large degree, photocatalytic properties, and good antibactericidal activity. By increasing refractive index, bactericidal inhibition of the coated samples increases. Films whose optical parameters approach those of stoichiometric TiO_2 were obtained best results. The results showed a substantial enhancement of the bactericidal activity of UV irradiation for the surfaces modified with the presented process [29].

The aim of the work of Sobczyk-Guzenda et al. [30] was to use the RF PECVD method to place a photocleaning titanium dioxide coating onto the surface of a cotton woven fabric.

The power of RF discharge was varied from 100 to 300 W. The optimum value found to be 200 W. The coating uniformly covers each fiber with no signs of splintering and stability of the coating remains unchanged after washing the fabric in a detergent solution even after 18 months of storage. By increasing the power, single cracks appear in the coatings and their number substantially increases after washing. A very thin film with low strength was deposited on the surface of cotton in lower used power (100 W).

It was concluded that deposition of TiO_2 coatings significantly enhanced the photocleaning properties of cotton under UV light as compared with untreated cotton. In addition, change of the surface character from hydrophobic to strongly hydrophilic under UV illumination is observed in TiO_2-coated fabric under optimum conditions. Both the above effects of photocleaning and photowetting were introduced to the cotton fabric by means of its modification with the RF PECVD-deposited TiO_2 coating [30].

In the work of Potock'y et al. [31], polyvinyl alcohol (PVA) fibers were used as a polymer matrix containing ultradispersed diamond (UDD) nanoparticles. The growth of diamond fiber-like structures and films by microwave PECVD was studied as a function of UDD concentration in the PVA matrix. The composite PVA polymer nanofiber with UDD powder is a suitable pretreatment layer for diamond growth by the microwave PECVD method.

By careful and mechanical separation of polypropylene (PP) cloth from the PVA nanofiber textile, the composite PVA and UDD nanofiber textile was transferred from PP cloth to Si substrates. From the PP cloth pattern, the fingerprint was maintained and then the PVA composite was transferred to the Si substrate by tweezers. In a CVD chamber (cavity resonator reactor), these substrates were loaded.

The composite PVA polymer nanofibers with UDD nanoparticles is a quick, easy, and economical method for the pretreatment of different substrate materials with three-dimensional geometry in large scale [31].

Struszczyk et al. [32] described the studies on the surface modification of so-called ballistic materials (materials commonly used to protect the human body against firearms; i.e., fragments or bullets). Two materials, an ultrahigh molecular weight polyethylene (UHMWPE) composite and aramid fabric, were investigated. The surfaces of these fibrous materials were modified using plasma-assisted CVD (PACVD) to examine the effects of the modification on the material properties, which are important for designing ballistic protections.

The aim of their study was to investigate the effects of low-temperature plasma treatment in the presence of two low molecular mass organic derivatives containing fluorine or silicon on the most critical properties of two types of ballistic raw materials, a p-aramid woven fabric and UHMWPE fiber composite [32].

They investigated the influence of the surface modification of two different materials by PACVD based on their properties (especially those critical for ballistic applications). Specifically, the materials studied were Kevlar and the Dyneema SB51 composite (ultrahigh molecular weight polyethylene); both materials are usually used to design soft ballistic inserts for bullet- and fragment-proof vests.

The surfaces of these materials were modified using a low temperature RF (13.56 MHz) plasma generator. The PACVD system (CD 400 PLC R/R model; Europlasma, Belgium) consisted of two parallel rectangular aluminum electrodes. The right one was connected to an alternative voltage (RF 13.56 MHz), whereas the left one was grounded. The surface to be modified was placed between the electrodes, facing the right one. During the process, the material to be deposited was injected (in the gas phase) into the chamber simultaneously with the gas used for the plasma initiation.

The surface modification of both Kevlar and Dyneema was performed in five stages: (a) proper evacuation of the chamber, (b) surface cleaning, (c) surface activation, (d) deposition of the material on the surface, and (e) conditioning of the deposited layer.

The results indicate that the use of PACVD to deposit a polymer layer on the fiber (Kevlar) and UHMWPE composite (Dyneema) surfaces might significantly improve the water resistance of the p-aramid fibers, and consequently, the ballistic protections can be used for a longer period of time without a significant loss of product performance and safety. The surface properties of both studied ballistic materials can be changed in a controlled way, which is helpful, for example, for the fabrication of a hybrid composite consisting of more than one type of ballistic or nonballistic material. The combination of materials with different performances might significantly improve the main functionality of the fabricated composites and provide them with new functionalities.

They concluded that PACVD is promising for ergonomics studies and the transformation of textiles, such as Kevlar and Dyneema, into materials used for protection against firearms [32].

2.2. Physical vapor deposition (PVD)

To deposit nanometer to thick layers of organic, inorganic, or metallic material at fiber surface, plasma is used. Several methods of plasma-assisted deposition such as sputtering, thermal evaporation, or anodic arc evaporation at low pressure to vaporize metals and polymers and to condense the gasified material onto a substrate (PVD) on textiles have been proposed. [33]. Vaporization and subsequent condensation of a coating species on a surface involves PVD, which is an atomistic deposition process. With the random (heterogeneous) nucleation of crystallites, growth and coating deposition begin. Growth and nucleation process will depend on different parameters, such as nucleation sites availability, energy of the substrate surface, atomic structure of the coating as a bulk material, and mobility of the deposited species.

Microcrystallites have a random variety of orientations possibly which influenced by epitaxy with the substrate crystal lattice, during the initial stages of growth. These regions of condensation will act as preferential nucleation sites from which will grow columns of textured material as the film grows [34].

Physical vapor deposition PVD coating is a product that is currently being used to enhance a number of products, including automotive parts such as wheels and pistons, surgical tools, drill bits, and guns [35, 36].

The PVD technique is based on the formation of vapor of the material to be deposited as a thin film. The material in solid form is either heated until evaporation (thermal evaporation) or sputtered by ions (sputtering) [37].

2.2.1. Different types of PVD

Cathodic arc deposition: In cathodic arc deposition process, a high-power electric arc discharged at the target or source material blasts away some into highly ionized vapor to be deposited on the sample or workpiece.

Electron beam PVD: In electron beam PVD, by electron bombardment in high vacuum, material to be deposited is heated to a high vapor pressure and is transported by diffusion to be deposited by condensation on the cooler workpiece.

Evaporative deposition: In evaporative deposition, by electrically resistive heating in "low" vacuum, the material to be deposited is heated to a high vapor pressure.

Pulsed laser deposition: In pulsed laser deposition, a high-power laser ablates material from the target into a vapor.

Sputter deposition: In sputter deposition process, glow plasma discharge, which is usually localized around the target by a magnet, bombards the material sputtering some away as a vapor for subsequent deposition [38, 39].

2.2.2. Application of PVD in different branches of industry

PVD technology is very versatile, enabling one to deposit virtually every type of inorganic materials, such as metals, alloys, compounds, and mixture, as well as some organic materials. Wear resistance, hardness, and oxidation resistance can be improved by PVD coatings. These coatings are used in surgical/medical, automotive, aerospace, dies and moulds for all manners of material processing, cutting tools, optic, watches, thin films, window tint, food packaging, firearms, and textile industry. This chapter is a review of the science and technology related to plasma-enhanced PVD (PEPVD) use in the textile industry.

Cathodic arc plasma deposition is one of the oldest and most modern emerging technologies. Cathodic arc plasma deposition is a coating technology with great potential, because cathodic arc plasmas are fully ionized with very energetic ions, promoting the adhesion and formation of dense films.

According to Hsu et al. [40], undoped ZnO films were successfully deposited on the glass substrate at a low temperature (<70°C) using cathodic arc plasma deposition. ZnO films were deposited onto the glass substrate in a cathodic arc plasma deposition (CAPD) system. As a cathode target, 100-mm diameter metallic Zn 99.99% purity in an alumina ceramic tube was held, and as the reactant gas, 99.99% purity O_2 gas was used. Glass substrates were washed by alcohol before deposition and then ultrasonically cleaned in alcohol for 10 min. In the depositions of ZnO films, the base pressure was kept at 3×10^{-4} torr. Substrate rotation of 2 rpm and substrate-anode electrode distance of approximately 21 cm remained constant during the deposition work. Without extra heating, the depositions of ZnO films were performed at room temperature. The result of the confirmed experiment showed that the transmittance increases and the resistivity decreases, indicating that the result of the final confirmed experiments could be certainly improved [40].

However, sputter deposition is a PEPVD method, in which a glow plasma discharge (usually localized around the "target" by a magnet) bombards the material sputtering some away as a vapor for subsequent deposition. Now, the sputter deposition is able to deposit most materials.

DC (diode) sputtering, RF (radiofrequency) sputtering, magnetron sputtering, and reactive sputtering are the basic techniques in sputter deposition. DC sputtering is the simplest sputtering technology. In DC sputtering, the sputter voltage is typically −2 to −5 kV; also, the substrate is being bombarded by electrons and ions from target and plasma. Sputtering film has neutral atoms deposit independently and put negative bias on the substrate to control this; also, it can significantly change film properties during deposition.

RF sputtering works well with insulating targets. Plasma can operate at lower Ar pressures (1–15 mtorr). In RF sputtering, for frequencies less than 50 kHz, electrons and ions in plasma are mobile, and it follows the switching of the anode and cathode.

For frequencies above 50 kHz, ions (heavy) can no longer follow the switching and there are enough electrons to ionize gases (5–30 MHz). Typically, 13.56 MHz is used.

To confine the glow discharge plasma to the region closest to the target plate, magnetron sputtering technique uses powerful magnets; ions with higher density can make the electron or gas molecule collision process much more efficient and improve the deposition rate [41].

Sputter deposition is a widely used technique for depositing thin metal layers on semiconductor wafers. The range of applications of sputtering and the variations of the basic process are extremely wide.

In another point of view, PVD, especially sputtering technology, has been regarded as an environment-friendly technique for the functionalization of textile materials. By sputter coatings of polymers, metals, and metal oxides, different functions can be achieved. By using co-sputtering methods, composite coatings can also be achieved. The functional textiles obtained include magnetic, optical, conductive, and biocompatible materials and can be used in different applications and industries [42].

Antimicrobial properties, electrical conductivity, or a shiny metallic appearance for decorative purposes by textile metallization show a growing interest for textile applications. Ag-coated yarns using electroless Ag plating technique are produced increasingly. Only on nylon fibers, good adhesion of the Ag layers could be achieved and washing fastness is an important technological challenge.

Hegemann et al. [43] described an alternative technique to deposit Ag on textiles, namely, plasma sputtering, which allows cleaning and deposition in a one-step process. They obtained excellent adhesion on polyester fibers with smooth coatings. Also, electrical conductivity was obtained with low amounts of deposited Ag maintaining their textile properties while showing metallic appearance and antimicrobial activity [43].

Fungal and bacterial infestation can cause material damage to fabrics, where damage becomes visible through discoloration and stains. Some fungi and bacteria are pathogenic. Antimicrobial active coatings might be a possible protection against an infestation of the fabrics. Fabrics consisting of SiO_2 fibers were coated with precious metal PVD layers using the magnetron sputtering technique by Scholz et al. [44]. Layers of silver, copper, platinum, platinum/rhodium, and gold were characterized according to their bonding strength and antimicrobial effectiveness. They found that copper was most effective against fungi and bacteria. But silver is effectiveness against fungi but it is effective against bacteria. The other tested metals did not have antibacterial properties. Due to a plasma treatment, the bonding strength between the coatings and the substrates was sufficient [44, 45].

Improving the fastness properties and antibacterial activity of dyed cotton samples was the main goal of Shahidi's research work. Cotton fabrics were dyed with various types of dyestuffs and then dyed samples were sputtered using plasma sputtering system for 15 s by silver and copper. A DC magnetron sputtering system for the deposition of metal nanolayer on the surface of samples has been used.

Silver or copper particles were deposited on the surface of cotton samples, and the antibacterial has been developed through the incorporation of metal nanoparticles onto fabric surfaces. It was concluded that the change in properties induced by sputtering can cause an improvement in certain textile products. Also, the results show that the sputtering technique can be a novel method for improving the fastness properties of dyed cotton samples with different classes of dyestuffs [46].

Also, the antibacterial activities of cellulose fabrics were easily achieved with a DC magnetron sputtering device, by coating copper on the surface of a fabric, without any chemical or wet process. The antibacterial activity of the fabrics remained even after laundering for at least 30 cycles [47].

Textiles were treated for the food industry with an original deposition technique based on a combination of plasma-enhanced chemical and physical vapor deposition to obtain nanometer-sized silver clusters incorporated into a SiOCH matrix by Brunon et al. [48]. The plasma tool used in this study was an industrial equipment with a chamber volume of 350 L for transparent and antimicrobial coating (Figure 7).

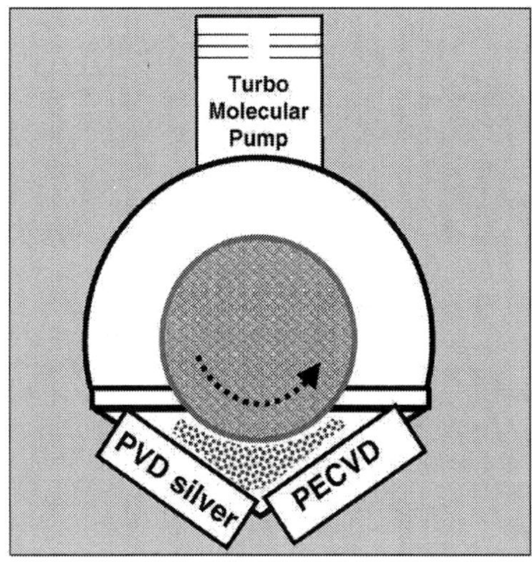

Figure 8. Thin film deposition equipment configuration [48].

The PECVD electrode and the sputtering cathode were installed next to each other to get a material made from both sources. The plasma was a mixed contribution of PECVD source and PVD sourced that are close enough together at an intermediate location. These sources were mounted with a relative angle of 130° and were not in the same plane (Figure 7).

Because of separate generators, both plasma discharges can be run alternately or simultaneously. Because of being uniformly coated, different gasses were mixed in a common pipe and injected through holes made in a vertical manifold installed next to the plasma sources in the chamber. With a digital liquid mass flow meter, HMDSO injection in the vacuum chamber before being vaporized was controlled.

To quantify the presence at the top surface and inside of the coating of the antimicrobial agent (Ag) as ~10 nm size particles and composition of antimicrobial coatings with acceptable transparency properties, the combination of spectroscopic and microscopic surface analysis was used. Microbiological characterization indicated that the level of antimicrobial activity directly influenced by the variable silver content on the surface [48].

3. Conclusion

This chapter is a study on the plasma-enhanced chemical and physical vapor deposition in textile industry. PECVD is also known as glow discharge CVD. It uses electron energy (plasma) as the activation method to enable deposition to occur at a low temperature and at a reasonable rate. PECVD is widely used in the textile industry. PVD techniques based on sputtering and cathodic arc methods, which are PEPVD, are widely used to deposit coating for various substrates and textiles. The overview concludes with a compilation of typical applications and textile materials.

PECVD has difficulty in depositing high-purity films. This is mostly due to the incomplete desorption of by-products and unreacted precursor at low temperatures, especially hydrogen that remains incorporated into the films. Such impurities, in some cases, are beneficial. PECVD at low frequencies is also prone to induce undesirable compressive stresses in the films. This is especially damaging in thick film for tribological applications, which could lead to cracking or spalling of the films. Toxic by-products and high cost of equipment are the top disadvantages of using PECVD. PECVD methods have some advantages in comparison with CVD methods. They are as follows:

- Flexibility for the microstructure of the deposition and film can be controlled separately

- Deposition can occur at relatively low temperatures on large areas

- To obtain the required film density, the ion bombardment can be substituted for deposition temperature. For temperature sensitive substrates, such low-temperature deposition is important [49].

High vapor pressure metalorganic precursors are available at reasonable prices and the commercial applications of metalorganic-assisted PECVD have been extended from metallic films, dielectric, and semiconductors to new applications, such as powder coating, biomaterials, diamond deposition diffusion barriers, abrasion-resistant coatings on polymer, optical filters, and fiber coatings [50].

In general, the advantages of using plasma-enhanced vapor deposition are low operation temperature, lower chances of cracking deposited layer, good dielectric properties of deposited layer, good step coverage, less temperature-dependent disadvantages of using PECVD, toxic by-products, and high cost of equipment.

Author details

Sheila Shahidi[1*], Jakub Wiener[2] and Mahmood Ghoranneviss[3]

*Address all correspondence to: sh-shahidi@iau-arak.ac.ir

1 Department of Textile, Arak Branch, Islamic Azad University, Arak, Iran

2 Department of Textile Chemistry, Faculty of Textile, Technical University of Liberec, Liberec, Czech Republic

3 Plasma Physics Research Center, Science and Research Branch, Islamic Azad University, Tehran, Iran

References

[1] D. Hegemann, M.M. Hossain, and D.J. Balazs. Nanostructured plasma coatings to obtain multifunctional textile surfaces. Progress in Organic Coatings 2007; 58: 237–240.

[2] www.corrosion-doctors.org.

[3] A. Surženkov. Aurustussadestatud pinded/Vapour Deposited Coatings/Loengukonspekt, Tallinn University of Technology, ppt.

[4] M. Esen, I. Ilhan, M. Karaaslan, E. Unal, F. Dincer, and C. Sabah. Electromagnetic absorbance properties of a textile material coated using filtered arc-physical vapor deposition method. Journal of Industrial Textiles, published online 12 May 2014. DOI: 10.1177/1528083714534710.

[5] www.reference.com.

[6] https://en.wikipedia.org/wiki/Chemical_vapor_deposition.

[7] P. Aminayi and N. Abidi. Imparting super hydro/oleophobic properties to cotton fabric by means of molecular and nanoparticles vapor deposition methods. Applied Surface Science 2013; 287: 223–231.

[8] M. Gorjanc, V. Bukosek, M. Gorensek, and A. Vesel. The influence of water vapor plasma treatment on specific properties of bleached and mercerized cotton fabric. Textile Research Journal 2010; 80(6): 557–567. DOI: 10.1177/0040517509348330.

[9] S. Vihodceva and S. Kukle. Low-pressure air plasma influence on cotton textile surface morphology and evaporated cooper coating adhesion. American Journal of Materials Science and Technology 2013; 2: 1–9. DOI: 10.7726/jac.2013.1001.

[10] Y. Chu, X. Chen, D.W. Sheel, and J.L. Hodgkinson. Surface modification of aramid fibers by atmospheric pressure plasma-enhanced vapor deposition. Textile Research Journal 2014; 84(12): 1288–1297.

[11] M. Mattioli-Belmonte, G. Lucarini, L. Virgili, G. iagini, L. Detomaso, P. Favia, R. DAgostino, R. Gristina, A. Gigante, and C. Bevilacqua. Mesenchymal stem cells on plasma-deposited acrylic acid coatings: an in vitro investigation to improve biomaterial performance in bone reconstruction. Journal of Bioactive and Compatible Polymers 2005; 20: 343–360.

[12] www.plasmatherm.com/pecvd.html. Received on 2 August 2015.

[13] www.plasmaequip.com

[14] www.plasmatherm.com/pecvd.html. Received 2 August 2015.

[15] https://en.wikipedia.org/wiki/Plasma-enhanced_chemical_vapor_deposition.

[16] S.B. Jin, J.S. Lee, Y.S. Choi, I.S. Choi, and J.G. Han. High-rate deposition and mechanical properties of SiO_x film at low temperature by plasma enhanced chemical vapor deposition with the dual frequencies ultra high frequency and high frequency. Thin Solid Films 2011; 519: 6334–6338.

[17] M.P. Hughey and R.F. Cook. Massive stress changes in plasma-enhanced chemical vapor deposited silicon nitride films on thermal cycling. Thin Solid Films 2004; 460: 7–16.

[18] D.T. Dias, V.C. Bedeschi, A. Ferreira da Silva, O. Nakamura, M.V. CastroMeira, and V.J. Trava-Airoldi. Photoacoustic spectroscopy and thermal diffusivity measurement on hydrogenated amorphous carbon thin films deposited by plasma-enhanced chemical vapor deposition, Diamond & Related Materials 2014; 48: 1–5.

[19] K. Yu, J. Zou, Y. Ben, Y. Zhang, and C. Liu. Synthesis of NiO-embedded carbon nanotubes using corona discharge enhanced chemical vapor deposition. Diamond & Related Materials 2006; 15: 1217–1222.

[20] N. De Vietro, C. Annese, L. D'Accolti, F. Fanelli, C. Fusco, and F. Fracassi. A new synthetic approach to oxidation organocatalysts supported on Merrifield resin using plasma-enhanced chemical vapor deposition. Applied Catalysis A: General 2014; 470: 132–139.

[21] J. Musschoot, J. Dendooven, D. Deduytsche, J. Haemers, G. Buyle, and C. Detavernier. Conformality of thermal and plasma enhanced atomic layer deposition on a nonwoven fibrous substrate. Surface and Coatings Technology 2012; 206: 4511–4517.

[22] R. Mossotti, G. Lopardo, R. Innocenti, G. Mazzuchetti, F. Rombaldoni, A. Montarso-lo, and E. Vassallo. Characterization of plasma-coated wool fabrics. Textile Research Journal 2009; 79(9): 853–861.

[23] M. Jimenez, H. Gallou, S. Duquesne, C. Jama, S. Bourbigot, X. Couillens, and F. Pero-ni. New routes to flame retard polyamide 6,6 for electrical applications. Journal of Fire Sciences 2012; 30(6): 535–551.

[24] G. Rosace, R. Canton, and C. Colleoni. Plasma enhanced CVD of $SiO_xC_yH_z$ thin film on different textile fabrics: Influence of exposure time on the abrasion resistance and mechanical properties. Applied Surface Science 2010; 256: 2509–2516.

[25] Q. Wei, Y. Wang, Q. Yang, and L. Yu. Functionalization of textile materials by plas-ma enhanced modification. Journal of Industrial Textiles 2007; 36(4): 301–309.

[26] N. Deng and C. Zhu. Influence of resident time and proportion of reactive gas on the growth of diamond films. Journal of Wide Bandgap Materials 2002; 10(1): 71–75.

[27] S. Mori and M. Suzuki. Characterization of carbon nanofibers synthesized by micro-wave plasma-enhanced CVD at low-temperature in a $CO/Ar/O_2$ system. Diamond & Related Materials 2009; 18: 678–681.

[28] H. Szymanowski, A. Sobczyk-Guzenda, A. Rylski, W. Jakubowski, M. Gazicki-Lip-man, U. Herberth, and F. Olcaytug. Photo-induced properties of thin TiO_2 films de-posited using the radio frequency plasma enhanced chemical vapor deposition method. Thin Solid Films 2007; 515: 5275–5281.

[29] H. Szymanowski, A. Sobczyk, M. Gazicki-Lipman, W. Jakubowski, and L. Klimek. Plasma enhanced CVD deposition of titanium oxide for biomedical applications. Sur-face and Coatings Technology 2005; 200: 1036–1040.

[30] A. Sobczyk-Guzenda, H. Szymanowski, W. Jakubowski, A. Błasińska, J. Kowalski, and M. Gazicki-Lipman. Morphology, photocleaning and water wetting properties of cotton fabrics, modified with titanium dioxide coatings synthesized with plasma en-hanced chemical vapor deposition technique. Surface and Coatings Technology 2013; 217: 51–57.

[31] ˇS. Potock´y, T. Iˇzák, B. Rezek, P. Tesárek, and A. Kromka. Transformation of poly-mer composite nanofibers to diamond fibers and films by microwave plasma-en-hanced CVD process. Applied Surface Science 2014; 312: 188–191.

[32] M.H. Struszczyk, A.K. Puszkarz, B. Wilbik-Hałgas, M. Cichecka, P. Litwa, W. Urban-iak-Domagała, and I. Krucinska. The surface modification of ballistic textiles using plasma-assisted chemical vapor deposition (PACVD). Textile Research Journal 2014; 84(19): 2085–2093.

[33] R. Morent, N. De Geyter, J. Verschuren, K. De Clerck, P. Kiekens, and C. Leys. Non-thermal plasma treatment of textiles. Surface and Coatings Technology 2008; 202: 3427–3449.

[34] J.P. Feist, A.L. Heyes, and J.R. Nicholls. Phosphor thermometry in an electron beam physical vapour deposition produced thermal barrier coating doped with dysprosium. Proceedings of the Institution of Mechanical Engineers, Part G: Journal of Aerospace Engineering 2001; 215: 333–341.

[35] http://en.wikipedia.org/w/index.php?oldid=619209469.

[36] M. Aliofkhazraei and N. Ali. PVD technology in fabrication of micro- and nanostructured coatings. Comprehensive Materials Processing 2014; 7: 49–84.

[37] P. Lu and B. Ding. Nano-modification of textile surfaces using layer-by-layer deposition methods. Surface Modification of Textiles 2009: 214–237.

[38] M. Polok-Rubiniec, K. Lukaszkowicz, and L.A. Dobrzański. Comparison of nanostructure and duplex PVD coatings deposited onto hot work tool steel substrate. Journal of Achievements in Materials and Manufacturing Engineering 2009; 37(2): 125–192.

[39] http://en.wikipedia.org.

[40] S.-F. Hsu, J.-H. Chou, C.-H. Fang, and M.-H. Weng. Optimization of the cathode arc plasma deposition processing parameters of ZnO film using the grey-relational Taguchi method, Hindawi Publishing Corporation. Advances in Materials Science and Engineering 2014. Article ID 187416, 6 pp. DOI: http://dx.doi.org/10.1155/2014/187416.

[41] www.angst romsciences.com.

[42] Q. Wei. Textile surface functionalization by physical vapor deposition (PVD). Surface Modification of Textiles 2009; woodhead publishing ISBN: 9781845694197

[43] D. Hegemann, M. Amberg, A. Ritter, and M. Heuberger. Recent developments in Ag metallised textiles using plasma sputtering. Materials Technology 2009; 24(1): 41–45.

[44] J. Scholz, G. Nocke, F. Hollstein, and A. Weissbach. Investigations on fabrics coated with precious metals using the magnetron sputter technique with regard to their anti-microbial properties. Surface and Coatings Technology 2005; 192(2–3): 252–256.

[45] www.chemweb.com.

[46] S. Shahidi. Plasma sputtering as a novel method for improving fastness and antibacterial properties of dyed cotton fabrics. The Journal of the Textile Institute 2015; 106(2): 162-172

[47] S. Shahidi, M. Ghoranneviss, B. Moazzenchi, A. Rashidi, and M. Mirjalili. Investigation of antibacterial activity on cotton fabrics with cold plasma in the presence of a magnetic field. Plasma Processes and Polymers 2007; 4: S1098–S1103.

[48] C. Brunon, E. Chadeau, N. Oulahal, C. Grossiord, L. Dubost, F. Bessueille, F. Simon, P. Degraeve, and D. Leonard. Characterization of plasma enhanced chemical vapor deposition-physical vapor deposition transparent deposits on textiles to trigger vari-

ous antimicrobial properties to food industry textiles. Thin Solid Films 2011; 519: 5838–5845.

[49] J. Pan. Chemical vapor deposition of one dimensional tin oxide nanostructures: structural studies. Surface Modifications and Device Applications 2011.

[50] K.L. Choy. Chemical vapour deposition of coatings. Progress in Materials Science 2003; 48: 57–170.

Special Plasma Sources

Chapter 9

Laser-Produced Heavy Ion Plasmas as Efficient Soft X-Ray Sources

Takeshi Higashiguchi, Padraig Dunne and Gerry O'Sullivan

Additional information is available at the end of the chapter

http://dx.doi.org/10.5772/63455

Abstract

We demonstrate extreme ultraviolet (EUV) and soft x-ray sources in the 2- to 7 -nm spectral region related to the beyond extreme ultraviolet (BEUV) question at 6.x nm and a water window source based on laser-produced high-Z plasmas. Strong emissions from multiply charged ions merge to produce intense unresolved transition array (UTA) toward extending below the carbon K-edge (4.37 nm). An outline of a microscope design for single-shot live- cell imaging is proposed based on a high-Z UTA plasma source, coupled to x-ray optics. We will discuss the progress and Z-scaling of UTA emission spectra to achieve lab-scale table-top, efficient, high-brightness high-Z plasma EUV-soft x-ray sources for *in vivo* bio-imaging applications.

Keywords: High-Z, unresolved transition array (UTA), extreme ultraviolet (EUV), soft x-ray, water window

1. Introduction

Laboratory- scale source development of shorter- wavelength spectral regions in the extreme ultraviolet (EUV) and soft x-ray has been motivated by their applications in a number of high-profile areas of science and technology. One such topic is the challenge of three-dimensional imaging and single-shot flash photography of microscopic biological structures, such as macromolecules and cells, *in vivo*. For x-ray microscopy, the x-ray source should emit a sufficient photon flux to expose the image of the biosample on the detector. Recently, the most practical source of high-power, high-brightness x-rays has been radiation from synchrotrons and x-ray– free electron lasers (XFEL) [1]. Compact sources using liquid nitrogen droplets are being developed for the use of the zone plates for the transmission microscopy. Recently, the wavelength at 2.48- nm narrowband emission from a liquid-nitrogen-jet laser-plasma [2] was successfully combined with the latest normal-incidence multilayer condenser optics and 20-nm zone-plate optics to work laboratory water-window x-ray microscopy [3] with resolution less than 25 nm and synchrotron-like image quality on biological and environment science samples. The development of a high-brightness source based on a focused electron beam impacting a liquid water jet resulting in 2.36-nm emission has also been studied [4]. The total collected energy, however, is low, when one combines the narrowband line emission with the low reflection coefficient of the collector mirror. As a result, long exposures are needed to take a picture, and there is not yet published evidence of single-shot, flash exposures by the use of a laboratory-scale source. In order to overcome the low efficiency imposed by line emission sources, we propose the use of high-power water-window emission from laser-produced high-Z plasmas, analogous to the extending scheme of efficient, high-volume manufacturing EUV sources.

High-power EUV sources with high efficiency for semiconductor lithography at 13.5 [5] and 6.7 nm [6–8] based on laser-produced plasmas (LPP) have been demonstrated in high-volume manufacturing of integrated circuits (IC) having node sizes of 22 nm or less [9, 10]. The EUV emission at the relevant wavelength may be coupled with La/B_4C or Mo/B_4C multilayer mirror with a reflectivity of 40% to provide a source at 6.5–6.7 nm. Recently, a reflection coefficient of about 60–70% was shown to be feasible in a theoretical study [11]. Consequently, the development of a new wavelength EUV source for the next-generation semiconductor lithography, which can be coupled with an efficient B_4C multilayer mirror, is particularly timely.

High-Z element plasmas of Sn and Gd produce strong resonant band emission due to $4d$–$4f$ and $4p$–$4d$ transitions around 13.5 nm and 6.7 nm, respectively, which are overlapped in adjacent ion stages to yield the intense unresolved transition arrays (UTAs) in their spectra. The in-band high-energy emissions are attributable to hundreds of thousands of near-degenerate resonance lines lying within a narrow wavelength range. Rare earth elements gadolinium (Gd) and terbium (Tb) produce strong resonant emission in an intense UTA around 6.5–6.7 nm [6–8]. The choice of these elements was prompted by the use of UTA radiation in tin (Sn) for the strong 13.5-nm emission, where $n = 4$–$n = 4$ transitions in Sn ions overlap to yield an intense UTA [12, 13], as the optimum source for 13.5 nm and the scaling of this emission to the shorter wavelength with increasing Z. Because the emitting ions in Gd and Tb plasmas have largely a similar electronic structure to Sn, they are expected to have a similar spectral behavior and emit an intense UTA due to $4d$–$4f$ and $4p$–$4d$ transitions at shorter wavelengths.

Plasmas of the rare earth elements gadolinium (Gd) and terbium (Tb) produce strong resonant emission due to the presence of an intense UTA around 6.5–6.7 nm in the spectra of their ions [6]. In tin (Sn), the presence of the corresponding feature at 13.5 nm prompted its selection as the optimum source material at that wavelength. The UTA emission scales to shorter wavelength with increasing atomic number, Z. Because the emitting ions in Gd (Z = 64) and Tb (Z = 65) plasmas have an electronic structure largely similar to Sn, they are expected to exhibit a similar spectral behavior and emit an intense UTA due to $4d$–$4f$ and $4p$–$4d$ transitions at shorter wavelengths. Recently, the suitability of Nd:yttrium-aluminum-garnet (Nd:YAG) LPP EUV sources based on Gd and Tb has been demonstrated for high-power operation [6]. Since at high plasma electron densities, the opacity effects reduce the intensity of the resonance lines thereby limiting the output power, methods of reducing the effects of reabsorption (opacity) were evaluated to achieve high energy conversion efficiency (CE) from the incident laser energy to the EUV emission energy and the spectral purity. The effect of optical thickness was evaluated by changing the laser wavelength to alter the plasma electron density [7, 14]. In order to increase the EUV energy CE and spectral efficiency (purity), the optical thickness in the dominant region of the EUV emission of high-Z highly charged plasmas should be controlled. To enhance the EUV emission from Gd plasmas, it is important to reduce reabsorption by the resonance lines and the emission from satellite lines that attribute to the long wavelength side of the array around 6.7 nm to improve the spectral purity as well as increase the resonance emission intensity [7]. In order to achieve this, we used low initial –density targets for the Nd:YAG LPPs [8]. In low-density, optically thin plasmas, a suppression of the reabsorption effect and the satellite emission, which originates from the high electron and ion density region, is expected, similar to the results obtained with low-density Sn targets used to optimize the emission from the Nd:YAG LPP EUV sources at 13.5 nm [15, 16]. It is known that optically thick plasmas can strongly self-absorb resonance emission. Optically, thin plasmas provide more efficient sources. Therefore, systematic LPP UTA source studies with up-to-date intense picosecond pulse lasers [17] or middle infrared laser, such as the CO_2 laser [14], are needed to determine available light source wavelengths for future applications.

In this chapter, we show the efficient EUV and soft x-ray sources in the 2- to 7- nm spectral region related to the beyond extreme ultraviolet (BEUV) question at 6.x nm and a water window source based on laser-produced high-Z plasmas. Resonance emission from multiply charged ions merges to produce intense UTA spectral structure, extending below the carbon K-edge (4.37 nm). An outline of a microscope design for single-shot live cell imaging is proposed based on a high-Z plasma UTA source, coupled to x-ray optics. We discuss the progress and Z-scaling of UTA emission spectra to achieve lab-scale table-top, efficient, high-brightness high-Z plasma EUV–soft x-ray sources for *in vivo* bioimaging applications.

2. Characteristics of the Gd plasmas for BEUV source applications

In order to increase the energy CE from the incident laser energy to the interested wavelength emission energy with the defined bandwidth, it is important to suppress not only the reabsorption by assurance of the plasma is optically thin but also plasma hydrodynamic expansion loss, while maintaining a plasma electron temperature of T_e = 100–120 eV [6, 17]. Lateral expansion of the plasma causes kinetic energy losses, which reduce the energy available for radiation and is particularly important for small focal spot diameters [6]. For practical EUV source development, it is important to establish the optimum plasma condition related to laser irradiation condition and construct a database of properties of the UTA plasma EUV sources. In addition, to compare with one-dimensional (1D) numerical simulation, it is important to produce 1D expansion plasmas by irradiating multiple laser beams based on the laser inertial confinement fusion (ICF) geometry [18]. Laboratory- scale experiments have, to date, only been studied under 2D conditions due to the use of a single laser beam and small focal spot diameters. Under multiple laser irradiation, it is expected that the highest CE will be achieved as plasma expansion loss can be neglected in plasmas from targets irradiated by solid-state laser pulses. In the database point of view, we demonstrate high CE for the EUV emission around 6.7 nm from multiple laser beam –produced 1D spherical plasmas of rare earth elements of Gd and Tb. The maximum in-band EUV CE at 6.7 nm within a 0.6% bandwidth (0.6%BW) in a solid angle of 2π sr was observed to be 0.8%, which is twice as large as that obtained by the use of a Joule-class laboratory- scale single laser beam with 2D or 3D plasma expansion losses. The CE value was one of the highest ever reported due to the reduction of the plasma expansion loss applying 12 laser beams under 1D plasma expansion condition.

A Nd:glass laser system, GEKKO-XII at the Institute of Laser Engineering (ILE) in Osaka University was used to produce the 1D expanding uniform plasma [19]. The GEKKO-XII laser facility consists of 12 laser beams each at a wavelength of 1.053 μm and a constant 1 J pulse energy, irradiating a total energy of 12 J, with a temporal Gaussian- shaped pulse width of 1.3 ns [full width at half maximum (FWHM)]. The 12 laser beams were located at 12 faces of a regular dodecahedron to irradiate spherical targets uniformly. A thick metallic layer of 2 μm was coated onto spherical polystyrene balls for providing targets. The laser power imbalance was monitored to be within ± 6.3% of the average. Then, the laser beams were uniformly irradiated onto the target, to provide a 1D plasma expansion with low expansion loss.

Figure 1 shows the temporal history of the in-band emission around 6.7 nm with the bandwidth of 0.6% from Gd plasmas observed by the x-ray streak camera to provide 1D time-resolved imaging. The red and blue lines are the EUV emission at the optimum intensity of 1×10^{12} W/cm^2 and the maximum intensity of 3×10^{13} W/cm^2, respectively. Under optimum irradiation conditions with the highest CE, the temporal profile of the EUV emission was similar of that of the laser pulse shown by the dashed line and reached a maximum a little later. On the other hand, the behavior of the EUV emission profile at 3×10^{13} W/cm^2 initially rose faster, but the peak was delayed by comparison with that obtained under optimum conditions. The initial steep rise indicates that the electron temperature quickly reaches a value necessary for the in-band EUV emission. The final electron temperature is expected to be higher than optimum, so that higher charge state ions higher than q = 28 are produced, which predominantly emit shorter- wavelength out-of-band emission around 2–4 nm. After the maximum electron temperature is attained, plasma recombination proceeds accompanied by adiabatic expansion, resulting in cooling. The in-band emission from ionic charge states of $q \approx 20$ arises in the recombination phase. Then, the time- resolved emission consists of a fast rising component and a delayed peak relative to the laser pulse. This measurement suggests that the temporal shape of the in-band emission should essentially behave similarly to the laser pulse shape under optimum laser irradiation conditions.

The in-band EUV CEs were evaluated at λ = 6.7 nm within a bandwidth of 0.6% for Gd and Mo and at λ = 6.5 nm with the bandwidth of 0.6% for Tb. The CEs were maximized at 0.8% in both Gd and Tb at $I_L = 1 \times 10^{12}$ W/cm^2, and the observed maximum CE was almost in agreement with the theoretical evaluation of 0.9% obtained from a collisional–radiative (CR) and modified 1D hydrodynamic code numerical simulation [21]. It is noted that the wavelength of 6.6 nm, predicted in the work, is slightly different compared to our spectral peaks at 6.5 and 6.7 nm.

A decrease in CE was also observed at the laser intensity higher than 1×10^{12} W/cm². Around these intensities, the rare earth highly charged plasmas are overheated, the average ionization stage increases and the population of relevant ions with $q \approx 20$ decreases. Then, the CE decreases due to the increase in electron temperature [20].

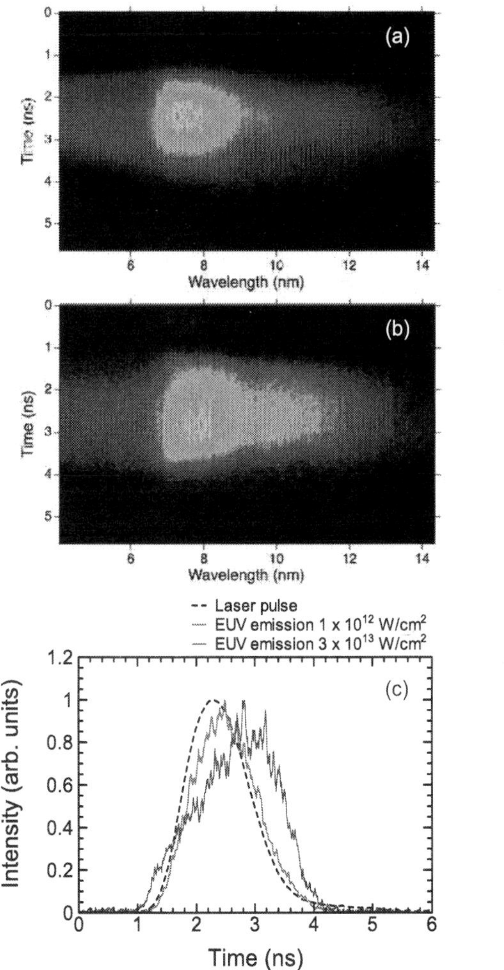

Figure 1. Time-resolved spectral images at two different laser intensities of (a) 1×10^{12} W/cm² and (b) 3×10^{13} W/cm², respectively. (c) Temporal histories of the EUV emission at 6.7 nm from Gd plasmas at two different laser intensities of 1×10^{12} W/cm² (red) and 3×10^{13} W/cm² (blue), together with a temporal profile of the laser pulse (dashed). At an optimum laser intensity of 1×10^{12} W/cm², the temporal behavior of the in-band emission is essentially the same as that of the laser pulse. It should be noted that intensities are normalized for timing comparison [20].

In addition, it is important to understand the physics of the EUV emission and transport in laser-produced dense high-Z plasmas. In order to achieve an efficient light source, or to diagnose complex highly charged ion (HCI) plasmas, the evaluation of plasma parameters is of fundamental importance in order to benchmark radiation hydrodynamic simulation codes. One matter of fundamental physics is the relationship between the electron density profile and the dominant EUV emission region. In general, dense high-Z plasmas are optically thick in the EUV spectral region, and the EUV emission originates from regions of reduced electron density where there is not only sufficient emissivity but also lower effects from opacity. We describe the results of measurements of the electron density profile of a laser produced isotropically expanding spherical Gd plasma using a Mach-Zehnder interferometer, as shown in Figure 2 [22]. The interferometry was performed at a wavelength of 532 nm to enable penetration of the plasmas to a high -density region, which has a maximum density close to the critical density of 1×10^{21} cm^{-3} as set by the plasma initiating laser wavelength of 1.053 μm. The EUV emission was observed using a monochromatic EUV pinhole camera. We present benchmark data for the electron density profile with the dominant EUV emission at 6.7 nm occurring in a region with an electron density close to 10^{19} cm^{-3} [14], which was corresponded to the critical density of the CO_2 (carbon dioxide) laser LPP, as shown in Figure 3 [22].

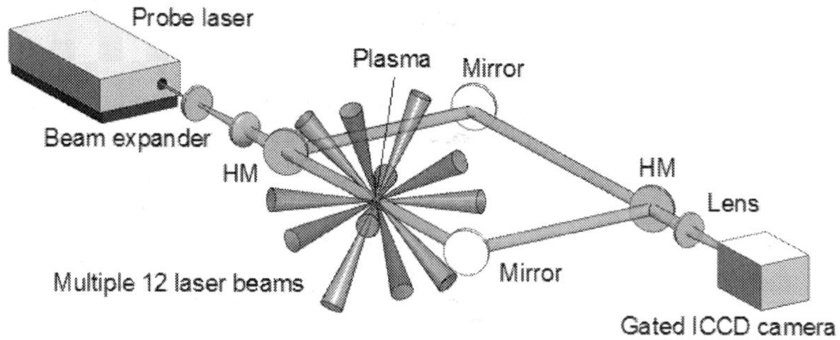

Figure 2. Schematic diagram of the experimental setup. Interferograms were produced by a Mach-Zehnder interferometer by the use of a Nd:YAG laser at a wavelength of 532 nm with a pulse duration of 6 ns (FWHM) [22].

The production of low-density plasma by the use of CO_2 LPPs has been proposed, because the critical electron density n_{ec} depends on the laser wavelength, λ_L, i.e., $n_{ec} \propto \lambda_L^{-2}$. The critical density at a laser wavelength of λ_L = 10.6 μm for a CO_2 laser is two orders of magnitude smaller than at λ_L = 1.06 μm for the solid-state laser. Then, a suppression of reabsorption and satellite emission in the wavelength region longer than 6.x nm is expected in CO_2 LPPs due to the lower plasma electron density. By extending efficient CO_2 laser–produced Sn plasma EUV sources around 13.5 nm, the CE and spectral efficiency, which is important when considering out-of-band spectral suppression, should be increased in an optically thin plasma. In order to ascertain the applicability of a CO_2 LPP EUV source at 6.x nm, its behavior needs to be clarified in a manner similar to the work performed on CO_2 LPP EUV sources at 13.5 nm.

We characterize the EUV emission from CO_2 laser–produced plasmas (CO_2-LPPs) of the rare earth element of Gd. The energy CE and the spectral purity in the CO_2-LPPs were higher than that for solid-state LPPs at 1.06 µm, because the plasma produced is optically thin due to the lower critical density, resulting in a maximum CE of 0.7% at 6.76 nm with 0.6% bandwidth in the solid angle of 2π sr. The peak wavelength was fixed at 6.76 nm for all laser intensities. The plasma parameters at a CO_2 laser intensity of 1.3×10^{11} W/cm^2 was also evaluated using the hydrodynamic simulation code to produce the EUV emission at 6.76 nm.

Figure 4(a) shows time-integrated EUV emission spectra from the Nd:YAG-LPPs at different laser intensities ranging from 9.7×10^{11} to 6.6×10^{12} W/cm^2. The peak wavelength shifts from 6.7 to 6.8 nm and is mainly due to $n = 4-n = 4$ ($\Delta n = 0$) transitions in HCIs with an open $4f$ or $4d$ outermost subshell. The sharp peak at 6.65 nm and the dip structure below 6.59 nm first appear at a laser intensity of 2.4×10^{11} W/cm^2. The emission at wavelengths less than 6 nm, increases with increasing laser intensity, and according to numerical evaluation, lines in the λ = 2.5–6 nm (hv = 207–496 eV) spectral region originate from Gd ionic charge states between Gd^{19+} and Gd^{27+} and arise from $n = 4-n = 5$ ($\Delta n = 1$) transitions [14].

In the case of CO_2-LPPs, the main spectral behaviors near 6.7 nm, on the other hand, are narrower than for Nd:YAG laser irradiating plasma, as shown in Figure 4(b). The CO_2 laser intensity was varied from 5.5×10^{10} to 1.2×10^{11} W/cm^2. The spectral structure was dramatically different to that from the Nd:YAG-LPPs. The peak wavelength of 6.76 nm remains constant with the increase of the laser intensity. Moreover, the emission intensity of the peak at 6.76 nm increases more rapidly with laser intensity than the emission in the ranges of λ = 3–6.6 nm and λ = 6.8–12 nm, respectively. Under the optically thin plasma conditions imposed by the CO_2-LPPs, this peak, which is mainly due to the $4d^{10}$ $^1S_0-4d^94f$ $^1P_1$ transition of Pd-like Gd^{18+} overlapped with $^2F-^2D$ lines of Ag-like Gd^{17+}, known to lie around 6.76 nm shows that these ions are indeed present in the plasma. Similar structure has been also observed in a discharge-produced plasma, which has low density and is optically thin like the CO_2-LPP. It is noted that

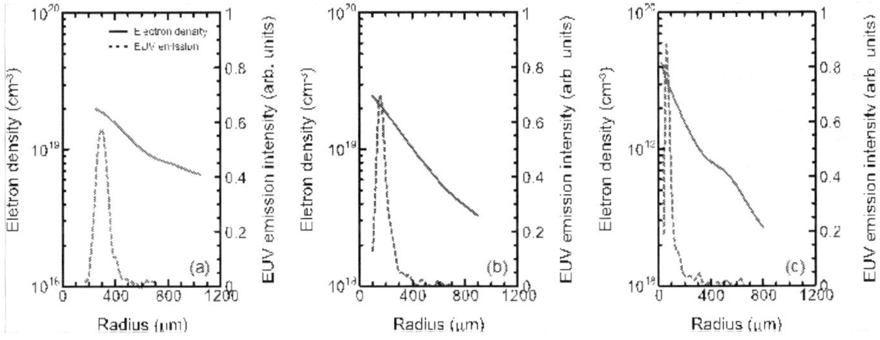

Figure 3. Profiles of the radial electron density (solid line) and radial EUV emission (dashed line) at the time of three different peak laser intensities of (a) 1×10^{12} W/cm^2, (b) 7×10^{12} W/cm^2, and (c) 1×10^{14} W/cm^2, corresponding to laser focal spot and target diameters of (a) 500 µm, (b) 200 µm, and (c) 50 µm [22].

the peak wavelength of 6.76 nm was constant with high spectral efficiency (purity) and energy CE in optically thin CO_2-LPPs of Gd [14].

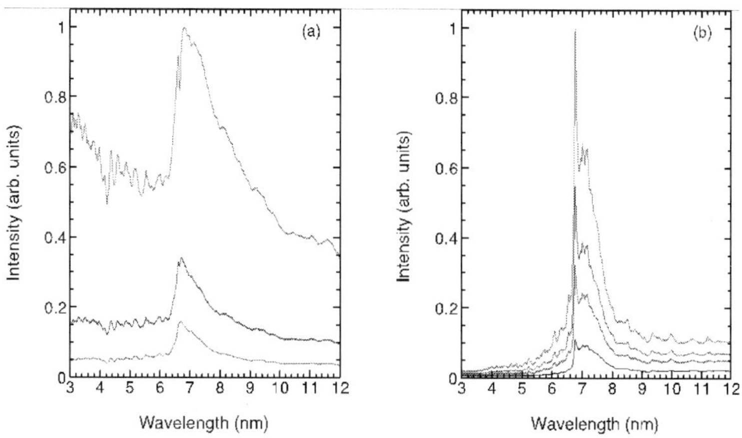

Figure 4. (a) Time-integrated EUV emission spectra from the Nd:YAG LPPs at different laser intensities of 9.7×10^{11}, 2.2×10^{12}, and 6.6×10^{12} W/cm^2, respectively. The peak wavelength shifts from 6.7 to 6.8 nm with increasing the laser intensity. (b) Time-integrated EUV emission spectra from the CO_2 LPPs at different laser intensities of 5.5×10^{10}, 8×10^{10}, 9.8×10^{10}, and 1.3×10^{11} W/cm^2, respectively. The peak wavelength of 6.76 nm remains constant with increasing the laser intensity [14].

In order to infer the laser parameters that maximize $6.x$-nm Gd-LPP emission, direct comparison between emissions from a laser-produced Gd plasma and that of Gd ions from well-defined charge states is necessary, as the charge state dependence of the emission at $6.x$ nm is defined by the electron temperature. We present a study of the charge state–defined emission spectra to explain the laser power density dependence of the Gd-LPP spectra and to evaluate the charge states contributing to the $6.x$-nm emission.

The profile of the intense emission at $6.x$ nm becomes broader, and its peak wavelength shifts to longer wavelength with increasing laser power density, as shown in Figure 5(a). However, the range of wavelengths involved is quite small, and the peak lies between 6.7 and 6.8 nm over this entire range of power densities. The emission from each of these peak wavelengths within a 0.6%BW becomes more intense with increasing laser flux. This behavior causes difficulty in fixing the precise wavelength of $6.x$ nm and optimization of the spectral efficiency while simultaneously maximizing the CE. The spectral efficiency denotes the ratio of the in-band energy at 6.70 nm within a 0.6%BW to that in the spectral ranging from 3 nm to 12 nm. An increase in laser power density raises the electron temperature which, in turn, implies an increase of both the highest charge state and the abundance of higher charge states. This change in the ion population must cause the observed shift of the peak wavelength for Gd-LPPs. Up to now, there was no direct experimental evidence that changes in emitting ion populations were responsible for this shift.

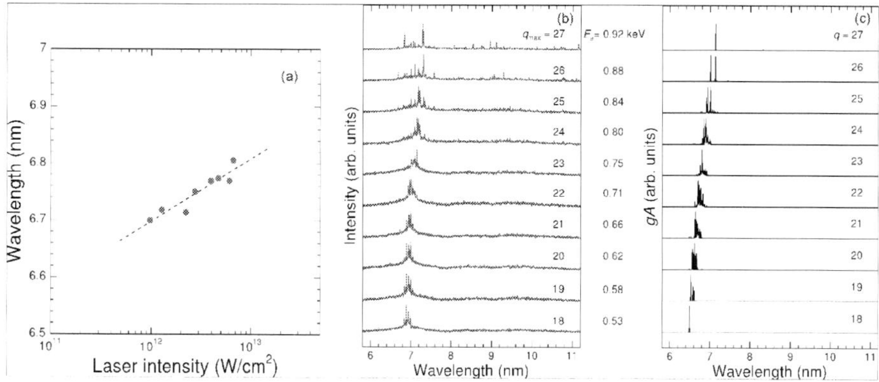

Figure 5. (a) The wavelength of the emission peaks near 6.x nm as a function of the Nd:YAG laser power density. The dotted line is a fitted curve. (b) EUV emission spectra of Gd ions with electron beam energies (E_e) of 0.43–0.92 keV. In the case of E_e = 0.43 keV, the compact EBIT with lower resolution was employed, while the Tokyo-EBIT was used in other cases. (c) Calculated gA values for $4d$–$4f$ transitions of the corresponding highest charge states (q_{max}) from Figure 5(b). The ground configuration of Gd^{18+} is $[Kr]4d^{10}$ [23]. Note that q is the charge state of Gd in Figures 5(b) and 5(c).

To verify the above explanation, charge-defined emission spectra were measured with the EBITs for different highest charge states. EUV emission spectra from EBIT experiments are shown in Figure 5(b), and calculated gA values of $4d$–$4f$ transitions for corresponding highest charge states are shown in Figure 5(c) to compare the charge state dependence of the emission near 6.x nm. The gA values are the transition probabilities from excited states multiplied by their statistical weights and thus are proportional to the emission intensities of the transitions. Note that the EBIT spectra include a subset of all possible radiative transitions that are predominantly resonant transitions to the ground state. For Pd-like Gd^{18+}, only one strong line is predicted corresponding to the $4d^{10}\,{}^1S_0$–$4d^94f\,{}^1P_1$ at 6.7636 nm, and this is clearly seen in the spectrum. In the absence of the configuration interaction (CI), according to the UTA model, the position of the intensity-weighted peak of the $4d^N$–$4d^{N-1}4f$ array depends directly on the occupancy of the $4d$ subshell, N, and the Slater-Condon $F^k(4d,4f)$ and $G^k(4d,4f)$ parameters. In the present case, the values of F^k and G^k change little with ionization stage, and therefore, the position of the array moves to lower energy with decreasing N. The presence of CI causes this shift to be reduced, but nevertheless, the overall trend is to move to longer wavelength with increasing ionization stage. The dominant emissions around 7 nm in the EBIT spectra indeed move to longer wavelengths with an increase of the highest charge state. The EBIT can thus generate charge- defined emission spectra, which are essential for both analysis of plasma emission spectra and the benchmarking of theoretical calculations [23].

3. Quasi-Moseley's law for the UTA emission

In this section, we show that the strong resonance UTAs of Nd:YAG LPPs for elements with Z = 50–83 obey a quasi-Moseley's law [24]. A 150-ps Nd:YAG laser with a maximum energy of 250 mJ at $\lambda_L = 1.064$ μm and an 8-ns Nd:YAG laser giving 400 mJ at $\lambda_L = 1.064$ μm were employed to provide the desired variation of laser intensity. The laser beam was incident normally onto planar high-Z metal targets in vacuo. The expected focal spot size, produced by an anti–reflection- coated plano-convex BK7 lens with a focal length of 10 cm, had a FWHM of approximately 50 μm. The laser was operated in single shot mode, and the target surface was translated to provide a fresh surface after each laser shot. A flat-field grazing incidence spectrometer (GIS) with an unequally ruled 2400 grooves/mm grating was placed at 30° with respect to the axis of the incident laser. Time-integrated spectra were recorded by a Peltier-cooled back-illuminated charge-coupled device (CCD) camera and were corrected by its quantum efficiency. The typical resolution was better than 0.005 nm (FWHM). The Large Helical Device (LHD) is one of the largest devices for magnetically confined fusion research and is described in detail elsewhere. The LHD plasmas were produced by the injection of a small amount of target elements into the background hydrogen plasma. The plasma density is about 10^{13} cm^{-3}, much lower than that in a LPP, and guarantees an optically thin condition. Emission spectra were recorded by a 2-m grazing incidence Schwob-Fraenkel spectrometer with a 600 grooves/mm grating. The exposure time of the detector was set at 0.2 s, and the spectral resolution is about 0.01 nm (FWHM).

Figures 6(a) –6(k) show LPP emission spectra from high-Z metal targets. The main UTA peak at 8.17 nm in the case of Nd clearly shifts to shorter wavelength with increasing atomic number, 3.95 nm in the case of Bi. This movement indicates the availability of a wide wavelength range for a LPP light source. While the main UTA peaks correspond to $4p^64d^N–4p^64d^{N-1}4f$ transitions, the $4p^64d^N–4p^54d^{N+1}$ UTAs were also observed around them, at 4 nm for the LPP of Pt, in the case of 150-ps LPPs. Optically thinner LHD plasma spectra are shown in Figures 6(l) –6(q). It should be noted that the electron temperatures of LHD plasma were relatively low, ≤ 1 keV, but higher than in 150-ps LPPs [24].

As a result, we have not observed significant emission of the type $4f^N–4f^{N-1}5l$ from stages with open $4f$ valence subshells in LHD spectra. Comparing LPP and LHD spectra, the UTA widths in LHD spectra are relatively narrower than in LPPs especially for lighter elements. This arises as a result of a number of factors: the increased contributions from ions with an outermost $4d^{10}4f^N$ configuration from transitions of the type $4d^{10}4f^N–4d^94f^{N+1}$ in LPP spectra and the differences in opacity that reduce the intensity of the strongest lines and the increased contribution from satellite emission. In addition, earlier research demonstrated that if the majority of radiation originates from open $4f$ subshell ions, whose complexity inhibits the emission of strong isolated lines, then no strong isolated lines are expected to appear through-out the EUV emission, which is clearly seen for the LPP spectra in Figure 6. Moreover, self-absorption effects are clearly observed in the case of 10-ns LPP for Nd due to optical thickness. Although the $n = 4–n = 4$ UTA transition peak was observed at 8.05 nm in the LHD spectrum, the strongest $4d–4f$ transitions essentially disappear in the 10-ns LPP owing to self-absorption. Because of their large transition probabilities, resonant lines that are strong in emission also strongly absorb in underdense ($n_e < n_{ec}$) or optically thick plasma conditions. An optically thinner plasma reduces the self-absorption effects and increases the spectral efficiency of $n = 4–n = 4$ UTA emissions.

Figure 7 shows the atomic number dependence of the observed peak wavelength of $n = 4$–$n = 4$ UTAs. The solid line is an approximated curve for 150-ps LPPs with a power-law scaling of the peak wavelength given by $\lambda = aR_0^{-1}(Z - s)^{-b}$ in nm, where $a = 21.86 \pm 12.09$, $b = 1.52 \pm 0.12$, $s = 23.23 \pm 2.87$ is the screening constant while Slater's rule gives $s = 36$–39.15 for $4d$ electrons, and R_0 is the Rydberg constant. This empirical law is surprisingly similar to Moseley's law, where $a = 4/3$, $b = 2$, and $s = 1$ were used to give the transition wavelength of the $K\alpha$-line of characteristic x-rays. It is noted that the Moseley's law derived from the Bohr model gives $\lambda = 0$ for $\Delta n = 0$ transitions in terms of the energy difference. It can, however, be fitted as a quasi-Moseley's law because there are energy differences between $\Delta n = 0$ levels due to different angular momentum quantum numbers [24].

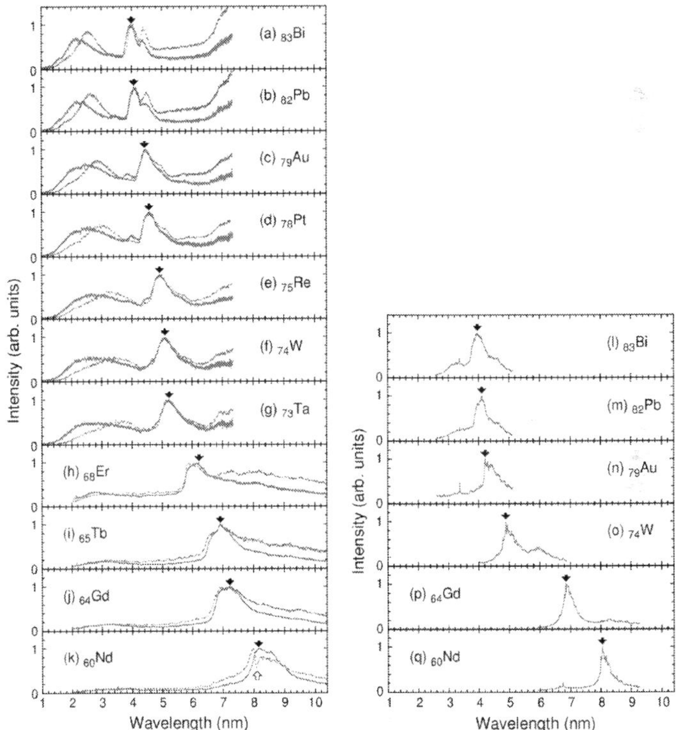

Figure 6. Time-integrated EUV emission spectra of the Nd:YAG LPPs for (a) $_{83}$Bi, (b) $_{82}$Pb, (c) $_{79}$Au, (d) $_{78}$Pt, (e) $_{75}$Re, (f) $_{74}$W, (g) $_{73}$Ta, (h) $_{68}$Er, (i) $_{65}$Tb, (j) $_{6}$4Gd, and (k) $_{6}$0Nd targets with 150-ps laser (red, solid line) and 10-ns laser (blue, dotted line), respectively. Typical laser power densities were 2.5×10^{14} W/cm^2 for ps-laser illumination and 5.6×10^{12} W/cm^2 for 10-ns laser irradiation. The measured LHD spectra (green, solid) for (l) Bi, (m) Pb, (n) Au, (o) W, (p) Gd, and (q) Nd targets, respectively. An emission line at 3.4 nm is from impurity carbon ions. Intensities were normalized at each maximum of the $n = 4$–$n = 4$ UTAs. Solid arrows indicate peak position of $n = 4$–$n = 4$ UTAs of 150-ps LPP and LHD spectra. An open arrow indicates structure due to self-absorption [24].

We propose here a pathway to produce feasible laboratory-scale high-Z LPP sources for a wide range of applications. For efficient UTA emission, plasmas of higher-Z elements need high-electron temperatures to produce higher charge state ions contributing to the $4p^64d^N$–$4p^64d^{N-1}4f$ UTAs. The electron temperature, T_e, rises with increasing laser intensity, such as $T_e \propto (I_L\lambda_L^2)^{0.4}$, where I_L and λ_L are the laser intensity and wavelength, respectively [25]. On the other hand, an optically thin plasma has a low electron density, n_e, which decreases with increasing λ_L. In terms of these features, use of a longer laser wavelength is necessary to generate the brightest LPP, such as a CO_2 laser operating at 10.6 μm due to the low critical density of 1×10^{19} cm^{-3} attainable with a pulse duration sufficiently short to give a laser intensity of the order of 10^{13} W/cm^2 but sufficiently long to permit excitation to the appropriate ionization stages, i.e., ~ 1 ns. Moreover, we can also obtain longer wavelengths, > 10.6 μm, with a Raman conversion system.

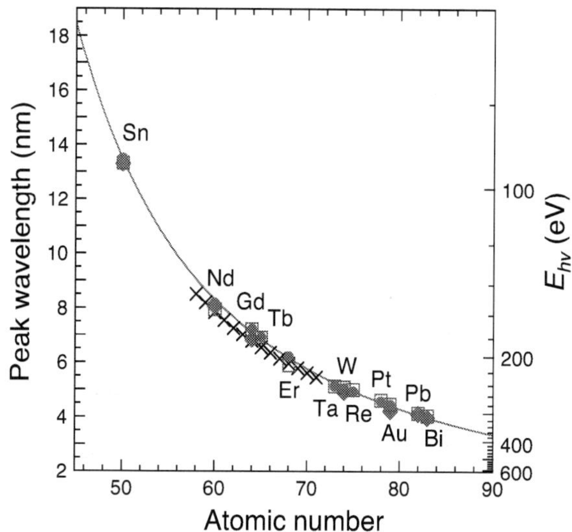

Figure 7. Atomic number dependence of the peak wavelength of $n = 4$–$n = 4$ UTAs in 150-ps LPP (red, circles), 10-ns LPP (blue, squares), and LHD (green, diamonds) spectra. Calculated peak wavelengths with GRASP are also shown (black, crosses). Sn spectra are not shown in Figure 6. The solid line is an approximated curve for $n = 4$–$n = 4$ UTAs in 150-ps LPPs with a power-law scaling [24].

4. Water window soft x-ray source by high-Z ions

4.1. Spectroscopy of low electron temperature in lab-scale laser-produced ions

According to the quasi-Moseley's law in Figure 7, the elements from $_{79}$Au to $_{83}$Bi are one of the candidates for high-flux UTA source in water window soft x-ray sources for single-shot (flash) bio-imaging in the laboratory size microscope, because the UTA emission is essentially high-power emission due to much resonant lines around the specific wavelength (photon energy). The UTA peak wavelengths of $_{79}$Au, $_{82}$Pb, and $_{83}$Bi reach the water window soft x-ray spectral region.

Figures 8(a)–8(c) show time-integrated spectra from Au, Pb, and Bi plasmas at a laser intensity of the order of 10^{14} W/cm^2 with a pulse duration of 150 ps (FWHM). The time-integrated soft x-ray spectra between 1 and 6 nm from each element display strong broadband emission around 4 nm, which is mainly attributed to the $n = 4$–$n = 4$ transitions from HCIs with an open $4f$ or $4d$ outermost subshell with the broadband emission of 2–4 nm originating from the $n = 4$–$n = 5$ transitions from HCIs with an outermost $4f$ subshell. The intensity of the $n = 4$–$n = 4$ UTA emission was higher than that of the $n = 4$–$n = 5$ transition emission. The atomic number dependence of the spectral structure is shown in Figure 8(d). The predicted emission photon energy of each peak photon energy was shifted to higher photon energy with the increase of the atomic number. Neither the emission spectra nor the plasma electron temperatures, however, have been optimized, as shown below. However, the emission intensity of the $n = 4$–$n = 5$ transitions was compared with that of the $n = 4$–$n = 4$ transitions of the UTA emission [26].

We compared the results of numerical calculation for some different experimental tempera-tures with the observed spectra as shown in Figure 9(a). Four regions corresponding to emission peaks were identified. The emission in the region of "1" results primarily from the $4f$–$5g$ transitions in HCIs with an open 4f subshell, i.e., the stages lower than 35+ Bi ions. The emission in regions of "2" and "3" originates from $4p$–$4d$ and $4d$–$4f$ transitions with an open $4d$ subshells of Bi^{36+}–Bi^{45+}, and numerical calculations show that the higher- energy region results from the more highly ionized species higher than Bi^{42+}. The emission in the region of "4" was also associated mainly with the $4d$–$4f$ transition emission from lower ionic charge stages with an open $4f$ outmost subshell. As a result, the bulk of the emission, especially from regions of "1" and "4", was associated with the recombining phase of the expanding plasma plume. We evaluate for comparison spectra calculated for steady-state electron temperatures of 180 and 700 eV, while the higher temperatures were required to produce the emission in the region of "2", the calculations verify that both the longer and shorter wavelength features were consistent with much lower plasma electron temperatures [26].

In Figure 9(b), evaluated spectra at different electron temperatures higher than 900 eV were shown. Numerical calculations show that high-Z plasmas at an electron temperature lower than 700 eV, as shown in Figure 6(a), radiate strongly around 3.9 nm. In the case of higher electron temperatures from 800 to 1500 eV, the strongest emission, however, is expected at around 3.2 nm, suitable for coupling with Sc/Cr multilayer mirrors. Therefore, for an optimized source, we should produce a plasma at high electron temperature of around 1 keV. The emission intensity of the Bi plasma was compared with 2.48-nm nitrogen line emission from a Si$_3$N$_4$ planar target, in the same experimental setup, and was observed to be 1.2 times higher within a bandwidth of 0.008 nm (FWHM) even though the plasma electron temperature was much lower than the optimum value [26].

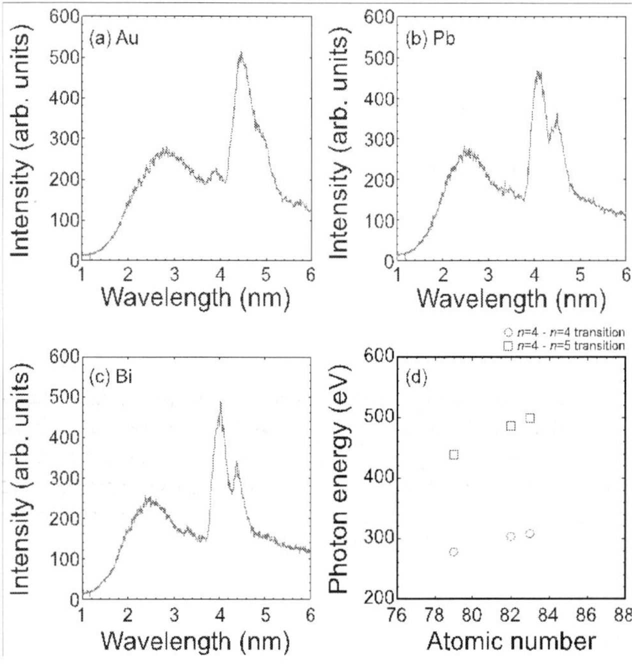

Figure 8. Time-integrated spectra from the picosecond-laser–produced high-Z plasmas by the use of Au (a), Pb (b), and Bi (c), and the atomic number dependence of the photon energies of the peak emission of the $n = 4$–$n = 4$ transition (circles) and the $n = 4$–$n = 5$ transition (rectangles) (d) [26].

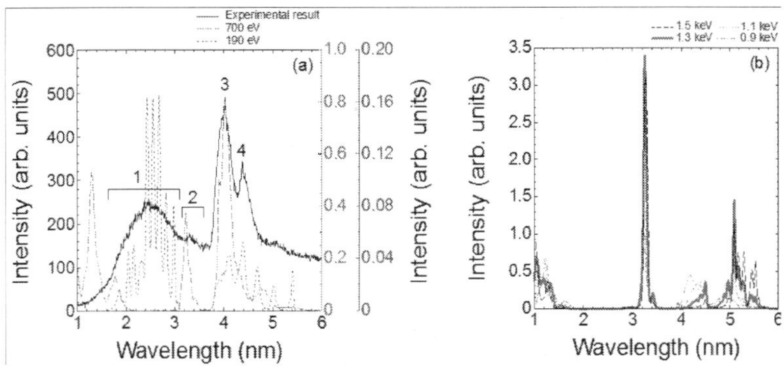

Figure 9. (a) The comparison between the observed spectrum with numerical calculation under assuming steady-state electron temperatures of 190 and 700 eV. (b) Calculated spectra for electron temperatures higher than 900 eV. [26]

4.2. Toward the laboratory water window soft x-ray microscope

Because of the broadband features of the emission, the zone plate components cannot be used, so one of the possible solutions would be to use a transmission planar x-ray nano-waveguide to image the sample. In order to achieve high resolution in the recorded image, we should also replace the recording device from the x-ray CCD camera to the sensitive EUV resist to overcome the resolution limitation of the CCD pixel size, coupling with the Schwarzschild optics, consisting of Sc/Cr multilayer mirrors. Although our proposal is based on a simple microscope construction, the key component is the UTA emitted from a hot dense Bi plasma point source, combined with Sc/Cr MLMs and sensitive EUV resists based on the photochemical reaction [26].

5. Summary

We have shown EUV and soft x-ray sources in the 2- to 7- nm spectral region related to the BEUV question at $6.x$ nm and a water window source based on laser-produced high-Z plasmas. The efficient $6.x$-nm BEUV sources have been demonstrated at the CE of 0.7% due to the high spectral purity by the optically thin plasmas after the database experiments. According to the atomic number dependence of the UTA emission, so- called quasi-Moseley's law, the Bi HCI plasma source is one of the solutions in the laboratory single-shot (flash) bio-imaging by extending the UTA light source feature.

Author details

Takeshi Higashiguchi[1*], Padraig Dunne[2] and Gerry O'Sullivan[2]

*Address all correspondence to: higashi@cc.utsunomiya-u.ac.jp

1 Department of Electrical and Electronic Engineering, Faculty of Engineering, Utsunomiya University, Utsunomiya, Tochigi, Japan

2 School of Physics, University College Dublin, Belfield, Dublin, Ireland

References

[1] Gorniak T, Heine R, Mancuso A. P, Staier F, Christophis C, Pettitt M. E, Sakdinawat A, Treusch R, Guerassimova N, Feldhaus J, Gutt C, Grübel G, Eisebitt S, Beyer A, Gölzhäuser A, Weckert E, Grunze M, Vartanyants I. A, Rosenhahn A: X-ray holographic microscopy with zone plates applied to biological samples in the water win-

dow using 3rd harmonic radiation from the free-electron laser FLASH. Optics Express. 2011;19:11059–11070. DOI: 10.1364/OE.19.011059]

[2] Jansson P. A. C, Vogt U, Hertz H. M: Liquid nitrogen jet laser plasma source for compact soft x-ray microscopy. Review of Scientific Instruments. 2005;76:043503. DOI: 10.1063/1.1884186

[3] Takman P. A. C, Stollberg H, Johansson G. A, Holmberg A, Lindblom M, Hertz H. M: High-resolution compact X-ray microscopy. Journal of Microscopy. 2007;226:175–181. DOI: 10.1111/j.1365-2818.2007.01765.x

[4] Skoglund P, Lundström U, Vogt U, Hertz H. M: High-brightness water-window electron-impact liquid-jet microfocus source. Applied Physics Letters. 2010;96:084103. DOI: 10.1063/1.3310281

[5] Bakshi V, editor. EUV Sources for Lithography. Bellingham: SPIE Press Book; 2006. 1094 p. DOI: 10.1117/3.613774

[6] Otsuka T, Kilbane D, White J, Higashiguchi T, Yugami N, Yatagai T, Jiang W, Endo A, Dunne P, O'Sullivan G: Rare-earth plasma extreme ultraviolet sources at 6.5–6.7 nm. Applied Physics Letters. 2010;97:111503. DOI: 10.1063/1.3490704; and references therein.

[7] Otsuka T, Kilbane D, Higashiguchi T, Yugami N, Yatagai T, Jiang W, Endo A, Dunne P, O'Sullivan G: Systematic investigation of self-absorption and conversion efficiency of 6.7 nm extreme ultraviolet sources. Applied Physics Letters. 2010;97:231503. DOI: 10.1063/1.3526383

[8] Higashiguchi T, Otsuka T, Yugami N, Jiang W, Endo A, Li B, Kilbane D, Dunne P, O'Sullivan G: Extreme ultraviolet source at 6.7 nm based on a low-density plasma. Applied Physics Letters. 2011;99:191502. DOI: 10.1063/1.3660275

[9] Meiling H, Boeij W, Bornebroek F, Harned N, Jong I, Meijer H, Ockwell D, Peeters R, Setten E, Stoeldraijer J, Wagner C, Young S, Kool R, Kürz P, Lowisch M: From performance validation to volume introduction of ASML's NXE platform. Proceedings of SPIE. 2012;8322:83221G. DOI: 10.1117/12.916971

[10] Wagner C, Harned N: EUV lithography: Lithography gets extreme. Nature Photonics. 2010;4:24–26. DOI: 10.1038/nphoton.2009.251

[11] Platonov Y, Rodriguez J, Kries M, Louis E, Feigl T, Yulin S. Status of Multilayer Coatings for EUV Lithography. In: Proceedings of 2011 International Workshop on EUV Lithography; 13-17 June 2012; Maui (Hawaii). Austin: EUV Litho, Inc.; 2012. p. 49 (P25).

[12] O'Sullivan G, Carroll P. K: $4d$–$4f$ emission resonances in laser-produced plasmas. Journal of the Optical Society of America. 1981;71:227–230. DOI: 10.1364/JOSA. 71.000227

[13] Carroll P. K, O'Sullivan G: Ground-state configurations of ionic species I through XVI for Z = 57–74 and the interpretation of $4d$–$4f$ emission resonances in laser-produced plasmas. Physical Review A. 1982;25:275–286. DOI: 10.1103/PhysRevA.25.275

[14] Higashiguchi T, Li B, Suzuki Y, Kawasaki M, Ohashi H, Torii S, Nakamura D, Takahashi A, Okada T, Jiang W, Miura T, Endo A, Dunne P, O'Sullivan G, Makimura T: Characteristics of extreme ultraviolet emission from mid-infrared laser-produced rare-earth Gd plasmas. Optics Express. 2013;21:31837–31845. DOI: 10.1364/OE. 21.031837; and references therein.

[15] Okuno T, Fujioka S, Nishimura H, Tao Y, Nagai K, Gu Q, Ueda N, Ando T, Nishihara K, Norimatsu T, Miyanaga N, Izawa Y, Mima K: Low-density tin targets for efficient. Applied Physics Letters. 2006;88:161501. DOI: 10.1063/1.2195693

[16] Higashiguchi T, Dojyo N, Hamada M, Sasaki W, Kuobdera S: Low-debris, efficient laser-produced plasma extreme ultraviolet source by use of a regenerative liquid microjet target containing tin dioxide (SnO_2) nanoparticles. Applied Physics Letters. 2006;88:201503. DOI: 10.1063/1.2206131

[17] Cummins T, Otsuka T, Yugami N, Jiang W, Endo A, Li B, O'Gorman C, Dunne P, Sokell E, O'Sullivan G, Higashiguchi T: Optimizing conversion efficiency and reducing ion energy in a laser-produced Gd plasma. Applied Physics Letters. 2012;100:061118. DOI: 10.1063/1.3684242

[18] Yamanaka C, Kato Y, Izawa Y, Yoshida K, Yamanaka T, Sasaki T, Nakatsuka M, Mochizuki T, Kuroda J, Nakai S: Nd-doped phosphate glass laser systems for laser-fusion research. IEEE Journal of Quantum Electronics. 1981;17:1639–1649. DOI: 10.1109/JQE.1981.1071341

[19] Shimada Y, Nishimura H, Nakai M, Hashimoto K, Yamaura M, Tao Y, Shigemori K, Okuno T, Nishihara K, Kawamura T, Sunahara A, Nishikawa T, Sasaki A, Nagai K, Norimatsu T, Fujioka S, Uchida S, Miyanaga N, Izawa Y, Yamanaka C: Characterization of extreme ultraviolet emission from laser-produced spherical tin plasma generated with multiple laser beams. Applied Physics Letters. 2005;86:051501. DOI: 10.1063/1.1856697

[20] Yoshida K, Fujioka S, Higashiguchi T, Ugomori T, Tanaka N, Ohashi H, Kawasaki M, Suzuki Y, Suzuki C, Tomita K, Hirose R, Ejima T, Nishikino M, Sunahara A, Scally E, Li B, Yanagida T, Nishimura H, Azechi H, O'Sullivan G: Efficient extreme ultraviolet emission from one-dimensional spherical plasmas produced by multiple lasers. Applied Physics Express. 2014;7:086202. DOI: 10.7567/APEX.7.086202

[21] Masnavi M, Szilagyi J, Parchamy H, Richardson M. C: Laser-plasma source parameters for Kr, Gd, and Tb ions at 6.6 nm. Applied Physics Letters. 2013;102:164102. DOI: 10.1063/1.4802789

[22] Yoshida K, Fujioka S, Higashiguchi T, Ugomori T, Tanaka N, Kawasaki M, Suzuki Y, Suzuki C, Tomita K, Hirose R, Ejima T, Ohashi H, Nishikino M, Sunahara A, Li B,

Dunne P, O'Sullivan G, Yanagida T, Azechi H, Nishimura H: Density and x-ray emission profile relationships in highly ionized high-Z laser-produced plasmas. Applied Physics Letters. 2015;106:121109. DOI: 10.1063/1.4916395

[23] Ohashi H, Higashiguchi T, Li B, Suzuki Y, Kawasaki M, Kanehara T, Aida Y, Torii S, Makimura T, Jiang W, Dunne P, O'Sullivan G, Nakamura N: Tuning extreme ultraviolet emission for optimum coupling with multilayer mirrors for future lithography through control of ionic charge states. Journal of Applied Physics. 2014;115:033302. DOI: 10.1063/1.4862441; and references therein.

[24] Ohashi H, Higashiguchi T, Suzuki Y, Arai G, Otani Y, Yatagai T, Li B, Dunne P, O'Sullivan G, Jiang W, Endo A, Sakaue H. A, Kato D, Murakami I, Tamura N, Sudo S, Koike F, Suzuki C: Quasi-Moseley's law for strong narrow bandwidth soft x-ray sources containing higher charge-state ions. Applied Physics Letters. 2014;104:234107. DOI: 10.1063/1.4883475

[25] Colombant D, Tonon G. F: X-ray emission in laser-produced plasmas. Journal of Applied Physics.1973;44:3524-3537. DOI: 10.1063/1.1662796

[26] Higashiguchi T, Otsuka T, Yugami N, Jiang W, Endo A, Li B, Dunne P, O'Sullivan G: Feasibility study of broadband efficient "water window" source. Applied Physics Letters. 2012;100:014103. DOI: 10.1063/1.3673912; and references therein.

Chapter 10

Laser-Induced Plasma and its Applications

Kashif Chaudhary, Syed Zuhaib Haider Rizvi and Jalil Ali

Additional information is available at the end of the chapter

http://dx.doi.org/10.5772/61784

Abstract

The laser irradiation have shown a range of applications from fabricating, melting, and evaporating nanoparticles to changing their shape, structure, size, and size distribution. Laser induced plasma has used for different diagnostic and technological applications as detection, thin film deposition, and elemental identification. The possible interferences of atomic or molecular species are used to specify organic, inorganic or biological materials which allows critical applications in defense (landmines, explosive, forensic (trace of explosive or organic materials), public health (toxic substances pharmaceutical products), or environment (organic wastes). Laser induced plasma for organic material potentially provide fast sensor systems for explosive trace and pathogen biological agent detection and analysis. The laser ablation process starts with electronic energy absorption (~fs) and ends at particle recondensation (~ms). Then, the ablation process can be governed by thermal, non-thermal processes or a combination of both. There are several types of models, i.e., thermal, mechanical, photophysical, photochemical and defect models, which describe the ablation process by one dominant mechanism only. Plasma ignition process includes bond breaking and plasma shielding during the laser pulse. Bond breaking mechanisms influence the quantity and form of energy (kinetic, ionization and excitation) that atoms and ions can acquire. Plasma expansion depends on the initial mass and energy in the plume. The process is governed by initial plasma properties (electron density, temperature, velocity) after the laser pulse and the expansion medium. During first microsecond after the laser pulse, plume expansion is adiabatic afterwards line radiation becomes the dominant mechanism of energy loss.

Keywords: Laser Induced Plasma, Plasma Diagnostics, Plasma Applications

1. Introduction

Laser-induced plasma (LIP) formation is a rapid process that is under investigation for several decades due to its versatile and complex nature. Intense laser pulse delivers energy to the target surface for a very short interval of time that instantly excites, ionizes and vaporizes the material

into an extremely hot vapour plume also called as 'plasma plume'. It has three main regions as depicted in Figure 1. Near the target surface is the hottest and the densest part of the plasma called core; in this region, the material is mostly found in the ionized state because of high temperatures. Within the core adjacent to the target surface there exists a so-called *Knudsen layer* having a thickness equal to a few mean free paths. *Knudsen layer* is defined as the region in which a particle achieves an equilibrium velocity distribution from non-equilibrium distribution within a few mean-free paths [1]. In the mid-region of plasma, ions and neutrals (atoms + molecules) coexist due to the ongoing ionization and recombination processes. The outermost region of the plasma is relatively cold, where the population of neutrals dominates and may absorb the radiations coming out of the core and mid-regions of the plasma. Beyond it, there is a shock wave front produced due to the explosive expansion of the plasma, and travels ahead of the plasma plume.

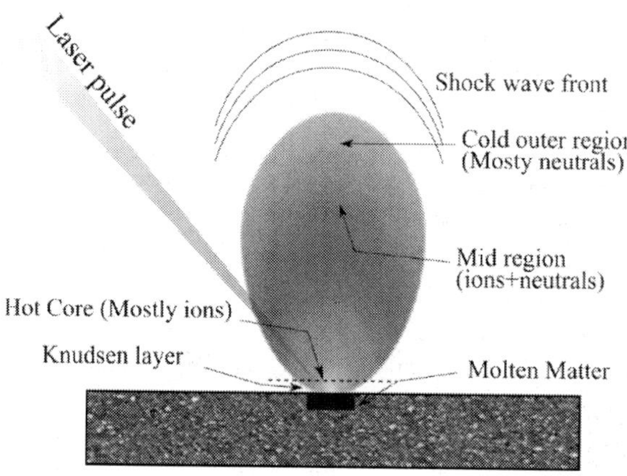

Figure 1. Schematic illustration of laser-induced plasma

In this chapter, the interaction of short (~nanosecond (ns)) and ultrashort (~picosecond (ps) and femtosecond (fs)) laser pulses with a variety of media, pulsed laser ablation (PLA), laser-induced plasma, plasma diagnostic techniques and applications of LIP is discussed. Out of the large canvas, the description about standard plasma diagnostic techniques encompassing optical emission spectroscopy (OES), laser-induced fluorescence (LIF), Langmuir probe, Faraday's cups and solid-state nuclear track detectors (SSNTDs) is provided. As the last section of the chapter, extensive literature review is presented that elaborates applications of the laser-induced plasma. Applications of laser-induced plasma in two main fields, that is, laser-induced breakdown spectroscopy (LIBS) and thin-film or pulsed laser deposition (PLD), are discussed in diverse fields of science and technology.

2. Laser–matter interaction

Lasers are unique energy sources characterized by their spectral purity, spatial and temporal coherence and high intensity. When laser interacts with matter, it can reflect, scatter, absorb or transmit depending upon the material characteristics (i.e. composition, physical, chemical and optical properties) and laser parameters. Laser energy, wavelength, spatial and temporal coherence, exposure time/pulse duration, etc. are the key influential parameters in laser–matter interaction.

The laser basically interacts with electrons in the material by energy transfer and excitation. Free electrons absorb energy and accelerate, whereas bound electrons excite to higher energy levels if laser photons match the bandgap energy of atoms. This absorbed energy is later released by the excited/accelerated electrons as electromagnetic radiation or dissipate into lattice as heat energy.

Different types of interactions of laser with the matter have found various applications. However, this chapter focuses on the laser-induced plasma that is, primarily, the consequence of laser energy absorption in the material. The interaction of laser is different for different materials.

In *metals*, there are a large number of free electrons available; the laser energy excites these electrons to various energies. The excitation of free electrons is taken as the increase in their kinetic energy. As free electrons are excited, the collision rate increases and the energy is transferred from the excited free electrons to lattice phonons.

Since *semiconductors* and *insulators* have very few free electrons, the laser interacts with bound electrons in atomic/ionic shells and generates electron–hole pairs. The generation of electron–hole pair depends upon the laser photon energy. If it matches the material's bandgap 'E_g', that is, the energy difference between the highest level in valence band and lowest level in the conduction band, an electron–hole pair is generated. If it is lower than the E_g, the photon may not absorb or cause low-energy intra-band electronic excitations in vibrational states.

In *dielectrics*, the laser energy causes ionization and liberates electrons within the material. Once the free electrons are available, collisional ionization starts that is termed as impact or avalanche ionization. It forms a dense cloud of ions near the surface that is opaque to the laser and absorbs later part of the laser pulse. In dense *ceramics*, the laser interaction phenomenon is same as in dielectrics; however, in the porous *ceramics*, the laser is mostly scattered due to the random distribution of particle size.

3. Pulsed laser ablation

Pulsed laser ablation is a high-photon-yield sputtering phenomenon in which the material is removed from a solid surface by the interaction of high-intensity short laser pulses. It is consequence of the conversion of laser-induced electronic and vibration excitation into the kinetic energy of nuclear motion. The material removal rate typically exceeds one-tenth of a monolayer per pulse that alters the surface composition at mesoscale. If particles are ejected out of the surface but there is no detectable modification in the surface composition, it is commonly referred to as *laser-induced desorption*. Being a mesoscale phenomenon, PLA is influenced by bulk properties of the material, for example, elasticity and compressibility. Moreover, the laser ablation need not involve massive, catastrophic destruction of the surface.

In pulsed laser ablation,

- The total yield of ejected particles is strongly proportional to the density of electronic and vibrational excitation.

- There is a threshold for photon fluence, below which only desorption can take place with negligible material damage.

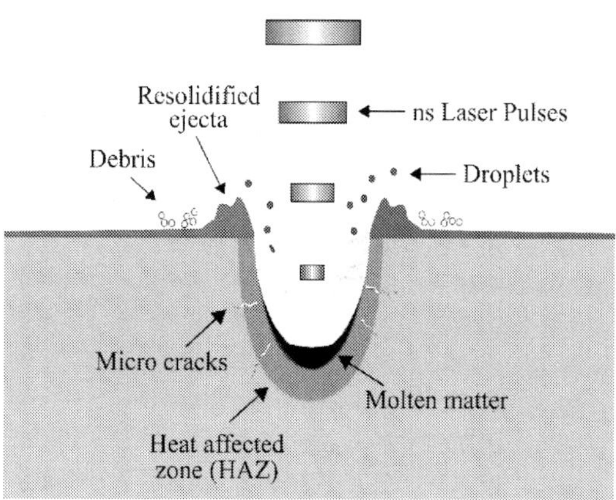

Figure 2. Illustration of nanosecond pulsed-laser ablation

PLA is a useful technique for surface machining. Although microcracks and heat-affected zone (HAZ) are formed as a result of thermal stresses and heat conduction into the bulk medium as illustrated in Figure 2, PLA provides an advantage of minimizing the heat-affected zone under certain circumstances. In the case of nanosecond laser pulses, the heat-affected zone will be minimal if the ablation depth per pulse Δh is comparable to the thermal penetration depth $l_t \approx 2\sqrt{D\tau_l}$ or the optical penetration depth $l_\alpha \approx \alpha^{-1}$ [2]:

$$\Delta h \approx \max\{l_\alpha, l_t\}$$

(1)

where D represents thermal diffusivity of the material, τ_l is the laser pulse duration and α is the optical absorption of the material. In case of ablation with ultrashort laser pulses, HAZ is nearly absent or is negligible due to non-thermal mechanisms governing the ablation.

3.1. Mechanisms of laser ablation

Laser ablation mechanisms are different for short (~ns) and ultrashort (~ps and fs) laser pulses due to the difference in laser coupling with matter at different timescales involved. Short pulse, that is, nanosecond pulse ablation, can be characterized through thermal, non-thermal and combination of both mechanisms. These are also referred to as photo-thermal, photo-chemical and photo-physical mechanisms, respectively. However, for certain experimental parameters and materials, the ablation can be described through one dominant mechanism.

If the excited material quickly converts the absorbed laser energy into heat, high temperature can build up near surface stresses in the material, which can lead to material ablation off the surface. Such an ablation is governed by the photo-thermal mechanism.

If the incoming laser photons have sufficiently high energy, the absorption can induce defects in the material or cause direct bond breaking of atoms, ions or molecules. Defects and bond breaking can individually cause ablation of the material. Such an ablation is termed as photo-chemical ablation. However, if the ablation occurs as a result of both thermal and non-thermal mechanisms, it is termed as photo-physical ablation. Figure 3 graphically illustrates the laser ablation process through thermal, non-thermal and photo-physical processes.

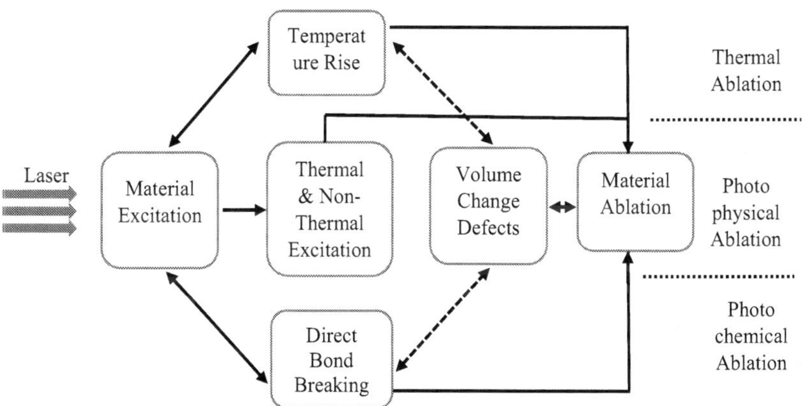

Figure 3. Graphical illustration of mechanisms leading to pulsed laser ablation

In case of ultrashort laser pulses, the mechanisms of photon absorption, heat transfer within the material and hydrodynamic ablation take place at different timescales independent of each other [3]. The phenomena that are not important or less significant in short-pulse laser ablation become dominant and significant because of ultrashort timescales and extreme laser intensities involved. Nonlinear absorption, Coulomb explosion and overcritical heating are the dominant phenomena in ultrashort pulsed laser ablation. At low laser intensities, close to the ablation threshold, Coulomb explosion governs the ablation. The absorbed laser energy accelerates electrons, and if the energy is sufficient to overcome the binding energy and work function of the material, electrons eject from the surface and form a charged cloud (of electrons) above the material surface leaving behind a charge density of ions on the surface. It creates a space charge separation; the electric field within the charge separation can be so strong that it can pull ions out of the material resulting in removal of top few monolayers from the surface [4]. Because of enormous intensities, non-linear multiphoton absorption phenomenon increases the laser energy absorption in the material that leads to cascade ionization. At ultrashort timescales (~pico- and femtoseconds), the material cannot either transfer energy to the lattice or evaporate instantly, rather the absorbed laser energy causes extreme excitation and heating of the material above critical temperatures. Consequently, a rapidly expanding mixture of liquid droplets and vapours is observed above the target surface [5].

4. Laser-induced plasma formation

4.1. Plasma ignition

Laser ablation can be divided into three phases

- Bond-breaking and plasma ignition
- Plasma expansion and cooling
- Particle ejection and condensation

The laser ablation process starts with electronic energy absorption (~fs) and ends at particle recondensation (~ms). Plasma ignition process includes bond breaking and plasma shielding during the laser pulse interaction with the material surface. Bond-breaking mechanisms influence the quantity and form of energy (kinetic, ionization and excitation) that atoms and ions can acquire.

For nanosecond laser pulses of irradiance $< 10^8$ W/cm^2, the dominant mechanism of plasma ignition is thermal vaporization [6], whereas for picosecond laser pulses with irradiances between 10^8 and 10^{13} W/cm^2 both thermal and non-thermal mechanisms contribute to the plasma ignition. However, with femtosecond laser pulses of irradiance $> 10^{13}$W/cm^2, the main bond-breaking (plasma ignition) mechanism is non-thermal, that is, Coulomb's explosion [6].

4.2. Plasma expansion and cooling

Plasma expansion depends on the initial mass and energy in the plume. The process is governed by initial plasma properties (electron density, temperature and velocity) after the laser pulse and the expansion medium. During the first microsecond after the laser pulse, plume expansion is adiabatic; afterwards, line radiation becomes the dominant mechanism of energy loss.

4.3. Particle formation and condensation

Condensation or particle formation takes place during the decay/cooling process of plasma. It starts when plasma temperature reaches the boiling point of the material and stops at the condensation temperature of the material. Exfoliation can also be observed as a result of high thermal stresses within the material due to fast heating. Thermal stresses produced in the material can break it into irregularly shaped particles that may eject from the surface. This phenomenon is called as exfoliation.

5. Plasma diagnostic techniques

Plasma is a rich source of radiations. A wide spectrum of electromagnetic radiation ranging from IR through X-rays is emitted by excited species of the plasma that is of practical interest. Various specialized techniques are utilized for the diagnostics of specific radiations emitting from the plasma. Diagnostic techniques can largely be categorized in optical diagnostic techniques, electrical diagnostic techniques and diagnostics using solid-state detectors. Optical diagnostics of the plasma gives indirect estimation of plasma conditions based on characteristics of plasma emission spectrum. However, electrical diagnostic techniques provide direct measurements of plasma characteristics and X-rays, electron and ion emissions from the plasma. For energy distribution of particles emitting from plasma, solid-state detectors are brought into use.

5.1. Optical diagnostics

Due to fast process and short lifetime, not all optical diagnostic techniques are useful to study laser-induced plasma that are used for persistent plasmas like inductively coupled plasma (ICP) and plasma focus (PF). The most common technique used for optical diagnostics of laser-induced plasma is emission spectroscopy. Radiations emitted from the plasma are collected by a spectrometer or a monochromator depending on the nature of study. Spectrometer captures a range of electromagnetic spectrum by a charge-coupled device (CCD) detector or with intensified CCD for even better results, whereas in monochromator a specific wavelength or a band is studied in high resolution with a CCD detector of a photomultiplier tube (PMT).

The spectrum is obtained as a combination of characteristic line emissions and continuum emissions. These spectra carry loads of information about the plasma conditions and composition. Various features of the spectrum can be utilized to obtain specific information about the plasma. For example, Doppler broadening of an emission line is related to velocity of the emitting particle, intensity of the emission line is proportional to the quantity of the emitter, Stark broadening of the line tells about electron density in the plasma while the ratio between intensity of emission line and continuum can provide us with temperature of the plasma, etc.

Another optical technique that is used for highly localized study of laser-induced plasma diagnostics is the laser-induced fluorescence. A tunable laser tuned to a specific wavelength, which is passed through the plasma to excite specific elements for certain transitions. In turn, upon de-excitation, radiations of a different frequency are obtained that help in studying the species of interest in the plasma. However, such a system is less popular because of the difficult set-up and expensive equipment. A combined system for plasma diagnostics through optical emission spectroscopy and laser-induced fluorescence is schematically shown in Figure 4.

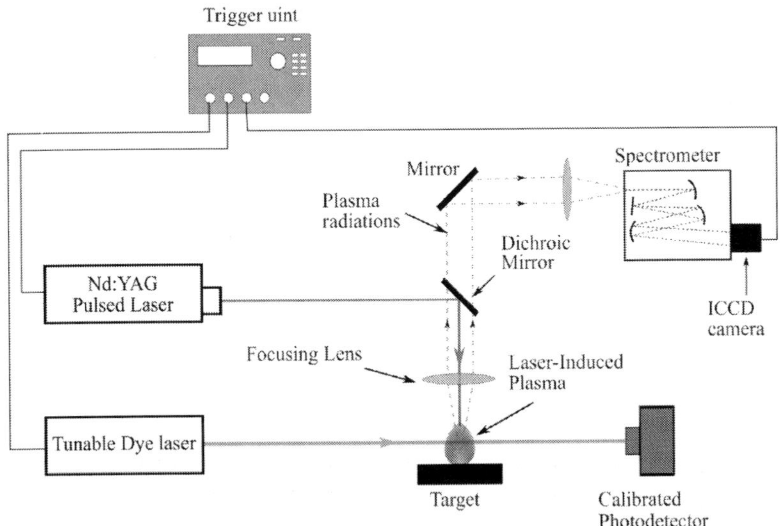

Figure 4. Schematic illustration of a combined set-up of laser-induced fluorescence and optical emission spectroscopy of laser-induced plasma

5.1.1. Electron density measurements

Electron density in the plasma can be measured through Stark broadening of an emission line or through the intensity ratio of two different emission lines of the same element. Utilizing the Stark-broadening method is considered more reliable because broadening of the emission line due to the Stark effect is a direct consequence of the presence of charged particles around the emitter.

For estimation of electron density in the plasma by making use of Stark broadening of an emission line, correct measurement of the Stark broadening is important. The emission line is normally broadened by a combination of three broadening mechanisms, that is, natural broadening, Doppler broadening that is caused by thermal motion of the emitter and Stark broadening caused by splitting of energy level because of the electric field strength of charged particles near the emitter. Doppler broadening is prominent at high temperatures, whereas the Stark broadening, which is also called as collision or pressure broadening, dominates at high densities of charged particles in the plasma.

Natural line broadening is related to naturally existing split in an energy level that makes an energy band. It can be related to the uncertainty in the energy of an excited state 'ΔE' for a limited excitation time 'Δt' of an electron through Heisenberg's uncertainty principle as $\sim \Delta E \Delta t > \hbar$. Normally, natural broadening is so small that it is undetectable by the spectrometers typically used in plasma diagnostics.

Doppler line broadening is a consequence of thermal motion of the emitter along the line of observation. The variation in the wavelength is explained on the basis of the Doppler effect. If movement of the emitter is towards the detector, a slightly shorter wavelength is recorded, and if the movement is away from the detector, a slightly longer wavelength is observed by the detector. Consequently, a broader emission line with a Gaussian profile is observed. For an emitter of atomic mass M, Doppler broadening $\Delta\lambda_D$ of an emission line at wavelength λ for a particular electron temperature T can be calculated as

$$\Delta\lambda_D = 2\lambda\sqrt{\frac{2kT\ln 2}{Mc^2}} \tag{2}$$

In addition to the above-described broadenings, the emission line is superimposed by another broadening contributed by the spectrometer itself that is referred to as *instrumental broadening*. It can be determined by using a narrow line laser beam. Typically, Doppler and Stark are the main competing broadening mechanisms. At high temperatures, Doppler broadening can be dominant, whereas at high densities Stark can be a dominant broadening mechanism. When Doppler broadening is negligible and Stark broadening mechanism is the main broadening mechanism, Stark broadening $\Delta\lambda_{\text{stark}}$ can be isolated from the instrumental broadening $\Delta\lambda_{\text{stark}}$ by deconvolution:

$$\Delta\lambda_{\text{stark}} = \Delta\lambda_{\text{total}} - \Delta\lambda_{\text{instumental}} \tag{3}$$

Stark broadening is directly linked with the electron density through electron impact parameter w_{FWHM}, by the following relation:

$$\Delta\lambda_{\text{stark}} = 2w_{\text{FWHM}}\left(\frac{n_e}{10^{16}\,\text{cm}^{-3}}\right) \tag{4}$$

Once the Stark broadening is obtained, the electron density n_e can be estimated by using this relation (eq. 4) if the electron impact parameter w_{FWHM} is known, which can be found in the literature. Electron density in the plasma depends upon a number of experimental parameters, for example, laser energy, background gas, ambient pressure and characteristics of the target itself. However, it represents a temporal profile that follows an exponential decay as a function of plasma lifetime [7, 8].

5.1.2. Plasma temperature measurements

Plasma temperature can be determined by several spectroscopic methods, for example, the line-pair intensity ratio method, Boltzmann plot, line-to-continuum ratio method, etc. One method may be more suitable than others under specific conditions. For example, diagnostics of the early-state plasma line-to-continuum ratio method is more suitable because during initial few hundred nanoseconds, both continuum and line intensities are of comparable strength. However, after a few microseconds, the continuum fades away and line intensities dominate in the spectrum; in such a situation, the line-pair intensity ratio or Boltzmann plot method would be more suitable. These methods are described as follows.

5.1.2.1. Line-pair intensity ratio method

Assuming that the plasma is in local thermodynamic equilibrium (LTE), the temperature of the plasma can be calculated through the intensity ratio of a pair of spectral lines of atom or ion of same ionization stage. In LTE, the level population supposedly obeys the Boltzmann distribution. Then the integrated intensity of a transition $(j \rightarrow i)$ can be represented as [9]:

$$I_{ij} = n_i^s A_{ij} \tag{5}$$

Here, n_i^s represents the population density of species 's' in level 'i' given as

$$n_i^s = \frac{g_i}{U^s(T)} n^s e^{E_i/kT} \tag{6}$$

Therefore, intensity I_{ij} can be written as

$$I_{ij} = \frac{g_i A_{ij}}{U^s(T)} n^s e^{E_i/kT} \tag{7}$$

where g_i is the partition function of level 'i', A_{ij} represents the transition probability of $i \rightarrow j$ transition, n^s and $U^s(T)$ are total number density of species in the plasma and partition function of the species 's', respectively. E_i, k and T are energies of the upper level 'i', Boltzmann constant and plasma temperature, respectively.

Now, consider another spectral line of the same species but originated from a different transition, that is, $m \rightarrow n$. Such that the upper energy level is different (i.e. $E_i \neq E_m$) and the lower energy level may or may not be the same. By taking intensity ratio of these two spectral lines, the plasma temperature can be calculated as follows:

$$\frac{I_{ij}}{I_{mn}} = \frac{\frac{g_i A_{ij}}{U^s(T)} n^s e^{E_i/kT}}{\frac{g_m A_{mn}}{U^s(T)} n^s e^{E_m/kT}} \tag{8}$$

Rearranging

$$T = \frac{E_i - E_m}{k \ln\left(\dfrac{I_{mn} A_{ij} g_i}{I_{ij} A_{mn} g_m}\right)} \tag{9}$$

It is advisable to choose lines that are as close in wavelength and as far in upper-level energy as possible. It will limit the device response variation in measurement of spectral line intensities. Assuming that the experimental error influences only intensities of the spectral lines, error in the temperature measurement is given as [10]

$$\frac{\Delta T}{T} = \frac{kT}{\Delta E}\frac{\Delta R}{R} \tag{10}$$

Here, $\Delta E = E_i - E_m$ energy difference between upper energy levels and R measures the intensity ratio, that is, $R = I_{ij}/I_{mn}$. ΔR is the uncertainty associated with ratio and ΔT is the corresponding uncertainty in temperature measurement.

5.1.2.2. Boltzmann plot method

Boltzmann plot is a reliable method for calculation of plasma temperature that has been trusted by many researchers in the latest literature.

The emission intensity of a spectral line is represented as [11]

$$I_{ij} = \frac{hc}{4\pi}\frac{A_{ij} g_i}{\lambda_{ij} U(T)} n e^{-E_i/kT} \tag{11}$$

Here, h, c, k, T and $U(T)$ are the Plank's constant, speed of light, Boltzmann constant, plasma temperature and partition function, respectively, whereas A_{ij}, g_i, E_i, λ_{ij} and n are the transition probability, degeneracy of upper level, upper-level energy, emission wavelength and total population density of the emitting species, respectively.

Taking natural logarithm and re-arranging the eq. (17), we obtain

$$\ln\left(\frac{I_{ij}\lambda}{g_i A_{ij}}\right) = \frac{-E_i}{kT} + \ln\left(\frac{hcn}{4\pi U(T)}\right) \tag{12}$$

This is the equation of a straight line that arises as a result of plot between $\ln\left(\dfrac{I_{ij}\lambda}{g_i A_{ij}}\right)$ and upper-level energy E_i. The slope of this plot is equal to $-1/kT$. From this slope, plasma temperature 'T' can easily be estimated [12, 13]. The further apart the upper-level energies of the selected lines, the better would be the measurement of the slope. The Boltzmann plot method is considered to be more precise for using several lines that averages out uncertainties involved in measurement than the intensity ratio method, which makes use of a pair of emission lines only [9]. The value of plasma temperature depends upon laser–target interaction, laser energy and characteristics of the ambient environment. However, it shows exponential decay during the lifespan of laser-induced plasma [7, 8].

5.2. Electrical diagnostics

A number of electrical diagnostic tools are used to characterize the laser-induced plasma. Few of them are briefly discussed in this section.

5.2.1. Langmuir probe

The Nobel laureate Irving Langmuir has invented the Langmuir probe to determine the electron temperature, density and the space potential V_{sp} in cold low-density plasmas. The local plasma parameters can be measured using stationary or slow time-varying electric (and/or magnetic) fields to collect or emit charged particles from the plasma. The charge particles, electrons, ions or both, are collected through the electric field established between the bulk plasma and the metallic surface of the probe. The simplest Langmuir probe consists of a metallic electrode, a bare wire or metal disk, with well-defined dimensions such as planar, cylindrical or spherical as shown in Figure 5. Langmuir probe is electrically biased with respect to a reference electrode to collect electron and/or positive ion currents. The dimensions of the probe as length, diameter and the thickness (or spatial extension) of the plasma sheath attached to the collecting surface play an important role in the collection of charge particles [14, 15].

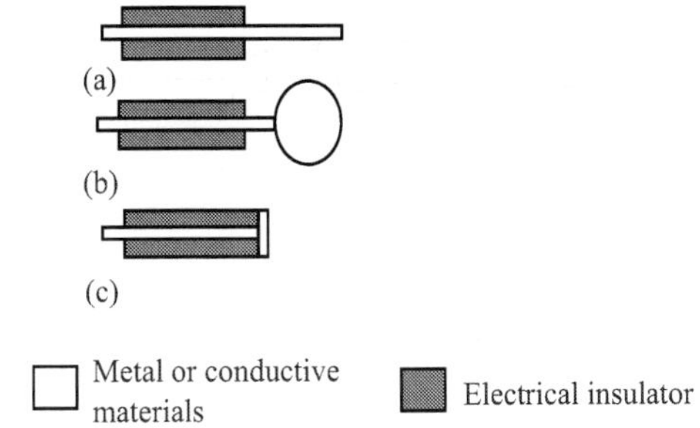

Figure 5 Schematic diagram of the Langmuir probe with different dimensions (a) cylindrical, (b) spherical and (c) planar

The metallic probe is immersed into the plasma and a potential V_p is provided to the probe using an external circuit, which bias the probe $V = V_p - V_{sp}$ with respect to the local space plasma potential V_{sp}. The drained current called probe current $I_p = I(V_p)$ for different probe potentials V_p is utilized to measure the different plasma parameters via voltage–current (IV) characteristic curves [14]. Figure 6 shows typical voltage–current (IV) curve of the Langmuir probe. The potential V_{sp} presents the localized electric potential at the point within plasma where the probe is immersed. As the probe does not emit particles, the drained current 'I_p' is the combination of ion current 'I_i' and electron current 'I_e', that is, '$I_p = I_i + I_e$'.

For very negative bias voltages where V_p is much less than V_{sp} (i.e. $V_p \ll V_{sp}$ at the left of point A as shown in Figure 6), the electrons are repelled, while ions are attracted by the probe. In this region, the characteristic curve is designated as the 'ion accelerating region'. The probe potential is sufficiently negative that only positive ions contribute to the probe current. The drained ion current from the plasma is limited by the electric shielding of the probe and I_p decreases slowly. The current $I_p \simeq I_{is}$ is the denominated ion saturation current.

The voltage corresponding to zero value of current is termed as the 'floating potential'. The bias potential V_F where $I_p = 0$ is the floating potential (point B) where the contributions of the ion and electron currents are equal.

After the floating potential, the I–V trace takes a sharp turn upwards. The bend is also referred to as the 'knee'. At this point, probe has the plasma potential. In the region, where V_p is much higher than V_{sp} (i.e. $V_p \gg V_{sp}$ at the right of point C as shown in Figure 6), the ions are repelled and the electrons are the attracted charges. In this case, the electrons are responsible for the electric shielding of the probe, and $I_p \simeq I_{es}$ is called the electron saturation current. As the positive bias is increased on the probe, eventually the collected electron current reaches a saturated value, that is, maximum electrons collected by the probe [14].

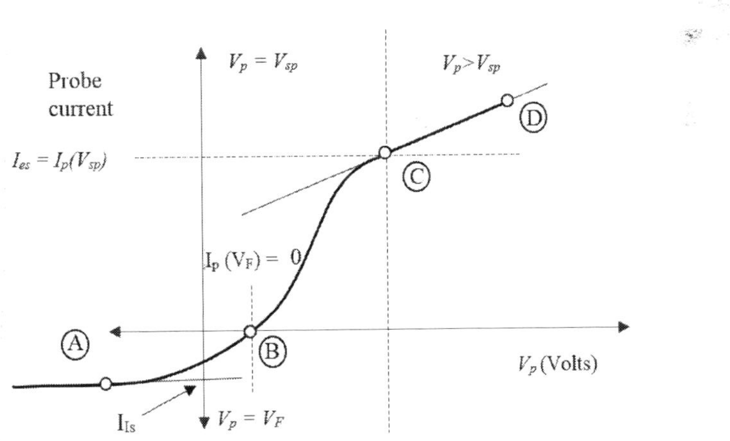

Figure 6. Typical voltage current (IV) curve of the Langmuir probe

For the qualitative interpretation, an idealized non-equilibrium collisionless, Maxwellian and unmagnetized plasma conditions are considered. Under the Maxwellian distribution, current density follows the Boltzmann law [16]:

$$J = J_0 \exp(-eV / kT_e) \tag{13}$$

where 'k' is the Boltzmann constant, 'T' is the temperature and 'J' is the current density at any potential 'V' and J_o is the current density at zero potential:

$$I = I_0 \exp(-eV / kT_e) \tag{14}$$

If 'T_e' is the electron temperature, 'V_a' is the applied biasing voltage; 'I_p' is the probe current at any biasing voltage and 'I_o' is the probe current at zero biasing:

$$I_p / I_o = \exp\left(eV / kT_e\right) \tag{15}$$

$$\ln(I_p / I_o) = eV / kT_e$$
$$T_e = eV / k. \, 1 / \ln(I_p / I_o) \tag{16}$$

Electron density of the laser-induced plasma can be calculated by the relation

$$n_e = I_0 / A_e.[m_e / 2\varepsilon_0 kT_e]^{\frac{1}{2}} \tag{17}$$

where T_e is the electron temperature; I_o is the probe current at $V_a = 0$; A_e is the probe surface area; n_e is the electron density; m_e is the mass of electron; k is the Boltzmann constant; and ε_0 is the permittivity of free space.

Debye's length can be calculated as

$$\lambda_D = [\varepsilon_0 kT_e / e^2 n_e]^{\frac{1}{2}} \tag{18}$$

Plasma frequency can be calculated as

$$\omega_p = (n_e e^2 / \varepsilon_0 m_e)^{\frac{1}{2}} \tag{19}$$

5.2.2. Faraday cup

Faraday cup (FC) is a metal cup with small aperture and a deep collector, which is designed to collect charged particles. When charged particles hit the metal, the metal gains a small net charge. The metal can then be discharged to determine a small current equivalent to the number of impinging charged particles as ions or electrons. A Faraday cup is used as a measurement tool for charged-particle-beam parameters such as current and current density profile. In the laser-induced plasma, the FC is used to measure the charged-particle parameters such as charge particle density, energy (using time of flight technique), the plasma angular distribution and distance charge dependence during the plasma-free expansion into the vacuum [17, 18].

The typical FC configuration is schematically shown in Figure 7, which mainly consists of two major parts: an inner cup for the charged particle collection and the grounded shield to measure the current I_c produced by the primary charged particle striking the cup. The electric current in the FC depends on the incident particle beam [19, 20].

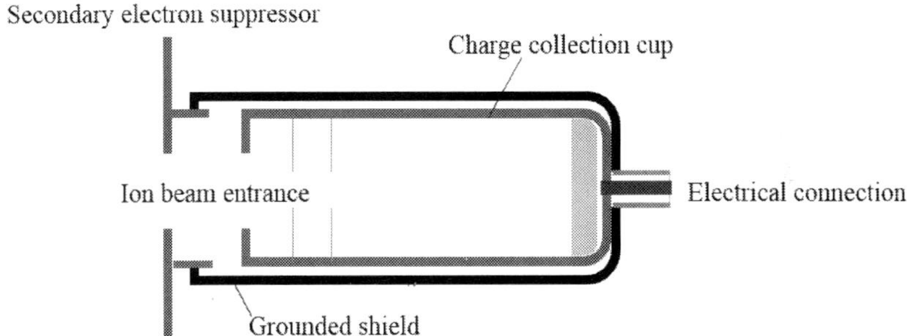

Figure 7. Schematic sketch of the Faraday cup

A bias voltage is applied to the cup to prevent secondary emission and electrons from escaping. The FC is connected to the measuring device as ammeter or oscilloscope to display the signal across a resistor from the conducting lead to ground [21, 22].

5.2.3. Scintillators and photomultiplier tubes

Radiation detectors play an important role in the non-destruction inspection and analysis in different field of studies. Photomultipliers are used to identify low-energy photons in UV–Vis range; high-energy photons (X-rays and gamma rays) and ionizing particles are identified using scintillators. The combination of photomultiplier tube and scintillator is one of the effective tools for the non-destruction characterization, which has several advantages such as fast time response, high detection efficiency, wide detection area and availability of different types of detective materials over number of other detection methods. A photomultiplier with scintillator converts radiations to an electrical signal, which is amplified to a useful level by emission of secondary electrons [23]. Due to high selectivity of photomultiplier tube, it is used to characterize the X-ray radiations ranging from 10 to 10 keV (soft and hard X-rays) [24, 25].

Figure 8 shows the schematic sketch of the photomultiplier system, which consists of a photocathode (scintillator) to convert radiation flux into electron flux, electron-optical input system to focus and accelerate the electron flux, electron multiplier comprises a series of secondary-emission electrodes (dynodes) to amplify the electric signal and an anode to collect the electron flux from the multiplier and displays the output signal.

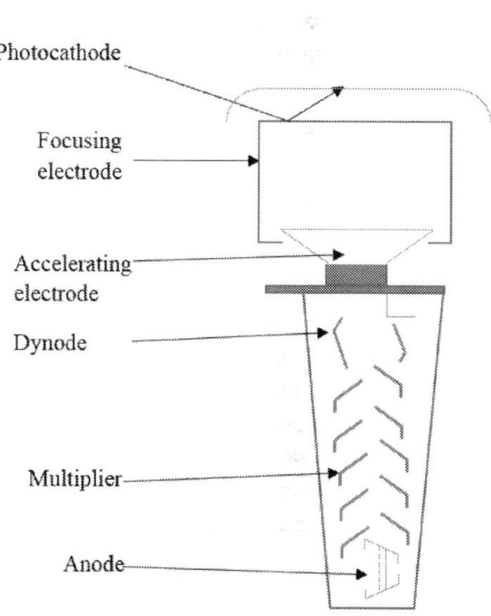

Figure 8. Schematic sketch of the photomultiplier system

The two fundamental phenomena to operate a photomultiplier are photoemission and secondary emission. Photoemission is due to a fraction of the incident photons that impart all their energy to bonded electrons of the photocathode material, giving some of the sufficient energy to escape, which are then amplified by dynodes and displayed as the output [23, 26].

5.3. Solid-state detectors

Solid-state nuclear track detectors (SSNTDs) are used to detect and estimate the energy of the charged particles in various disciplines. The solid-state nuclear track detection is an important and useful method to detect charged particles with energy ranging from tens of kilo-electron volts up to few hundred mega-electron volts. For instance, the detector of CR-39 is suitable for the detection of protons, deuterons and alpha detection due to high sensitivity of CR-39 material towards these particles. Once the SSNTD material is exposed to charged particles, these particles leave narrow damage trails on their passage through SSNTDs [27]. These damage trails represent regions of enhanced chemical activity as compared with bulk material due to disorders in the structure caused by the charged particles, which in turn are associated with large free energy. These trails are too small to be seen even with the help of a microscope. Therefore, the exposed detector material containing damage trails is subjected to the etching process in next stage to remove the material from the track.

To make the tracks visible, the exposed detectors are etched in the chemical solution to a certain temperature and for specific time. The etched detectors were then examined using an optical microscope or scanning electron microscope [28]. An optical micrograph for the carbon ions on the CR-39 material is shown in Figure 9.

On treating with suitable chemical reagents, the detector material along the damage trails interact much faster as compared with the undamaged bulk material. The resulting etch pits can be approximated by geometrical cones with the damage trails as axes. The study of etch-pit geometry and determination of the range of particles in the detector provides the identification of the particles creating the tracks and energies of particles [27].

The diameter of the etched tracks is used to evaluate the energy of the ion using diameter energy relationship given as

$$D = 1.8409E^{0.1624} \tag{20}$$

where D is the track diameter in micrometers and E is the energy in kilo-electron volt [29].

Figure 9. Optical micrograph of carbon ions track on CR-39 material

6. Applications of laser-induced plasma

Due to versatility of laser-induced plasma, it has been found to be suitable for numerous applications in various fields of science and technology. The utility of powerful lasers in remote, physically inaccessible locations makes it a special tool for applications in hostile environments. Laser-induced plasma has been utilized for lateral and depth elemental analysis, thin-film deposition, X-ray production, ion production, identification of materials, etc. Further details on applications of different aspects of laser-induced plasma are described in the following sections.

6.1. Laser-induced breakdown spectroscopy

The optical emission spectroscopy of laser-induced plasma is named as laser-induced breakdown spectroscopy (LIBS) or sometimes laser-induced plasma spectroscopy (LIPS). It is a powerful and versatile analytical technique for elemental analysis, which could be utilized for an enormous range of materials. Spectral features such as emission lines, peak intensity and integrated intensity, etc. are used for the determination of elemental concentration of the target or discriminate one material, organic or inorganic, from another through their unique spectral signatures. The basic set-up of LIBS is very simple (Figure 10); however, for higher signal intensities and lower limit of detections (LOD), double-pulse configuration of LIBS is preferred that has two variants, that is, collinear and orthogonal as shown in Figure 11. Reheating or preheating modes of operation can be utilized in both of these configurations. In the reheating mode, the laser pulse 1 ignites plasma plume, whereas laser pulse 2 (after a certain time delay) further excites plasma species to obtain larger amount of radiations. While in preheating mode, the laser pulse 1 heats the atmosphere above target surface and lets the plasma, generated by laser pulse 2, expand in hot and rare ambience.

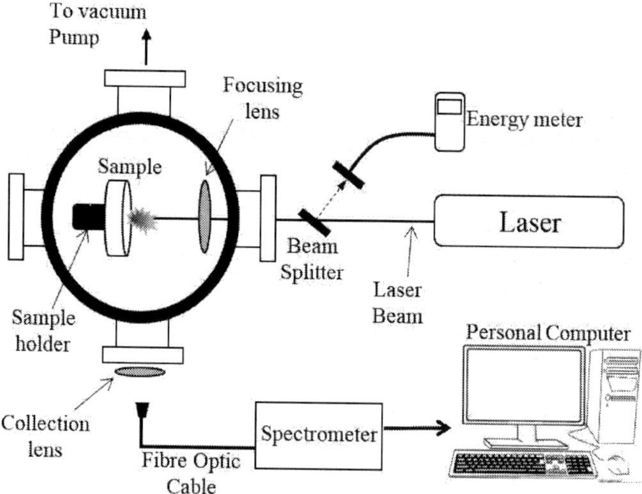

Figure 10. Schematic diagram of a typical laser-induced breakdown spectroscopy set-up

LIBS provides a huge advantage of requiring no specific sample preparation procedure to follow. Ideally, it can be utilized for solid, liquid and gaseous types of materials. Unique features of LIBS are stand-off and remote analyses that are typically useful when dealing with hazardous materials, toxic environments and humanly inaccessible locations. These characteristics are unique for LIBS because other chemometric techniques do not offer such utility. In the following sections, we discuss versatile applications of LIBS in various fields of science in the recent years.

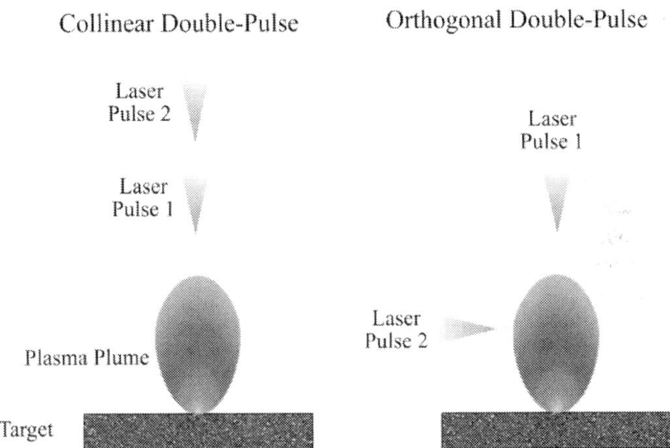

Figure 11. Two different modes of double-pulse laser-induced breakdown spectroscopy (DP-LIBS)

6.1.1. Geochemical applications

LIBS has proved to be useful in *geochemical* studies. The spectral signatures obtained from laser-induced plasma of targets have made investigations of geological materials possible. It has been taken up as a new technique for geological studies to determine the origin of petrogenetically important minerals and archeologically important rocks and stones on the basis of their physical and chemical properties. It includes not only the typical elemental measurements but also material identification with and without using machine-learning software [30-34].

With the capability of highly localized analysis (on the scale of μm), LIBS has made the spatial profiling of minerals possible in geological samples. Prochazka et al. [35] utilized LIBS for 3D profiling of minerals in geological samples performed at a resolution as high as 100 μm. They practised the advantage of no pretreatment of samples and found very good results.

6.1.2. Environmental applications

Another aspect of LIBS applications is *environmental studies*. It has widespread applications in environmental sciences for real-time elemental detection and analysis outside the laboratory. Its portability, simple execution, no requirement of sample preparation and simultaneous multi-element identification are attractive attributes for on-site analysis in terms of environmental health and safety. It covers a wide range of samples from industrial waste water to aerosols in the air.

LIBS has been successfully used for the identification and quantification of various pollutants and elements including carbon (C) in samples of diverse nature and origin [36-39]. Carbon is a good indicator of CO_2 gas in the environment, which is a greenhouse gas and influences the climate change. Since CO_2 cycles between atmosphere, vegetation and soil, it is quite possible that it increases the content of C in terrestrial carbon sinks, for example, soil. Therefore, the increase in the C content of terrestrial carbon sinks is an indicator of increasing CO_2 in the environment. Several studies have been performed that demonstrate the inherent advantages and capabilities of LIBS for quantification of the C content in soil from various origins [40-42]. Martin *et al.* [43] successfully attempted to differentiate the organic and inorganic C in the selected set of soil samples by the application of chemometric techniques of partial least squares (PLS) and principal component analysis (PCA) on LIBS spectra. By obtaining the correct signal intensity, linearity of their calibration curves significantly improved, that is, indicator of improvement accuracy of measurements. Nguyen *et al.* [44] determined the C concentration as an indicator of CO_2 greenhouse gas. The challenge regarding overlapping of C emission lines was handled by optimizing the time window of data collection for minimum broadened lines to mitigate interference effects. An accuracy of 95% was ensured in measurements of C in forest soils, wetland soils and sediment samples.

In addition to greenhouse gases, heavy metals are of major concern for human health and safety. The presence of toxic heavy metals like arsenic, lead, antimony, zinc and cadmium in natural resources of the environment, for example, water, soil, air etc. or synthetic products that come in direct human contact, such as paint, are of risk for living organisms. Because of the toxicity of Pb, which could be poisonous for the human body [45], its detection and achieving low limits of detection have always been the matter of interest. Analysis of liquid samples with LIBS is challenging; short lifetime of plasma, liquid–laser coupling and splashes resulting from plasma shockwaves raise the difficulty of LIBS measurement and increase the limit of detection for liquid samples. However, new techniques for LIBS are improving the analysis of liquid samples. Järvinen *et al.* [46] incorporated evaporation pre-concentration for the measurement of Zn and Pb in water and at limits of detection as low as 0.3 and 0.1 ppm, respectively.

6.1.3. Agricultural applications

LIBS has a prominent potential for agricultural studies. Portability and no sample preparation features again provide LIBS an advantage over conventional analytical techniques. Soil, crops, plants, fruits, grains and irrigation water can be tested in field. Although the portable systems for agricultural applications are not very mature, the pace of technological advancement suggests that some decent portable LIBS systems for a variety of applications would be available soon.

LIBS has successfully been utilized for agricultural studies from various aspects. Nutrients and minerals in the soil are of great importance from agricultural point of view. There are many studies referring to the employment of LIBS for measurement of macro- and micronutrients in soil and crops. Naturally, organic samples are not very homogeneous in elemental composition. It is sometimes difficult to analyse heterogeneous samples with LIBS because it only provides information about a highly localized region on the sample surface. The determination of Fe, Mn, Mg, Ca, Na and K in heterogeneous soil samples was performed by Nguyen [44] by using laser ablation–LIBS (LA–LIBS) without sample preparation. Unnikrishnan *et al.* [47] demonstrated the measurement of trace amounts of nutrients Cu, Zn and Ca in soil samples with limits of detection of 4.8, 4.4 and 6 ppm, respectively. In addition to mare elemental analysis, a recent publication by Ferreira *et al.* [48] shows that LIBS can potentially be effective in determination of pH of soil.

LIBS has been employed for studying diverse plants for practical applications and for exploration and development [49-51]. It is well known that the deficiency of macro-/micronutrients can adversely affect the crop yield and degrade the nutritious quality of the product. Therefore, monitoring of essential elements in crop plants, fruits and vegetables is vital for maintaining the nutritional status and food security. Numerous studies can be listed in which LIBS is featured as an investigation tool for nutritious quality of diverse range of organic samples. For instance, Arantes de Carvalho *et al.* [52] studied macro- (Ca, Mg, P) and micronutrient (Cu, Fe, Mn and Zn) concentrations in a variety of crop samples that include sugarcane, soy, citrus, coffee, maize, eucalyptus, mango, bean, banana, lettuce, brachiaria, pearl millet, grape, rubber tree and tomato. Kim *et al.* [53] investigate the quantity of macronutrients (Mg, Ca, Na and K) in spinach and rice samples and apply discriminatory operations upon LIBS spectra to discriminate the samples contaminated with pesticides from clean ones. LIBS has also been utilized for the determination of toxic heavy metals and contaminations residing on the surface of fruits because of environmental pollution and pesticide sprays [54, 55]. The list of publications regarding LIBS applications for investigations of agricultural products encompassing diverse variety of samples is long. One can simply deduce its applicability in almost all parts associated with agriculture, ranging from irrigation water to ripe crops, fruits and vegetables.

6.1.4. Stand-off and remote applications

Stand-off and remote analysis capabilities are unique for LIBS. In stand-off application, a powerful laser pulse is fired at a distant (approximately several meters) target and emissions from the microplasma at the target surface are collected through a telescope. Schematic diagram of a typical stand-off LIBS system is provided in Figure 12. In remote studies, a fibre-optic probe is commonly utilized to reach remote and humanly inaccessible locations. This probe delivers the laser to the target and transfers the plasma radiations from target to the spectrometer back in the system. Schematic illustration of a typical remote LIBS system is depicted in Figure 13. Portability of LIBS system has elevated the potential of LIBS capability in the investigation of quarantined regions, potentially hostile and physically inaccessible locations. Applications of stand-off and remote LIBS can be found in identification of biomaterials at far distance, investigation of deep sea objects, studying radioactive and explosive materials from a safe distance, monitoring the composition of alloys in production line, Mars and exploration, etc.

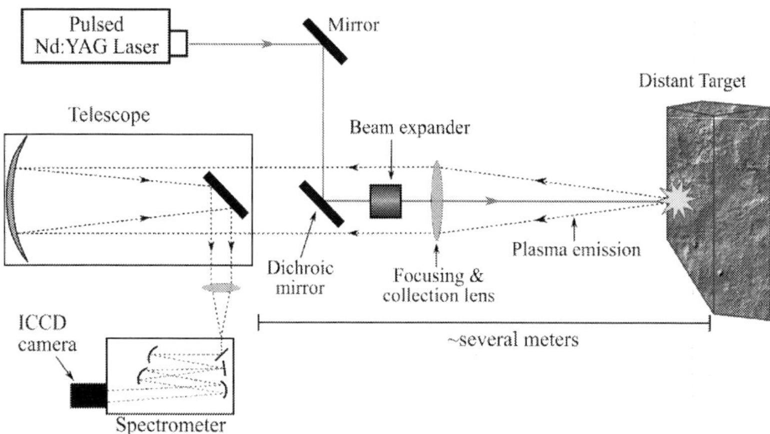

Figure 12. Schematic illustration of a typical stand-off LIBS system

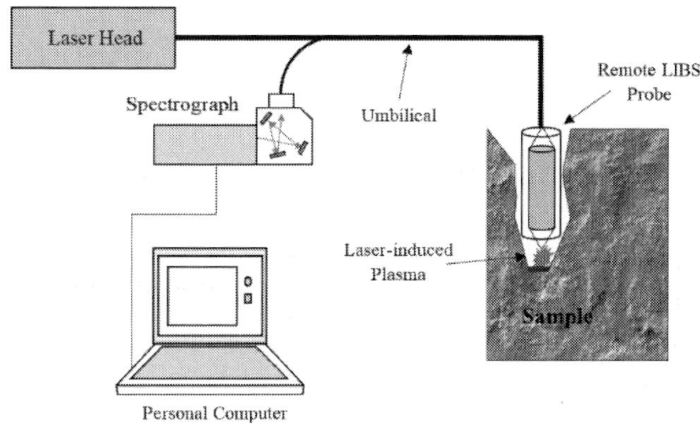

Figure 13. Schematic illustration of a typical remote LIBS system

At a distance of ~6.2 m, Vítková *et al.* [56] performed LIBS analysis on a variety of biological samples that included mortar, sea shell, human tooth, swine bone, soil pellet and a brick. Each of these samples was classified on the basis of their spectral signatures by complementing LIBS with supervised machine-learning methods, such as linear discriminant analysis (LDA) and artificial neural networks (ANN). In another study, Vítková *et al.* [57] demonstrated the LIBS-based discrimination of historically important bricks according to their archaeological localities with the help of LDA at a stand-off distance of ~6.2 m. Another application of stand-off LIBS is presented for the analysis of soil and vegetation powders for studying the content of heavy metals and toxic trace elements in targets [58].

An interesting underwater application of stand-off LIBS is demonstrated by Fortes *et al.* [59]. To test the feasibility of underwater LIBS at a stand-off distance of 0.8 m from sensor to the target deeply submerged in the water, fundamental investigations were performed. A similar study has reported [60] the application of remote LIBS for studying underwater shipwreck. In such applications, the advantage of remote over stand-off approach is that the laser beam interaction with ocean water is minimal and the laser energy losses are likewise minimum. However, the advantage of stand-off approach is that it is not required to bring the LIBS probe to the target in deep waters, rather, measurements can be taken from far.

Stand-off LIBS is capable of investigating nuclear materials from a safe distance, for instance, Gong *et al.* [61] has demonstrated the investigations made on cerium oxide (CeO_2) and potassium chloride (KCl) at a stand-off distance of 1.45 m through the protection shield. Stand-off LIBS is particularly suitable for industrial applications like real-time monitoring of product quality in production line without interfering the production process [62].

LIBS has gained popularity among the geologists since the NASA has equipped the Curiosity Mars rover with a ChemCam LIBS system. A field test of this system with a remote analysis probe was published as a meeting abstract [63]. The ChemCam LIBS system has proved useful and reliable in Mars exploratory mission since the Curiosity has landed the surface of Mars in year 2012 [64].

6.2. Pulsed laser deposition

Pulsed laser deposition is a versatile technique. It makes use of the laser's ability to ablate and excite almost every type of material and, therefore, has been utilized to deposit an extensive range of thin films may it be metallic, polymeric or composite.

A high-power laser pulse, typically $\geq 10^8$ W/cm^2, is utilized to evaporate and ionize the material into a plasma plume that expands in the ambient environment perpendicular the target surface. This evaporated and highly accelerated material is collected on an appropriately positioned substrate upon which it condenses and grows a thin film. The size and expansion velocity of the plasma plume depend largely upon nature and density of the ambient atmosphere along with laser pulse duration and irradiation. The plasma plume has found to expand with velocities of the order of 10^6 cm/s under vacuum conditions (~10^{-6} mbar) for typical laser fluences ~10^9–10^{10} W/cm^2 and increases for higher fluences. Plume dimensions and expansion velocity are also inverse functions of the ambient pressure and time [65, 66]. When the plasma species strike surface of the substrate with such high velocities, sufficient adhesion is obtained to keep particles sticking with the surface for gradual growth of the film. The film growth and quality depend upon characteristics of substrate, substrate temperature and energy of the plasma species, that is, atoms, ions and clusters striking the surface of substrate that in turn depends upon laser parameters. Surface morphology and texture of thin films are strongly influenced by substrate temperature, nature and pressure of the ambient gas. By controlling the ambient atmosphere, we can actually control mechanical, optical, surface morphological properties of the deposited thin film. The choice of laser and laser parameters are vital for PLD. The selection of wavelength and energy of laser depends upon the coupling efficiency with target material to be used. Nanosecond Nd:YAG and excimer lasers are the most widely utilized lasers for PLD as found in the published literature.

PLD is principally a simple technique and versatile in terms of utilizing target and substrate materials. A typical PLD arrangement is schematically illustrated in Figure 14. Laser ablation is considered to be a stoichiometric phenomenon, which means that plasma plume has the same composition as that of the target that is delivered onto the substrate surface as a thin film. It is normally referred to as 'stoichiometry transfer' from target to film on substrate surface that is difficult to achieve with other techniques. It, therefore, permits the control of thin-film composition by appropriate selection of target composition. Multilayers of different materials can be deposited on a single substrate by using multiple targets during the PLD procedure. Since only small substrate areas (typically ~1 × 1 cm^2) can be targeted by the laser-induced plasma, it allows the preparation of complex samples enriched with isotopes for laboratory-scale research purposes. Being a short-pulsed phenomenon makes PLD a flexible operation, because operational parameters can be varied from pulse to pulse according to the requirement of the application.

Despite numerous advantages of the PLD technique, it is not yet well adopted in the industrial manufacturing process because of the uncontrolled plasma process and small-area deposition. In addition, the deposition may not be of uniform thickness because of strong forward peaking nature of the laser-induced plasma. However, it has gained popularity among the scientists who are convinced and exploring its enormous potential in laboratories and discovering new possibilities and procedures for extensive applications in diverse fields of science and technology. Amorphous, polycrystalline, epitaxial thin films can be grown, depending on the nature of the substrate and its temperature. There are a large number of applications in which PLD has shown great promise, for example, photovoltaics, superconductor technology, nanostructured thin films, optical coatings, etc.. Following is given a brief review of versatile applications of PLD.

PLD has been found to be very suitable for deposition of thin films for *photovoltaic* applications. Sekiguchi *et al.* [67] were perhaps the first to demonstrate the use of PLD for epitaxial growth of CZTS films on GaP substrates heated at 350–400 °C. Stoichiometric crystalline thin films with an appropriate optical band gap of ~1.5 eV were observed. However, the first-ever publication about Sn-rich CZTS thin film deposition by PLD, particularly for solar cell applications was reported by Moriya *et al.* [68]. Films showed an optical band gap of ~1.5 eV, consequently prepared solar presented a conversion efficiency of 1.74% for an active area of 0.092 cm^2.

Pulsed laser deposition has also found application in depositing high-quality superconducting thin films. Superconductivity has found to be thickness dependent; therefore, precise control of thin films is crucial. Here, PLD provides an advantage of controlled growth of thin films through variation in experimental parameters [69, 70], which can improve the quality of thin films in individual applications. There are many recent studies found in literature that demonstrate the exploration of experimental parameters of PLD for diverse types of high-quality superconducting thin films. Oshima [70] report the deposition of LiTi$_2$O$_4$ thin films of various thicknesses by employing the PLD method using krypton fluoride laser (KrF) excimer

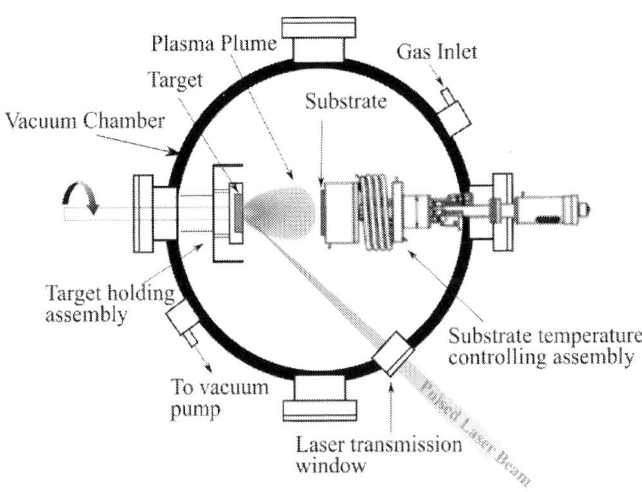

Figure 14. Schematic illustration of pulsed-laser deposition system

laser. Superconducting transition was observed around 12 K at a film thickness of 9 nm. One of the benefits of PLD method is that quality and thickness of thin films can be controlled by varying deposition time, growth rate (determined by pulse repetition rate) and substrate temperature during the film growth [69, 70].

PLD method is highly suitable for growing *composite thin films* of a wide range of materials over diverse substrates. Sharma *et al.* [71] report on composite thin films $0.9BiFeO_3$-$0.1YCrO_3$ on Pt, TiO_2, SiO_2 and Si substrates by sequential deposition of $BiFeO_3$ and $YCrO_3$ targets by the PLD method. Consequently obtained thin films represented fine, particle-free smooth surfaces with a crystal structure as a mixture of $BiFeO_3$ and $YCrO_3$ crystals. Chowdhury *et al.* [72] deposited SnO_2-Fe_2O_3 composite thin films by the PLD method with the aim to investigate structural, morphological, optical and electrical properties that revealed their suitability for optoelectronic applications.

An interesting possible application of PLD can be the coating of surgical instruments with antimicrobial composite thin films. Several studies have demonstrated the growth of antifungal and antibacterial composite thin films on various substrates. For instance, Ag-SiO_2 antibacterial composite thin films were grown on Si (111) substrate by the PLD method using KrF Excimer laser [73] at room temperature in the oxygen environment. In another study, Pradhaban *et al.* [74] demonstrated the deposition of antibacterial Ag:ZrO_2 composite films on stainless steel substrates by the PLD technique using Nd:YAG laser. Similary, Eraković *et al.* [75] utilized PLD to grow antifungal Ag:HA (hydroxyapatite) composite thin films on titanium (Ti), anodized Ti and Si (111) substrates.

PLD is still a laboratory-scale method for thin-film deposition. Large amount of data published on investigation of the PLD method for a variety of applications is an evidence of intensive efforts being made by the scientists. However, it will take time to see the PLD method being used for industrial-scale applications.

7. Conclusions

This chapter encompasses fundamentals of laser-induced plasma from laser–matter interaction to radiation emission from plasma and deposition of the ablated material on a distant substrate. LIP is a rapid process and fairly complex in nature. It starts with absorption of pulsed-laser energy by the material that breaks it down and evaporates it in the form of plasma that eventually decays after few tens of microseconds. During its lifetime, the plasma temperature, particle density and continuum and line emissions show temporal fluctuations. Characteristics and emissions of laser-induced plasma strongly depend upon laser parameters, nature of the target and ambient conditions. There are several methods used for estimation of plasma parameters based on spectroscopic data. Among them, the Boltzmann plot method for plasma temperature measurements and the Stark broadening method for estimation of electron density are the most commonly used methods. Besides spectroscopic diagnostics, the most well-known electrical diagnostic tools include Langmuir probe, Faraday cups and scintillators and photomultipliers, which have been widely used to study plasma parameters, energy of charged particles and X-ray emissions from plasma, respectively. A physical method of measuring the density, energy and distribution of charged particles is the use of solid-state nuclear track detectors. By investigating the density and dimensions of the tracks produced on SSNTDs, the density, energy of the particles emitting at various angles of the plasma can be obtained.

Being a rich source of radiation and particle emission, LIP has found a variety of applications in diverse areas. Line emissions from the laser-induced plasmas inform about elemental composition of the material that has opened up several fronts of applications. It is termed as LIPS (laser-induced plasma spectroscopy). It has been utilized for numerous laboratory and in-field applications from food safety to explosive detection, soil studies to Mars exploration, archaeology to artefact studies, micro to bulk exploration and many more. The samples that could be studied with LIPS have virtually no limits, from solids to gasses and aerosols have been investigated that encompass organic materials to nuclear and explosive materials. In addition to spectroscopic investigations, the excited material of LIP has been utilized for thin-film deposition on a wide range of substrates for diverse applications. It is commonly termed as pulsed-laser deposition. Through PLD, films of various materials including conductors, insulators, dielectrics, in general, have been deposited on a variety of substrates. Deposition of complex hybrid and multilayer films, a combination of multiple materials, have also been successfully demonstrated. The quality, morphology and structure of films depend upon the ambient atmosphere and substrate temperature besides laser parameters.

Because of numerous intrinsic benefits of LIP for various applications over other conventional techniques, some commercial products, specifically LIPS systems, can be seen in the market. However, the uncontrolled nature of LIP hinders its wider deployment for industrial applications. Most of the potential applications are still in infancy and are being explored at the laboratory level. Keeping in mind the current pace of advancement in technology and enormous potential of LIP, in near future, we can expect to see some products in the market that are based on LIP for innovative applications.

Author details

Kashif Chaudhary*, Syed Zuhaib Haider Rizvi and Jalil Ali

*Address all correspondence to: kashif@utm.my

Laser Center, Ibnu Sina Institute for Scientific & Industrial Research (ISI-SIR), Universiti Teknologi Malaysia (UTM), Malaysia

References

[1] Wu, B. and Shin, Y.C. (2006), Modeling of nanosecond laser ablation with vapor plasma formation. Journal of Applied Physics. 99(8): 084310.

[2] Bäuerle, D.W., Laser Processing and Chemistry. 2011: Springer Berlin Heidelberg.

[3] Phipps, C., Laser ablation and its applications. Vol. 129. 2007, New Mexico: Springer.

[4] Harilal, S.S., Freeman, J.R., Diwakar, P.K., and Hassanein, A., Femtosecond laser ablation: fundamentals and applications, in Laser-Induced Breakdown Spectroscopy. 2014, Springer. p. 143-166.

[5] Leitz, K.-H., Redlingshöfer, B., Reg, Y., Otto, A., and Schmidt, M. (2011), Metal ablation with short and ultrashort laser pulses. Physics Procedia. 12: 230-238.

[6] Singh, J.P. and Thakur, S.N., Laser-Induced Breakdown Spectroscopy. 2007: Elsevier Science.

[7] Hegazy, H., Abd El-Ghany, H.A., Allam, S.H., and El-Sherbini, T.M. (2014), Spectral evolution of nano-second laser interaction with Ti target in nitrogen ambient gas. Applied Physics B. 117(1): 343-352.

[8] Aragón, C., Aguilera, J.A., and Manrique, J. (2014), Measurement of Stark broadening parameters of Fe II and Ni II spectral lines by laser induced breakdown spectroscopy using fused glass samples. Journal of Quantitative Spectroscopy and Radiative Transfer. 134: 39-45.

[9] Miziolek, A.W., Palleschi, V., and Schechter, I., Laser Induced Breakdown Spectroscopy: Fundamentals and Applications. 2006: Cambridge University Press.

[10] Harmon, R.S., DeLucia, F.C., McManus, C.E., McMillan, N.J., Jenkins, T.F., Walsh, M.E., and Miziolek, A. (2006), Laser-induced breakdown spectroscopy – An emerging chemical sensor technology for real-time field-portable, geochemical, mineralogical, and environmental applications. Applied Geochemistry. 21(5): 730-747.

[11] Cremers, D.A. and Radziemski, L.J., Handbook of Laser-Induced Breakdown Spectroscopy. 2006: Wiley.

[12] Aragón, C. and Aguilera, J.A. (2008), Characterization of laser induced plasmas by optical emission spectroscopy: A review of experiments and methods. Spectrochimica Acta Part B: Atomic Spectroscopy. 63(9): 893-916.

[13] Unnikrishnan, V., Mridul, K., Nayak, R., Alti, K., Kartha, V., Santhosh, C., Gupta, G., and Suri, B. (2012), Calibration-free laser-induced breakdown spectroscopy for quantitative elemental analysis of materials. Pramana. 79(2): 299-310.

[14] Conde, L. (2011), An introduction to Langmuir probe diagnostics of plasmas. Madrid: Dept. Física. ETSI Aeronáut ngenieros Aeronáuticos Universidad Politécnica de Madrid. 1-28.

[15] Merlino, R.L. (2007), Understanding Langmuir probe current-voltage characteristics. American Journal of Physics. 75(12): 1078-1085.

[16] Tufail, K. (2008), Magnetic field effect on electron emission from plasma, Master of Science Thesis, University of Engineering and Technology.

[17] Lorusso, A., Velardi, L., and Nassisi, V. (2008), Characterization of laser-produced plasma of metal targets. Radiation Effects & Defects in Solids. 163(4-6): 429-433.

[18] Kohno, S., Lisitsyn, I., Kawauchi, T., Katsuki, S., and Akiyama, H. *Characterization of laser-produced plasma in a plasma opening switch.* (1997) 2, 1186-1191: IEEE.

[19] Roshani, G., Habibi, M., and Sohrabi, M. (2011), An improved design of Faraday cup detector to reduce the escape of secondary electrons in plasma focus device by COMSOL. Vacuum. 86(3): 250-253.

[20] Ghadikolaee, M.R.B. and Ghadikolaee, E.T. (2012), Design of a new Faraday cup to measure the beam current of an ion source with residual gas. Journal of fusion energy. 31(6): 569-572.

[21] Naieni, A.K., Bahrami, F., Yasrebi, N., and Rashidian, B. (2009), Design and study of an enhanced Faraday cup detector. Vacuum. 83(8): 1095-1099.

[22] Guethlein, G., Houck, T., McCarrick, J., and Sampayan, S. (2000), Faraday Cup Measurements of Ions Backstreaming into a Electron Beam Impinging on a Plasma Plume*. arXiv preprint physics/0008165.

[23] Giulietti, D. and Gizzi, L.A. (1998), X-ray emission from laser-produced plasmas. La Rivista del Nuovo Cimento. 21(10): 1-93.

[24] Serbanescu, C., Fourmaux, S., Kieffer, J.-C., Kincaid, R., and Krol, A. *K-alpha x-ray source using high energy and high repetition rate laser system for phase contrast imaging.* (2009) San Diego, CA 7451, 745115-745115-9: International Society for Optics and Photonics.

[25] Bukhari, M., Iqbal, S.J., Rafique, M.S., and Iqbal, M. (2013), PRODUCTION OF HARD X-RAY BY LASER PRODUCED PLASMA. Science International-Lahore. 25(4): 687-691.

[26] Flyckt, S.-O., Photomultiplier tubes: principles and applications. 2002: Photonis.

[27] Dey, S., Gupta, D., Maulik, A., Raha, S., Saha, S.K., Syam, D., Pakarinen, J., Voulot, D., and Wenander, F. (2011), Calibration of a solid state nuclear track detector (SSNTD) with high detection threshold to search for rare events in cosmic rays. Astroparticle Physics. 34(11): 805-808.

[28] Bhagwat, A. (1993), Solid State Nuclear Track Detection: Theory and Applications. Published by Indian Society for Radiation Physics, Kalpakkam chapter, ISRP (K)-TD-2.

[29] Bhatti, K., Rafique, M., Khaleeq-ur-Rahman, M., Latif, A., Hussain, K., Hussain, A., Chaudhary, K., Tahir, B., and Qindeel, R. (2011), Characterization of Platinum and Gold ions emitted from laser produced plasmas using solid state nuclear track detectors. Vacuum. 85(10): 915-919.

[30] Iqbal, J., Mahmood, S., Tufail, I., Asghar, H., Ahmed, R., and Baig, M.A. (2015), On the use of laser induced breakdown spectroscopy to characterize the naturally existing crystal in Pakistan and its optical emission spectrum. Spectrochimica Acta Part B: Atomic Spectroscopy. 111: 80-86.

[31] Zhu, X., Xu, T., Lin, Q., Liang, L., Niu, G., Lai, H., Xu, M., Wang, X., Li, H., and Duan, Y. (2014), Advanced statistical analysis of laser-induced breakdown spectroscopy data to discriminate sedimentary rocks based on Czerny–Turner and Echelle spectrometers. Spectrochimica Acta Part B: Atomic Spectroscopy. 93: 8-13.

[32] Dyar, M.D., Tucker, J.M., Humphries, S., Clegg, S.M., Wiens, R.C., and Lane, M.D. (2011), Strategies for Mars remote laser-induced breakdown spectroscopy analysis of sulfur in geological samples. Spectrochimica Acta Part B: Atomic Spectroscopy. 66(1): 39-56.

[33] Gottfried, J.L., Harmon, R.S., De Lucia, F.C., and Miziolek, A.W. (2009), Multivariate analysis of laser-induced breakdown spectroscopy chemical signatures for geomaterial classification. Spectrochimica Acta Part B: Atomic Spectroscopy. 64(10): 1009-1019.

[34] Alvey, D.C., Morton, K., Harmon, R.S., Gottfried, J.L., Remus, J.J., Collins, L.M., and Wise, M.A. (2010), Laser-induced breakdown spectroscopy-based geochemical fin-

gerprinting for the rapid analysis and discrimination of minerals: the example of garnet. Applied Optics. 49(13): C168-C180.

[35] Prochazka, D., Kynický, J., Prochazková, P., Zikmund, T., Novotný, J., Novotný, K., Kizek, R., Petrilak, M., and Kaiser, J. (2014), 3D CHEMICAL MAPPING OF SELECTED GEOLOGICAL SAMPLES USING COMBINATION OF LASER-INDUCED BREAKDOWN SPECTROSCOPY AND uCT.

[36] Ferreira, E.C., Milori, D.M.B.P., Ferreira, E.J., dos Santos, L.M., Martin-Neto, L., and Nogueira, A.R.d.A. (2011), Evaluation of laser induced breakdown spectroscopy for multielemental determination in soils under sewage sludge application. Talanta. 85(1): 435-440.

[37] Díaz, D., Hahn, D.W., and Molina, A. (2012), Evaluation of Laser-Induced Breakdown Spectroscopy (LIBS) as a Measurement Technique for Evaluation of Total Elemental Concentration in Soils. Applied Spectroscopy. 66(1): 99-106.

[38] Ayyalasomayajula, K.K., Yu-Yueh, F., Singh, J.P., McIntyre, D.L., and Jain, J. (2012), Application of laser-induced breakdown spectroscopy for total carbon quantification in soil samples. Applied Optics. 51(7): B149-B154.

[39] Dell'Aglio, M., Gaudiuso, R., Senesi, G.S., De Giacomo, A., Zaccone, C., Miano, T.M., and De Pascale, O. (2011), Monitoring of Cr, Cu, Pb, V and Zn in polluted soils by laser induced breakdown spectroscopy (LIBS). Journal of Environmental Monitoring. 13(5): 1422-1426.

[40] Nicolodelli, G., Marangoni, B.S., Cabral, J.S., Villas-Boas, P.R., Senesi, G.S., dos Santos, C.H., Romano, R.A., Segnini, A., Lucas, Y., Montes, C.R., and Milori, D.M.B.P. (2014), Quantification of total carbon in soil using laser-induced breakdown spectroscopy: a method to correct interference lines. Applied Optics. 53(10): 2170-2176.

[41] Martin, M.Z., Wullschleger, S.D., Garten, C.T., and Palumbo, A.V. (2003), Laser-induced breakdown spectroscopy for the environmental determination of total carbon and nitrogen in soils. Applied Optics. 42(12): 2072-2077.

[42] Martin, M.Z., Wullschleger, S., and Garten, C. *Laser-induced breakdown spectroscopy for environmental monitoring of soil carbon and nitrogen*. (2002) 4576, 188-195.

[43] Martin, M.Z., Mayes, M.A., Heal, K.R., Brice, D.J., and Wullschleger, S.D. (2013), Investigation of laser-induced breakdown spectroscopy and multivariate analysis for differentiating inorganic and organic C in a variety of soils. Spectrochimica Acta Part B: Atomic Spectroscopy. 87: 100-107.

[44] Nguyen, H.-M., Moon, S.-J., and Choi, J. (2015), Improving the application of laser-induced breakdown spectroscopy for the determination of total carbon in soils. Environmental Monitoring and Assessment. 187(2): 1-11.

[45] Flora, G., Gupta, D., and Tiwari, A. (2012), Toxicity of lead: A review with recent updates. Interdisciplinary toxicology. 5(2): 47-58.

[46] Järvinen, S.T., Saarela, J., and Toivonen, J. (2013), Detection of zinc and lead in water using evaporative preconcentration and single-particle laser-induced breakdown spectroscopy. Spectrochimica Acta Part B: Atomic Spectroscopy. 86: 55-59.

[47] Unnikrishnan, V., Nayak, R., Aithal, K., Kartha, V., Santhosh, C., Gupta, G., and Suri, B. (2013), Analysis of trace elements in complex matrices (soil) by Laser Induced Breakdown Spectroscopy (LIBS). Analytical Methods. 5(5): 1294-1300.

[48] Ferreira, E.C., Gomes Neto, J.A., Milori, D.M.B.P., Ferreira, E.J., and Anzano, J.M. (2015), Laser-induced breakdown spectroscopy: Extending its application to soil pH measurements. Spectrochimica Acta Part B: Atomic Spectroscopy. 110: 96-99.

[49] Haider, Z., Munajat, Y.B., Kamarulzaman, R., and Shahami, N. (2015), Comparison of Single Pulse and Double Simultaneous Pulse Laser Induced Breakdown Spectroscopy. Analytical Letters. 48(2): 308-317.

[50] Barefield, J.E., Judge, E., Clegg, S.M., Berg, J., Colgan, J.P., Kilcrease, D.P., Johns, H.M., Le, L.A., Lopez, L.N., and Trujillo, L.A., Laser-Induced Breakdown Spectroscopy (LIBS): Applications to Analysis Problems from Nuclear Material to Plant Nutrients for Sustainable Agriculture, (2014). Los Alamos National Laboratory (LANL).

[51] Sankaran, S., Ehsani, R., and Morgan, K.T. (2015), Detection of Anomalies in Citrus Leaves Using Laser-Induced Breakdown Spectroscopy (LIBS). Applied Spectroscopy. 69(8): 913-919.

[52] Arantes de Carvalho, G.G., Moros, J., Santos Jr, D., Krug, F.J., and Laserna, J.J. (2015), Direct determination of the nutrient profile in plant materials by femtosecond laser-induced breakdown spectroscopy. Analytica Chimica Acta. 876: 26-38.

[53] Kim, G., Kwak, J., Choi, J., and Park, K. (2012), Detection of nutrient elements and contamination by pesticides in spinach and rice samples using laser-induced breakdown spectroscopy (LIBS). Journal of agricultural and food chemistry. 60(3): 718-724.

[54] Li, W., Huang, L., Yao, M., Liu, M., and Chen, T. (2014), Investigation of Pb in Gannan Navel Orange with Contaminant in Controlled Conditions by Laser-Induced Breakdown Spectroscopy. Journal of Applied Spectroscopy. 81(5): 850-854.

[55] Yao, M., Huang, L., Zheng, J., Fan, S., and Liu, M. (2013), Assessment of feasibility in determining of Cr in gannan navel orange treated in controlled conditions by Laser Induced Breakdown Spectroscopy. Optics & Laser Technology. 52: 70-74.

[56] Vítková, G., Novotný, K., Prokeš, L., Hrdlička, A., Kaiser, J., Novotný, J., Malina, R., and Prochazka, D. (2012), Fast identification of biominerals by means of stand-off laser-induced breakdown spectroscopy using linear discriminant analysis and artificial neural networks. Spectrochimica Acta Part B: Atomic Spectroscopy. 73: 1-6.

[57] Vítková, G., Prokeš, L., Novotný, K., Pořízka, P., Novotný, J., Všianský, D., Čelko, L., and Kaiser, J. (2014), Comparative study on fast classification of brick samples by combination of principal component analysis and linear discriminant analysis using

stand-off and table-top laser-induced breakdown spectroscopy. Spectrochimica Acta Part B: Atomic Spectroscopy. 101: 191-199.

[58] Fang, X. and Ahmad, S.R. (2014), Elemental analysis in environmental land samples by stand-off laser-induced breakdown spectroscopy. Applied Physics B. 115(4): 497-503.

[59] Fortes, F., Guirado, S., Metzinger, A., and Laserna, J. (2015), A study of underwater stand-off laser-induced breakdown spectroscopy for chemical analysis of objects in the deep ocean. Journal of Analytical Atomic Spectrometry. 30(5): 1050-1056.

[60] Guirado, S., Fortes, F.J., and Laserna, J.J. (2015), Elemental analysis of materials in an underwater archeological shipwreck using a novel remote laser-induced breakdown spectroscopy system. Talanta. 137: 182-188.

[61] Gong, Y., Choi, D., Han, B.-Y., Yoo, J., Han, S.-H., and Lee, Y. (2014), Remote quantitative analysis of cerium through a shielding window by stand-off laser-induced breakdown spectroscopy. Journal of Nuclear Materials. 453(1–3): 8-15.

[62] Sturm, V., Fleige, R.d., de Kanter, M., Leitner, R., Pilz, K., Fischer, D., Hubmer, G., and Noll, R. (2014), Laser-induced breakdown spectroscopy for 24/7 automatic liquid slag analysis at a steel works. Analytical chemistry. 86(19): 9687-9692.

[63] Wiens, R., Clegg, S., Barefield, J., and Maurice, S. *The Chemcam LIBS and Imaging Instrument Suite on the Curiosity Mars Rover, and Terrestrial Field Testing of LIBS*. (2014) 1, 3737.

[64] Gasda, P.J., Acosta-Maeda, T.E., Lucey, P.G., Misra, A.K., Sharma, S.K., and Taylor, G.J. (2015), Next Generation Laser-Based Standoff Spectroscopy Techniques for Mars Exploration. Applied Spectroscopy. 69(2): 173-192.

[65] Bogaerts, A. and Chen, Z. (2005), Effect of laser parameters on laser ablation and laser-induced plasma formation: A numerical modeling investigation. Spectrochimica Acta Part B: Atomic Spectroscopy. 60(9–10): 1280-1307.

[66] Albert, O., Roger, S., Glinec, Y., Loulergue, J.C., Etchepare, J., Boulmer-Leborgne, C., Perrière, J., and Millon, E. (2003), Time-resolved spectroscopy measurements of a titanium plasma induced by nanosecond and femtosecond lasers. Applied Physics A. 76(3): 319-323.

[67] Sekiguchi, K., Tanaka, K., Moriya, K., and Uchiki, H. (2006), Epitaxial growth of Cu2ZnSnS4 thin films by pulsed laser deposition. Physica Status Solidi (C) Current Topics in Solid State Physics. 3(8): 2618-2621.

[68] Moriya, K., Tanaka, K., and Uchiki, H. (2007), Fabrication of Cu2ZnSnS4 thin-film solar cell prepared by pulsed laser deposition. Japanese Journal of Applied Physics, Part 1: Regular Papers and Short Notes and Review Papers. 46(9 A): 5780-5781.

[69] Palonen, H., Huhtinen, H., and Paturi, P. (2015), Growth Condition Dependence of Microcracks in YBCO Thin Films Pulsed Laser Deposited on <inline-formula> <img

src="/images/tex/24016.gif" alt="\hbox {NdGaO}_{3}"> </inline-formula> (001) Substrates. Applied Superconductivity, IEEE Transactions on. 25(3): 1-4.

[70] Oshima, T., Yokoyama, K., Niwa, M., and Ohtomo, A. (2015), Pulsed-laser deposition of superconducting LiTi2O4 ultrathin films. Journal of Crystal Growth. 419: 153-157.

[71] Sharma, Y., Misra, P., Katiyar, R.K., and Katiyar, R.S. (2014), Photovoltaic effect and enhanced magnetization in 0.9 (BiFeO3)–0.1 (YCrO3) composite thin film fabricated using sequential pulsed laser deposition. Journal of Physics D: Applied Physics. 47(42): 425303.

[72] Chowdhury, M., Kumar Kunti, A., Kumar Sharma, S., Gupta, M., and Janay Chaudhary, R. (2014), Synthesis and characterization of pulsed laser deposited SnO2-Fe2O3 composite thin films for TCO application. The European Physical Journal - Applied Physics. 67(01): null-null.

[73] Lei, L., Liu, X., Yin, Y., Sun, Y., Yu, M., and Shang, J. (2014), Antibacterial Ag–SiO2 composite films synthesized by pulsed laser deposition. Materials Letters. 130: 79-82.

[74] Pradhaban, G., Kaliaraj, G., and Vishwakarma, V. (2014), Antibacterial effects of silver–zirconia composite coatings using pulsed laser deposition onto 316L SS for bio implants. Progress in Biomaterials. 3(2-4): 123-130.

[75] Eraković, S., Janković, A., Ristoscu, C., Duta, L., Serban, N., Visan, A., Mihailescu, I.N., Stan, G.E., Socol, M., Iordache, O., Dumitrescu, I., Luculescu, C.R., Janaćković, D., and Miškovic-Stanković, V. (2014), Antifungal activity of Ag:hydroxyapatite thin films synthesized by pulsed laser deposition on Ti and Ti modified by TiO2 nanotubes substrates. Applied Surface Science. 293: 37-45.

Plasma Diagnostics

Diagnostics of Magnetron Sputtering Discharges by Resonant Absorption Spectroscopy

Nikolay Britun, Stephanos Konstantinidis and Rony Snyders

Additional information is available at the end of the chapter

http://dx.doi.org/10.5772/61840

Abstract

The determination of the absolute number density of species in gaseous discharge is one of the most important plasma diagnostics tasks. This information is especially demanded in the case of low-temperature sputtering discharges since the time- and space-resolved behavior of the sputtered particles in the ground state determines the plasma kinetics and plasma chemistry in this case. Historically, magnetron sputtering is often implied when talking about sputtering discharges due to the popularity and the numerous advantages this technique provides for coating applications. The determination of the absolute density of various atomic and molecular species in magnetron sputtering discharges along with its time and space evolution may be important from several points of view, since it may help to estimate the total flux of particles to a virtual surface in the plasma reactor, to compare the throughputs of two different sputtering systems, to use the absolute particle concentrations as an input data for discharge modeling, etc. This chapter is intended to provide an overview on the advantages and main principles of resonant absorption spectroscopy technique as a reliable tool for in situ diagnostics of the particle density, as well as on the recent progress in characterization of magnetron sputtering discharges using this technique, when the role of reference source is played by another low-temperature discharge. Both continuous and pulsed magnetron sputtering discharges are overviewed. Along with the introduction covering the main principles of magnetron sputtering, the description of the basics of resonant absorption technique, and the selected results related to the particle density determination in direct current and high-power pulsed magnetron sputtering discharges are given, covering both space- and time-resolved density evolutions.

Keywords: Magnetron sputtering, resonant optical absorption spectroscopy, ROAS, atomic absorption, absolute number density

1. Introduction: Magnetron sputtering discharges

Among the families of plasma-related methods involved in thin films growth, those belonging to physical vapor deposition (PVD) are among the most commonly utilized. The PVD-related techniques cover wide range of the deposition methods, including evaporation, vacuum arc deposition, laser ablation, and sputtering. These methods are different from another class referred to as chemical vapor deposition (CVD) in the sense that the source of material is solid or liquid as opposed to a gaseous one in the case of CVD [1]. Thermal evaporation has been the most used PVD process for many years because of the easy handling and relatively high deposition rate comparing to the first known sputtering process, namely, the diode sputtering [2]. The latter process has been known since its first description in 1852 by W. R. Grove, who had performed sputtering using a massive inductive coil [3]. In diode sputtering, ions of the sputter gas, commonly argon, hit a negatively biased cathode (also known as target) with energy up to several hundred electron volts (eV). This energy is high enough to induce ejection of the superficial atoms, which is followed by their condensation on the chamber surfaces, including a potential substrate [4]. In diode sputtering, however, due to the relatively high process pressure needed to ensure the discharge stability (nearly 0.1 Torr), the mean free path of the ejected atoms is in the mm range, which is much lower than the typical substrate-target distance (cm range) resulting in relatively poor quality of the deposited coating.

The limitations of diode sputtering have been overcome in 196× with the works of E. Kay and W. Gill [5,6], who have proposed to utilize magnetic field to create efficient electron trapping above the cathode (target) in order to accelerate the ionization and sputtering itself. The electron trapping region is created due to the fact that the electrons propagate mainly along the magnetic field lines as a result of well-known gyration effect (with gyration radius being in the μm range). The implementation of magnetic field for sputtering, naturally producing the name "magnetron," has been shortly followed by the introduction of planar magnetron sources in 1974 [2,7], for which the presence of permanent magnets beneath the cathode was a distinctive feature, as schematically illustrated in Figure 1. As a result of the electron trapping, these innovations led to essential increase in the ionization degree of the sputter (bulk) gas in the cathode vicinity. As a result, the pressure necessary to maintain the discharge current in this case could be reduced by about one order of magnitude, i.e., down to the mTorr range [5]. Another critical point related to efficiency of magnetron sputtering is the topology of magnetic field. Indeed, the arrangement of magnets beneath the cathode affects significantly the degree of electron confinement above the target. In the so-called balanced magnetron sources, the dense plasma region above the target is roughly comparable to its radius (assuming a planar circular target) [8]. If the substrate is located outside this region, the bombardment of growing film by the plasma ions is essentially reduced (the resulting ion current is <1 mA/cm^2), limiting the benefit of the sputtering process. This effect is different in the so-called unbalanced sources, where the topology of the magnetic field near cathode is different, letting some magnetic field lines reach the substrate. In this case, plasma is not confined completely, and the ion current densities of about 2–10 mA/cm^2 can be reached, which is typically one order of magnitude higher than in the case of balanced magnetrons [9].

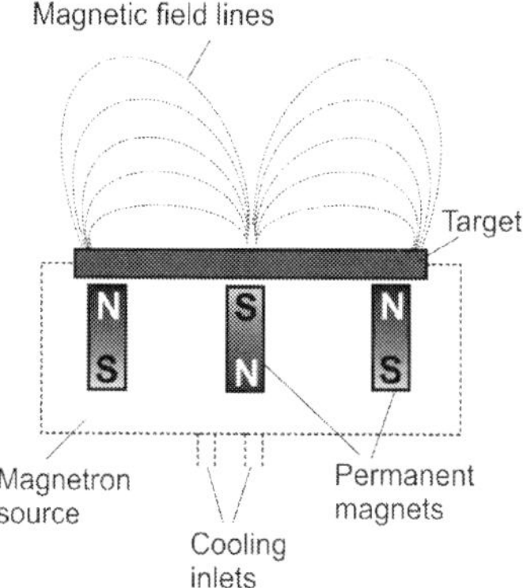

Figure 1. Side view of a typical magnetron source, including planar circular magnetron target, permanent magnets, and cooling system. The magnetic field lines are shown schematically.

In addition to magnetic field configuration, the voltage waveform and its repetition rate also play significant role in magnetron sputtering. The development of the plasma sources utilizing this effect has been mainly driven by the efficient deposition of the insulating compounds using reactive sputtering (see below). Thus, in addition to the well-known direct current magnetron sputtering (DCMS) devices, the radio frequency (RF) magnetron sources mainly working at 13.56 MHz have been introduced. Following the same trend, in the early 1990s, the idea of using pulsed-DCMS (also known as P-DCMS, or pulsed-DC) has been proposed (see, e.g., [10,11]). In most cases, the pulsed-DC power supplies operate successively alternating the negative (sputtering phase) and positive (charge dissipation phase) voltage cycles. The power supply in this case might be unipolar or bipolar, depending on polarity of the utilized pulses. Nowadays, most of magnetron sputtering processes for the synthesis of insulating compound coatings use pulsed magnetron sputtering approach [9].

Further development of the pulsed magnetron sputtering technology has triggered the new family of magnetron sputtering processes, so-called ionized physical vapor deposition (IPVD) techniques [12]. In a typical IPVD discharge, significant fraction of the sputtered atoms is ionized, reaching up to 100% in certain cases. The main idea behind the IPVD techniques is to generate denser plasmas than those appear in conventional magnetron sputtering in order to ionize the sputtered atoms more efficiently (up to the level of number densities $\sim10^{13}$–10^{14} cm^{-3} vs. 10^{8}–10^{11} cm^{-3} in the DCMS case). Among the different approaches targeted to increase the ionization degree, such as using an inductive coil [13], or hollow cathodes [14], the high-power impulse magnetron sputtering technique, or HiPIMS (also known as high-power pulsed magnetron sputtering—HPPMS), is the most notorious IPVD example. HiPIMS discharge uses the pulse duration ranged from few μs to few hundreds μs, while the pulse repetition frequency typically varies from ~10 Hz to ~10 kHz. Under these conditions, the peak current density may reach values of up to several A/cm^{-2} compared to a few mA/cm^{-2} in DCMS, but only during a short time, typically 1–2% of the repetition period [15]. As a result, a dense plasma is generated during the plasma on time enabling not only efficient target sputtering but also high ionization degree of metallic vapor [16], at the same time keeping the average applied power comparable to that of DCMS [15–17]. The schematic comparison of the DCMS, P-DCMS, and HiPIMS techniques in terms of the applied power is shown in Figure 2. Note that the level of time-averaged power applied to the sputtered cathode in each case is roughly the same, which is first of all due to the fact that the permanent magnets beneath the target should not be overheated and reach the Curie temperature (which is as low as about 580 K for widely used neodymium-based magnets). The actual applied power level is typically equal to several hundred Watt and often defined by the magnetron source cooling system (i.e., water temperature) used beneath the target. The interest to the pulsed magnetron discharges and, namely, the HiPIMS discharges from both scientific and application points of view has been continuously increasing since the introduction these techniques [18,19]. The main advantages of the HiPIMS technique, along with the achievements in characterization of these discharges, are described in the numerous works [15, 17,20–23].

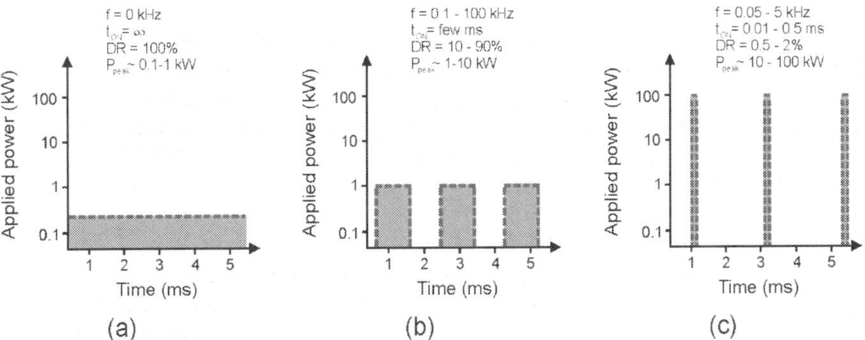

Figure 2. Comparison between the DCMS (a), P-DCMS (b), and HiPIMS (c) magnetron sputtering techniques in terms of the temporal distribution of the applied power. Typical range of the relevant discharge parameters and the values along both horizontal and vertical axes are given for reference only. The quantities f, t_{ON}, DR, and P_{peak} stand for repetition frequency, plasma on time, duty ratio, and peak power, respectively.

Why is the knowledge of the absolute density in plasma and particularly in magnetron sputtering discharges important? There are few main reasons making this parameter critical. First, *particle fluxes* in the discharge can be estimated knowing the density and velocity of each sort of the sputtered species, which is critical for any film growth process. Second, the discharge physics and chemistry can be significantly clarified having the absolute concentration of the relevant species in the discharge estimated properly. Third, the absolute concentration of the discharge species or their fluxes can be used as input information for further kinetic/chemical modeling of the discharge processes. Among the known plasma characterization approaches, resonant optical absorption spectroscopy (ROAS), also known as atomic absorption spectroscopy (AAS) [24,25], represents one of the straightforward ways for the determination of the absolute density of atomic and/or molecular species in gaseous discharges. Utilizing this technique, the density of the absorbers can be determined using an external (reference) light source by measuring the attenuation in its intensity after passing a volume with the absorbing species (see, e.g., Mitchell and Zemansky [24], p. 92). Along with the laser-based plasma diagnostic techniques, such as laser-induced fluorescence (LIF), ROAS technique has been playing a significant role in studying the plasma discharges during the last decades, involving both coherent (such as diode lasers) [26–28], and noncoherent (e.g., hollow cathode lamps— HCL) [29–31] radiation sources. The range of applications of this technique has been continuously increasing due particularly to the improvements of the tunable diode laser systems, namely, their stability, power level, tuning range, etc. [27,32].

In the domain of magnetron sputtering discharges, the determination of number density of the discharge particles by ROAS has been undertaken in the variety of systems, including time-resolved [33] and space-resolved [34] characterization. In particular, among the recent achievements, the works of various research groups devoted to absolute density of sputtered atoms in the DCMS discharges [31,34,35], including those amplified by an RF coil [30], as well as the works of devoted to flux measurements in DCMS [36], should be mentioned. What is related to the utilization of ROAS in the HiPIMS domain, the works dealing with the determination of time-resolved absorption line profile [33], which have been realized using tunable diode laser absorption spectroscopy (TD-LAS), should be emphasized. In addition, a detailed time evolution of the Ar metastable $1s_5$ states (Ar^{met}) has been investigated in a long-pulsed (200 μs) HiPIMS discharge by Vitelaru et al. [26], where the Ar^{met} density growth, followed by its rarefication after about 50 μs, and a gradual refill afterward have been clearly demonstrated. These results are generally in a good agreement with the recent studies in HiPIMS involving LIF [37] and ROAS [38] techniques, which are partially overviewed here. The last two works ([37,38]) are devoted to the short-pulsed HiPIMS discharges studying the propagation of the ground state and metastable sputtered particles, being mainly targeted to the systematic time-resolved characterization of HIPIMS discharges in terms of absolute density of species, 2-D imaging of the main atomic states corresponding to the studied discharge species (LIF+ROAS), as well as to the study of several optically available energy states of the discharge species and their sublevels.

Unifying the recent achievements in the domain of the absolute density measurements for two "extreme" magnetron sputtering cases, namely, DCMS and HiPIMS, this chapter is mainly focused on the particularities of the mentioned sputtering discharges in terms of the absolute density of species where resonant absorption is used as a main diagnostics technique. The resonant absorption method overviewed in this chapter is considered as a reliable tool for the determination of the absolute density of species in the discharge volume and for understanding the sputtering processes at the atomic level.

2. Particularities of resonant absorption

The main principles of the resonant absorption are described in details in the work of Mitchell and Zemansky [24]. This method is based on resonant absorption of radiation emitted by a reference source and absorbed (by atoms or molecules) in an optically thin gaseous discharge. At a definite spectral line the absolute density of states corresponding to a lower level of a chosen spectral transition can be determined by measuring so-called line absorption, i.e., an integral under spectral absorption line of interest, if the effective absorption lengths as well as the line width of both plasma and source spectral lines are known. This approach can be applied for the case when the spectral lines in the discharge are Doppler-limited [31], as well as can be generalized for more general cases [39]. The biggest advantage of ROAS technique is its nonintrusiveness. In this regard, this technique is different from the other methods used for density determination, which are based on the introduction of additional gases to the discharge (such as titration [40]), which may potentially change the electron energy distribution and the discharge kinetics. In addition, the fact ROAS uses external source of radiation and does not depend on the discharge emission itself is critical for characterization of the pulsed discharges, such as HiPIMS, especially during the afterglow time when optical emission spectroscopy (OES), cannot be applied.

As mentioned above, ROAS can be implemented in two different ways, namely, using a gaseous discharge emitting the spectral lines of interest, as well as using a monochromatic laser (e.g., a continuous tunable laser diode). Despite the fact that the laser-based diagnostics has several advantages [21,33], the discharge-based ROAS should be implemented. Among the main reasons for this are the following: (i) discharge-based ROAS setup is normally less expensive than the one based on a tunable laser; (ii) discharge-based ROAS depends on the line width ratio between the plasma and the reference source, which can be easily taken into account [24]; (iii) several spectral transitions can be studied in parallel in this case, as far as the emitters in the reference source coincide with the absorbers of interest, which is not possible with a monochromatic laser; and (iv) in case of a large spectral separation (few nm) between the lines of interest, laser-based absorption setup requires separate laser diodes for each transition, which involves additional calibration procedures. Due to the mentioned advantages, for example, the absolute density at the energy sublevels corresponding to a certain electronic state can be probed in a straightforward way, by simply following the changes in the emission intensity of the corresponding peaks from the reference source, as performed in the numerous studies [30,31,34,35,39]. Let us consider the main relations necessary to understand the ROAS method.

2.1. The basics of resonant absorption method

This chapter deals with the classical ROAS method described by Mitchell and Zemansky [24]. In the case of temperature-limited emission (absorption) lines in the reference source (and in the studied discharge), the application of the resonant absorption method is rather straightforward, and the details of its implementation in magnetron discharges can be found elsewhere [35,38]. In this section, only the essential relations important for understanding the absolute density determination using ROAS method are given. The number density of the absorbing species in the lower (often ground) state can be determined using the following relation:

$$N_{abs} = 1.2 \cdot 10^{12} k_0 \frac{\delta \sigma^p}{f_{ji}},$$

(1)

where N_{abs} (cm^{-3}) is the absolute number density of the absorbing species in the lower state, k_0 (cm^{-1}) is the absorption coefficient corresponding to the center of the absorption line, $\delta\sigma^p$ (cm^{-1}) is the full width at half maximum (FWHM) of the plasma emission line, f_{ji} is the absorption oscillator strength, and j (i) stands for lower (upper) level of a chosen transition, respectively. f_{ji} can be found using the relation [41]:

$$f_{ji} = 1.5 \cdot 10^{-14} \frac{g_i}{g_j} A_{ij} \lambda_{ij}^2 , \tag{2}$$

where g is the statistical weight of the corresponding energy level, A_{ij} is the emission probability corresponding to $i \rightarrow j$ transition, and λ_{ij} is the transition wavelength.

The absorption coefficient k_0 in Eq. (1) can be determined based on measurements of line absorption A_L, representing the integral under the absorption line profile. In the case of Doppler broadening of the source emission and plasma absorption lines, A_L can be expressed as follows [34,39]:

$$A_L = \frac{2}{\sqrt{\pi}} \int_0^\infty \mathrm{Exp}\left(-x^2\right)\left(1 - \mathrm{Exp}\left(-k_0 L \cdot \mathrm{Exp}\left(-\alpha^2 x^2\right)\right)\right) dx, \tag{3}$$

where L is the effective absorption length (~30 cm in our case), and α is the reference source-to-plasma line broadenings ratio. In the Doppler-limited case $\alpha^2 = T_S/T_P$, where T_S and T_P are the absolute gas temperatures in the reference source and in the plasma, respectively. A_L is normally determined experimentally by measuring the intensities of the corresponding emission lines:

$$A_L = 1 - \frac{\text{transmitted radiation}}{\text{incident radiation}} = 1 - \frac{I_{PS} - I_P}{I_S}, \tag{4}$$

where I_S is the light intensity from the reference source, I_P is the light intensity from the plasma, and I_{PS} is the (partially absorbed) light from the reference source passing through absorption region, along with the light from plasma, if any. The physical meaning of these quantities is illustrated in Figure 3.

After the experimental determination of line absorption A_L, according to Eq. (4), the absorption coefficient k_0 can be found as a result of solution of Eq. (3). This leads to a direct determination (according to Eq. (1)) of the absolute density of species corresponding to the lower atomic (molecular) state of a chosen spectral transition, if the effective absorption length L, as well as the oscillator strength f_{ji} are known. The critical role in the determination of the absorption

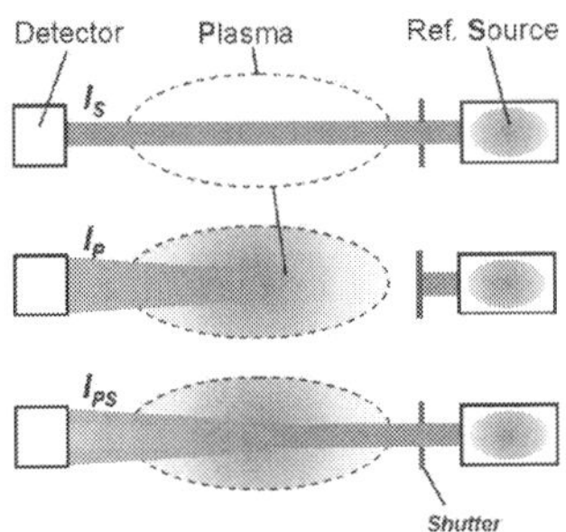

Figure 3. Schematic illustration of the absorption technique and the physical meaning of the quantities I_P, I_S, and I_{PS}.

coefficient is played by the coefficient α, representing the line broadening ratio of the spectral lines, as explained below.

2.2. Role of spectral line broadening

Several important remarks on the line broadening of plasma and reference source spectral lines should be made for proper interpretation of ROAS data. As mentioned above, this chapter deals with discharge-based ROAS when an HCL is primarily used as reference source. Since the gas pressure in HCL normally exceeds few Torr, and the mean free path of the gas particles is in the μm-range, the total particle thermalization can be assumed. The temperature in HCLs under the reviewed conditions, additionally measured by a high-resolution planar Fabry–Perot interferometer [42], is in the range of about 700–1100 K (see Table 1). Despite much lower pressure (~20–30 mTorr) in the magnetron discharges considered here (Table 1), similar considerations are also valid for the *bulk gas* atoms (Ar) in the studied DCMS and HiPIMS cases, which can be considered quasi-thermalized. The Ar temperature can be often assumed to be in the range of 300–500 K in this case, depending on the discharge type and its conditions [21,30,35,43].

Parameter	Value	Comment
	DCMS system parameters	
Target	99.99% Ti	With water cooling
Target diameter/thickness	5 cm/0.5 cm	10 cm/1 cm for Figure 9
Working gas	99.999% Ar	
Working pressure	20 mTorr	30 mTorr for Figure 9
Base pressure	~10^{-6} Torr or less	
Time-averaged power	~200 W	500 W for Figure 9
Monocromator used	Princeton Instrument SpectraPro-500i	Jobin-Yvon HR-460 for Figure 9

Parameter	Value	Comment
Monocromator slit width	10 μm	
Spectral resolution	0.05 nm	
Optical detector used	Princeton Instrument PI-MAX ICCD	Hamamatsu R928 photomultiplier tube
HiPIMS system parameters		
Target	99.99% Ti	With water cooling
Target diameter/thickness	10 cm/1 cm	
Working gas	99.999% Ar	
Working pressure	20 mTorr	
Base pressure	$<10^{-6}$ Torr	
pulse duration/frequency	20 μs/1 kHz	Unless stated otherwise
Supplied energy per pulse(at 20 μs pulse)	≈0.26 J	Corresponding to ~260 W of time-averaged power
Monocromator used	Andor Shamrock 750	
Monocromator slit width	20–100 μm	
Spectral resolution	0.04–0.1 nm typically	
Optical detector used	Andor iStar DH740-18F ICCD	With ~10^3 pulses averaged
Resonant absorption parameters		
Beam diameter	≈1 cm	
Distance above the target, z	≈5 cm	z = 0..10 cm for Figure 8, z = 8 cm for Figure 9.
Effective absorption length, L	30 cm	Variable for Figure 8, L=25 cm for Figure 9.
Reference source type	Ar-Ba and Ne-Ti hollow cathode lamps	
Reference source pulse duration	40–100 μs	Continuous for Figure 8, 150 μs for Figure 9.
Reference source temperature	≈1100 K (in pulsed mode [38])	950 K for Figure 8 [34], 630 K for Figure 9 [54].
DCMS plasma temperature	580 K (Figure 8), 300 K (Figure 9)	
HiPIMS plasma temperature (Ar)	≈500 K	Based on OES measurements [43]
HiPIMS plasma line width (Ti)	≈5.5 GHz	Based on LIF measurements [44]

Table 1. Main parameters of the DCMS and HiPIMS discharges described in this work and characterized by resonant optical absorption spectroscopy.

The broadening of spectral lines corresponding to the *sputtered* particles may differ from those of the bulk gas (Ar), however. In the DCMS case, despite the fact the sputtered particles are far from thermalization, their velocity distribution function (VDF) can be often approximated by a Gaussian profile. This VDF does not change in time, by definition of DCMS discharge [37], and often can be associated with the particle temperature in the range of 300–700 K for the conditions described here (see Table 1). Due to the directional motion of the sputtered particles in magnetron plasma, the spectral line in the direction of measurements (i.e., along the ROAS line-of-sight) might be significantly broadened, as compared to the one of the reference source. This situation is schematically illustrated in Figure 4. In this case, the resonant absorption will happen only for a part of the total line width, and the ratio between the mentioned broadenings has to be introduced ($\delta\sigma^S/\delta\sigma^P$) to take into account the broadening difference, as described in details by Mitchell and Zemansky [24]. The similar approach should also be utilized in case non-Doppler line broadening dominates in the discharge, as studied recently by Li et al. [39].

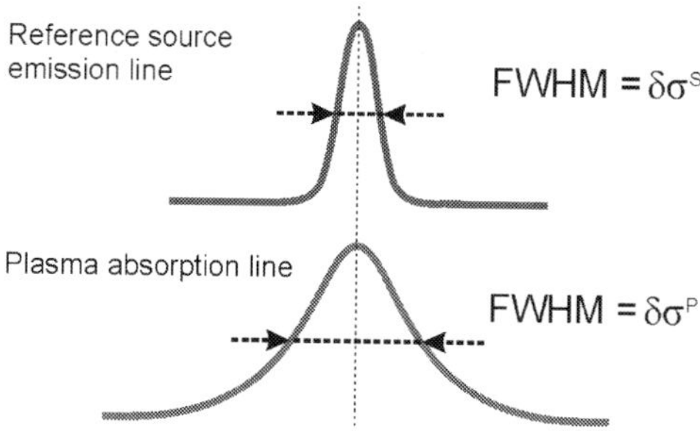

Figure 4. Illustration of the spectral line broadening (FWHM) of the reference source (emission) and plasma (absorption) lines along with their broadening notations.

The situation is yet different for HiPIMS discharges, which are far from thermalization during the plasma on time, especially in the target vicinity [37,43]. The distribution of the velocity component parallel to the target ($v_{||}$) in this case is often different from the Maxwell–Boltzmann one (even though it still can be roughly fitted by a Gaussian profile [33]), and the gas temperature in the classic sense cannot be applied. Because of this, the VDF line width corresponding to a particular velocity component and defining the line width of the emission (absorption) line should be determined separately. The dynamics of the line broadening in HiPIMS has been studied recently by LIF [44,45]. Based on these works, under the conditions described here (20 mTorr, $z = 5$ cm), the VDF width changes only by about 25%, which has rather minor impact on the final density of absorbers determined by Eq. (1). Due to this fact, the plasma absorption line width is assumed constant during the whole HiPIMS discharge period for simplicity, as described in [38,44] (see Table 1). These assumptions are additionally justified by the fact that the absorption coefficient k_0, determined based on Eq. (3) is weakly dependent on the plasma line width $\delta\sigma^P$.

2.3. Experimental setup for resonant absorption

The typical experimental arrangement for characterization of magnetron sputtering discharge by resonant spectroscopy is shown in Figure 5. The main parameters related to the discharges overviewed in this work are summarized in Table 1. Both in DCMS and HiPIMS cases, a cylindrical vacuum chamber with either horizontally or vertically placed balanced magnetron sputtering source holding a planar circular magnetron target has been used. The Ti magnetron targets, 5 or 10 cm in diameter, attached to a magnet-supporting water-cooled copper base, have been utilized. The commercial power supplies have been used to sustain the DCMS discharges, as described elsewhere [30,34]. In the HiPIMS case, the discharge current and voltage along with the pulse parameters have been controlled by the Lab.-made power supply, described by Ganciu et al. [46]. The HiPIMS discharge typically had 20 µs of the pulse duration and 1 kHz of the repetition frequency (thus having 980 µs of the plasma off time). The typical current–voltage waveforms can be found elsewhere [43,44,47]. The base pressure in the magnetron reactor has been kept $<10^{-6}$ Torr, whereas the bulk gas (99.999% Ar) pressure was normally fixed at 20 mTorr during all the ROAS measurements, unless stated otherwise. In order to avoid deposition on the quartz window surface, the special collimators (Ni tubes) with high length-to-radius ratio have been installed in front of each viewport inside the reactor.

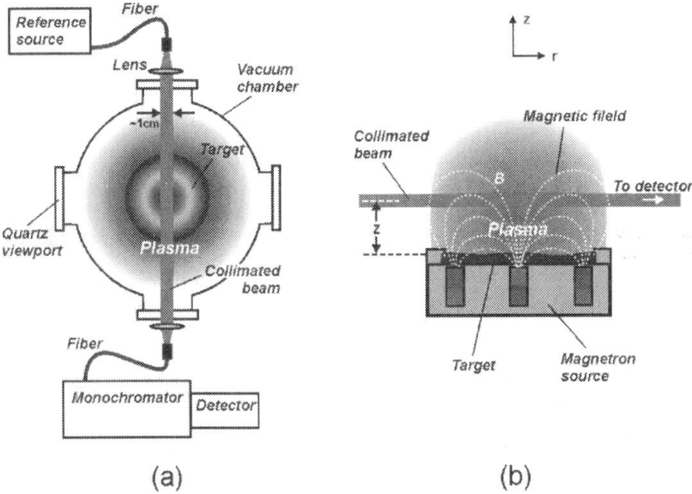

Figure 5. Top (a) and side (b) view of the experimental setup utilized for absolute density measurements in a magnetron sputtering discharge by ROAS. Role of the reference source is played by a hollow cathode lamp. Reproduced with permission from Britun et al. [38]. Copyright 2015 AIP Publishing.

The Perkin Elmer hollow cathode lamps (one with Ti cathode filled by Ne, and another one with Ba cathode filled by Ar) have been used as the reference sources. The HCLs have been running either in the continuous or pulsed regime (mainly in the HiPIMS case) in order to increase the emission intensity and make it comparable with the HiPIMS plasma emission, aiming at reducing the measurements error during the plasma on time [43,48]. Apart from this, the pulse regime of HCL source favors the ionization of the cathode material inside the lamp, which is critical for analysis of the Ti^+ states [30]. During the measurements, the HCLs have a pulse duration typically ranging from 40 to 100 μs. About 10^3 pulses from the reference source have been averaged by the detector during the measurements of the absorption coefficient. The HCL pulse sequence was triggered by the HiPIMS power supply using the external transistor-transistor logic (TTL) trigger and was time shifted relatively to the HiPIMS plasma pulse using an additional analog TGP-110 pulse generator.

In some cases, when the density of absorbers in the studied discharge is too low, an improved detection scheme, including a triple optical fiber and allowing simultaneous acquisition of the I_S, I_P, and I_{PS} signals, may significantly increase the reliability of the obtained data. Such a scheme, illustrated in Figure 6, allows to avoid the instabilities of the plasma discharge and/or of the reference source during the acquisition time (which may last for several minutes) because the signals necessary to calculate the line absorption are taken simultaneously in this case, as described by Britun et al. [48]. The diagnostics setup in this case contains a reference emission source, a multichannel optical fiber, a group of lenses for beam collimation before passing the vacuum chamber, and a monochromator with an optical detector, according to Figure 5(a).

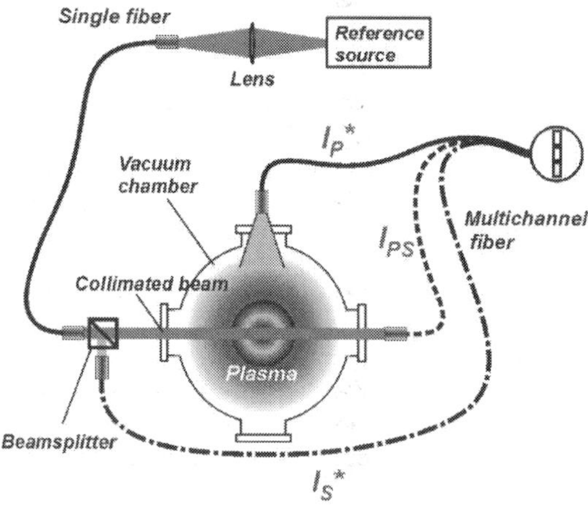

Figure 6. Scheme of the experimental setup for absorption measurements with increased reliability utilizing simultaneous acquisition of the plasma and source optical signals (I_P, I_S, and I_{PS}) necessary for calculation of the line absorption A_L.

A monochromator equipped with an intensified charge-coupled device (ICCD) camera and connected to the discharge reactor by an optical fiber has been used for spectral acquisition in most cases. The spectral resolution of the monochromator was typically ≈0.05–0.1 nm. The typical emission spectra acquired in Ar-Ti HiPIMS discharge at the end of plasma on time are presented in Figure 7. From the general emission spectrum (Figure 7(a)), the relevant spectral regions containing Ti and Ti$^+$ emission lines are visible (Figure 7(b–d)). The representative emission lines used in this work for resonant absorption are marked by arrows (see also Table 2). In most cases, the studied emission lines are well-resolved from the neighboring peaks. Due to this fast, rather low spectral resolution is normally sufficient for ROAS measurements. The total procedure for the number density determination consisted of the emission peaks intensities measurements (I_S, I_P, and I_{PS} values at each condition), A_L value determination by Eq. (4), determination of the absorption coefficient k_0 as a result of solving Eq. (3), followed by calculation of the absolute density of absorbers according to Eq. (1).

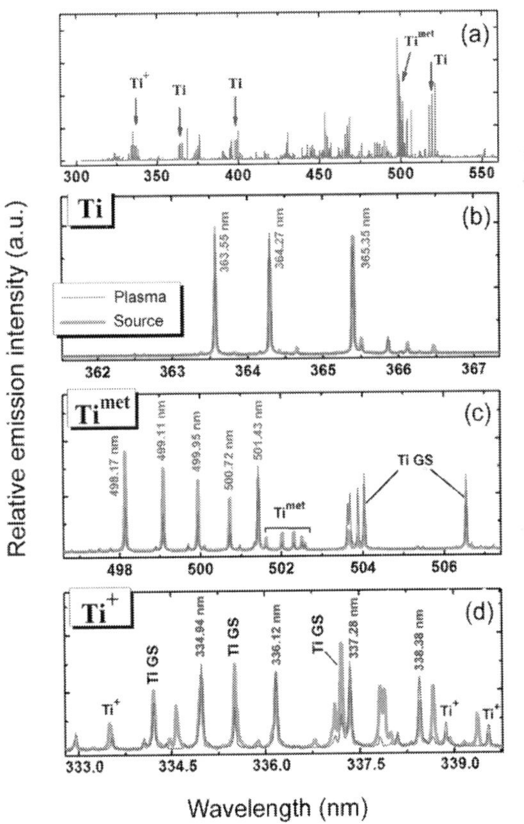

Figure 7. General emission spectrum (a) obtained in Ar-Ti HiPIMS discharge showing the spectral regions of interest used for the determination of the absolute density of different atomic species. The spectral regions corresponding to Ti ground state atoms (b), Ti metastables (c), as well as Ti ions (d) are shown separately. The HiPIMS and HCL spectra (b–d) are normalized arbitrary. Reproduced with permission from Britun et al. [38]. Copyright 2015 AIP Publishing.

Studied atom	Nature of the probed state	Energy of the probed state (eV)	Emission wavelength λ (nm)	Spectral transition (lower–upper level, j–i)	Statistical weight of the lower level, g_j	Oscillator strength, f_{ji}[a]
Ti	Ground	0.000	363.55	$3d\,^24s\,^2\,a^3F_2$–$3d\,^2(^3F)4s4p(^1P°)\,y^3G°_3$	5	0.20
		0.021	364.27	$3d\,^24s\,^2\,a^3F_3$–$3d\,^2(^3F)4s4p(^1P°)\,y^3G°_4$	7	0.16
		0.048	365.35	$3d\,^24s\,^2\,a^3F_4$–$3d\,^2(^3F)4s4p(^1P°)\,y^3G°_5$	9	0.14
Ti^{met}	Metastable	0.813	501.43	$3d\,^3(^4F)4s\,a^5F_1$–$3d\,^3(^4F)4p\,y^5G°_2$	3	0.38
		0.818	500.72	$3d\,^3(^4F)4s\,a^5F_2$–$3d\,^3(^4F)4p\,y^5G°_3$	5	0.26
		0.826	499.95	$3d\,^3(^4F)4s\,a^5F_3$–$3d\,^3(^4F)4p\,y^5G°_4$	7	0.22
		0.836	499.11	$3d\,^3(^4F)4s\,a^5F_4$–$3d\,^3(^4F)4p\,y^5G°_5$	9	0.17
		0.848	498.17	$3d\,^3(^4F)4s\,a^5F_5$–$3d\,^3(^4F)4p\,y^5G°_6$	11	0.22
Ti$^+$	Ground	0.000	338.38	$3d\,^2(^3F)4s\,a^4F_{3/2}$–$3d\,^2(^3F)4p\,z^4G°_{5/2}$	4	0.28
		0.012	337.28	$3d\,^2(^3F)4s\,a^4F_{5/2}$–$3d\,^2(^3F)4p\,z^4G°7/2$	6	0.25
		0.028	336.12	$3d\,^2(^3F)4s\,a^4F7/2$–$3d\,^2(^3F)4p\,z^4G°9/2$	8	0.26
		0.049	334.94	$3d\,^2(^3F)4s\,a^4F9/2$–$3d\,^2(^3F)4p\,z^4G°11/2$	10	0.27
Ti^{+met}	Metastable	0.574	376.13	$3d\,^2(^3F)4s\,a^2F_{5/2}$–$3d\,^2(^3F)4p\,z^2F°_{5/2}$	6	0.22
		0.607	375.93	$3d\,^2(^3F)4s\,a^2F_{7/2}$–$3d\,^2(^3F)4p\,z^2F°_{7/2}$	8	0.21
Ar^{met}	Metastable	11.548	811.53	$3s\,^23p\,^54s(1s\,_5)$–$3p\,^54p(2p\,_9)$b	5	0.51
		11.723	794.82	$3s23p54s(1s\,_3)$–$3p54p(2p\,_4)$b	1	0.56

[a]The values of the oscillator strength are taken from Crintea et al. [75] for Ar and Wiese and Fuhr [76] otherwise.

[b]Paschen notations are used for Ar energy levels [75].

Table 2. Spectral parameters of the optical transitions for Ti, Ti^{met}, Ti$^+$, and Ar^{met} atomic states used in this work for ROAS diagnostics. Spectroscopic data are taken from the online NIST spectral database [74] unless stated otherwise. Reproduced with permission from Britun et al. [38]. Copyright 2015 AIP Publishing.

3. Particle density behavior in DCMS discharges

3.1. Spatial characterization of the DCMS discharge

Even though the main plasma parameters of the DCMS discharges do not alter in time, the plasma density as well as the density of the sputtered species may vary significantly as a function of the spatial position in the discharge. The first parameter in this case defines mainly the excitation level of the bulk and sputtered species as well as their ionization degree, whereas the (ground state) density of the sputtered species is mainly affected by the applied power, sputtering yield, target size, diffusion processes in the discharge volume, as well as by the other processes. The effects of ground state density depletion also may take place, as recently shown for DCMS discharges by LIF [49], but these effects are rather minor, especially comparing with HiPIMS case [37,38], since the ionization degree in the DCMS discharges generally remains at the level of few percent [15].

The optical diagnostics applied to a DCMS discharge clearly visualizes the plasma region corresponding to the confined electrons, which is represented by the bright area near the target surface observable by a naked eye, as shown in Figure 8. This region clearly corresponds to a spatial segment formed by the magnetic lines in the target vicinity, as shown in Figure 8(b), where the simulated magnetron magnetic field lines are drawn. The target diameter in this case is equal to 5 cm, but nevertheless the obtained data can be compared to the typical absolute density measured with larger targets (see Table 1), as soon as the target materials are the same and the current density are known, due to the scalability of the magnetron discharges [15]. In addition to this, as shown by Britun et al. [34], the emission lines in the plasma region decay exponentially along z direction, where an exponential dependence with significantly lower decrement is found at higher z, which corresponds to the diffusion region.

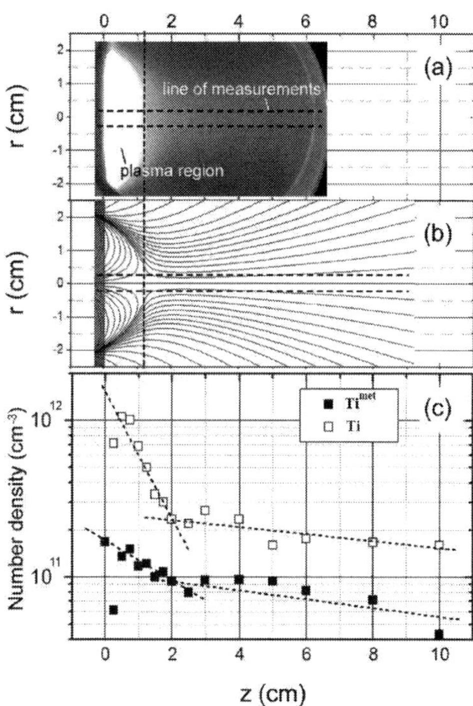

Figure 8. A photograph of the actual DCMS discharge operating at 4 mTorr of Ar pressure (a) along with the simulated magnetic field lines (b), and the distribution of the absolute density of Ti ground state atoms and Ti metastables measured by ROAS as a function of z at 30 mTorr of Ar pressure (c). Reproduced with permission from Britun et al. [34]. Copyright 2006 IOP Publishing.

In addition, the ground state density of the sputtered Ti as well as Ti metastables, measured by ROAS, follows the trend found for the emission lines by Britun et al. [34], as shown in Figure 8(c). As one can observe, the number density of Ti neutrals reaches nearly 10^{12} cm^{-3} near the target at the examined conditions (see Table 1), dropping about 5 times already at $z = 2$ cm, which is mainly due to the diffusion and the angular particularities of sputtering [50]. The Ti metastable atoms (the energy level is about 0.8 eV, see Table 2) reveal the same tendency having the total density 2–5 times smaller. The presented absolute values of Ti density are in the good agreement with related measurements performed in DCMS under the similar conditions [31]. We have to note that the found particularities of spatial distribution for the sputtered particles in DCMS should resemble those for HiPIMS discharges, as far as the magnetic field topology and the average power are similar. In HiPIMS discharge, however, the additional physical effects, such as the time-dependent broadening of the emission [22] and absorption spectral lines [44], cross-field ion transport [15,51], ionization-induced density depletion near the target [37], time-dependent current-induced changes of the magnetic field [52], etc., may not allow a direct comparison between the DCMS and the HiPIMS discharges in terms of the absolute density of particles. These differences should be even more essential with increasing of the plasma density, i.e., the magnetron current density, as briefly mentioned below.

3.2. Absolute density in reactive DCMS discharge

Another interesting possibility that ROAS provides is visualization and control of the physical processes during the so-called reactive sputtering. This process is mainly known for synthesis of oxide, nitride, and oxynitride thin films [21,53]. Reactive sputtering normally involves standard magnetron sputtering sources, where a certain percentage of, e.g., molecular oxygen or nitrogen, is added to the bulk gas (typically Ar) in order to finally synthesize oxide or nitride compounds. Physically, during the reactive sputtering, oxidation (nitriding, etc.) of the cathode superficial layer takes place, which is often referred to as "poisoned" regime. This process is characterized by a significant drop of the density of metallic species in the discharge volume [54]. Apart from this, a hysteresis effect (in terms of discharge voltage or metal atom density) depending on the direction of change of the reactive gas content is often considered to be among the main characteristics of reactive sputtering [15,55–57].

The example of measurements of the absolute number density of Ti atoms sputtered in a DCMS discharge at $z = 8$ cm away from the target using resonant absorption is given in Figure 9. Molecular nitrogen has been added to the bulk gas in this case in order to synthesize nitride films, as described in details in [54]. As we can see, as a result of nitrogen addition, a clear drop in the density of the metal (Ti) atoms can be detected. Due to the target poisoning, the density of sputtered Ti drops by a factor of about 4 when the nitrogen content in the discharge exceeds ≈5%. The relative density of Ti ions in this case is measured using mass-spectrometry, demonstrating the same effect. Based on this example, we can conclude that the (real-time) measurements of the absolute density using absorption spectroscopy, contrary to the well-known plasma emission monitoring method [58], may have additional benefits for reactive sputtering since the emission is always defined by electron excitation and may be weak at the high distances from the target. In addition, the spatial selectivity of ROAS may also be better than that of OES, since the collimated beam used in this case can be highly focalized and thin, especially if a laser is used as a reference source [27].

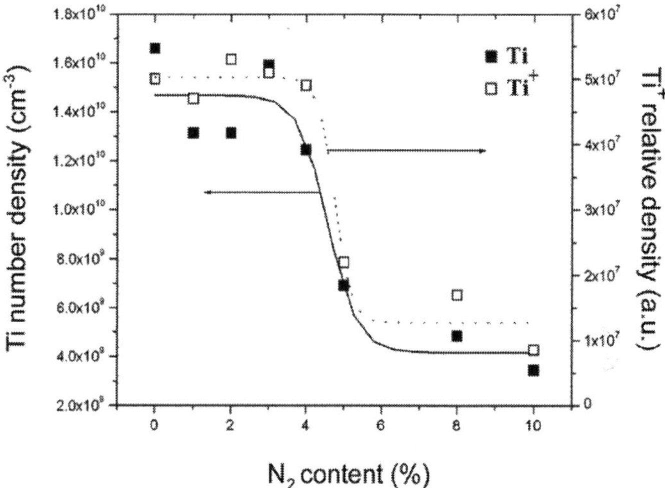

Figure 9. Evolution of the absolute Ti ground state density measured by ROAS (filled squares) and Ti⁺ relative density obtained by mass spectrometry (open squares) in a reactive DCMS discharge as a function of the nitrogen content. Ar pressure = 30 mTorr, applied power = 500 W, z = 8 cm. Reproduced with permission from Konstantinidis et al. [54]. Copyright 2005 Elsevier Publishing.

4. Particle density dynamics in HiPIMS discharges

HiPIMS discharges, possessing very short plasma on time, as illustrated in Figure 2, demonstrate different physics from that known for the DCMS discharges. The main reason for this is significantly elevated plasma density as well as the ionization degree related to both the bulk gas and the sputtered particles in HiPIMS, as mentioned in the Introduction and analyzed in the numerous works. Despite the fact that DCMS and HiPIMS sputtering processes can be realized using the same magnetron source, the direct comparison between these discharges is rather difficult, especially at the high values of the peak power applied to HiPIMS (or cathode current density), even though the time-averaged level of the applied power might be the same. In this section, for the sake of illustration of the dynamics of HiPIMS discharge in terms of the absolute density of species, the time-resolved evolution of the main discharge species in an Ar-Ti short-pulsed HiPIMS discharge analyzed by ROAS is presented. The experimental details on the HiPIMS system used for diagnostics are listed in Table 1, whereas the corresponding spectroscopic data are available in Table 2, or can be found elsewhere [31,38].

4.1. Time-resolved particle density evolution in HiPIMS

The time-resolved evolution of the absolute densities of Ti and Ti^+ atomic species sputtered in a 20-μs pulse Ar-Ti HiPIMS discharge measured at $z = 5$ cm above the target is shown in Figure 10. The total density in this case represents the sum of populations of the corresponding ground state sublevels (see Table 2). As we can observe from this figure, after a small density decrease (depletion) during the plasma on time, the Ti density is peaking at the time delay $\Delta t \approx 100$ μs and reaches the value of about 6×10^{11} cm^{-3}. The number density is found to be proportional to the pulse duration in this case, as far as the other discharge parameters are the same, which should be due to the change in the pulse energy (0.13 vs. 0.26 J). The pulse duration effect is also valid for the Ti ions in HiPIMS (not shown). The decrease in Ti density at the end of the plasma on time (roughly twice) is likely related to the VDF broadening of the sputtered particles during this time, neglected during the density calculation, as explained in [38]. In this case, an increase in the VDF width during the plasma on time observed by Palmucci et al. [44] results only in ~20% of the total density drop, which is less than the depletion visible in Figure 10(a) for Ti. This implies a presence of the additional factors responsible for density depletion, such as ionization of the sputtered particles, as discussed by Britun et al. [37]. Quantitative estimation of the different contributions is barely possible in this case, primarily due to the strong plasma emission in this time interval, which results in high deviation of the obtained data. The error bar shown in Figure 7(a) represents a standard deviation of the data points corresponding to several independent ROAS measurements, whereas the density data correspond to the values averaged among several measurements, as well as among three Ti ground state energy sublevels listed in Table 1.

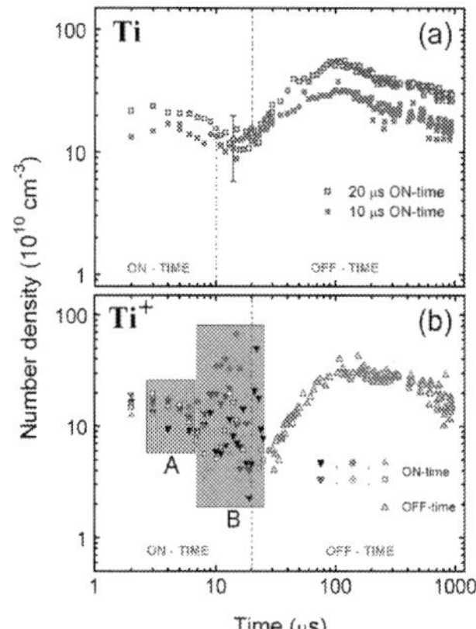

Figure 10. Time-resolved evolution of the Ti (a) and Ti+ (b) number density in Ar-Ti HiPIMS discharge measured by ROAS. The A and B time intervals correspond to the density depletion and the current self-organization, respectively. Six independent measurements have been undertaken in the case of Ti+ continuously showing the density instabilities in the interval B. Ar pressure = 20 mTorr, $z = 5$ cm. Reproduced with permission from Britun et al. [38]. Copyright 2015 AIP Publishing.

The evolution of the Ti ion density shown in Figure 10(b) reveals similar tendencies, especially during the plasma off time, where the density has a maximum at 100–200 µs, reaching ~3 × 10^{11} cm^{-3}. During the plasma on time, two phenomena can be observed: the first one is the beginning of density depletion interval, which is similar to that found for Ti neutrals (called interval A), and the second one which characterized by the strong instabilities of the measured Ti$^+$ density at the end of the plasma on time, resembling the stochastic density behavior (called interval B). Note that six independent density measurements made during the plasma on time always lead to the different Ti$^+$ density evolution. Since Ti ionization is likely defined by the electrons presented in the discharge volume during this time ($Ti + e \rightarrow Ti^+ + 2e$), the latter phenomenon is probably related to the discharge current self-organization at the end of the plasma on time, resulting in a nonperiodic rotation of the excitation zones ("spokes") along the target race-track area, as reported by Kozyrev et al. [59] and studied by the other groups [60–62]. Apparently, the uncertainties in the measured Ti$^+$ density during the interval B persist even after the averaging of the signal over 10^3 pulses by the ICCD detector. This implies an existence of the additional factors responsible for the strong uncertainty of Ti$^+$ density, such as the high plasma-to-source emission peak ratio at the end of plasma on time. Indeed, a strong domination of the plasma emission intensity over the emission from the reference source dramatically decreases the stability of ROAS measurements. In our case, the typical emission intensity from plasma may surpass the one coming from the reference source by a factor of about 100 for Ti ions, and of about 30 for Ti neutrals, as illustrated in Figure 11. As follows from the recent analysis performed by Britun et al. [48], the relative error of the ROAS measurements in this case may easily exceed 100%, even if the relative error of the emission intensity measurements is only about 5% (typical for ICCD detectors), which may explain the instabilities observed during the HiPIMS on time for ions. Because of the mentioned instabilities, the plasma parameters such as Ti ionization degree, which can be defined for Ti as

$$\eta = \frac{\left[Ti^+ \right]}{\left[Ti^+ \right] + \left[Ti \right]} \tag{5}$$

(where $[X]$ stand for the absolute number density of specie X), are rather difficult to determine during the interval B. In the plasma off time, however, η reaches the level of about 0.4, which is typical for HiPIMS off time [63], taking into account rather low pulse energy used in this work ($E_P = 0.26$ J).

The summary of the time evolution of several atomic species studied in a 20-µs Ar-Ti HiPIMS discharge at $z = 5$ cm is given in Figure 12. From this figure, the main stages of HiPIMS discharge can be distinguished. First, after the discharge ignition, the excitation of Ar and Ti atoms remaining from the previous plasma pulse is taking place, as a result of propagation of the secondary electrons away from the cathode. The background Ti density is found to be ~2 × 10^{11} cm^{-3}, whereas the density of Ti metastables is much lower (~10^{10} cm^{-3}) for the considered conditions. The propagation of the hot electrons [64,65] results in formation of the Ar and Ti metastable states (Ar^{met}, Ti^{met}) as well as in the intensive Ti ionization during the plasma on

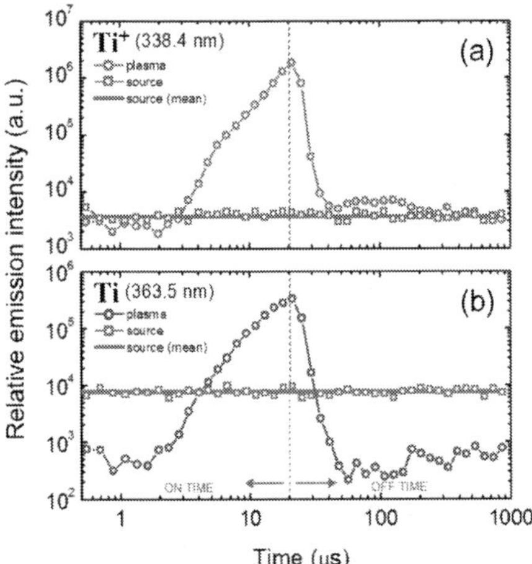

Figure 11. Typical example of the time evolution of Ti⁺ (a) and Ti (b) emission line intensity in Ar-Ti HiPIMS discharge comparing to the intensity of the same emission lines in the reference source (HCL) showing a dramatic increase in the emission intensity at the end of the HiPIMS pulse. Ar pressure = 20 mTorr, $z \sim 1$ cm.

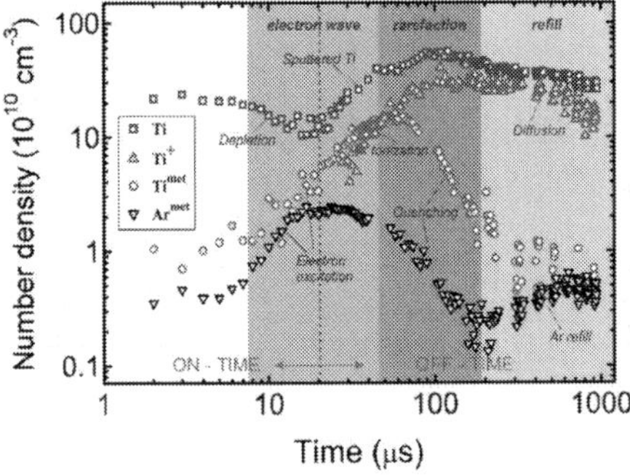

Figure 12. Summary of time evolution of the absolute density corresponding to the main discharge species in Ar-Ti HiPIMS discharge measured by ROAS. Three main time intervals corresponding to the secondary electron wave propagation, rarefaction, and refill are indicated. Ar pressure = 20 mTorr, z = 5 cm. Reproduced with permission from Britun et al. [21] (Copyright 2014 IOP Publishing) and Britun et al. [22] (Copyright 2015 WILEY-VCH Publishing).

time–beginning of the plasma off time, which is accompanied by the inversion of population of the Ti energy sublevels, as discussed below. At the beginning of the plasma off time, the density of metal ions increases (red triangles in Figure 12), corresponding to their transport toward the volume analyzed by ROAS. Second, the depletion of Ti density at the end of the plasma on time is also visible. This depletion can be induced by the VDF broadening of Ti neutrals at the end of the plasma pulse [44]. Apart from this, a strong ionization of the sputtered neutrals in this region should also contribute to the observed density depletion [37,38]. Third, the strong excitation and ionization above the target are followed by rarefication of the bulk gas, resulted by the sputtered Ti species incoming to the studied volume, which is clearly observable together with a decrease in the Ar^{met} density, and also supported by the recent 2D density mapping results [37]. This process should be accompanied by electron cooling, as a result of collisions with the incoming heavy particles [64]. The ionization of sputtered Ti is taking place during the same time interval (20–100 μs), presumably as a result of the direct electron impact $(Ti + e \rightarrow Ti^+ + 2e)$. The role of Penning ionization of Ti $(Ti + Ar^{met} \rightarrow Ti^+ + Ar + e)$ in this case is supposed to be negligible, due to the much higher measured increase in Ti^+ density ($\Delta[Ti^+] \approx 2 \times 10^{11}$ cm^{-3}) comparing to a small decrease in the density of Ar^{met} ($\Delta[Ar^{met}] \approx 2 \times 10^{10}$ cm^{-3}) during this time. Finally, in addition to the gas rarefication above the target, the quenching of both Ti and Ar metastable states by the sputtered Ti should also contribute in the observable decrease of Ti^{met} and Ar^{met} density during the 50- to 200-μs time interval. Afterward (at $\Delta t > 200$ μs), the refill of Ar to this area is observed, which is consistent with the Ar refill time estimated in the related works [26,66]. The Ti and Ti^+ ground state densities gradually decrease during this time, due mainly to the diffusion of these species. The described processes are summarized in Table 3 where the main kinetic reactions along with the possible reasons for the observed processes are given.

Time	Particle	Processes	Main reaction(s)	Evidence
0–5 μs	Electrons	Generation		
	Ar	Ionization + acceleration toward the cathode	$Ar + e \rightarrow Ar^+ + 2e$	Related works [15]
	Ti			
	Ti^{met}	Have their "background" density level, remaining from previous plasma pulse.		
	Ti^+			
5–20 μs	Electrons	**Electron wave** propagation (secondary hot electrons— e^{sec}) (few km/s)		LIF imaging data [37], Langmuir probe measurements [77]
	Ar^{met}	Generation by e^{sec}	$Ar + e \rightarrow Ar^{met} + e$	LIF imaging data [37]
	Ti	VDF broadening and strong ionization, resulting in density depletion		LIF imaging data [37], measured VDF width [44]
	Ti^{met}	Generation by e^{sec}	$Ti + e \rightarrow Ti^{met} + e$	Presence of Ar^{met} during this interval [37]
	Ti^+	Generation by e^{sec} (?)	$Ti + e \rightarrow Ti^+ + 2e$	Not visible, probably due to the discharge current instabilities [59,60]
20–50 μs	Electrons	**Electron wave** (gradual cooling)		LIF imaging data [37], inversion of sublevel populations [67] (Figures 13 and 14).

Time	Particle	Processes	Main reaction(s)	Evidence
	Ar^{met}	1. Saturation, rarefication 2. Quenching by incoming Ti neutrals	2. $Ar^{met} + Ti \rightarrow Ar + Ti$	1. Decrease in $[Ar^{met}]$, Related studies [15,26] 2. LIF imaging data [37]
	Ti	1. Propagation of sputtered Ti 2. Ti VDF relaxation		1. LIF imaging data [37]2. Related studies [44,45]
	Ti^{met}	Follow the growth of [Ti]	$Ti + e \rightarrow Ti^{met} + e$	Same growth rate for Ti and Ti^{met} (Figure 12)
	Ti^+	Generation (following Ti wave)	1. $Ti + e \rightarrow Ti^+ + 2e$ 2. $Ar^{met} + Ti \rightarrow Ti^+ + Ar + e$ (Penning ionization)	1. Faster growth of $[Ti^+]$ and slower growth of [Ti] at the same time 2. Negligible, due to very small drop in $[Ar^{met}]$ (Figure 12)
50–200 μs	Electrons	Thermalization	$X + e^{hot} \rightarrow X + e^{cold}$, where X is mainly Ti	Loss of the inversion of sublevel populations (Figure 14), Langmuir probe data [64,65], modeling [78]
	Ar^{met}	**Rarefication**		Figure 12
	Ti	Diffusion		Gradual decrease of [Ti] (Figure 12)
	Ti^{met}	Quenching, diffusion	$Ti^{met} + X \rightarrow Ti + X$, where X is a heavy particle	Permanent density drop, as well as $[Ti^{met}]/[Ti]$ ratio drop (Figure 12)
	Ti^+	Density saturation (production-loss balance)		Figure 12
200–1000 μs	Electrons	Thermalized		Related works [64,65,78]
	Ar^{met}	**Refill**		Increase in $[Ar^{met}]$, refill time estimations [26,66]
	Ti	Continuing diffusion		Gradual decrease of [Ti] (Figure 12)
	Ti^{met}	Reaching background $[Ti^{met}]$ level $\sim 10^{10}$ cm^{-3}.		Figure 12
	Ti^+	Slight drop after 600 μs, might be due to:	1. $X^+ + Ar + e \rightarrow X + Ar^{met}$ 2. $X^+ + e \rightarrow X$ where $X = Ar$, Ti	1. $[Ti^+]$ drop + $[Ar^{met}]$ increase 2. $[Ti^+]$ drop (Figure 12)

Table 3. Summary of the processes related to the absolute density evolution of the discharge species measured by ROAS in a 20-μs Ar-Ti HiPIMS discharge (see Figure 12). Partially reproduced with permission from Britun et al. [38]. Copyright 2015 AIP Publishing.

4.2. Inversion of the energy sublevel populations

An interesting effect can be observed if the measured density is represented separately for each energy sublevel at the studied electronic state, as shown in Figure 13. As one can observe, a clear inversion among the studied energy states is achieved for five Ti sublevels corresponding to the metastable Ti a^5F_J states (see Table 2). An inversion of the considered energy sublevels (i.e., when the population of the state corresponding to higher energy is found to be higher) takes place in Figure 13(a) during the 20- to 100-µs time interval. The maximum of the mentioned inversion for Ti^{met} case is located around $\Delta t \approx 60$ µs. Similar but much weaker effect for Ar metastables (only two levels are considered) is also visible in Figure 13(b).

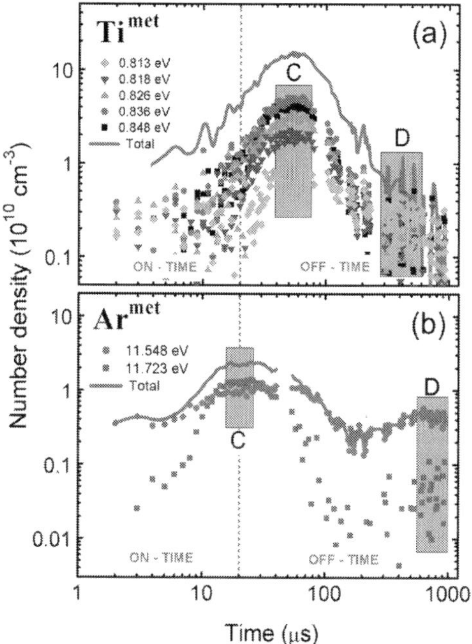

Figure 13. Time-resolved number density evolution for Ti (a) and Ar (b) metastable atoms measured by ROAS in Ar-Ti HiPIMS showing the populations of the corresponding energy sublevels (see Table 2). The total density evolution is given by gray lines for comparison. Ar pressure = 20 mTorr, z = 5 cm. Reproduced with permission from Britun et al. [38]. Copyright 2015 AIP Publishing.

Despite the fact that the inversion effect is rather new for HiPIMS, it is well known for the other low-temperature discharges, such as Ti hollow cathode plasma [67], and RF-amplified DCMS discharge [30]. In the former case, it is explained by the (partial) equilibrium of the heavy discharge species with the energetic electrons during the short plasma on time interval. Besides this, the inversion of the sublevel populations might be also responsible for the apparent density depletion in the magnetron target vicinity detected in the DCMS discharges

[30,49,68,69]. The sublevel inversion can be explained based on the Boltzmann distribution of the corresponding energy sublevel populations. Assuming the presence of the dense flux of hot electrons during certain time in the discharge, the heavy particles should be intensively excited by the electrons during this time. Under these conditions, for the close energy sublevels, the factor $\Delta E / kT$ in the Boltzmann distribution may be negligible due to rather high effective (excitation) temperature of the involved particles. This should result finally in the proportionality between the level population N_i and its statistical weight g_i, as shown by the following expressions:

$$\frac{N_i}{g_i} = \frac{N_j}{g_j} \cdot \mathrm{Exp}\left(-\frac{E_i - E_j}{kT_{exc}}\right) \equiv \frac{N_j}{g_j} \cdot \mathrm{Exp}\left(-\frac{\Delta E}{kT_{exc}}\right) \tag{6}$$

where i (upper) and j (lower) stand for two close levels, $N_{i,j}$ is the level populations, $g_{i,j}$ - is their statistical weights, ΔE is the energy difference between two levels, and T_{exc} is the excitation temperature. Under $\Delta E \ll kT_{exc}$, we obtain the following:

$$\frac{N_i}{g_i} \approx \frac{N_j}{g_j}\left(1 - \frac{\Delta E}{kT_{exc}}\right) \approx \frac{N_j}{g_j} \tag{7}$$

which, after normalization to $\frac{N_j}{g_j}$, results in $N_i \sim g_i$. Note that this effect has the same nature as the "saturated LIF" mode described by Amorim et al. [70].

The sublevel inversion is even more clear if a "stabilized" modification of the resonant absorption method is implemented, when a pulsed reference source is synchronized with the time gate of the ICCD detector, as discussed above. The relative error of the ROAS measurements can be significantly reduced in this case, and fine spectral effects in the discharge, such as inversion of the energy sublevels, can be easily visualized. The example of these measurements using three Ti ground state energy sublevels is given in Figure 14. In Figure 14(a), the experimental ROAS results, clearly showing the presence of the Ti ground state sublevel inversion (roughly between 20 and 300 µs), are shown. At the same time, a sketch of the same level populations drawn in accordance to the inversion model described by Eqs. (6) and (7) is given in Figure 14(b) for the sake of clarification of the experimental data. We have to note that before the inversion interval, the studied Ti sublevels do not perfectly follow the Boltzmann distribution, as the middle level is mostly populated in this case. This fact might be due to the complicated HiPIMS discharge kinetics, which is not well studied yet. Let us also note that the inversion interval might be significantly shifted in time depending on position of the ROAS line of sight, and the inversion should happen much earlier if the measurements are performed closer to the target surface.

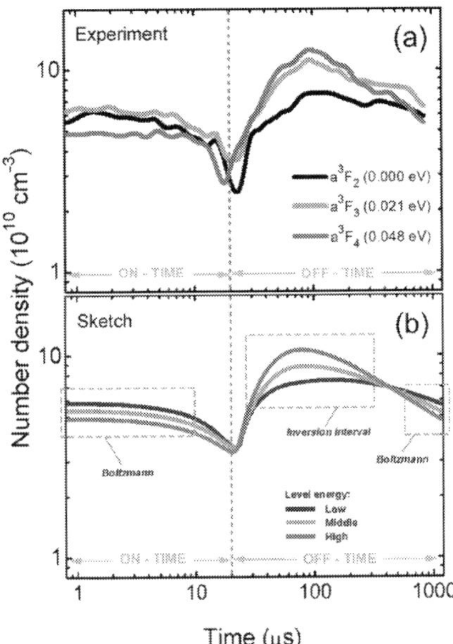

Figure 14. The time-resolved populations of three ground state Ti energy sublevels (a) measured in Ar-Ti HiPIMS discharge by ROAS using the dynamic source triggering. Ar pressure = 20 mTorr, z = 5 cm. A sketch (b) illustrating the dynamics of the Ti sublevels population as a function of time in HiPIMS. The beginning of the inversion interval should depend on the observation point (z) and should start earlier if the ROAS line-of-sight is located closer to the target.

Fitting the experimental data corresponding to Ti sublevels, the Ti *excitation* temperature can be obtained, which is a measure of excitation of the corresponding energy sublevels by the hot electrons. In our case, the estimations lead to T_{exc} ~0.25 eV for Ti, and to the close value for Ti^{met}. In general, a precise ROAS measurement of the energy sublevel populations for a certain atomic state should allow a reliable determination of the excitation temperature T_{exc} for the heavy discharge particles after their interaction with hot electrons, as well as the determination of its temporal evolution in the discharge. Representing a measure of excitation of the heavy particles by electrons in the discharge volume, T_{exc} should be proportional to the HiPIMS pulse energy, secondary electron emission coefficient, and should be generally higher in the target vicinity [22].

4.3. Density dynamics in reactive HiPIMS

It is also interesting to consider some selected phenomena in the case of reactive HiPIMS discharge. The behavior of the number density of the sputtered species in reactive HiPIMS should follow the same tendency as in the reactive DCMS case (shown in Figure 9), i.e., a density drop by several times in the poisoned regime of sputtering should be expected. Let us consider the behavior of the absolute density of *reactive* species here. The presented example is related to the time-behavior of the absolute density of the atomic oxygen measured by ROAS. This question has been particularly studied by Vitelaru et al. [71] using a tunable diode laser and giving the dynamics of the atomic oxygen in the HiPIMS discharge, however, without measuring its absolute density, which is a very important parameter, as mentioned in Section 1. In the considered case, a discharge-based ROAS with simultaneous signal acquisition (see Figure 6) has been applied for measurements of the absolute density of O atoms in the HiPIMS discharge volume. Due to the spectral limitations, however, the ground state O atoms are not accessible by ROAS (the corresponding transitions are located in the deep UV range [72]), so only the O *metastable* atoms (O^{met}, 3s $^5S°_2$ level with the energy = 9.15 eV) are analyzed in this case. A microwave surfaguide discharge [73] operating at 2.5 GHz and using Ar–7% O_2 gas mixture has been utilized as a reference source in this case.

Analyzing the O^{met} atoms in any discharge, one should bear in mind that these are excited species which keep the dynamics of the plasma electrons as well as the dynamics of the ground state O_2 and O. The time evolution of O^{met} atoms measured by ROAS at $z = 5$ cm in the HiPIMS discharges working with Ar–20% O_2 gas mixture and using Ti and Ag targets is shown in Figure 15. The typical stages of nonreactive sputtering discharge can be observed in this case (compare to Figure 12). The first noticeable processes are the O^{met} density depletion at the beginning of the plasma on time (visible in the Ti target case only) followed by the density increase at the end of the plasma on time. These processes are followed by the rarefication of O^{met} and the gas refill afterward (exists both for Ti and Ag target cases). We would like to emphasize that both in Ti and Ag cases the O^{met} density starts to increase already at the end of the plasma on time, unlike in the nonreactive case. Such an increase correlates well with the time-resolved dynamics of the Ar metastables in the nonreactive case, as shown in Figure 12. This similarity, as well as the proximity of the excitation thresholds for these species (11.55 eV for Ar^{met} vs. 9.15 eV for O^{met} [74]) point out on the role of the secondary electrons in formation of the O^{met} state, which should take place already during the plasma on time (i.e., $O_2 + e \rightarrow O + O + e$, followed by $O + e \rightarrow O^{met} + e$). The correlation with Ar^{met} is additionally confirmed by the gas rarefication and refill processes observed both for Ar and atomic O during the plasma off time. The different density values in the density maxima of O^{met} ($\approx 5 \times 10^8$ cm^{-3} for Ti vs. $\approx 3 \times 10^9$ cm^{-3} for Ag) may be related to the differences in the ion-induced secondary electron emission (ISEE) coefficients in these cases, assuming the electron impact to be a primary mechanism for O^{met} formation. Interestingly, the absolute number density below 10^8 cm^{-3} has been detected in these measurements, approving the high reliability of ROAS with simultaneous signal detection. (see Figure 6).

Since in the case of Ar–20% O_2 gas mixture magnetron target is supposed to be completely oxidized [47], the sputtering of oxide material from (i.e., M_xO_y molecules, where M is a metal atom) followed by formation of O atoms and finally O^{met} should be another possible mechanism of the observed O^{met} density elevation at the end of the plasma on time. This mechanism, however, is supposed to be rather minor since the wave of the sputtered species arrives in the volume of measurements much later, having the maximum at $\Delta t \approx 100$ µs, as shown in Figure 12.

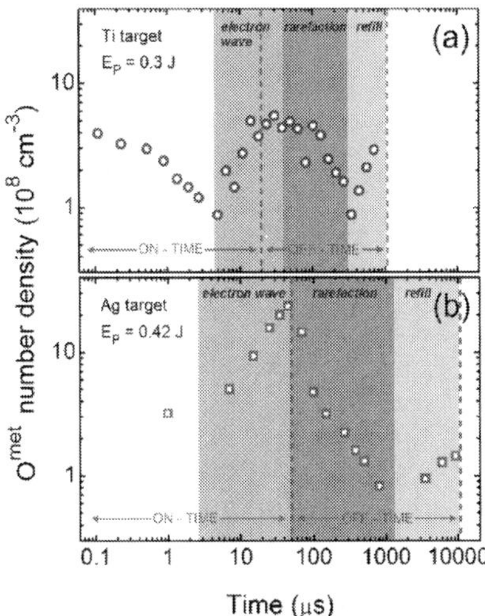

Figure 15. Time-resolved evolution of the absolute density of O metastable atoms (3s $5S^0_2$ state) measured by ROAS in HiPIMS discharge using Ar–20% O_2 gas mixture for the case of Ti (a) and Ag (b) cathode. Ar pressure = 20 mTorr, z = 5 cm. Pulse duration, discharge frequency, and pulse energy (E_P) are, respectively, 20 ms, 1 kHz, 0.3 J (a) and 50 ms, 0.1 kHz, 0.42 J (b). Reproduced with permission from Britun et al. [22]. Copyright 2015 WILEY-VCH Publishing.

5. Summary and conclusions

5.1. DCMS discharges

Resonant absorption analysis of the direct current sputtering discharges shows that the absolute density of the sputtered particles is ranged in a relatively narrow interval, being always in the range of 10^{10}–10^{12} cm^{-3} in the Ti case (assuming the average power level to be several hundred Watt). These values may be somewhat different varying the cathode materials with essentially different sputtering yield and/or in the sputtering systems, which use significantly higher power levels. The spatial distribution of the sputtered particles is mainly determined by their diffusion (at high pressure), by the angular directivity of sputtering (low pressure), as well as by the magnetic field topology (critical mainly for the sputtered ions), which lead to nearly exponential decay of the observed density of species as a function of the distance from cathode, so the density might decay at least by one order of magnitude already at few cm away from the target surface (for 5 cm diameter round target).

One of the advantages of using ROAS technique for diagnostics of sputtering processes is its ability to probe the species in a wide range of absolute density, starting from roughly 10^7–10^8 cm^{-3}, where the signal-to-noise ratio of the used detector is a limiting factor, and finishing by the values several orders of magnitude higher, when the increase of the optical thickness of the volume of interest becomes the limiting factor. The bottom sensitivity level may be additionally affected by the type of the absorption scheme used (e.g., single-pass, multipass, with or without simultaneous signal acquisition, etc.). The variety of the ROAS implementation schemes allow real time monitoring of the sputtered species also during *reactive* sputtering, which is very demandable, as the synthesis of the compound (oxides, nitrides, etc.) coatings is one of the main sputtering workflow nowadays. Since the density drop between the "metallic" and "reactive" mode of sputtering normally drop by about one order of magnitude, the absolute density levels in both modes can be easily real time monitored by absorption technique, which might be useful for research-oriented as well as for industry applications. This conclusion is valid both for DCMS (as shown in this work), pulsed-DC, as well as for HiPIMS discharges.

5.2. HiPIMS discharges

The physical phenomena in the HiPIMS discharges are rather different from those observed in DCMS and P-DCMS cases, first of all in terms of the plasma density and much shorter timing (µs scale). As a result, numerous dynamic effects should be taken into account explaining the properties of the HiPIMS discharges. Among these effects, the following time-dependent phenomena should be mentioned: the propagation of the fast electron wave the dynamic gas rarefaction and refill, the ionization and excitation of various discharge particles, quenching of the excited states as a result of collisions, the anomalous electron transport, etc. ROAS technique, being essentially limited by only the time-resolution of the used optical detector, suits very well for systematic characterization of these discharges in terms of the absolute density. In addition, the modern ICCD detectors typically provide ns time resolution, which surpasses the typical HiPIMS time scale by few orders of magnitude.

In terms of the time-resolved evolution of the discharge particles density, there are several processes in HiPIMS discharge, which can be visualized using ROAS technique. The time-resolved summary of these processes is given in Table 3, considering the main discharge species studied in the Ar-Ti nonreactive HiPIMS case. The presence of such dynamic effects as the density depletion, gas rarefaction, gas refill, quenching of the metastable states, ionization, as well as the propagation of the electron wave away from the target surface can be concluded based on the undertaken ROAS analysis. Despite a rather high HiPIMS repetition rate considered in this work (1 kHz), the gradual relaxation for the majority of the listed processes till the end of the plasma off time can be observed. Even though the total range of the measured absolute density of the discharge species reaches nearly three orders of magnitude in the considered case, the detected densities are still well above the detection threshold (which is estimated to be roughly 10^7–10^8 cm^{-3} with the detection scheme used). The sensitivity of the used ROAS method can be additionally enhanced using the lock-in detection technique, dynamic triggering of reference source, and related approaches.

One of these approaches, namely, the simultaneous acquisition of the emission signals using a triple optical fiber, has been utilized for time-resolved detection of the oxygen metastable atoms during the reactive HiPIMS process. The results reveal the presence of the same discharge phases (such as gas rarefaction and refill, as well as the density depletion in some cases) as detected in the nonreactive HiPIMS case. The absolute density of the O metastable atoms detected in this case is still above the mentioned ROAS sensitivity threshold, being in the range of 10^8–10^9 cm^{-3}. The optical diagnostics of the reactive HiPIMS discharges, however, is a relatively new area and the obtained results require additional verification by the other methods, such as LIF, in order to be understood completely.

5.3. General remarks

The resonant absorption technique represents a powerful diagnostic tool suitable for the determination of the absolute density of the relevant species in sputtering discharges, such as DCMS, P-DCMS, HiPIMS, as well as the other types of magnetron discharges. This technique has several modifications allowing its optimization depending on the discharge geometry etc. For example, using a plasma discharge as a reference source, ROAS setup can be rather simple, requiring only the additional optics for reference source beam collimation, and (sometimes) additional synchronization between the source and the discharge of interest. Both direct current and pulsed reference sources can be used for resonant absorption purposes. In addition to this, in case a broad range spectral detector, such as ICCD, is utilized, a parallel detection of several spectral transitions as well as the corresponding absolute densities becomes possible. These advantages open new possibilities to study the basic effects in the discharge volume. Among these effects are the population of the metastable levels of the various discharge species, the population of the atomic energy sublevels of the atoms and their time-resolved dynamics, the study of the excitation temperature across the discharge volume, etc. These possibilities are important for studying magnetron sputtering processes at the fundamental level, whereas rather basic configuration of the ROAS setup may be helpful in industry, e.g., for real-time density monitoring of the relevant species during a certain process.

The sensitivity of the absorption method can also be significantly improved by varying the detection schemes, such as lock-in amplification (improving the sensitivity by 1–2 orders of magnitude typically), multipass methods, as well as by simultaneous signal acquisition, dynamic source triggering, etc. Using the simultaneous signal acquisition for example, the sensitivity below 10^8 cm^{-3} has been achieved for O metastable atoms in a reactive HiPIMS discharge case, as illustrated in this chapter. This value is far below the typical concentrations of the sputtered atoms in the discharge (10^{10}–10^{12} cm^{-3}) achieved under the laboratory conditions, which is enough for most applications. For very weakly populated energy states, like those appearing in molecular discharges, the multipass absorption, including the ring-down spectroscopy method, could be suggested, which are, being out of the scope of this chapter, still based on the same fundamental principles.

Acknowledgements

This work is supported by the Belgian Government through the "Pôle d'Attraction Interuniversitaire" (PAI, P7/34, "Plasma-Surface Interaction," Ψ). N. Britun is a postdoctoral researcher; S. Konstantinidis is a research associate of the Fonds National de la Recherche Scientifique (FNRS), Belgium.

Author details

Nikolay Britun[1*], Stephanos Konstantinidis[1] and Rony Snyders[1,2]

*Address all correspondence to: nikolay.britun@umons.ac.be

1 Chimie des Interactions Plasma–Surface (ChIPS), CIRMAP, Université de Mons, Mons, Belgium

2 'Materia Nova' Research Center, Parc Initialis, Mons, Belgium

References

[1] S. M. Rossnagel, J. Vac. Sci. Technol. A. 21, S74 (2003).

[2] G. Bräuer, B. Szyszka, M. Vergöhl, and R. Bandorf, Vacuum 84, 1354 (2010).

[3] W. R. Grove, Philos. Trans. R. Soc. London 142, 87 (1852).

[4] P. Sigmund, *Sputtering by Particle Bombardment I. Physical Sputtering of Single-Element Solids* (Springer, Berlin, 1981).

[5] W. D. Gill and E. Kay, Rev. Sci. Instrum. 36, 277 (1965).

[6] E. Kay, J. Appl. Phys. 34, 760 (1963).

[7] J. S. Chapin., Res. Dev. 25, 37 (1974).

[8] R. D. Arnell and P. J. Kelly, Surf. Coat. Technol. 112, 170 (1999).

[9] P. J. Kelly and R. D. Arnell, Vacuum 56, 159 (2000).

[10] S. Schiller, K. Goedicke, J. Reschke, V. Kirchhoff, S. Schneider, and F. Milde, Surf. Coat. Technol. 61, 331 (1993).

[11] J. Musil, J. Lestina, J. Vlcek, and T. Tolg, J. Vac. Sci. Technol. A. 19, 420 (2001).

[12] U. Helmersson, M. Lattemann, J. Bohlmark, A. P. Ehiasarian, and J. T. Gudmundsson, Thin Solid Films 513, 1 (2006).

[13] S. M. Rossnagel and J. Hopwood, Appl. Phys. Lett. 63, 3285 (1993).

[14] A. Tonegawa, H. Taguchi, and K. Takayama, Nucl. Instrum. Methods Phys. Res. B. 55, 331 (1991).

[15] J. T. Gudmundsson, N. Brenning, D. Lundin, and U. Helmersson, J. Vac. Sci. Technol. A. 30, 030801 (2012).

[16] K. Sarakinos, J. Alami, and S. Konstantinidis, Surf. Coat. Technol. 204, 1661 (2010).

[17] J. T. Gudmundsson, Vacuum 84, 1360 (2010).

[18] I. K. Fetisov, A. A. Filippov, G. V. Khodachenko, D. V. Mozgrin, and A. A. Pisarev, Vacuum 53, 133 (1999).

[19] V. Kouznetsov, K. Macák, J. M. Schneider, U. Helmersson, and I. Petrov, Surf. Coat. Technol. 122, 290 (1999).

[20] U. Helmersson, M. Lattemann, J. Bohlmark, A. P. Ehiasarian, and J. T. Gudmundsson, Thin Solid Films 513, 1 (2006).

[21] N. Britun, T. Minea, S. Konstantinidis, and R. Snyders, J. Phys. D. Appl. Phys. 47, 224001 (2014).

[22] N. Britun, S. Konstantinidis, and R. Snyders, Plasma Process. Polym. DOI: 10.1002/ppap.201500051 (2015).

[23] K. Sarakinos, J. Alami, and S. Konstantinidis, Surf. Coat. Technol. 204, 1661 (2010).

[24] A. C. Mitchell and M. W. Zemansky, *Resonance Radiation and Excited Atoms* (Plenum Press, Cambridge, 1971).

[25] G. F. Kirkbright and M. Sargent, *Atomic Absorption and Fluorescence Spectroscopy* (Academic Press, London, 1974).

[26] C. Vitelaru, D. Lundin, G. D. Stancu, N. Brenning, J. Bretagne, and T. Minea, Plasma Sources Sci. Technol. 21, 025010 (2012).

[27] A. Zybin, J. Koch, H. D. Wizemann, J. Franzke, and K. Niemax, Spectrochim. Acta Part B At. Spectrosc. 60, 1 (2005).

[28] D. S. Baer and R. K. Hanson, J. Quant. Spectrosc. Radiat. Transf. 47, 455 (1992).

[29] L. De Poucques, J.-C. Imbert, C. Boisse-Laporte, J. Bretagne, M. Ganciu, L. Teulé-Gay, and M. Touzeau, Plasma Sources Sci. Technol. 15, 661 (2006).

[30] S. Konstantinidis, A. Ricard, M. Ganciu, J. P. Dauchot, C. Ranea, and M. Hecq, J. Appl. Phys. 95, 2900 (2004).

[31] M. Gaillard, N. Britun, Y. M. Kim, and J. G. Han, J. Phys. D. Appl. Phys. 40, 809 (2007).

[32] TOPTICA Photonics Webpage: http://www.toptica.com/products/research_grade_diode_lasers.html

[33] C. Vitelaru, L. de Poucques, T. M. Minea, and G. Popa, J. Appl. Phys. 109, 053307 (2011).

[34] N. Britun, M. Gaillard, L. Schwaederlé, Y. M. Kim, and J. G. Han, Plasma Sources Sci. Technol. 15, 790 (2006).

[35] N. Britun, S. Ershov, A.-A. El Mel, S. Konstantinidis, A. Ricard, and R. Snyders, J. Phys. D. Appl. Phys. 46, 175202 (2013).

[36] C. Vitelaru, L. de Poucques, T. M. Minea, and G. Popa, Plasma Sources Sci. Technol. 20, 045020 (2011).

[37] N. Britun, M. Palmucci, S. Konstantinidis, and R. Snyders, J. Appl. Phys. 117, 163302 (2015).

[38] N. Britun, M. Palmucci, S. Konstantinidis, and R. Snyders, J. Appl. Phys. 117, 163303 (2015).

[39] L. Li, A. Nikiforov, N. Britun, R. Snyders, and C. Leys, Spectrochim. Acta Part B At. Spectrosc. 107, 75 (2015).

[40] A. Ricard, V. Monna, and M. Mozetic, Surf. Coat. Technol. 174–175, 905 (2003).

[41] W. Lochte-Holtgreven, editor, *Plasma Diagnostics* (North-Holland publishing company, Amsterdam, 1968).

[42] N. Britun, J. G. Han, and S.-G. Oh, Plasma Sources Sci. Technol. 17, 045013 (2008).

[43] N. Britun, M. Palmucci, S. Konstantinidis, M. Gaillard, and R. Snyders, J. Appl. Phys. 114, 013301 (2013).

[44] M. Palmucci, N. Britun, S. Konstantinidis, and R. Snyders, J. Appl. Phys. 114, 113302 (2013).

[45] N. Britun, M. Palmucci, and R. Snyders, Appl. Phys. Lett. 99, 131504 (2011).

[46] M. Ganciu, M. Hecq, S. Konstantinidis, J.-P. Dauchot, M. Touzeau, L. de Poucques, and J. Bretagne, 2007/0034498 A1 (2007).

[47] M. Palmucci, N. Britun, T. Silva, R. Snyders, and S. Konstantinidis, J. Phys. D. Appl. Phys. 46, 215201 (2013).

[48] N. Britun, M. Michiels, and R. Snyders, Rev. Sci. Instrum. 86, (2015) In press.

[49] N. Britun, M. Gaillard, and J. G. Han, J. Phys. D. Appl. Phys. 41, 185201 (2008).

[50] A. Goehlich, D. Gillmann, and H. F. Döbele, Nucl. Instrum. Methods Phys. Res. B. 164–165, 834 (2000).

[51] D. Lundin, P. Larsson, E. Wallin, M. Lattemann, N. Brenning, and U. Helmersson, Plasma Sources Sci. Technol. 17, 035021 (2008).

[52] J. Bohlmark, U. Helmersson, M. VanZeeland, I. Axnäs, J. Alami, and N. Brenning, Plasma Sources Sci. Technol. 13, 654 (2004).

[53] J. Musil, P. Baroch, J. Vlček, K. H. Nam, and J. G. Han, Thin Solid Films 475, 208 (2005).

[54] S. Konstantinidis, A. Ricard, R. Snyders, H. Vandeparre, J. P. Dauchot, and M. Hecq, Surf. Coat. Technol. 200, 841 (2005).

[55] D. Depla, J. Haemers, G. Buyle, and R. De Gryse, J. Vac. Sci. Technol. A. 24, 934 (2006).

[56] M. Audronis and V. Bellido-Gonzalez, Thin Solid Films 518, 1962 (2010).

[57] D. Depla and S. Mahieu, editors, *Reactive Sputter Deposition* (Springer, 2008).

[58] P. P. Woskov, K. Hadidi, P. Thomas, K. Green, and G. Flores, Waste Manag. 20, 395 (2000).

[59] A. V. Kozyrev, N. S. Sochugov, K. V. Oskomov, A. N. Zakharov, and a. N. Odivanova, Plasma Phys. Rep. 37, 621 (2011).

[60] A. Hecimovic, M. Böke, and J. Winter, J. Phys. D. Appl. Phys. 47, 102003 (2014).

[61] A. P. Ehiasarian, A. Hecimovic, T. De Los Arcos, R. New, V. Schulz-Von Der Gathen, M. Boke, J. Winter, and M. Bke, Appl. Phys. Lett. 100, 114101 (2012).

[62] J. Andersson, P. Ni, and A. Anders, Appl. Phys. Lett. 103, 054104 (2013).

[63] A. Ehiasarian, A. Vetushka, A. Hecimovic, and S. Konstantinidis, J. Appl. Phys. 104, 083305 (2008).

[64] A. Pajdarová, J. Vlček, P. Kudláček, and J. Lukáš, Plasma Sources Sci. Technol. 18, 025008 (2009).

[65] P. Poolcharuansin and J. W. Bradley, Plasma Sources Sci. Technol. 19, 025010 (2010).

[66] D. Lundin, N. Brenning, D. Jädernäs, P. Larsson, E. Wallin, M. Lattemann, M. A. Raadu, and U. Helmersson, Plasma Sources Sci. Technol. 18, 045008 (2009).

[67] D. Ohebsian, N. Sadeghi, C. Trassy, and J. M. Mermet, Opt. Comm. 32, 81 (1980).

[68] N. Nafarizal, N. Takada, K. Shibagaki, K. Nakamura, Y. Sago, and K. Sasaki, Jpn. J. Appl. Phys. 44, L737 (2005).

[69] N. Nafarizal, N. Takada, and K. Sasaki, Jpn. J. Appl. Phys. 48, 126003 (2009).

[70] J. Amorim, G. Baravian, and J. Jolly, J. Phys. D. Appl. Phys. 33, R51 (2000).

[71] C. Vitelaru, D. Lundin, N. Brenning, and T. Minea, Appl. Phys. Lett. 103, 104105 (2013).

[72] R. Payling and P. Larkins, *Optical Emission Lines of the Elements (CD-ROM)* (Wiley & Sons Inc., New York, 2000).

[73] T. Godfroid, J. P. Dauchot, and M. Hecq, Surf. Coat. Technol. 174–175, 1276 (2003).

[74] NIST Atomic Spectra Database Lines Form: http://physics.nist.gov/PhysRefData/ASD/lines_form.html

[75] D. L. Crintea, U. Czarnetzki, S. Iordanova, I. Koleva, and D. Luggenhölscher, J. Phys. D. Appl. Phys. 42, 045208 (2009).

[76] W. L. Wiese and J. R. Fuhr, J. Phys. Chem. Ref. Data 4, 263 (1975).

[77] K. B. Gylfason, J. Alami, U. Helmersson, and J. T. Gudmundsson, J. Phys. D. Appl. Phys. 38, 3417 (2005).

[78] T. Kozák and A. Dagmar Pajdarová, J. Appl. Phys. 110, 103303 (2011).

A Technique for Time-Resolved Imaging of Millimeter Waves Based on Visible Continuum Radiation from a Cs-Xe DC Discharge — Fundamentals and Applications

Mikhail S. Gitlin

Additional information is available at the end of the chapter

http://dx.doi.org/10.5772/61843

Abstract

The chapter presents a review of a highly sensitive technique for time-resolved imaging and measurement of the 2D intensity profiles of millimeter waves (MMW) based on the use of visible continuum radiation (VCR) from the positive column (PC) of a medium pressure Cs-Xe DC discharge (VCRD technique). The review focuses on the operating principles, fundamentals, and applications of this new technique. The design of a discharge tube and an experimental setup which were used to create a wide homogeneous PC plasma slab are described. The MMW effects on the plasma slab are studied. The mechanism of microwave-induced variations in the VCR intensity and the causes of violation of the local relation between the visible continuum emissivity and the MMW intensity are discussed. The main characteristics, e.g., spatial and temporal resolution, and sensitivity of the VCRD technique have been evaluated. Experiments on imaging of the field patterns of horn antennas and quasioptical beams demonstrated that the VCRD technique can be used for a good-quality imaging of the MMW beams in the entire MM-wavelength band. The VCRD technique was applied for imaging of output field patterns of the MMW electron tubes and determination of some of their characteristics, as well as for active real-time imaging and nondestructive testing using MM waves.

Keywords: Microwaves, Imaging, Electric discharges, Continuum

1. Introduction

Millimeter waves (MMWs) are electromagnetic (EM) waves with wavelengths ranging from 1 to 10 mm in free space. Millimeter waves are widely used in radar, navigation, telecommunications, remote sensing, plasma heating and diagnostics, material processing, spectroscopy, etc. Imaging and nondestructive testing (NDT) with millimeter waves is also of great interest

for scientific, industrial, security, and biomedical applications [1–10]. In many MMW applications, as well as for engineering of MMW devices and systems, it is necessary to measure the spatial profiles of MMW radiation in real time. Electronic, electro-optic (EO), and thermographic techniques are commonly used for measuring and imaging the MMW field profiles. The electronic techniques employ receivers with semiconductor [4–9], superconducting [10], and glow discharge [11–13] MMW detectors. Electro-optic techniques use EO sensors for MM waves [14, 15]. Mechanically scanned receiver systems are widespread. However, a mechanically scanned receiver cannot be applied for a time-resolved imaging of MMW field profiles. Time-resolved images of MMW beams can be obtained with two-dimensional (2D) antenna arrays with MMW detectors [4–8, 12, 13]. The images of MMW beams obtained using 2D antenna arrays are of rather poor quality, because of the element-to-element gain and sensitivity variations, large element-to-element spacing, and small number of sensor elements. Because of the strong scattering and reflection of millimeter waves from antenna arrays, they can hardly be used for near-field imaging. Moreover, 2D antenna arrays are expensive and sophisticated systems. A live electro-optic camera has recently been used for imaging of traveling millimeter waves at video frame rates [15]. However, a live 2D electro-optic imaging system is a very complex system as well. Small aperture of the modern live EO sensor is another disadvantage of this technique. In thermographic techniques of measuring of the MMW beam profiles, thermo-sensitive phosphor or liquid crystal films, thermal paper, and dielectric sheets are used as screens, which are irradiated with millimeter waves. Thermographic techniques, which use thermo-sensitive phosphor [16, 17] or liquid crystal films [18], provide images of MMW intensity profiles in the visible region. A very simple way of obtaining MMW beam images in the visible region is based on the utilization of thermal papers. However, only rough time-integrated images of the field profiles of high-energy MMW beams can be obtained by using thermal paper [19]. In another variant of the thermographic technique, a flat sheet of dielectric material (e.g., paper, polyvinylchloride) is used as an absorbing screen for MMWs [20–23]. A spatial distribution of the microwave-induced variation in the screen temperature is measured by an infrared (IR) camera. This technique has proven itself well in installation, adjustment, and evaluation of the characteristics of the MMW transmission lines and sources of CW and long-pulse high-power millimeter waves. The main drawbacks of all thermographic methods of the MMW imaging are their low temporal resolution, which can be 10 ms at best, and low energy flux sensitivity, which is no better than 1 mJ/cm^2. Moreover, distortions, which are caused by heat conduction, can significantly impair the quality of thermographic images of MMW beams. These distortions increase with increasing MMW pulse duration or camera exposure time.

Light emission from microwave-induced gas breakdowns in space and in gas-filled tubes (luminescent lamps, neon indicator lamps, etc.), as well as breakdowns near a dielectric surface coated with a thin layer of metal powder, have been exploited to analyze the field profiles of millimeter waves [24, 25]. However, the microwave-induced breakdown yields rough distorted images of the field pattern of high-power pulsed MM waves. This technique operates only for narrow ranges of MMW intensities and pulse durations. It is impossible to obtain good-quality images by using this technique because of the properties both of the microwave-breakdown plasma and the light emission from such a plasma. No plasma glow is observed when the intensity of millimeter waves is below the breakdown threshold, or the MMW pulse duration is shorter than the time to breakdown. When the breakdown threshold is exceeded, a filamentary array with subwavelength spacing is formed in space if a gas pressure is higher than a few Torr [26]. The electron density in the filaments is higher than the critical plasma density; therefore, millimeter waves are reflected and scattered by the filaments. The boundary of the region occupied by the breakdown plasma propagates towards the MMW beam. The main contribution to the light emission from the breakdown plasma is given by a bound-bound transition between electronic levels of atoms and molecules. The dependences of the intensity of the atomic and molecular emission lines on the MMW intensity are sophisticated and nonlinear. Moreover, the dependences are different for different spectral lines.

The above-specified drawbacks of the conventional techniques for imaging and measurement of the MMW field profiles require their further improvement, as well as the development of alternative techniques. Recently, a technique for imaging of MMWs based on visible continuum (VC) radiation from a slab of the positive column (PC) of a medium-pressure DC discharge in a mixture of cesium vapor and xenon (a Cs-Xe DC discharge) was proposed and developed [27–35]. The idea behind this MMW imaging technique is to use the effect of the increase in intensity of the VC radiation from the PC of a Cs-Xe discharge due to MMW electron heating. By means of this technique, MMW beam profiles are converted into visible images, thus allowing a conventional digital camera with an optical filter to acquire the images. This review is focused on the operating principles, fundamentals, and applications of this new technique for imaging of MMWs. The chapter is organized as follows. Section 2 provides methods for a plasma slab generation using a Cs-Xe DC discharge and its diagnostics. Section 3 focuses on the experimental evaluation of some important characteristics of the technique for MMW imaging, which is based on using visible continuum radiation from the discharge (VCRD) in a mixture of Cs-Xe. We shall call this technique for imaging of MMWs a VCRD technique. In Section 3, experiments on imaging of the field profiles at the output of MMW horn antennas and quasioptical MMW beams are also discussed. The subjects of Section 4 are the fundamentals of the VCRD technique for imaging of MMWs. In particular, it discusses the nature of the visible continuum radiation (VCR) from the PC of a Cs-Xe discharge, the mechanism of microwave-induced variations in the VCR intensity, and the causes of violation of the local relation between the VC emissivity and the MMW intensity. Section 5 reviews applications of the VCRD technique for imaging of output field patterns of short-wavelength MMW vacuum electron tubes and determination of some characteristics of these tubes. Applications of the VCRD technique for real-time imaging and nondestructive testing using MMWs are also the subject of Section 5.

2. Plasma slab generation using a Cs-Xe DC discharge and experimental study of its characteristics

A sealed discharge tube (DT) was used to generate a slab of the positive column of a Cs-Xe DC discharge in the experiments on imaging of MM waves [27]. Figures 1(a) and 1(b) show side and top views of the discharge tube. A hollow rectangular parallelepiped was located at the center of the tube. It was glued from fused quartz plates. Two square quartz windows with 10 × 10 cm² apertures and 0.65 cm thick were set at a distance of 2 cm from each other. They were used to form the plasma slab and input the MMW beam into the discharge tube. Two plane anodes and two heated cathodes were sealed in glass cylinders 10 cm in diameter. The glass cylinders were glued to the quartz cell. The distance between the anodes and cathodes was 30 cm. Each pair of electrodes was powered by a separate power supply. Two pairs of thin quartz plates with a size of 5 × 9 cm² were installed inside the cylinders parallel to each other at a distance of 2 cm. These two pairs of quartz plates restricted the region in the cylinders occupied by the positive column and increased the length of a homogeneous part of the PC plasma. The longitudinal electric field E in the positive column was determined using the difference in potentials between the electrical probes. The discharge tube was filled with 45 Torr xenon. The tube had side arms in which drops of cesium metal were placed. Cesium was a readily ionized seed. To obtain the required density of cesium vapor, the discharge tube was heated by a hot air flow in an oven (see Figure 1(c)). The DT wall temperature was measured using thermocouples. The oven has one or two quartz windows 20 cm in diameter. The spectrum of visible radiation emitted by the PC of a Cs-Xe discharge was recorded by a monochromator (see Figure 1(c)). A lens focused light from the PC onto the input slit of the monochromator. A photomultiplier tube was attached to the exit slit of the monochromator. A signal from the photomultiplier was recorded by a digital oscilloscope. The data from the oscilloscope were saved to a computer and processed. The relative calibration of the spectroscopic system in a spectral region from 350 to 700 nm was performed with a tungsten ribbon lamp.

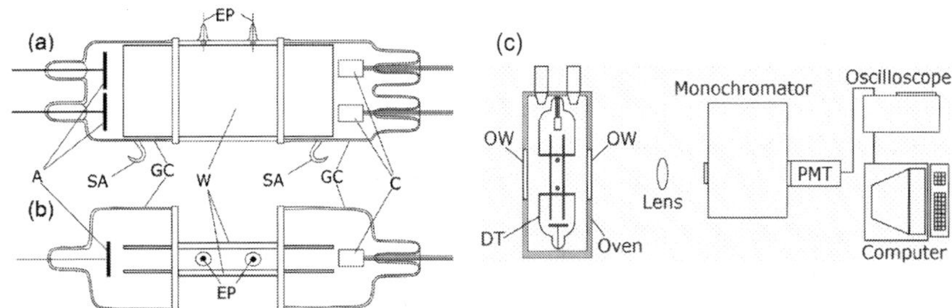

Figure 1. (a) Side and (b) top views of the discharge tube used to generate a slab of the positive column of a Cs-Xe DC discharge; A: anodes; C: cathodes; W: quartz windows; GC: glass cylinder; SA: sidearm; EP: electrical probes. (c) Schematic diagram of the experimental setup (top view) used for a study of emission spectra of light from the PC of a Cs-Xe discharge; DT: discharge tube; OW: oven window; PMT: photomultiplier tube.

The characteristics of the positive column of a Cs-Xe discharge and the spectrum of the light from the PC plasma were studied when there were no incident MM waves [27, 28]. The discharge current J was 1.5 A and the DT wall temperature varied from 80°C to 120°C, which corresponded to the partial pressures of Cs vapor from about 10^{-4} to $2\,10^{-3}$ Torr. For a fixed value of the discharge current, three modes of the positive column were observed depending on the tube temperature [27, 28, 36], specifically, a constricted PC, a wide homogeneous PC, and a PC with a filament. The first PC mode occurred at a high DT wall temperature, i.e., under a relatively high density of Cs vapors. At the tube wall temperature exceeding 110°C the positive column of a Cs-Xe discharge was constricted and occupied only part of the cross section of the discharge tube. For the first PC mode, the DC electric field in the positive column E_0 was less than 0.8 V/cm. Emission lines of cesium atoms dominated in the spectrum of visible light from the constricted PC. As the tube temperature decreased, the width of the positive column and the DC electric field increased. The second mode of the PC was observed for the tube wall temperature ranging from 100°C to 80°C. At these temperatures, the homogeneous positive column filled the entire cross section of the discharge tube. The aperture of the homogeneous plasma slab was approximately equal to the operating aperture of the discharge tube windows 10 × 8 cm² (see Figure 2(a)). The DC electric field in the homogeneous PC increased from 0.9 to 1.7 V/cm when the tube temperature decreased from 100°C to 80°C. If the discharge tube was cooled further, a bright filament 5 mm in diameter appeared. The filament was directed along the discharge current. The rest of the tube cross section was filled with a homogeneous positive column. The positive column with the filament was the third PC mode [27, 36–38]. Xenon was excited and ionized in the filament, and bright xenon atom lines along with the cesium atom lines were observed in its emission spectrum.

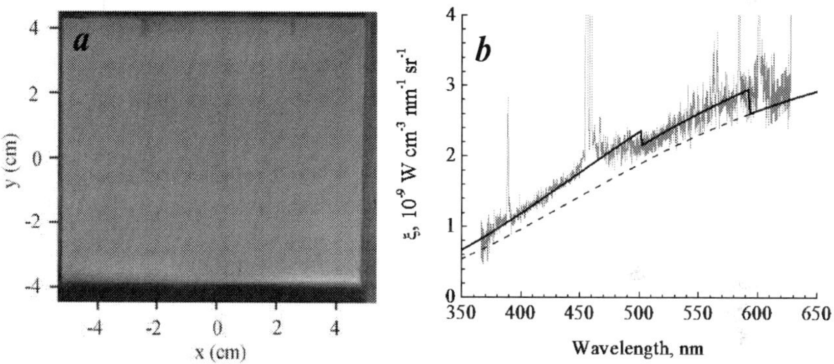

Figure 2. (a) A photo of the homogeneous positive column of a Cs-Xe discharge (the second PC mode). (b) Spectrum of the light emitted from the homogeneous positive column of a Cs-Xe discharge in the visible region.

The decrease in the cesium density with decreasing DT temperature was the main reason for the deconstriction of the PC of a Cs-Xe discharge and the transition from the first to the second PC mode. The electron temperature in the first and second PC modes is too low for a considerable ionization of xenon, and the main ion species in these PC modes are cesium atomic ions Cs^+ [27, 28, 35]. The ionization degree of cesium in the Cs-Xe discharge plasma is high and the number density of positive cesium ions is more than 80% of cesium [35]. Therefore, a decrease in the cesium density with decreasing tube temperature results in an almost proportional decrease in the density of cesium atomic ions. Owing to the plasma quasineutrality, the electron density in the PC is equal to the density of cesium atomic ions; hence, a decrease in the tube temperature also leads to a decrease in the electron density in the PC of the Cs-Xe discharge. For a fixed value of the discharge current, a decrease in the electron number density in the PC plasma leads to an increase in the transverse dimension of the current flow region, i.e., deconstriction of the PC. The positive column ceased to become wider with decreasing tube temperature only when the PC plasma filled entirely the internal cross section of the quartz cell.

Gray thin line in Figure 2(b) shows emission spectrum of the homogeneous positive column (second PC mode) in the visible region for the discharge current density j=0.1 A/cm^2 and DC electric field E_0 = 1.1 V/cm. This spectrum clearly reveals the continuum and the cesium atom emission lines. The electron temperature T_e in the thermal nonequilibrium PC plasma was measured using the spectra of visible continuum radiation from the PC of a Cs-Xe discharge [27, 28, 39–41]. In the homogeneous positive column T_e rose from 0.4 ± 0.05 to 0.5 ± 0.05 eV as the DC electric field increased from 1 to 1.4 V/cm. For the second PC mode, the DC electric field, electron temperature, and density vary in space only weakly, except for the narrow layers near the tube walls: $E_0(x, y) \approx$ const, $T_{e0}(x, y) \approx$ const, and $N_{e0}(x, y) \approx$ const, where the subscript zero denotes the values of the parameters in the absence of MMW exposure. Using the results of E_0 and T_{e0} measurements, the electron number density N_{e0} in the positive column was determined by the equation $N_{e0} = j / e\mu(T_{e0})E_0$ [42], where e is the electron charge, $\mu(T_e)$ is the DC electron mobility, and j is the discharge current density equal to $j = J / S$; here S is the internal cross-sectional area of the quartz cell.

The wide homogeneous plasma slab, i.e., the second PC mode, was used as an imager for MM waves. In the experiments on imaging of MMWs, the discharge current was 1.5 A, and the DT temperature was about 90°C. When there were no incident MMWs, the DC electric field in the PC was equal to E_0 = 1.15 ± 0.05 V / cm, the electron temperature was T_{e0} = 0.47 ± 0.03 eV, and the electron density was N_{e0} = (2.7 ± 0.3)•10^{12} cm^{-3}. The electron density in the homogeneous PC of a Cs-Xe discharge was much less than the critical density for MMW frequency band, so reflection of the MM waves from the plasma was weak (less than 1%) [42]. The absorption coefficients $\gamma_0 = \gamma(T_{e0})$ of the PC plasma for Ka and D band MM waves were equal to 0.3 cm^{-1} and 0.01 cm^{-1}, respectively [27, 29].

3. Evaluation of the performance of the VCRD technique

The MMW effects on a slab of the PC of a Cs-Xe discharge have been experimentally studied to develop the VCRD technique for imaging of MMWs and determine its basic characteristics. The model experiments[1] on imaging of the field patterns of horn antennas and quasioptical beams have been performed. Figure 3 shows a layout of the experiments on imaging of the near-field patterns at the output of a MMW horn antenna. The images of the plasma slab were captured by a black and white charged coupled device (CCD) camera. The data from the CCD camera were processed with a computer. A set of optical filters rejecting atomic emission lines and transmitting the continuum in the visible region was placed in front of the camera lens.

---

1 In the model experiments, the MMW field profiles have been known from calculations or measurements by conventional techniques.

The filter set had a transmissivity of less than 0.1% in the near IR range and in the region from 450 to 460 nm, where the bright lines of the second resonance doublet of cesium occur (see Figure 2(b) and refs. [27, 28]). The contribution of atomic-line intensity into the wavelength-integrated intensity of the transmitted light was a few percent only.

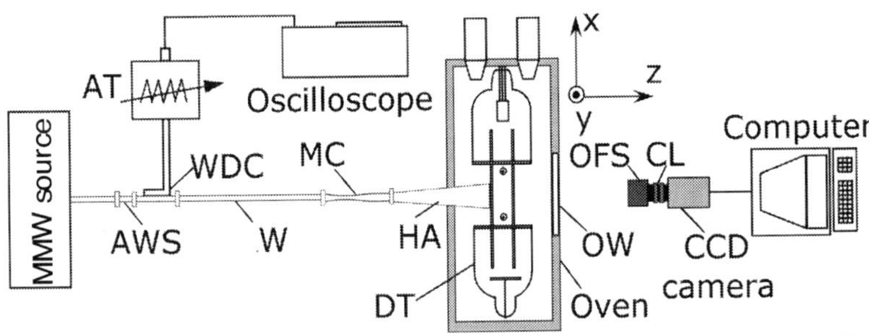

Figure 3. Schematic diagram of the experimental setup (top view) for imaging of the field patterns at the output of a horn antenna; W: waveguide; AWS: waveguide section with MMW absorber; WDC: waveguide directional coupler; AT: MMW attenuator; MC: mode converter; HA: horn antenna; DT: discharge tube; OW: oven window; OFS: optical filter set; CL: camera lens.

To evaluate the microwave-induced variation in the VCR energy flux $\Delta S_C(x, y)$, the background VCR energy flux $S_{C_0}(x, y)$ was subtracted from the VCR energy flux $S_C(x, y)$: $\Delta S_C(x, y) = S_C(x, y) - S_{C_0}(x, y)$. The VCR energy flux $S_C(x, y)$ was obtained using the CCD camera frame simultaneous with the MMW pulse, and the background VCR energy flux $S_{C_0}(x, y)$ was obtained using the CCD camera frame preceding the MMW pulse. When the VC emissivity is time-independent during the camera exposure time τ_e, the VCR energy flux is equal to $S_C(x, y) = \tau_e I(x, y)$, where $I(x, y)$ is the intensity of the visible continuum radiation, and the relative variation in the VCR intensity is equal to the relative variation in the VCR energy flux $\Delta I(x, y)/I_0(x, y) = \Delta S_C(x, y)/S_{C_0}(x, y)$, where $\Delta I(x, y) = I(x, y) - I_0(x, y)$; here $I_0(x, y)$ is the background VCR intensity.

Study of the millimeter wave effect on the homogeneous positive column of a Cs-Xe discharge was performed using conical horn antennas (see Figure 3) excited by the TE_{11}^O mode of a circular waveguide [27]. A 35.4-GHz magnetron was used as a source of MM waves. The magnetron supplied coherent radiation with an output power of up to 20 W in the long-pulse mode. The MMW pulse length was up to 100 ms, and the pulse leading edge was less than 0.1 µs. Mode converters were used to transform the TE_{10} mode of a rectangular waveguide into the TE_{11} mode of a circular waveguide [43, 44]. The conical horns had aperture radii from 10 to 30 mm. The MMW electric field polarization in the vicinity of the beam axis was directed along the x coordinate (see Figure 3). The discharge tube window was attached directly to the output of a horn antenna, and the spatial distribution of the MMW intensity was measured in the near-field region. A set of rectangular waveguide sections with MMW absorber was used to decrease the power of millimeter waves that are incident on the PC plasma. By using the waveguide section with a MMW absorber, the MMW power can be decreased up to 15 dB with discrete steps of about 3 dB. The MMW power was also varied smoothly by changing voltage and current supplied to the magnetron. For relative measurements of the radiated power, a small part of the MMW radiation was branched to the calibrated attenuator by a waveguide directional coupler. The output signal from the attenuator was detected using a MMW detector and measured using an oscilloscope.

When the MMW intensity W exceeded the breakdown threshold W_{Br}, a microwave-induced breakdown of the PC slab occurred, and bright thin filaments were observed in the plasma area affected by a MMW beam. The filaments were elongated in the direction of the MMW electric field. Figure 4(a) shows the spatial distribution of the VCR energy flux variation when there was a microwave-induced breakdown of the plasma slab. The threshold value of the microwave breakdown for the Ka-band was equal to about 4 W/cm² for $E_0 = 1.1$ V/cm. It decreased to approximately 2 W/cm² for $E_0 = 1.5$ V/cm. When $W > W_{Br}$, MMW heating increases the electron temperature so that excitation and ionization of xenon become significant. Ionization of xenon is the major cause of ionization instability of the homogeneous PC and the appearance of filaments with a high electron density, i.e., a microwave breakdown of the plasma slab [35].

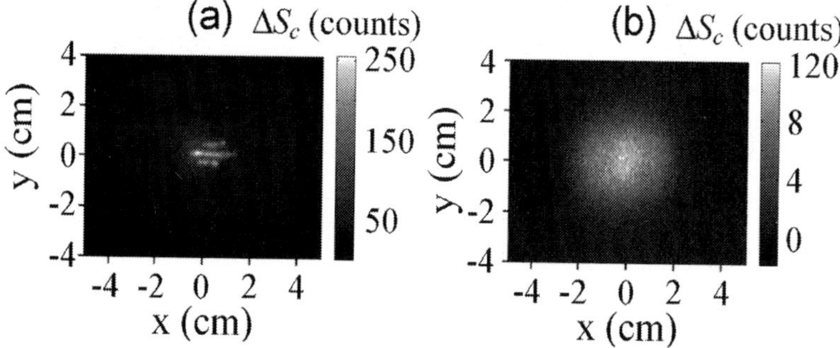

Figure 4. Spatial distribution of the VCR energy flux variation at the output of a conical horn antenna excited by the TE_{11}^{O} mode for the MMW intensities (a) above and (b) below the microwave-induced breakdown threshold of the plasma slab.

Figure 4(b) shows a 2D distribution of the VCR energy flux variation $\Delta S_c(x, y)$ at the output of a conical horn antenna,[2] when the MMW intensity in the field maximum was less than the microwave breakdown threshold. The horn antenna with the taper angle of 5° and output aperture radius $R_h = 22.5$ mm was excited by the TE_{11}^{O} mode. The CCD camera exposure time was equal to 200 μs. The spatial distribution of the VCR energy flux variation is smooth, and

---

2 The coordinates are denoted x and y in the reference frame, whose origin coincides with the center of the DT window, and the coordinates are denoted x' and y' in the reference frame, whose origin coincides with the MMW beam center.

it resembles the intensity profile for the TE_{11}^O mode (see Figure 4(b)). Using the same conical horn excited by the TE_{11}^O mode, the dependences of the relative VCR intensity variation $\Delta I(0, 0)/I_0(0, 0) = \Delta S_C(0, 0)/S_{C_0}(0, 0)$ on the intensity of millimeter waves incident on the plasma slab $W(0, 0)$ was measured. The CCD camera exposure time was 100 and 200 μs. The measured dependence of $\Delta I/I_0$ on W/W_{Br} is shown in Figure 5 by diamonds. In the range of MMW intensities from 0 to W_{Br}, the second-order polynomial function fits the dependence $\Delta I(W)/I_0$ with an accuracy of up to a few percent (black solid line in Figure 5). For the MMW intensity range $0 < W < W_{Br}/3$, the value of ΔI could be considered approximately directly proportional to the intensity of millimeter waves (with an accuracy of about 10%). If the MMW intensity is below the breakdown level, MMW beam profile can be obtained from the distribution of the VCR intensity variation using the dependence of $\Delta I/I_0$ on W.

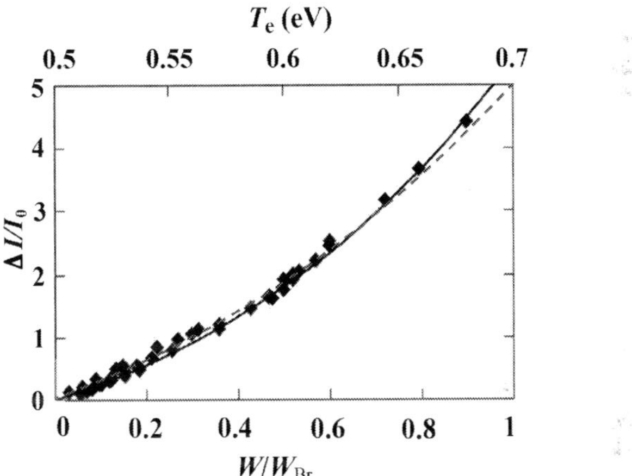

Figure 5. Experimental dependence of the relative VCR intensity variation as a function of the intensity of MM waves incident on the plasma slab (black diamonds), and calculated dependence of the relative VCR intensity variation as a function of the electron temperature (blue dashed line).

The data of the experiment on the imaging MMW intensity profile at the output of the conical horn antenna with an aperture radius of 22.5 mm excited by the TE_{11}^O mode (see Figure 4(b)) were compared with the calculated spatial distribution of the EM wave intensity for the TE_{11}^O mode of a circular waveguide of the same radius. For $x' = 0$, the calculated EM wave intensity decreases as a function of $|y'|$ without discontinuities from the maximum value W_m at the center of the waveguide to zero near the waveguide wall, i.e., for $y' = \pm 22.5$ mm [43, 44]. The measured dependences of the VCR intensity variation vs. coordinate y' for $x' = 0$ were in good agreement with the calculated spatial distribution of the EM wave intensity $W(x' = 0, y')$. For $y' = 0$, the measured dependence of the VCR intensity variation vs. the coordinate x' was different from the calculated spatial distribution of the EM wave intensity for the TE_{11}^O mode $W(x', y'=0)$ in the regions near $x' = \pm 22.5$ mm. For $y' = 0$ the sharp edges of the MMW intensity distribution near the waveguide wall were blurry at the image. The blurring effect appeared due to the diffraction of MMWs at the edge of a horn antenna and electron heat conduction (see Section 4.2).

The experiments on imaging of the MMW field profiles at the output of the conical horn antennas excited by the axially symmetric TE_{01}^O mode of a circular waveguide were carried out in ref. [27]. The same 35.4 GHz long-pulse magnetron was used in these experiments. The TE_{10} mode of a rectangular waveguide was transformed into the TE_{01}^O mode of a circular waveguide by a Marier transducer (see Figure 3). Figure 6(a) shows a 2D distribution of the MM wave-induced variation of the VCR intensity $\Delta I(x, y)$ at the output of a conical horn with a taper angle of 6° and output aperture radius $R_h = 28$ mm. The MMW intensity at the field maximum was $W_m \approx 1$ W/cm². The CCD camera exposure time was 100 μs. The obtained pattern of the VC glow corresponds to the TE_{01}^O mode with a small admixture of other modes[3] (see Figure 6(a)). Black lines in Figures 6(b) and 6(c) show the dependence of the VCR intensity variation under the MMW effect vs. the coordinates y for $x' = 0$ and x for $y' = 0$, respectively. For the TE_{01}^O mode, the dependence of the MMW intensity on the radius $r = \sqrt{(x')^2 + (y')^2}$ is $W(r) \approx W_m \cdot [1.72 J_1(3.832 r / R_h)]^2$ for $r < R_h$, where J_1 is a first-order Bessel function and R_h = is the radius of a circular waveguide or the aperture of a conical horn [43, 44]. Gray lines in Figures 6(b) and 6(c) show the function $W(r)$. The function $W(r)$ fits well with the measured dependences $\Delta I(x, y'=0)$ and $\Delta I(x'=0, y)$. However, the intensity variation of the VC radiation from the plasma region near the axis of the MMW beam $\Delta I(x'=0, y'=0)$ was equal to 15% of the VC intensity variation at the field maximum ΔI_m (see Figures 6(b) and 6(c)), while the MMW intensity at the beam center $W(0, 0)$ is equal to zero. Thin solid line in Figure 6(d) shows the dependence of the VCR intensity variation under the MMW effect vs. the coordinate x' for $y' = 0$, which was obtained for a conical horn with an aperture radius $R_h = 20$ mm. For a horn antenna of this aperture size, $\Delta I(0, 0)/\Delta I_m$ was about 25%. When conical horns with smaller aperture radii were used, the relative value of the VCR intensity variation in the beam center $\Delta I(0, 0)/\Delta I_m$ increased, i.e., the central deepening became shallower. Thus, a violation of the local relation between the intensity of the plasma VC radiation and the MMW intensity was observed in these experiments. The causes of the nonlocality of microwave-induced variations in the intensity of the VC radiation from a medium-pressure cesium-xenon DC discharge are discussed in Section 4.2.

An experiment on imaging of the Ka-band quasioptical millimeter-wave beam using the VCRD technique is described in ref. [27]. The quasi-Gaussian MMW beam was focused by a double

---

3 An experiment showed that the dependence of the VCR intensity on the azimuthal coordinate (see Figure 6(a)) is caused by the admixture of other waveguide modes excited in a conical horn, but not the spurious plasma effect. In this experiment, a rectangular waveguide, which was twisted through an angle of approximately 90°, was installed before the Marier transducer. The image of the spatial distribution of the MMW intensity was also rotated through an angle 90° in the same direction.

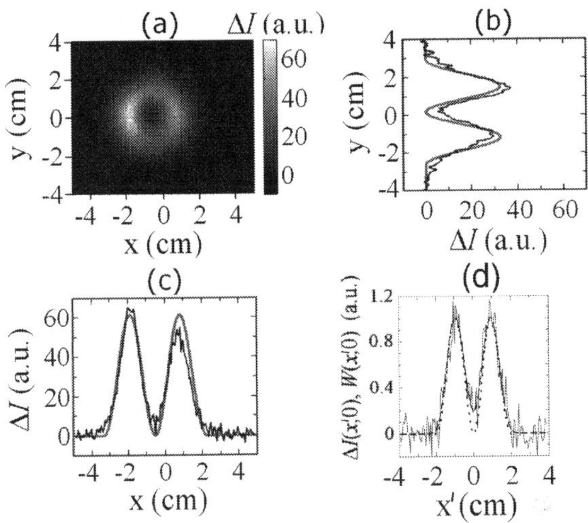

Figure 6. Spatial distributions of the VCR intensity variation at the output of the conical horn antennas excited by the TE_{01}^O mode. (a) Two-dimensional distribution of the VCR intensity variation. (b) and (c) Measured dependence of the VCR intensity variation (black lines) and calculated dependences of the MMW intensity variation (gray lines) vs. the coordinates y for $x' = 0$ (b) and x for $y' = 0$ (c) for a horn with the aperture radius $R_h = 28$ mm. (d) Measured (solid gray line) and calculated (dashed black line) dependences of the VCR intensity variation vs. the coordinates x' for $y' = 0$ for a horn with the aperture radius $R_h = 20$ mm. Calculated dependence of the MMW intensity variation vs. x for $y' = 0$ (dotted line).

convex Teflon lens 20 cm in diameter and a focal length of 32 cm at the center of the plasma slab. The width of the MMW beam (FWHM) in the focus was approximately 2.5 cm. The distribution of the VCR intensity variations in the focal plane of the lens was compared with the MMW field profile. The MMW intensity profile was measured by a mechanically scanned receiver with calibrated MMW detector. The patterns measured by the two techniques coincided within the limits of the random experimental error. Thus, model experiments on imaging of the field profiles of horn antennas and quasioptical beams have demonstrated that the VCRD technique can be used for good-quality imaging of Ka-band MMW beams.

The response time of the VCR intensity to the variation in the MMW intensity was measured using a photomultiplier tube (PMT) [27]. In these experiments, the photomultiplier tube was set in place of the CCD camera (see Figure 3). The optical filter set rejecting atomic emission lines and transmitting the VC radiation from the PC of a Cs-Xe discharge was mounted before the PMT input window. The temporal resolution of the setup used for recording of the VCR intensity time history was less than 50 ns. After the MMW radiation was switched on, the measured waveform of VCR intensity fit well with the exponential time dependence:

$$\Delta I = \Delta I_\infty \left\{ 1 - \exp\left[-\left(t - t_0 \right) / \tau \right] \right\}, \tag{1}$$

where ΔI_∞ is the VCR intensity variation, t_0 is the time point at which the MMW pulse was switched on, and τ is the characteristic time of VCR intensity variation. The VCR response time was about $\tau = 0.8 \pm 0.1$ μs. This time is approximately equal to the time of the electron temperature variation in the PC plasma (see Section 4.2 and refs. [27, 35]). Thus, it was shown that the VCRD technique for MMW imaging is very rapid and its temporal resolution was less than 1 μs.

In the experiments on imaging of Ka-band MM waves using the VCRD technique, a single-shot signal to noise ratio (SNR) was about one when the MMW intensity was $W \approx 1$ W/cm² and the CCD camera exposure time was set at 10 μs [27]. Thus, in the Ka-band, a single-shot threshold energy flux sensitivity (SNR = 1) of about 10 μJ/cm² was achieved. In the performed experiments, the energy flux sensitivity was restricted by the noise performance of the CCD camera. The actual sensitivity of the VCRD technique is determined by fluctuations of the parameters of the PC of a Cs-Xe discharge, primarily, by the fluctuations of the electron temperature in the plasma [45]. Estimated sensitivity of the VCRD technique is higher by at least an order than the sensitivity achieved in the experiments. Further study is required for more accurate determination of the VCRD technique sensitivity.

The VCRD technique can be used to image the intensity patterns of the EM waves in a wide band from short-wavelength centimeter to submillimeter waves. When the MMW frequency increases, the plasma becomes more transparent for MMWs due to a decrease in the MMW absorption coefficient, and the MMW effect on the plasma parameters decreases [27, 29]. At the xenon pressure of tens of Torr, the electron-Xe atom collision frequency for the momentum transfer was much less than the angular frequency ω of MM waves [27, 29]. In this case, the scaling law for the efficiency of electron heating under the MMW effect is $\Delta T_e \propto W / \omega^2$, where ΔT_e is the electron temperature variation. Therefore, on passing from long-wavelength to short-wavelength millimeter waves, the sensitivity of the technique decreases if the identical systems are used for VC radiation recording. The threshold of the microwave breakdown of a plasma slab is significantly increased in a short MMW band. For example, on passing from Ka-band to D-band the threshold of the microwave breakdown of a plasma slab increases to the value of the order of 100 W/cm², but the energy sensitivity of the VCRD technique deteriorates to the level of hundreds of μJ/cm².[4]

The ability to image the intensity profiles of short-wavelength MM waves using the VCRD technique was demonstrated in refs. [29, 30]. Thus, the wide frequency band of the technique has been confirmed. In paper [29], the intensity profile of the D-band MMW beam was measured at the output of a corrugated conical horn antenna. The arrangement of this model experiment was similar to the arrangement of the experiment on imaging of Ka-band MMWs at the output of conical horn antennas (see Figure 3). A pulsed F- and D-band orotron OR-180 was used as a source of MM waves (see Section 5.1 and ref [46]). In this experiment, the orotron radiation frequency was 140 GHz. The maximum orotron power was about 0.6 W. The MMW pulse length was 3 ms, and the pulse repetition rate was 10 Hz. The corrugated conical horn was fed by the orotron through an E-band rectangular waveguide and a down taper. The

---

4 In the D band, a single-shot energy flux sensitivity of about 200 μJ/cm² was demonstrated [29, 30].

output diameter of the corrugated horn was 16 mm, and its length was about 90 mm. The discharge tube window was attached to the output of the horn. The corrugated horn forms a high-quality axially symmetric Gaussian beam [47]. The width (FWHM) of the Gaussian MMW beam in the middle plane of the plasma slab was equal to 7.8 mm. The MMW electric field vector at the output of the horn was directed along the y coordinate. Figure 7(a) shows the 2D distribution of the VCR intensity variation induced by the MMW beam, which was measured at the output of the corrugated conical horn. The solid lines in Figures 7(b) and 7(c) show the VCR intensity variation as functions of the x' and y' coordinates, respectively. The CCD camera exposure time was 2 ms. To improve the SNR, the images of the MMW field pattern (see Figure 7) were averaged using 300 camera frames. The image was broader than the MMW beam by approximately 1 mm, and it had no axial symmetry in the region of the beam tail.[5] The causes of the broadening of the MMW beam images and their axial asymmetry are discussed in Section 4.2. The MMW intensity profile in the focal region of a Teflon conical lens (axicon) was imaged by the VCRD technique in refs. [29, 30]. A 140 GHz Gaussian EM beam, which is radiated from a corrugated conical horn and collimated by a Teflon lens, was incident on the axicon. The corrugated horn was fed from orotron OR-180. The 2D intensity profiles of the focused MMW beam were imaged at various distances from the axicon. It has been shown that the axicon forms a beam with the width (FWHM) of a focal spot of less than 3 mm and a focal length of about 10 cm. The imaging of the profiles of the MMW beams, which were formed by a corrugated horn antenna and an axicon, demonstrates that the VCRD technique can be used for imaging of MMW beams having a width of more than several millimeters.

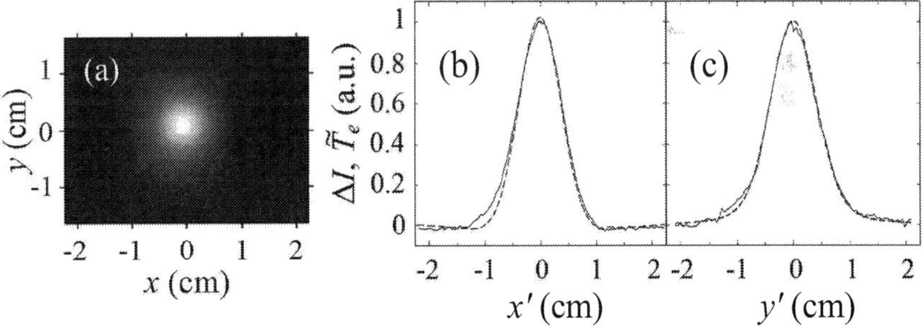

Figure 7. (a) 2D distribution of the VCR intensity variation at the output of a D-band corrugated conical horn. (b) and (c) Measured dependences of the VCR intensity variation ΔI (solid lines) at the output of a corrugated conical horn and calculated electron temperature perturbation $\widetilde{T}_e$ (dashed lines) on the coordinates x' for $y' = 0$ (b) and y' for $x' = 0$ (c).

---

5 When a section of the D-band rectangular waveguide twisted by an angle of 90° was installed before the corrugated conical horn, the horn rotated around its axis by 90°, and the MMW polarization in the Gaussian beam also changed from direction along the y-axis to direction along the x-axis. However, this rotation does not change the shape of the image of the MMW beam.

4. Fundamentals of the VCRD technique

This section discusses the nature of the visible continuum radiation from the PC of a Cs–Xe discharge and the mechanism responsible for the VCR intensity variation under the MMW effect. The relationship of the spatial distributions of the VC emissivity variation and intensity of MMW radiation incident on the plasma slab is also studied. The factors determining the spatial and temporal resolution of the VCRD technique are discussed.

4.1. Mechanism of the MMW effect on the intensity of the VC radiation from the PC of a Cs-Xe discharge

The nature of the visible continuum radiation from the positive column of a Cs-Xe discharge for plasma parameters that are typical for the MMW imaging experiments was studied in refs. [27, 28]. It was shown that the electron-Xe atom bremsstrahlung continuum (e-Xe BC) and the cesium recombination continuum (RC) give the main contributions to the visible continuum radiation from the PC of a Cs-Xe discharge. The VC emission coefficient $\xi(\lambda)$ is equal to the sum of emission coefficients for the electron-Xe atom bremsstrahlung continuum $\xi_{e-Xe}^{BC}(\lambda)$ [39–41, 48–50] and the cesium recombination continuum $\xi^{RC}(\lambda)$ [39, 51]:

$$\xi(\lambda) = \xi_{e-Xe}^{BC}(\lambda) + \xi^{RC}(\lambda). \tag{2}$$

Visible RC radiation originates from recombination of the electrons and Cs^+ ions into $6P$ and $5D$ electronic states of cesium (blue and yellow cesium RC, respectively) [51]. Thus, the emission coefficient for the recombination continuum of cesium in the visible region is equal to $\xi^{RC}(\lambda) = \xi_{6P}^{RC}(\lambda) + \xi_{5D}^{RC}(\lambda)$, where $\xi_{6P}^{RC}(\lambda)$ and $\xi_{5D}^{RC}(\lambda)$ are the emission coefficients for $6P$- and $5D$-recombination continua of cesium, respectively. The emission coefficients of the blue and yellow cesium recombination continua for $\lambda < hc/E_n$ are equal to [27, 39, 51]

$$\xi_n^{RC}(\lambda) = A\lambda^{-3}\sqrt{\varepsilon}\,\sigma_n(\varepsilon)N_+N_e f_0(\varepsilon), \tag{3}$$

where n means $6P$ or $5D$, E_n is the binding energy of atomic state n, h is the Planck constant, c is the speed of light, λ is the wavelength of the emitted light, A is a constant which is equal to approximately $6\bullet10^{-36}$ $kg^{3/2}m^6s^{-4}sr^{-1}$, ε is the electron energy, $\sigma_n(\varepsilon)$ is the cross section of radiative recombination into the nth electronic state of a cesium atom, N_+ is the number density of the Cs^+ ions ($N_+ \approx N_e$), and $f_0(\varepsilon)$ is an isotropic component of the electron energy distribution function (EEDF). For RC radiation, the relationship between the free electron energy and the light wavelength is $\varepsilon = hc/\lambda - E_n$. For $\lambda > hc/E_n$, $\xi_n^{RC}(\lambda)$ is equal to zero. The emission coefficient of the e-Xe BC is given by [27, 28, 40]

$$\xi_{e-Xe}^{BC}(\lambda) = C\frac{N_e N_{Xe}}{\lambda^2}\int_{\frac{hc}{\lambda}}^{\infty}\sqrt{\varepsilon - \frac{hc}{\lambda}}\left\{\varepsilon\sigma_t\left(\varepsilon - \frac{hc}{\lambda}\right) + \left[\varepsilon - \frac{hc}{\lambda}\right]\sigma_t(\varepsilon)\right\}f_0(\varepsilon)d\varepsilon, \tag{4}$$

where C is a constant which is equal to approximately 0.9 $sr^{-1}kg^{-1/2}m$, N_{Xe} is the Xe atom number density, and $\sigma_t(\varepsilon)$ is the momentum transfer cross section for the electron scattering from a xenon atom. Thick solid line in Figure 2(b) shows the VC emission coefficient $\xi(\lambda)$, which was calculated by Eqs. (2)–(4) for $N_{e0} = 2.5\ 10^{\cdot\ 12}$ cm^{-3} and EEDF with the effective electron temperature $T_{e0} = 0.47$ eV [28]. The results of calculations agree well with the measured VC emission spectra. Dashed line in Figure 2(b) shows calculated e-Xe BC emission coefficient for the same plasma parameters. In the visible region, the intensity of e-Xe BC radiation is several times greater than the maximum intensity of the cesium RC radiation.

The intensity of the visible continuum from the PC of a Cs-Xe discharge rises when the millimeter waves incident on the plasma slab or the DC electric field in the PC plasma increases. It follows from Eqs. (3) and (4) that the rise in the VCR intensity can be caused by an increase in the electron number density or by EEDF variation. The electron-density variation induced by the MMW or DC electric field cannot explain the observed increase in the VCR intensity. Firstly, due to a high ionization degree of cesium in the homogeneous positive column, the microwave-induced variation in the electron density is relatively small ($\Delta N_e / N_{e0} <<1$) [35], whereas the effect of watt-scale Ka-band MMWs on the PC plasma increases the VCR intensity significantly ($\Delta\xi / \xi \sim 1$) (see Figure 5). Secondly, the characteristic time of the electron density variation [35, 36] is several orders of magnitude longer than the VCR response time (see Section 3). Thirdly, when the discharge tube temperature was decreased, the electron number density N_{e0} in the homogeneous PC also decreased, whereas the VCR intensity and the DC electric field E_0 increased [28]. Hence, the VCR intensity variation is caused by the EEDF variation resulting from the MMW or DC electron heating. Using Eqs. (2)–(4), the VC emission coefficient were calculated for different electron temperatures in the homogeneous PC plasma. The electron density was assumed constant. Solid lines in Figure 8 show the continuum emission coefficient $\xi(\lambda)$ in the visible region calculated for 45 Torr xenon, $N_e = 2.5\bullet10^{12}$ cm^{-3} and the EEDF with the effective electron temperatures $T_e = 0.45, 0.5, 0.55,$ and 0.6 eV. Dashed lines in Figure 8 show the e-Xe BC emission coefficient $\xi_{e-Xe}^{BC}(\lambda)$ calculated for the same plasma parameters. In the visible region, the wavelength-integrated emission coefficient for the e-Xe BC radiation exceeds the wavelength-integrated emission coefficient for the cesium recombination continuum by about one order of magnitude. Figure 8 demonstrates that the increase in the VC emissivity is caused by the e-Xe BC emissivity increase due to an additional electron heating and is not caused by an increase in the cesium recombination continuum emissivity as was believed earlier [31]. As the electron temperature rises from $T_e = 0.45$ to 0.6 eV, the intensity of the e-Xe BC in the visible region increases more than fourfold.

Explanation of the phenomenon of the increase in the e-Xe BC radiation intensity in the visible region induced by the electron heating is quite obvious. Bremsstrahlung photons with wavelength λ can be emitted only by electrons with kinetic energies greater than the energy of a light quantum, i.e., $\varepsilon > hc / \lambda$. The visible light wavelengths ranging from 620 nm to 410 nm correspond to photon energies from 2 to 3 eV. The bremsstrahlung photons with such an energy can be emitted by the electrons from the tail of the EEDF ($\varepsilon > 2$–3 eV) because the electron temperature in the homogeneous PC of a Cs-Xe discharge is from 0.4 to 0.8 eV. Thus, the

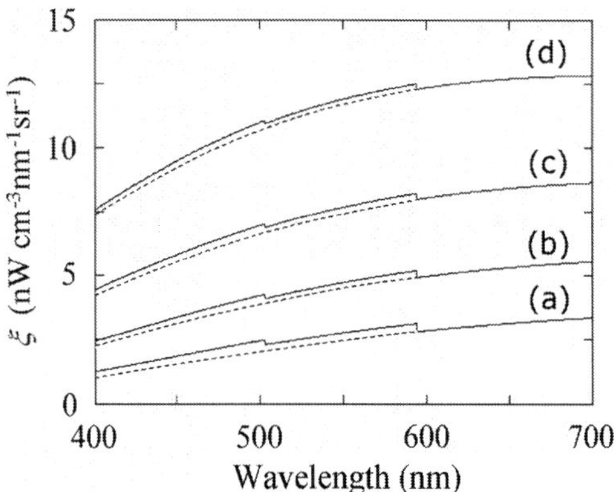

Figure 8. The visible continuum (solid line) and e-Xe bremsstrahlung continuum (dashed line) emission coefficients vs. wavelength for effective electron temperatures (a) 0.45, (b) 0.5, (c) 0.55, and (d) 0.6 eV.

increase in the e-Xe bremsstrahlung radiation intensity in the visible region is caused primarily by an increase in the fast electron number density, which results from the additional electron heating induced by the MMW or DC electric field. The bremsstrahlung cross section for electron-atom collisions increases with increasing electron energy (see Eq. (4) and refs. [40, 50]). This is another cause of the increase in the VC brightness due to DC- or microwave-induced electron heating.

Verification of the previously discussed mechanism of the electromagnetic field effect on the brightness of the visible continuum radiation from the PC of a Cs-Xe discharge was carried out in ref. [28], where the effects of the DC electric field strength and electron density on the intensity of VC radiation were studied experimentally. There was no MMW radiation incident on the plasma. The electron density was varied by changing the discharge current. The DC electric field strength was varied by changing the tube temperature for a fixed value of the discharge current. Experimental data are in a good agreement with the results of calculations done using the above model [28]. In ref. [27], the dependence of the relative intensity variation of the VC radiation from the PC of a Cs-Xe discharge, $\Delta I / I_0$, on the electron temperature was calculated. The best agreement between the results of calculations and the experimental data is achieved when it is assumed that the microwave breakdown of the plasma slab occurs at an electron temperature equal to $T_{eBr} = 0.7$ eV. Dashed line in Figure 5 shows the calculated dependence $\Delta I(T_e) / I_0$ for $T_{eBr} = 0.7$ eV.

4.2. Relation between spatial profiles of the VC emissivity and the MMW intensity. Spatial and temporal resolution of the VCRD technique

The causes of violation of the local relation between the visible continuum emissivity and the MMW intensity, as well as the main factors which limit the spatial resolution of the VCRD technique, were explored in ref. [29]. The emissivity of the e-Xe bremsstrahlung continuum depends locally on the effective electron temperature $T_e(x, y, z)$ and electron density $N_e(x, y, z)$ in the PC of a Cs-Xe discharge (see Eqs. (2)–(4)). The electron density variation is relatively small in the PC plasma, and its effect on the brightness of VC radiation is not significant. The main cause of the increase in the VC intensity is the microwave-induced increase in T_e (see Section 4.1). The value of ΔI is directly proportional to the local value of the electron temperature variation ΔT_e when $\Delta T_e < 0.1$ eV. Spatial profiles of the microwave-induced perturbations in the electron temperature and density in the slab of the PC of a Cs-Xe DC discharge were analytically modeled [29]. The model considers the effects of the electron heating by the DC and MMW electric field, the electron energy loss due to electron-atom collisions, electron heat conduction, diffusion, and thermal diffusion of the charged particles. The perturbations of the electron temperature induced by MMW beams which had different shapes were described by the electron-energy balance equation. Perturbation technique was used to find the solutions for this equation. It was sufficient to calculate the first- and second-order terms for the electron temperature perturbation to explain the results of the experiments. For a one-dimensional (1D) MMW beam having a width much smaller than the size of the operating aperture of a plasma slab, the spatial distribution of the electron temperature perturbation of the first order is given by the convolution of the MMW intensity $W(\zeta)$ and the line spread function (LSF) for the VCRD technique $G_e(\zeta)$:

$$\tilde{T}_{e1}(\zeta) = \frac{2\gamma_0 \tau_\varepsilon}{3kN_{e0}} \int\limits_{-\infty}^{+\infty} W(\hat{\zeta}) G_e(\zeta - \hat{\zeta}) d\hat{\zeta}. \tag{5}$$

In Eq. (5), the coordinate ζ is a linear function of x and y, k is the Boltzmann constant, and $\tau_\varepsilon \approx (2P'_0/3k)^{-1}$ is the characteristic time of electron temperature variations due to the electron energy loss in electron-atom collisions, where $P' = \partial(Q/N_e)/\partial T_e$; here, Q is the power density of the electron energy loss due to electron-atom collisions. The line spread function for VCRD technique is in direct proportion to the electron temperature perturbation induced by an infinitely narrow 1D MMW beam with the intensity $W(\zeta) = W_m \delta(\zeta - \zeta_0)$, where $\delta(\zeta)$ is a delta function. In a first-order approximation, LSF is given by the Laplace distribution

$$G_e(\zeta) = \frac{1}{2\Lambda_T} \exp\left(-\frac{|\zeta - \zeta_0|}{\Lambda_T}\right); \tag{6}$$

here $\Lambda_T = (2\tau_\varepsilon \chi_0/3k)^{1/2}$, where χ is the electron heat conduction coefficient divided by the electron density. For a xenon pressure of 45 Torr and an electron temperature of 0.5 eV, Λ_T is equal to about 1.5 mm, so the width of the line spread function (FWHM) equals to approximately 2 mm. When the plasma slab is affected by a two-dimensional beam, whose width is significantly smaller than the aperture sizes of the discharge tube, the electron temperature perturbation of the first order is given by a 2D convolution of $W(x, y)$ with a modified Bessel function of the second kind of order zero $K_0(r/\Lambda_T)$,

$$\tilde{T}_{e1}(x,y) = \frac{\gamma_0}{2\pi N_{e0}\chi_0} \int\limits_{-\infty}^{\infty}\int\limits_{-\infty}^{\infty} W(\hat{x},\hat{y})K_0(\xi/\Lambda_T)d\hat{x}d\hat{y}; \tag{7}$$

where $\xi = \sqrt{(x-\hat{x})^2+(y-\hat{y})^2}$. It follows from Eqs. (5) to (7) that the value of spatial broadening of the MMW beam image is determined by a parameter Λ_T, which depends on the electron heat conduction coefficient and the time of electron energy relaxation due to electron-atom collisions. Therefore, at a first-order approximation, the nonlocality of the microwave-induced variation in the intensity of the VC radiation is caused by the electron heat conduction effect. The electron heat conduction restricts the spatial resolution of the VCRD technique. Using Eq. (7), one can find that the plasma imager with aperture dimensions of about 10 cm has an equivalent number of effective sensor elements of the order of one thousand.

The calculations of the microwave-induced electron temperature perturbation based on the above-described model were compared with the data of the experiments on imaging of 34.5GHz MMW field profiles at the output of conical horn antennas with different aperture radii, which were excited by the TE_{01}^O mode (see Section 3). The dashed line in Figure 6(d) shows the calculated dependence of the first-order perturbation of the electron temperature on transverse coordinate r for $R_h = 20$ mm. The calculated dependence fits well with the measured spatial distributions of the VC brightness variations. Figure 7 allows one to compare the data of the experiment on imaging of a D-band MMW beam at the output of the corrugated conical horn and the calculated profiles of the electron temperature perturbation induced by a Gaussian MMW beam with a width of 7.8 mm [29]. The dashed lines in Figures 7(b) and 7(c) show the calculated dependences $\tilde{T}_e = \tilde{T}_{e1} + \tilde{T}_{e2}$ on coordinates x and y, respectively, wher $\tilde{T}_{e2}$ is the second-order perturbation of the electron temperature. The calculations fit well with the experimental data. Figures 7(b) and 7(c) show that the calculated distributions of the electron temperature perturbation, as well as the measured variations in the VCR intensity, are not axisymmetric. Considering the second-order perturbation of the electron temperature, one can explain the violation of the axial symmetry in the images of axisymmetric beams. The difference in the VCR intensity profile along the x and y coordinates is due to the MM wave-induced spatial inhomogeneity of the electron heating by the DC electric field [29].

When a smooth and wide (i.e., wider than 2 cm) MMW beam affects the plasma slab, the influence of the electron heat conduction and other manifestations of nonlocality can be neglected in a first-order approximation. In this case, the perturbation in the electron temperature induced by a long pulse of MMW radiation with a sharp leading edge is given by [27]

$$\tilde{T}_e = \frac{2}{3}\frac{\gamma_0 W\tau_\varepsilon}{kN_{e0}}\left[1\text{-}\exp\left(-(t-t_0)/\tau_\varepsilon\right)\right]. \tag{8}$$

The time history of the electron temperature perturbation, calculated using Eq. (8), coincides well with the measured time history of the VCR intensity variation. Therefore, the response time τ of the VC radiation is approximately equal to the characteristic time τ_ε of electron temperature variations due to the electron energy loss in electron-atom collisions.

5. Applications of the VCRD technique

The study of the performance and fundamentals of the VCRD technique, which was reviewed above, shows that this technique has some advantages over conventional techniques for MM imaging. In contrast to the imaging techniques, which use the receiving antenna arrays, a continuous medium with a small MMW reflection coefficient is exploited in the VCRD technique as an imager. Hence, it is possible to perform imaging of MMW beam profiles in the near-field region via the VCRD technique. The advantages of the VCRD technique over the thermographic techniques include a much higher energy sensitivity and a microsecond temporal resolution. These and some other advantages of the VCRD technique give hope for its applications for determination of the characteristics of output radiation from moderate-power MMW sources as well as for imaging and nondestructive testing with MM waves. At the same time, the VCRD technique has some peculiarities; therefore, experimental verification of the possibility of its use for solving applied problems is required. In this regard, Section 5 is focused on demonstration of two application capabilities of the VCRD technique. Subsection 5.1 discusses the experiments, where this technique is successfully used to measure the parameters of MM waves at the output of new-designed electron tubes. These studies are very relevant because short-wavelength millimeter waves are supposed to be widely used in new advanced radar, communication, diagnostic, and spectroscopy systems. However, in these frequency bands, the power of most types of modern CW and pulsed sources of EM waves is insufficient for use of thermographic imaging techniques. Due to its higher sensitivity, the VCRD technique can be successfully used for measuring the output characteristics of such MM wave sources. Subsection 5.2 provides an overview of experiments on the use of the VCRD technique for real-time imaging and nondestructive testing in the millimeter-wave band. Time-resolved imaging and NDT using MM waves are required for many applications [1–9]. Millimeter waves can penetrate through many opaque dielectric materials, such as paper, wood, fabric, ceramics, polymers, semiconductors, and others. This fact gives MMW imaging and NDT systems some advantage over optical and infrared systems, even though the spatial resolution of MMW systems is much worse than that of optical and IR systems. MMW imaging also has some advantages over THz imaging: first, atmospheric absorption and scattering is lower in the millimeter-wave region, and second, many types of conventional MMW sources and components are widely available. In contrast to EM radiation of the centimeter-wave band, which is also used for imaging and nondestructive testing, millimeter waves have a better spatial resolution. X-ray imaging also has some disadvantages compared to MMW imaging: first, the X-ray is ionizing and requires the implementation of stringent safety rules, and second, some materials, e.g., water, produce only low-contrast images in the X-ray region.

5.1. Imaging of the beam profiles at the output of a D-band orotron and a W-band gyrotron

Orotrons are vacuum electron sources of coherent EM radiation of the short-wavelength band of the MMW range. They are used in the microwave and NMR spectroscopy, for MMW frequency standards. A pulsed orotron OR-180 has been designed and manufactured by a collaboration of Russian organizations [46]. The orotron frequency was tuned electromechanically in the F and D bands. Output coupling of MMW radiation from the orotron cavity occurred through an oversized E-band rectangular waveguide. The design of the orotron assumes that the TE_{10} mode of a rectangular waveguide will be excited predominantly in its output waveguide. Verification of this assumption and determination of the field profiles at the output of the orotron when it was tuned to the different frequencies were important steps towards the comprehension of the operating principle of this device and the improvement of its performance. For the solution of these problems, the VCRD technique for imaging of MM waves has been successfully used [30]. The scheme of this experiment is similar to that shown in Figure 3. Millimeter waves were generated by the orotron OR-180. The orotron pulse duration was 3 ms. A pyramidal horn antenna with aperture dimensions 14 × 14 mm² and 21 cm long was installed at the output of the E-band waveguide. The MMW electric field vector at the output of the horn was directed along the y coordinate. The output of the horn was placed very close to the discharge tube window. Near-field intensity profiles at the output of a pyramidal horn antenna were imaged when the frequencies orotron radiation were 130, 140, and 150 GHz. The maximum orotron output power was about 0.3, 0.6, and 0.2 W for the frequencies 130, 140, and 150 GHz, respectively. Thin red lines in Figure 9 show the dependences of the VCR intensity variations on x for $y' = 0$ and on y for $x' = 0$, which were measured at the output of a horn antenna when the orotron frequency was 130 GHz (see Figures 9(a) and 9(b)) and 150 GHz (see Figures 9(c) and 9(d)). The VCR intensity profiles, which are shown in Figure 9, were averaged using 200 CCD camera frames recorded when MMW pulses were incident on the plasma. Figure 9 illustrates that the tuning the orotron frequency changes the intensity profile of the output MMW beam. In the frequency range 130–150 GHz in the E-band rectangular waveguide, besides the fundamental mode $TE_{10}^{\square}$, several higher modes can also be excited [43, 44]. By comparing the measured spatial distributions of VCR intensity variation with the profiles of the microwave-induced electron temperature perturbation, which were calculated using Eq. (7), it is possible to analyze the mode content of the orotron output radiation. Thick blue lines in Figures 9(a) and 9(b) show distributions of the first-order perturbation of the electron temperature on the coordinate x for $y' = 0$ and on y for $x' = 0$, respectively. These dependences were calculated assuming that in the orotron waveguide excited only one fundamental waveguide mode, $TE_{10}^{\square}$. For an orotron frequency of 130 GHz, the profiles of the VCR intensity are approximated well by these calculated distributions of the electron temperature perturbation. The measured relative value of the VCR intensity was higher than calculated electron temperature perturbation at the "tails" of distribution on y for $x' = 0$ due to diffraction of the MM waves at the edges of a horn antenna, as well as the MM wave-induced nonuniformity of the electron heating by a DC electric field (see Section 4.2 and ref. [29]). Therefore, only fundamental waveguide mode is excited for an orotron radiation frequency of 130 GHz. The same situation occurred when the orotron frequency was 140 GHz. For orotron frequencies of 130 GHz and 140 GHz, the admixture of higher modes does not exceed 10%. When the orotron was tuned to 150 GHz, multimode excitation of its output waveguide was observed. This is confirmed, particularly, by strong asymmetry of the distribution of the VCR intensity along the y coordinate (see Figure 9(d)) and its broadening on x (see Figure 9(c)). The results of the experiments on imaging of the MMW intensity profile for an orotron frequency of 150 GHz can be explained assuming that in the output waveguide, a $TE_{11}^{\square}$ mode is excited in addition to the fundamental mode $TE_{10}^{\square}$. Thick blue lines in Figures 9(c) and 9(d) show distributions of the first-order perturbation of the electron temperature on the x coordinate for $y' = 0$ (see Figure 9(c)) and on y for $x' = 0$ (see Figure 9(d)). The distributions were calculated assuming that the electric field amplitude ratio for the $TE_{10}^{\square}$ and $TE_{11}^{\square}$ modes is equal to $E^{TE10} / E^{TE11} = 3{:}2$ and the phase difference between them is $\pi/4$. Figures 9(c) and 9(d) show that good agreement between the results of calculation and the experimental data occurs for such an amplitude ratio and a phase difference for the two modes. At an orotron frequency of 150 GHz, the power of the spurious mode was about 40% of the power of the fundamental mode.

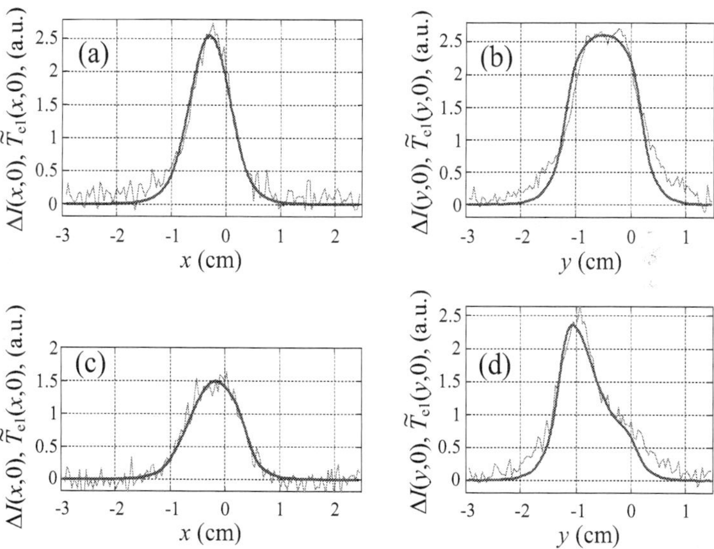

Figure 9. The dependences of the VCR intensity variations (thin red lines) on x for $y' = 0$ ((a) and (c)) and on y for $x' = 0$ ((b) and (d)), which were obtained at the orotron frequency 130 GHz ((a) and (b)) and 150 GHz ((c) and (d)). The first-order perturbation of the electron temperature (thick blue lines) vs. x for $y' = 0$ ((a) and (c)) and vs. y for $x' = 0$ ((c) and (d)).

The VCRD technique was applied for imaging of the field profiles at the output of a 110 GHz kilowatt-scale gyrotron with a pulsed magnetic field [32]. These experiments are of interest for the improvement of this type of gyrotrons and for their successful exploitation in various applications. The scheme of this experiment is similar to that shown in Figure 3. The millimeter waves were output from the gyrotron cavity without mode conversion via a circular hollow metal waveguide 16 mm in diameter and about 200 mm long. A circular hollow metal waveguide 16 mm in diameter and 640 mm long was attached to the gyrotron waveguide. A conical horn antenna 140 mm long and having an aperture diameter of 40 mm was set at the output of the circular waveguide. The gyrotron power can be varied by changing the electron beam voltage and current. The pulse duration of the gyrotron was 60 μs. The CCD camera exposure time was 64 μs. The gyrotron was in the stage of debugging and improvements at the time of the experiments on imaging of its output field profiles. In accordance with the gyrotron design, its operating mode was the rotating mode TE_{32}^{O} of a circular waveguide. In the first series of experiments, the length of the gyrotron cavity was by 1.5 times longer than optimal for its high-efficiency operation. The maximum output power of the gyrotron with that long cavity was about 3 kW. Figure 10(a) shows a 2D distribution of CVR intensity variation under the MMW effect, which was measured at the output of a conical horn for a gyrotron output power $P = 1$ kW. For such a power, the MMW intensity in the field maxima did not exceed the microwave breakdown threshold of the PC of a Cs-Xe discharge. The VCR image corresponds to the TE_{32}^{O} mode pattern. The standing- wave pattern in the azimuthal direction occurred due to the presence of an admixture of the counter-rotating TE_{32}^{O} mode. Figure 10(b) shows the image of VCR intensity variation recorded at the output of a horn antenna for the gyrotron power $P = 3$ kW. For such a power, the intensity of millimeter waves at their maxima exceeded the breakdown threshold of the plasma slab. Due to the strong increase of VCR intensity in the plasma breakdown areas, the image of the spatial distribution of the MMW intensity was distorted, i.e., it was much more contrasting than in reality. 2D distributions of the MMW intensity were calculated for different power ratios of the counter-rotating TE_{32}^{O} modes of a circular waveguide. The best agreement between experimental data (Figure 10(a)) and computed distribution was obtained for a power of the spurious counter-rotating mode equal to about 0.5% of the power of the fundamental mode TE_{32}^{O}. Figure 10(c) shows the calculated 2D distribution of MMW intensity for TE_{32}^{O} modes with such a power of the counter-rotating mode. The darker areas in Figure 10(c) correspond to the higher-intensity MM waves. In the second series of experiments, a new cavity was installed in the gyrotron. The length of this cavity was optimal for the high-efficiency gyrotron operation. As a result, the maximum gyrotron power increased up to about 10 kW. Figure 10(d) shows 2D distribution of the VCR intensity variation induced by MMW radiation from gyrotron with the optimal cavity. This pattern was recorded for reduced gyrotron power of about 1 kW. At such a gyrotron power, the MMW intensity did not exceed the breakdown threshold of the plasma. The image of MMW intensity profile (see Figure 10(d)) corresponds to the profile of the TE_{32}^{O} mode, which rotates in one direction only. The admixture of the counter-rotating TE_{32}^{O} mode at the output of the gyrotron with a new cavity was less than 0.1% in power.

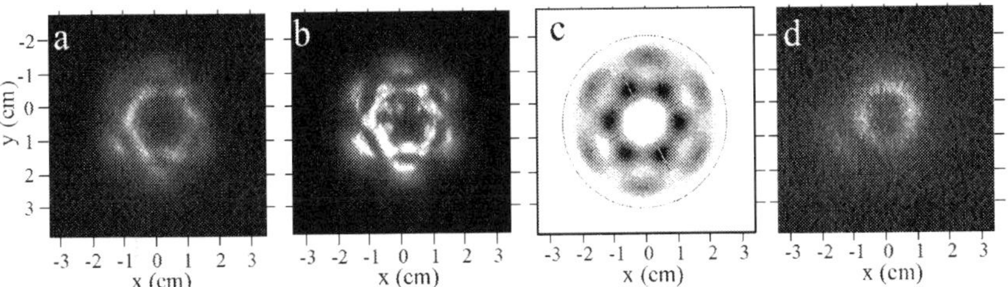

Figure 10. (a), (b), and (d) 2D distributions of the VCR intensity variation at the output of a 110 GHz gyrotron. (c) Calculated 2D distribution of MMW intensity for the TE_{32}^{O} mode.

Thus, the VCRD technique has been successfully applied for determining of the fundamental waveguide modes of moderate-power MMW electron tubes and evaluating the relative power of some spurious modes.

5.2. Applications of the VCRD technique for imaging and nondestructive testing with MM waves

The VCRD technique has been used for active near-field MMW imaging and NDT via shadow projection method [33, 34]. In active-mode MMW imaging and NDT systems MM wave sources are used for object illumination [1–8]. A slab of the positive column of a Cs-Xe DC discharge has been used as a rapid imager for millimeter waves. In the near-field shadow projection method of MMW imaging, the objects are located close to the plasma imager. This is a drawback of the method in comparison with the quasioptical camera-mode method of MMW imaging, in which the object is located far away from the receiver [4, 6, 8, 13]. Realization of the active camera-mode method of imaging based on the VCRD technique is also feasible with using short-wavelength MM waves for object illumination. However, in the camera-mode method distortions of the images occur due to the diffraction and aberrations by lenses and mirrors. Diffraction and aberration effects are very significant when long-wavelength MM waves are used for object illumination. In this case, the use of the shadow projection method may provide a better image quality.

Figure 11 shows schematic of the experimental setup (top view), which was used for active near-field shadow projection MMW imaging using the VCRD technique [34]. A 35.4 GHz magnetron was used as a MMW source. The parameters of the magnetron were specified in Section 3. The MMW pulse length was about 10 ms. The repetition rate of MMW pulses was 12.5 Hz, i.e., half of the frame rate of the CCD camera. Camera frames, which were simultaneous with the MMW pulses, alternated with frames with the background VC radiation from the plasma slab. The CCD camera exposure time was 1 ms. The MMW beam was radiated by a pyramidal horn antenna with a length of 50 cm and aperture dimensions 6×8 cm^2. A plane-convex Teflon lens collimated the MMW beam. The polarization of the MMW beam was directed along the x axis (see Figure 11). A test object was placed close to the window of the discharge tube. The width (FWHM) of the quasi-Gaussian beam in the object plane was about

8 cm. The MMW intensity in the center of the beam was approximately 0.3 W/cm². Time-averaged MMW intensity in the center of the beam was about 30 mW/cm². The MMW dose can be reduced by an order of magnitude by shortening the MMW pulse duration to the CCD camera exposure time. For this millimeter wave intensity, the MMW thermal effect on the majority of objects is not significant.[6] For the MMW intensity less than 1 W/cm², the VCR intensity $\Delta I(x, y)$ is directly proportional to the intensity of the MMW radiation incident on the plasma $W(x, y)$ (see Section 3); hence, the image of the VCR intensity variations corresponds to a 2D distribution of the MMW intensity behind the object. The effect of nonlocality of the MM wave-induced variation of the VCR intensity slightly blurs the MMW images.

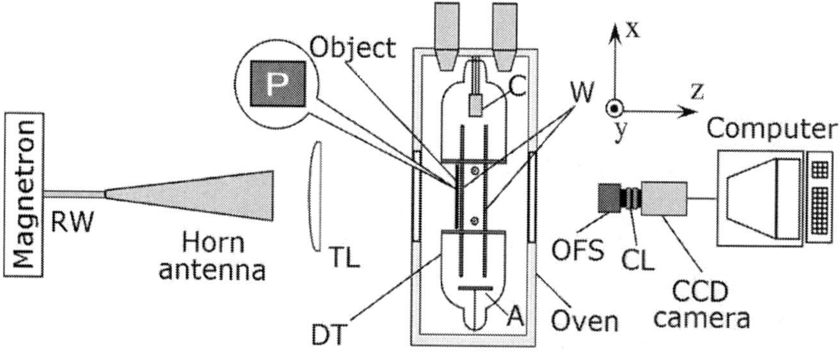

Figure 11. Schematic of the experimental setup (top view); RW: rectangular waveguide; TL: Teflon lens; DT: discharge tube; A: anode; C: cathode; W: quartz windows; OFS: optical filter set; CL: camera lens. Inset, front view of the test object placed before the tube window.

The near-field MMW images of the amplitude objects, which are made of materials that are opaque for MMWs, and the phase objects, which are transparent to the MMW radiation, have been obtained using the VCRD technique. At first, static amplitude and phase objects were imaged. The letters and numbers from IRMMW-THZ 08, an acronym for the name of the International conference, were used as the amplitude test objects. They were cut in the aluminum foil. The foils were glued on cardboard sheets. The letters and numbers were 50 mm high. The width of the rectangular strips transparent for MMWs was 12 mm. The images of the VCR intensity variation under the effect of the MM waves transmitted through the slit objects are shown in Figure 12. The single-shot signal-to-noise ratio for the images was about 20:1. The noise level of the images was given by the noise performance of the CCD camera. The obtained images were distorted by the diffraction because the plasma slab was located in the

---

6 The VCRD technique is well suited for inspection and MMW imaging of the inanimate objects. The time-averaged MMW intensity of 3mW/cm² is several times larger than the maximum MMW dose, which is permitted for general public use in the case of uniform MMW irradiation the whole human body [52]. However, this value is an order of magnitude less than the maximum permitted dose for local MMW irradiation of a part of the human body. Therefore, the VCRD technique cannot be used for entire-body personnel screening, but its applications for security screening and medical imaging with irradiation of a part of human body may be feasible.

Fresnel region. Nevertheless, all of the letters and numbers can easily be recognized. Figure 13(a) shows the image of letter E cut out of aluminum foil and glued on a cardboard sheet. The letter was 50 mm high and 45 mm wide. The width of the foil strips was 10 mm. The shape of this object, which is opaque for MM waves, can be easily determined from the image.

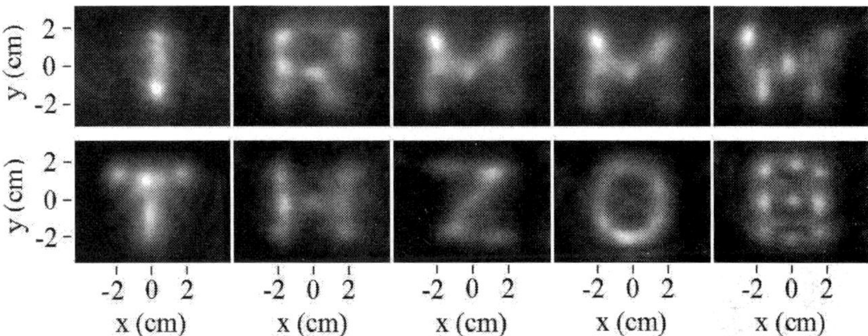

Figure 12. MMW transmission images of the letters and numbers from the acronym IRMMW-THZ 08 cut in aluminum foils.

Due to the peculiarities of the electromagnetic wave diffraction at the edge of transparent and opaque objects, the images of phase objects can have a higher quality as compared to the images of amplitude objects. The EM wave transmitted through a transparent dielectric slab with the thickness L_C acquires an additional phase shift $\Delta\varphi_M = 2\pi(\sqrt{\varepsilon_M} - 1)L_C/\lambda_M$, where ε_M is the permittivity of the material from which the object is made, λ_M is the wavelength of the MMW radiation. If the additional phase shift of the EM wave transmitted through the object differs from the multiple of 2π, then the diffraction gives rise to a narrow minimum of the EM wave intensity in the near-field region behind the edge [53, 54]. The minimum is located between two maxima of intensity, i.e., an edge contrast enhancement occurs in the image of a phase object. Exploiting the VCRD technique, a shadow projection MMW image of a rectangular parallelepiped from Teflon ($\varepsilon_M \approx 2$) was obtained. The length, width, and thickness of the parallelepiped were 180, 28, and 6.8 mm, respectively. In the experiment, the long edges of the parallelepiped were directed parallel to the y axis. When the MMW beam transmitted through Teflon parallelepiped, it acquired an additional phase shift which was equal to about 0.64π. Figure 13(b) shows the 2D MMW image of the parallelepiped. Gray line in Figure 13(c) shows the dependence of the VCR intensity variation on x for $y = 0$. The MMW intensity distribution in the near-field of the Teflon plate 28 mm width was calculated using the theory of Fresnel diffraction from a transparent thin plate [53, 54]. The black line in Figure 13(c) shows the dependence of the integral of the MMW intensity over the plasma slab thickness W_S vs. the coordinate x calculated for such a plate. Calculated dependence $W_S(x)$ is close to the experimental distribution of the VCR intensity variation. Images of more complex dielectric objects are shown in Figures 14 and 15. Figure 14(a) is a drawing of a rectangular Teflon plate with two intersecting grooves having the form of half a cylinder. In the drawing, the dimensions are given in millimeters. Figure 14(b) shows the MMW image of this Teflon plate. The plate edges and both grooves are clearly seen in the image given in Figure 14(b). Perforating circular holes with a diameter of 6 mm are invisible. Figure 15(a) shows the MMW transmission image of a fish-shaped pumice-stone. The height, width, and thickness of the pumice-stone fish were 95, 58, and 18 mm, respectively. In the MMW images of the fish, as well as the Teflon plates, an edge contrast enhancement occurs. Figure 15(b) shows the MMW transmission image of three Septogal pellets (made by JADRAN, Croatia). The pellets were disks 16 mm in diameter and about 3.5 mm thick. The pellets were arranged as a trefoil and placed inside an envelope. Although the diameter of the pills is only twice the MMW wavelength, they are fairly well seen on the image. The experiments have demonstrated that the VCRD technique can be used for a real-time security screening, including the screening of drugs hidden in envelopes.

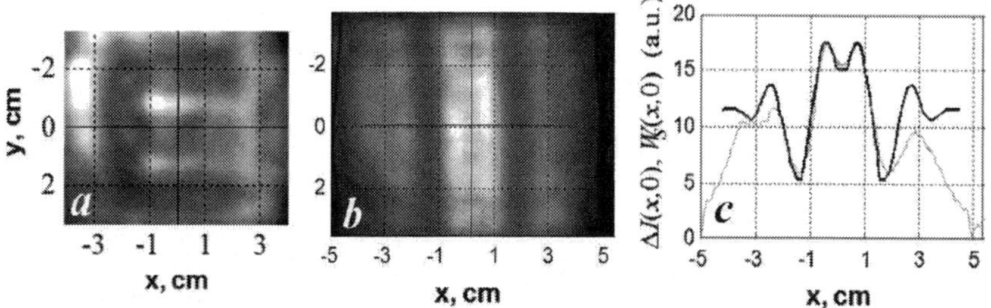

Figure 13. (a) MMW image of the letter E cut out of aluminum foil. (b) 2D MMW image of a rectangular parallelepiped from Teflon. (c) The dependences of the VCR intensity variation (gray line) and the MMW intensity integrated over the plasma slab thickness (black line) on x for $y = 0$.

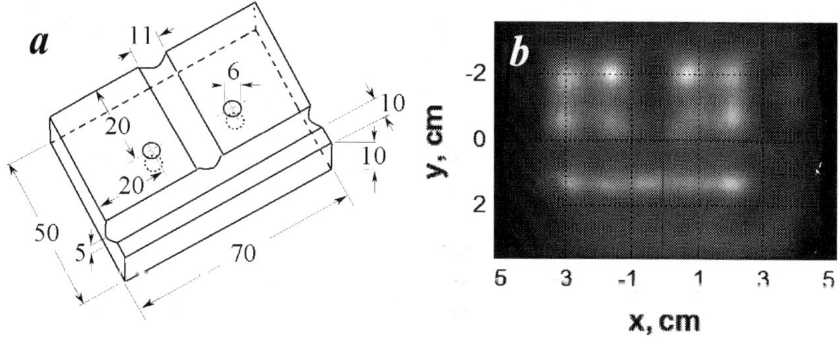

Figure 14. (a) A drawing of the rectangular plate with two intersecting grooves. (b) MMW transmission image of the Teflon plate.

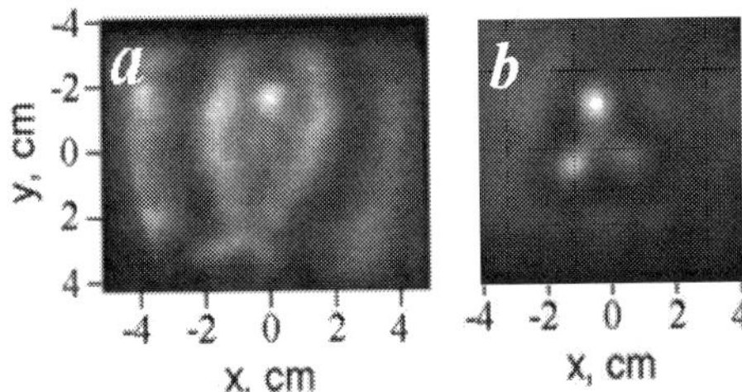

Figure 15. (a) MMW transmission image of a pumice-stone in the form of a fish. (b) MMW transmission image of three Septogal pellets.

The VCRD technique was used for time-resolved record of the MMW images of moving objects and registration of transient processes [34]. In particular, real-time shadow projection images of damped pendulum oscillation were obtained. The pendulum was a Teflon ring suspended by a thread about 15 cm long. The outer and inner ring diameters were 25 mm and 13 mm, respectively, and the ring was 8 mm thick. The ring affected the MMW beam as a three-zone phase filter which focused the MMW beam in the near field in a focal spot about 7 mm in diameter. The decay time of the pendulum oscillation was about 4 s, and the oscillation period was approximately 0.8 s. Figures 16(a) –16(e) show a sequence of MMW images of the pendulum motion, which was recorded at equal time intervals during half of the first period of pendulum oscillations. The process, by which a cylindrical glass tube filled with water became empty, was also imaged (see Figures 16(f) –16(j)). The inner and outer tube diameters were 6 and 8 mm, respectively. The tube filled with water was opaque for MMWs (see Figure 16(f)). After opening a valve which was located at the foot of the glass tube (outside the aperture), the water began to flow out of the tube. Figures 16(g) –16(i) show how the glass tube becomes transparent for MMWs as it is emptied. The water poured out of the tube for about 1 s. The MMW transmission image of the empty tube is shown in Figure 16(j). The tube wall is well defined in Figure 16(j) because of the considerable difference in the refractive indices of glass and air.

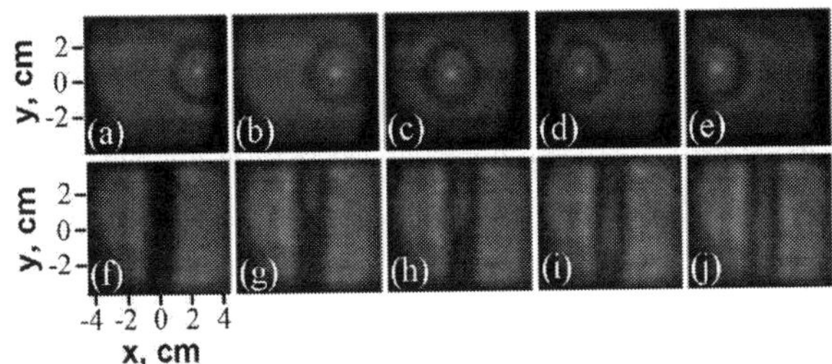

Figure 16. (a)–(e) Real-time MMW transmission images of pendulum motion obtained during one-half of the oscillation period. (f)–(j) MMW transmission movie showing how a glass tube filled with water was emptied.

6. Conclusions

The chapter presents a review of the VCRD technique for imaging and measurement of the 2D spatial profiles of MMW intensity based on the use of visible continuum radiation from the positive column of a Cs-Xe DC discharge. The design of the discharge tube and the experimental setup that were used to create a CW homogeneous slab of thermal nonequilibrium discharge in a mixture of medium pressure xenon and cesium vapor were described. Results of the study of the basic parameters of the positive column of a Cs-Xe discharge were presented. Three modes of the positive column of a Cs-Xe discharge were observed depending on the tube temperature for a fixed value of the discharge current: a constricted PC, a spatially homogeneous PC, and a PC with a filament. For imaging of MM waves, a wide homogeneous PC plasma slab was used. The MMW effects on the plasma slab have been experimentally studied. It was shown that brightness of the VC radiation from the PC plasma increases by several times when the MMW intensity changes from zero to the threshold of the microwave-induced plasma breakdown. The model experiments on imaging of the field patterns of horn antennas and quasioptical beams have been performed using the VCRD technique. These experiments demonstrate that the VCRD technique can be used for a good-quality time-resolved imaging of MMW beams with a width of about 10 mm or more. The temporal resolution of the technique is about 0.8 μs. It was shown that the VCRD technique is wideband and can be used for imaging of the intensity profiles of the EM waves in the entire MMW band. Energy flux sensitivities of about 10 $\mu J/cm^2$ in the Ka band and about 200 $\mu J/cm^2$ in the D band have been demonstrated.

The fundamentals of the VCRD technique for imaging of MMWs have been discussed. The studies of the nature of the visible continuum radiation from the PC of a Cs-Xe discharge and mechanism of microwave-induced variations in the VCR intensity were reviewed. It is shown that the electron-xenon atom bremsstrahlung is the dominant component of the visible continuum emitted by the homogeneous PC of a Cs-Xe discharge. The increase in the e-Xe bremsstrahlung intensity in the visible region is caused by an increase in the number density of electrons with the energy more than 2 eV, which is a result of the additional electron heating induced by the MM waves. The profiles of the microwave-induced variation in the intensity of the VC radiation from the plasma slab were analytically modeled. It was proved that the nonlocality of the microwave-induced variations in the VC radiation intensity, as well as the spatial resolution of the VCRD technique of MMW imaging, are primarily determined by the influence of the electron heat conduction. The line spread function for the VCRD technique has been calculated. It was found that its width is about 2 mm. It was shown that the main cause of the axial asymmetry of images of the axisymmetric MMW beams is the MM wave-induced spatial inhomogeneity of the electron heating by a DC electric field.

The final part of the paper reviews some applications of the CVRD technique for MMW imaging. The experiments on imaging of the output field pattern of a D-band watt-scale orotron and a W-band kilowatt-scale gyrotron were described. Operating modes for these vacuum tubes were identified and the power of spurious modes were determined. Applications of the VCRD technique for real-time imaging and nondestructive testing using MM waves are also the subject of the final part of the review. Near-field shadow projection images of the objects which are opaque and transparent for MM waves have been obtained using pulsed watt-scale MM waves for object illumination. Near video frame rate millimeter-wave shadowgraphy has been demonstrated. It was shown that this technique can be used for a single-shot security screening, including the screening of drugs hidden in envelopes, and for a real-time imaging of the transient processes.

Acknowledgements

The author acknowledges A. E. Fedotov, M. Yu. Glyavin, V. V. Golovanov, A. G. Luchinin, A. G. Spivakov, S. E. Stukachev, A. I. Tsvetkov, and V. V. Zelenogorsky for their contributions to this work. The author is also grateful to N. A. Bogatov, I. L. Epstein, Yu. A. Lebedev, V. I. Malygin, and V. E. Semenov for helpful discussions.

Author details

Mikhail S. Gitlin*

Address all correspondence to: gitlin@appl.sci-nnov.ru

Institute of Applied Physics, Russian Academy of Sciences, Nizhny Novgorod, Russia

References

[1] Bolomey J.-C. Recent European developments in active microwave imaging for industrial, scientific, and medical applications. IEEE Trans. Microwave Theory Tech. 1989;37:2109–2117. DOI: 10.1109/22.44129

[2] Kharkovsky S., Zoughi R. Microwave and millimeter wave nondestructive testing and evaluation—Overview and recent advances. IEEE Instrum. Meas. Mag. 2007;10:26–38. DOI: 10.1109/MIM.2007.364985

[3] Gomez-Maqueda I., Almorox-Gonzalez P., Callejero-Andres C., Burgos-Garcia M. A millimeter-wave imager using an illuminating source. IEEE Microwave Mag. 2013;14:132–138. DOI: 10.1109/MMM.2013.2248652

[4] Sato M., Mizuno K. Millimeter-wave imaging sensor. In: Minin I, editor. Microwave and Millimeter Wave Technologies from Photonic Bandgap Devices to Antenna and Applications. InTech; 2010. p. 331–349. ISBN: 978-953-7619-66-4

[5] Ahmed S. S., Schiessl A., Gumbmann F. F., Tiebout M., Methfessel S., Schmidt L.-P. Advanced microwave imaging. IEEE Microwave Mag. 2012;13:26–43. DOI: 10.1109/MMM.2012.2205772

[6] Tang A., Gu Q. J., Chang M.-C. F. CMOS receivers for active and passive mm-wave imaging. IEEE Communications Magazine. 2011;49:190–198. DOI: 10.1109/MCOM.2011.6035835

[7] Sheen D. M., McMakin D. L., Hall T. E. Three-dimensional millimeter-wave imaging for concealed weapon detection. IEEE Trans. Microwave Theory Tech. 2001;49:1581–1592. DOI: 10.1109/22.942570

[8] Watabe K., Shimizu K., Yoneyama M., Mizuno K. Millimeter-wave active imaging using neural networks for signal processing. IEEE Trans. Microwave Theory Tech. 2003;51:1512–1515. DOI: 10.1109/TMTT.2003.810132

[9] Greeney N. S., Scales J. A. Dielectric microscopy with submillimeter resolution. Appl. Phys. Lett. 2007;91:222909. DOI: 10.1063/1.2818674

[10] Ozhegov R., Gorshkov K., Vachtomin Y., Smirnov K., Finkel M., Goltsman G., Kiselev O., Kinev N., Filippenko L., Koshelets V. Terahertz imaging system based on superconducting heterodyne integrated receiver. In: Corsi C., Sizov F, editors. THz and Security Applications, Detectors, Sources and Associated Electronics for THz Applications. Netherlands: Springer; 2014. p. 113-125. DOI: 10.1007/978-94-017-8828-1_6

[11] Kopeika N. S., Fartiat N. H. Video detection of millimeter waves with glow-discharge tubes. IEEE Trans. on Electron Devices. 1975;22:534–548. DOI: 10.1109/T-ED.1975.18175

[12] Abramovich A., Kopeika N. S., Rozban D. Design of inexpensive diffraction limited focal plane arrays for millimeter wavelength and terahertz radiation using glow discharge detector pixels. J. Appl. Phys. 2008;104:033302-1-033302-4. DOI: 10.1063/1.2963714

[13] Brooker G., Johnson D. G. Low-cost millimeter wave imaging using a commercial plasma display. IEEE Sensors J. 2015;15:3557-3564. DOI: 10.1109/JSEN.2015.2393306

[14] Hisatake S., Nguyen Pham H. H., Nagatsuma T. Visualization of the spatial-temporal evolution of continuous electromagnetic waves in the terahertz range based on photonics technology. Optica. 2014;1:365–371. DOI: 10.1364/OPTICA.1.000365

[15] Tsuchiya M., Kanno A., Sasagawa K., Shiozawa T. Image and/or movie analyses of 100 GHz traveling waves on the basis of real-time observation with a live electrooptic imaging camera. IEEE Trans. Microwave Theory Tech. 2009;57:3373-3379. DOI: 10.1109/TMTT.2009.2033890

[16] Bazhulin A. P., Vinogradov E. A., Irisova N. A., Fridman S. A. Obtaining a visible image of radio emission in the millimeter band. JETP Lett. 1968;8:160–162.

[17] Cherkassky V. S., Gerasimov V. V., Ivanov G. M., Knyazev B. A., Kulipanov G. N., Lukyanchikov L. A., Merzhievsky L. A., Vinokurov N. A. Techniques for introscopy of condense matter in terahertz spectral region. Nuclear Instruments and Methods in Physics Research A. 2007;575:63–67. DOI: 10.1016/j.nima.2007.01.025

[18] Chang T. H., Yu C. F., Fan C. T. Polarization-controllable TE21 mode converter. Rev. Sci. Insrum. 2005;76:074703-1-074703-6. DOI: 10.1063/1.1942528

[19] Junchang L., Yanmei W. An indirect algorithm of Fresnel diffraction. Optics Communications. 2009;282:455–458. DOI: j.optcom.2008.10.060

[20] Thumm M. K., Kasparek W. Passive high-power microwave components. IEEE Trans. on Plasma Science. 2002;30:755–786. DOI: 10.1109/TPS.2002.801653

[21] Malygin V. I., Paveljev A. V. Determination of the mode content in spurious microwave radiation of the gyrotron with a straight axisymmetric output. Int. J. Infrared Millimeter Waves. 1999;20:33–56. DOI: 10.1023/A:1021747516720

[22] Jawla S., Hogge J-P., Alberti S., Goodman T., Piosczyk B., Rzesnicki T. Infrared measurements of the RF output of 170-GHz/2-MW coaxial cavity gyrotron and its phase retrieval analysis. IEEE Trans. on Plasma Sci. 2009;37:414-424. DOI: 10.1109/TPS. 2008.2011488

[23] Idehara T., Kosuga T., Agusu L., Ogawa I., Takahashi H., Smith M. E., Dupree R. Gyrotron FU CW VII for 300 MHz and 600 MHz DNP-NMR spectroscopy. J. Infrared Millim. Terahz Waves. 2010;31:763–774. DOI: 10.1007/s10762-010-9637-9

[24] Gold S. H., Fliflet A. W., Manheimer W. M., McCowan R. B., Lee R. C., Granatstein V. L., Hardesty D. L., Kinkead A. K., Sucy M. High peak power Ka-band gyrotron oscillator experiments with slotted and unslotted cavities. IEEE Trans. on Plasma Sci. 1988;16:142–148. DOI: 10.1109/27.3806

[25] Bratman V. L., Denisov G. G., Ofitserov M. M., Korovin S. D., Polevin S. D., Rostov V. V. Millimeter-wave HF relativistic electron oscillators. IEEE Trans. on Plasma Sci. 1987;15:2–15. DOI: 10.1109/TPS.1987.4316655

[26] Cook A., Shapiro M., Temkin R. Pressure dependence of plasma structure in microwave gas breakdown at 110 GHz. Appl. Phys. Letters. 2010;97:11504-1-011504-3. DOI: 10.1063/1.3462320

[27] Gitlin M. S., Golovanov V. V., Spivakov A. G., Tsvetkov A. I., Zelenogorskiy V. V. Time-resolved imaging of millimeter waves using visible continuum from the positive column of a Cs-Xe DC discharge. J. Appl. Phys. 2010;107:063301-1-063301-11. DOI: 10.1063/1.3327218

[28] Gitlin M. S., Spivakov A. G. Mechanism of the electric field effect on the intensity of visible continuum emission from the positive column of gas discharge in a cesium vapor-xenon mixture. Tech. Phys. Lett. 2007;33:205-208. DOI: 10.1134/ S1063785007030078

[29] Gitlin M. S., Fedotov A. E., Stukachev S. E., Tsvetkov A. I. Nonlocality of microwave-induced variations in the intensity of the visible continuum from a medium-pressure cesium-xenon dc discharge. Physics of Plasmas. 2012;19:033508-1-033508-11. DOI: 10.1063/1.3692077

[30] Fedotov A. E., Gitlin M. S., Golovanov V. V., Perminov A. O., Stukachev S. E. Imaging of short millimeter waves using the visible continuum emitted by the Cs-Xe DC discharge. In: Proc. Sixth Int. Kharkov Symposium on Physics and Engineering of Microwaves, Millimeter and Submillimeter Waves and Workshop on Terahertz Technologies; Kharkov. Kharkov, Ukraine: IRE NASU; 2007. 263–265.

[31] Gitlin M. S., Zelenogorsky V. V., Perminov A. O. Microwave beam imaging by recombination continuum from the positive column of gas discharge in a cesium vapor–xenon mixture. Tech. Phys. Letters. 2002;28:445–447. DOI: 10.1134/1.1490955

[32] Gitlin M. S., Glyavin M. Yu., Luchinin A. G., Zelenogorsky V. V. Imaging the output field pattern of a 110 GHz gyrotron with pulsed magnetic field using recombination continuum emitted by a slab of the Cs-Xe DC discharge. IEEE Trans. on Plasma Science. 2005;33:380-381. DOI: 10.1109/TPS.2005.845294

[33] Gitlin M. S., Golovanov V. V., Tsvetkov A. I. Real-time shadow projection millimeter-wave imaging using visible continuum from a slab of the Cs-Xe DC discharge. IEEE Trans. on Plasma Sci. 2008;36:1398–1399. DOI: 10.1109/TPS.2008.920901

[34] Gitlin M. S., Tsvetkov A. I. Real-time millimeter-wave shadowgraphy using the visible continuum from a slab of the Cs-Xe DC discharge. Appl. Phys. Lett. 2009;94:234102-1-234102-3. DOI: 10.1063/1.3152285

[35] Gitlin M. S., Epstein I. L., Lebedev Yu. A. Modeling of the positive column of a medium pressure Cs-Xe dc discharge affected by a millimeter wave pulse. J. Phys. D: Appl. Phys. 2013;46:415208-1-415208-11. DOI: 10.1088/0022-3727/46/41/415208

[36] Bogatov N. A., Gitlin M. S., Dikan D. A., Luchinin G. A. Cs-Xe dc gas discharge as a fast highly nonlinear volumetric medium for microwaves. Phys. Rev. Lett. 1997;79:2819–2822. DOI: 10.1103/PhysRevLett.79.2819

[37] Wazink J. H., Polman J. Cesium depletion in the positive column of Cs-Ar discharge. J. Appl. Phys. 1969;40:2403–2408. DOI: 10.1063/1.1658005

[38] van Tongeren H. Positive column of the Cs-Ar law-pressure discharge. J. Appl. Phys. 1974;45:89–96. DOI: 10.1063/1.1663024

[39] Batenin V. M., Chinnov V. F. Electron bremsstrahlung in the field of argon or helium atoms. Sov. Phys. JETP. 1972;34:30–33.

[40] Golubovskii Yu. B., Kagan Yu. M., Komarova L. L. Diagnostics of the discharge in argon at medium pressures by bremsstrahlung continuum. Opt. Spectrosc. (USSR). 1972;33:646–648.

[41] Park J., Henins I., Herrmann H. W., Selwyn G. S. Neutral bremsstrahlung measurement in an atmospheric-pressure radio frequency discharge. Phys. Plasmas. 2000;7:3141–3144. DOI: S1070-664X(00)03808-8

[42] Raizer Y. P. Gas Discharge Physics. Berlin: Springer-Verlag; 1991. 449 p.

[43] Harvey A. F. Microwave Engineering. London and New York: Academic Press; 1963. 1313 p.

[44] Marcuvitz N. Waveguide Handbook. London: P. Peregrinus; 1986. 434 p.

[45] Camparo J., Fathi G. Effects of rf power on electron density and temperature, neutral temperature, and Te fluctuations in an inductively coupled plasma. J. Appl. Phys. 2009;105: 103302-1-103302-9. DOI: 10.1063/1.3126488

[46] Bratman V. L., Dumesh B. S., Fedotov A. E., Makhalov P. B., Movshevich B. Z., Rusin F. S. Terahertz orotrons and oromultipliers. IEEE Trans. Plasma Sci. 2010;38:1466–1471. DOI: 10.1109/TPS.2010.2041367

[47] Clarricoats P. J. B., Olver A. D. Corrugated horns for microwave antennas. London: P. Peregrinus; 1984. 231 p.

[48] Rutscher A., Pfau S. On the origin of visible continuum radiation in rare gas glow discharges. Physica C. 1976;81:395–402. DOI: 10.1016/0378-4363(76)90078-4

[49] de Regt J. M., van Dijk J., van der Mullen J. A. M., Schram D. C. Components of continuum radiation in an inductively coupled plasma. J. Phys. D: Appl. Phys. 1995;28:40–46. DOI: 10.1088/0022-3727/28/1/008

[50] Burm K. T. A. L. Continuum radiation in a high pressure argon–mercury lamp. Plasma Sources Sci. Technol. 2004;13:387–394. DOI: 10.1088/0963-0252/13/3/004

[51] Agnew L., Reichelt W. H. Identification of the ionic species in a cesium plasma diode. J. Appl. Phys. 1968;39:3149–3155. DOI: 10.1063/1.1656749

[52] Wu T., Rappaport T. S., Collins C. M. Safe for generations to come. IEEE Microwave Mag. 2015;16:65–84. DOI: 10.1109/MMM.2014.2377587

[53] Anokhov S. P. Plane wave diffraction by a perfectly transparent half-plane. J. Opt. Soc. Am. A. 2007;24:2493–2498. DOI: 10.1364/JOSAA.24.002493

[54] Amiri M., Tavassoly M. T. Fresnel diffraction from 1D and 2D phase steps in reflection and transmission modes. Optics Communications. 2007;272:349–361. DOI: 10.1016/j.optcom.2006.11.048

Chapter 13

Optically Thick Laser-Induced Plasmas in Spectroscopic Analysis

Fatemeh Rezaei

Additional information is available at the end of the chapter

http://dx.doi.org/10.5772/61941

Abstract

Studies on the plasma physics has been grown over the past few decades as a major research field. The plasma can be produced by different sources such as acr, spark, electric discharge, laser and so on. The spectral radiation of the plasma which acts as its fingerprint, contains valuable information about plasma features. Characterization of plasmas by spectroscopic measurement is a powerful tool for increasing the knowledge and applications of these kinds of radiation sources. Therefore, the spectral diagnostics methods are proposed which are based on measurement of spectral lines intensity, estimation of continuous and absorption radiation, and as well as determination of shifts and halfwiths of the spectrum [1]. The fundamental characteristic parameters of the plasma, i.e., the number densities of plasma species, electron temperature, and as well as particle transport property at each plasma space can be determined by optical emission spectroscopy and utilizing appropriate methods [2]. For accurate evaluation of plasma parameters, its thickness must be thoroughly considered. Generally, the plasmas can be separated into two categories of thin and thick groups. In thin plasmas, the re-absorption of radiation is negligible. Consequently, in spectroscopic analysis, the non-self-absorbed spectral radiation is evaluated by considering the summation of all spectral emissions along the line of sight. In optically thick plasmas, the radiation trapping happens which leads to the self-absorption phenomenon in spectroscopic analysis that is explained with details in below section.

Keywords: Thick plasma, self absorption

1. Introduction

Studies on plasma physics have been grown over the past few decades as a major research field. Plasma can be produced by different sources such as arc, spark, electric discharge, laser, and so on. The spectral radiation of the plasma, which acts as its fingerprint, contains valuable

information about plasma features. Characterization of plasmas by spectroscopic measurement is a powerful tool for increasing the knowledge and applications of these kinds of radiation sources. Therefore, spectral diagnostics methods are proposed which are based on measurements of spectral line intensity, estimation of continuous and absorption radiation, as well as the determination of shifts and half-widths of the spectrum [1]. The fundamental characteristic parameters of the plasma, i.e., the number densities of plasma species, electron temperature, and the particle transport property at each plasma space can be determined by optical emission spectroscopy and by utilizing appropriate methods [2]. For accurate evaluation of plasma parameters, its thickness must be thoroughly considered. Generally, the plasmas can be separated into two categories of thin and thick groups. In thin plasmas, the reabsorption of radiation is negligible. Consequently, in spectroscopic analysis, the non-self-absorbed spectral radiation is evaluated by considering the summation of all spectral emissions along the line of sight. In optically thick plasmas, radiation trapping happens, which leads to the self-absorption phenomenon in spectroscopic analysis that is explained in details in the following section.

2. Self-absorption effect

In thick plasmas, when light is emitted from the interior hot parts of plasma and travels to the outside cold regions, the light may be absorbed by the same sort of emitting atoms and molecules. Consequently, the resultant spectrum in a spectrograph will be weakened so that the plasma itself absorbs its emission. This particular kind of absorption of a light source is called self-absorption. The main error that happens in the evaluation of the plasma parameters is the erosion of the spectral intensity due to self-absorption. This phenomenon results in peak height reduction and growth of spectral line widths. In some cases, absorption in the center of the spectral line is more severe than its sides, so that the self-absorption appears as a self-reversal [3]. Self-reversal happens especially in strong resonance lines and in inhomogeneous thick plasmas. In this case, a central dip is observed in spectral lines due to the cold absorbing atoms from the outer parts of the plasma plume. In most cases, the self-absorption is shown as a height reduction, which will not be well recognized from the shape of the spectrum. Self-absorption is mainly severe for atomic lines with low excitation energies of upper levels or spectral lines with high transition probabilities. Furthermore, resonant lines are particularly influenced by the self-absorption effect.

For the spectroscopic purposes in a reabsorbed plasma, the spectral intensity has a complicated relation with plasma parameters, as well as with emission coefficients. Here, the self-absorption effect is investigated for the laser-induced plasmas in local thermal equilibrium (LTE) condition. The details of optically thick plasma calculations in collisional radiative models is indicated in ref. [4].

Ref. [3] completely investigated the effects of different spectral distributions such as Doppler, resonance, and natural broadening on the magnitude of self-absorption. They focused mainly on self-absorption treatment in arcs and sparks sources. Moreover, they investigated the influences of uniformly excited sources and non-homogenous sources on the amount of line self-absorption. In this study, most of the attention is concentrated on laser as a plasma sources. In the laser-induced breakdown spectroscopy (LIBS) technique, it is observed that in nearly all strong lines of a spectrum and for concentrations more than approximately 3% in the sample, the plasma can behave as a thick medium [5].

It should be noted that self-absorption can be comprehensively studied for different spectro-scopic techniques in all intervals of electromagnetism emission, including gamma and X-ray spectroscopy to radiofrequency region, and it comprises the relativistic synchrotron emission as well. In this chapter, the attention is focused on spectral intervals from UV to IR region and on plasma produced by laser radiation.

Self-absorption coefficient of a particular line is usually defined as the ratio of spectral peak height in the presence of self-absorption to its peak magnitude in the condition without self-absorption as:

$$SA = \frac{I(\lambda_j)}{I_0(\lambda_j)}.$$

(1)

or it can be expressed as the ratio of line width as below:

$$\frac{\Delta\lambda}{\Delta\lambda_0} = (SA)^{-0.5}.$$

(2)

or as the ratio of integrated intensities by

$$\frac{\overline{I}}{\overline{I_0}} \approx \frac{I(\lambda_0)\Delta\lambda}{I_0(\lambda_0)\Delta\lambda_0} = (SA)^{0.5}.$$

(3)

Several research groups proposed different methods of line ratios [6, 7], duplicating mirror [8], curve of growth (COG) [9–12], and calculation models [13–17] for the identification and evaluation of the self-absorption of the considered spectrum. Then, after diagnostic stages, appropriate corrective methods corrected the self-absorbed spectral line intensities before utilizing them for analytical goals. Finally, they calculated the plasma parameters after suitable correction methods, but some groups used these self-absorbed lines straightly by applying appropriate theoretical models without any correction, in spite of its complicated calculations [4, 18–21].

3. Homogenous plasmas

In homogenous plasmas, it is assumed that the plasma parameters are uniform, i.e. they have the same magnitudes in the entire plasma volume. The main investigated parameters are electron and species temperatures T, and number densities of all different elements in the plasma n.

The intensity of a particular spectral line along the line profile related to transition between two ionic or atomic levels l and u (lower and upper levels), can be evaluated by radiation transfer equation. This equation describes the radiation intensity changes after passing a distance dl of a plasma by taking into account the contribution of the emission within this distance and emission reduction because of absorption along dl [4]:

$$dI_\lambda = \left(\varepsilon_\lambda - k(\lambda) I_\lambda \right) dl, \tag{4}$$

where, ε_λ is the emission coefficient in thin plasmas, and $k(\lambda)$ is the absorption coefficient. In this equation, the source function is expressed as:

$$S_\lambda = \frac{\varepsilon_\lambda}{k(\lambda)}. \tag{5}$$

The optical depth or optical thickness is defined as multiplication of absorption coefficient by the geometrical thickness of plasma as $\tau = k(\lambda)dl$.

In a two-level system, by neglecting the stimulated emission, the spectral emission can be calculated by considering the spontaneous emission coefficient (in SI units) as follows:

$$\varepsilon_{spec}(\lambda) = N_u A_{ul} h\nu \, V(\lambda). \tag{6}$$

In the above equation, N_u indicates the number of atoms in the upper level and by assumption of holding the local thermal equilibrium (LTE) condition, it is calculated by Boltzmann distribution function. h is the Plank constant, and ν is the spectral line frequency. Here, for more simplification, it is assumed that the line profile distribution for the emission coefficient and the absorption coefficient is similar as Voigt profile $V(\lambda)$.

The absorption coefficient of the two-level system, in SI units, is explained by taking into account the absorption and induced emission between low and high levels of u and l as:

$$k(\lambda) = \frac{h}{\lambda}(B_{ul}N_l - B_{lu}N_u)V(\lambda). \tag{7}$$

B_{ul} and B_{lu} are Einstein coefficients due to absorption and induced emission, respectively, and they are dependent to spontaneous probability coefficient, A_{ul} by:

$$A_{ul} = \frac{g_l}{g_u} B_{lu} \frac{8\pi h\nu^3}{c^3}, \, g_u B_{ul} = g_l B_{lu}. \tag{8}$$

where, g_l and g_u are the degeneracy of the lower and upper levels, respectively. c is the light velocity in the vacuum. N_l is atomic density in lower level. Therefore, by substituting the above equations in Eq. (7), the absorption coefficient of the mentioned spectral line can be expressed as:

$$k(\lambda) = \frac{g_u A_{ul} N_{tot} \lambda^2}{8\pi Z} e^{-\frac{E_l}{k_B T}} (1 - e^{-\frac{h\nu}{k_B T}}) V(\lambda). \tag{9}$$

here, E_l is the energy of lower level, N_{tot} is the total number densities of species, T is the plasma temperature, Z is the partition function, and K_B is the Boltzman constant.

It should be mentioned that in the collisional-radiative plasma, the calculation procedure is the same as in LTE model, except that the estimation of number densities are performed by rate equation as mentioned in ref. [4].

In this chapter, at first, some corrective methods will be explained for evaluation of self-absorption coefficient in laser-induced homogeneous plasmas by utilizing suitable experimental and numerical methods. Then, the effect of self-absorption on inhomogoneous plasmas will be discussed. Afterward, the impressive parameters on spectral lines charactristics affected by self-absorption will be expressed.

4. Different corrective methods in homogeneous plasmas

One of the methods for observation of self-absorption in the experiment is bending of calibration curve (or curve of growth) constructed by standard samples at high concentrations. Consequently, the self-absorbed spectral lines must be corrected to reach the condition of thin plasmas for prediction of the accurate magnitude of sample concentration without any reduction in intensity. Therefore, in the following sections, some of the corrective methods proposed by different research groups are mentioned.

4.1. Ratio of two spectral line features (width, peak, and surface)

Amamou et al. [6, 7] calculated the self-absorption for both Gaussian and Lorentzian line profiles with a Simplex algorithm program fitting method [22] in a homogenous laser-produced plasma. They fitted the experimental results with theoretical calculation and then, their expressions for Lorentzian profiles were used for quantification of the transition probabilities, ratios, and the ratios of optical thicknesses as well. They introduced different correction factors by considering the ratio of the peaks, line widths, and surfaces of two spectral lines for both of the considered line profiles. The correction factor for the line height can be evaluated by considering the ratio of peak intensities of spectral lines in case of non-self-absorbed atomic line to the case of the self-absorbed line as follows:

$$F^L_{max} = \frac{I^{Thin}_0}{I^{Thick}_0} = \frac{\tau_0}{1-\exp(-\tau_0)}. \tag{10}$$

As well as by taking into account the FWHM ratios, the correction factor for the line width can be calculated by the following equation:

$$F^L_{\Delta\lambda} = \frac{\Delta\lambda^{Thin}_L}{\Delta\lambda^{Thick}_L} = \left(-1 - \frac{\tau_0}{\log([1+\exp(-\tau_0)]/2)}\right)^{-1/2}. \tag{11}$$

For an optical thickness of less than 4, it is assumed that the line surface is proportional to the multiplication of line width by its height. Hence, the correction factor for the line surface can be estimated as:

$$F^L_S = \frac{S^{Thin}_L}{S^{Thick}_L} = \frac{\tau_0}{1-\exp(-\tau_0)}\left(-1 - \frac{\tau_0}{\log([1+\exp(-\tau_0)]/2)}\right)^{-1/2}. \tag{12}$$

here, τ_0 is the optical thickness at central line wavelength (or maximum of optical thickness). The results of these calculations are obtained from the plasma created by laser irradiation on a silicate solid sample placed in a xenon and hydrogen atmosphere for the various multiples of Si II lines. Figure 1 illustrates the evolution of self-absorption correction factors for the above-mentioned parameters in a Lorentzian distribution.

4.2. Simple theoretical equation

El Sherbini et al. [15] presented a simple relation for correcting the self-absorption effect in a homogenous plasma. This model is applicable when the Stark broadening parameter of selected spectral lines is known, as well as when the plasma electron density is available from experiments.

In this work, the intensity of a spectral line in thick condition (erg.s^{-1}.cm^{-3}) along the line profile due to the transition between two levels (j and i) is expressed by

$$I(\lambda) = \frac{8\pi hc^2}{\lambda^5_0}\frac{n_j}{n_i}\frac{g_i}{g_j}(1-e^{-k(\lambda)l}). \tag{13}$$

Generally, the absorption coefficient $k(\lambda)$ is described by a Voigt profile, which is convolution of a Lorentzian and a Gaussian distribution. In laser sources plasmas, Lorentzian width is associated with the Stark effect and the Gaussian line width is dominated by the Doppler broadening. In a typical LIBS experiment, the Gaussian contribution of a spectral line width is negligible compared to the Lorentzian component, hence, the optical depth $k(\lambda)l$ can be calculated by

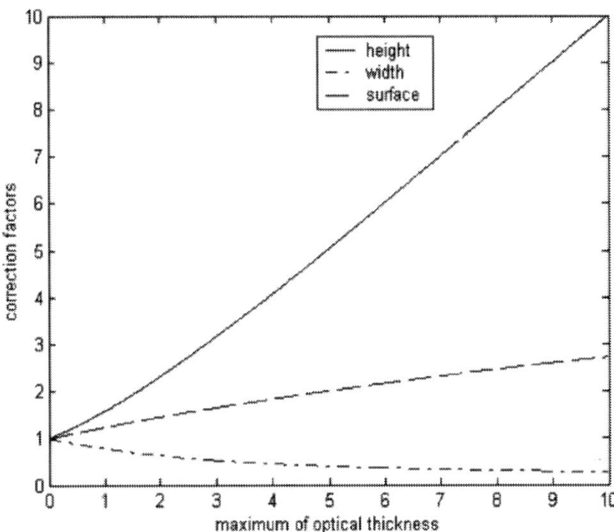

Figure 1. Self-absorption correction factors for: height "solid line", width "alternate dashed line", and surface "long dashed line", in Lorentzian line profile.

$$k(\lambda)l \approx K\frac{\Delta\lambda_0}{4(\lambda - \lambda_0)^2 + \Delta\lambda_0^2},\tag{14}$$

Here, $K = 2\dfrac{e^2}{mc^2}n_i f\lambda_0^2 l$ and the line width is $\Delta\lambda_0 = \Delta\lambda_L$. In the condition of thin plasmas with $k(\lambda)l \ll 1$, the above equation can be approximated as

$$I_0(\lambda) \approx \frac{8\pi hc^2}{\lambda_0^5}\frac{n_j}{n_i}\frac{g_i}{g_j}k(\lambda)l.\tag{15}$$

Consequently, $I_0(\lambda)$ illustrates the line intensity in case of negligible self-absorption. According to Eqs. (13) and (15), it is clearly seen that, in the case of self-absorption, the line intensity at its peak (i.e. for $\lambda = \lambda_0$) has less value compared to that in the case of thin plasmas. Then, the self-absorption coefficient, SA, can be expressed as

$$\frac{I(\lambda_0)}{I_0(\lambda_0)} = \frac{\left(1 - e^{-k(\lambda_0)l}\right)}{k(\lambda_0)l} = \Delta\lambda_0\frac{\left(1 - e^{\frac{-K}{\Delta\lambda_0}}\right)}{K} = SA.\tag{16}$$

SA is a coefficient between 0 and 1. SA equals to 1 in the condition of thin plasma and 0 in highly absorbing plasma. Moreover, the line width magnitude is influenced by the self-absorption effect. For obtaining an equation relating FWHM of an optically thick line $\Delta\lambda$ to thin spectral line width $\Delta\lambda_0$ and SA, the following equation can be used:

$$\frac{I(\lambda)}{I(\lambda_0)} = \left(\frac{1 - e^{-K\frac{\Delta\lambda_0}{4(\lambda-\lambda_0)^2 + \Delta\lambda_0^2}}}{1 - e^{-\frac{K}{\Delta\lambda_0}}} \right). \tag{17}$$

By numerically solving the above equation for $\Delta\lambda$ and taking into account the definition of FWHM as $\lambda = \lambda_0 \pm \Delta\lambda / 2$, the intensity of $I(\lambda)$ equals to $I(\lambda_0)/2$. Hence, the exact equation between the measured spectral width ($\Delta\lambda$) and related non-self-absorbed line width ($\Delta\lambda_0$) can be evaluated. Then, after appropriate calculation, provided that $\Delta\lambda$ and n_e are measured from the experiment and w_s magnitude is inserted from the literature, the SA coefficient can be obtained as:

$$SA = (\frac{\Delta\lambda}{2\omega}\frac{1}{n_e})^{1/\alpha}. \tag{18}$$

In the above equation, n_e can be measured from the non-self-absorbed spectral line of hydrogen H_α at 656.27 nm. For evaluation of the mentioned method, the experiment is performed on several Al spectral lines radiated from pure aluminum (99.9%) samples. The experiments are done with different equipment, one, at the Physics Department of Cairo University (Egypt) and another, at the Applied Laser Spectroscopy Laboratory in Pisa (Italy). At Cairo University, the experiment is performed with using a single pulse Nd:YAG laser with 160 mJ laser energy, 6 ns pulse duration, and 1064 nm laser wavelength. At Pisa Laboratory, the measurement is done utilizing a mobile double-pulse laser with 8 ns FWHM and laser energy of 80 + 80 mJ with a 2 µs delay between the pulses in collinear configuration.

In Figure 2, the temporal evolutions of the self-absorption coefficients SA for three spectral lines of Al I at 394.4 nm, Al II line at 281.6 nm, and Al II line at 466.3 nm are shown. In this figure, it is seen that the Al I spectral line at 394.4 nm exposes to a higher self-absorption (in the spectra taken at Cairo) compared to the other two cases. This is probably because of the higher electron density (produced by the higher laser energy) so that, based on the Saha equation, it provides a larger amount of neutral atoms in the plasma. Moreover, it is clearly observed that the ionized aluminum lines illustrate a low to moderate self-absorption at later delay times, but they are approximately optically thin for delay times lower than 3 µs. Furthermore, the increase of plasma optical thickness at longer delay times is proved for the Al I spectral line at 394.4 nm and is likely because of the plasma plume cooling which induces a growth in the population of the atomic and ionic lower energy levels.

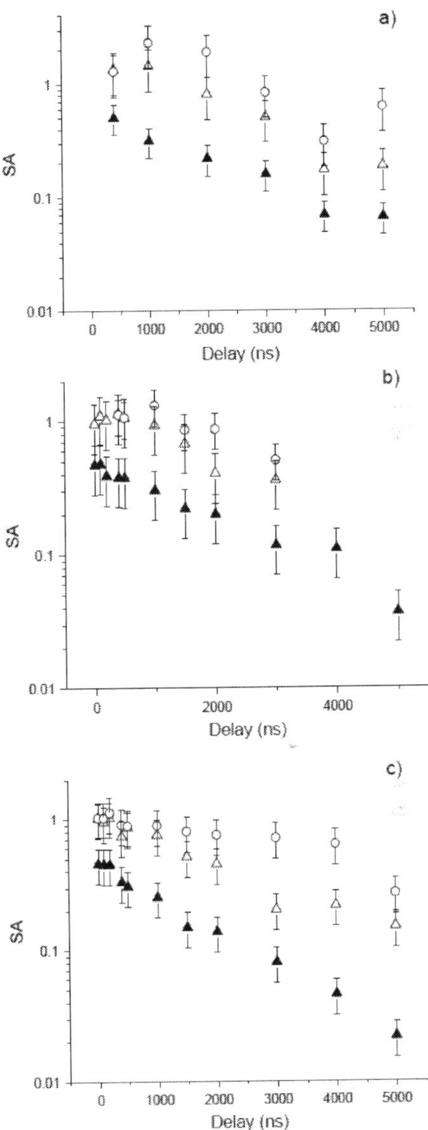

Figure 2. Temporal evolution of self-absorption coefficient SA for Al I spectral line at 394.4 nm (solid triangles), Al II line at 281.6 nm (open triangles), and Al II line at 466.3 nm (open circles). Measurements performed at Cairo University by impact of a single pulse with laser energy of 760 mJ.

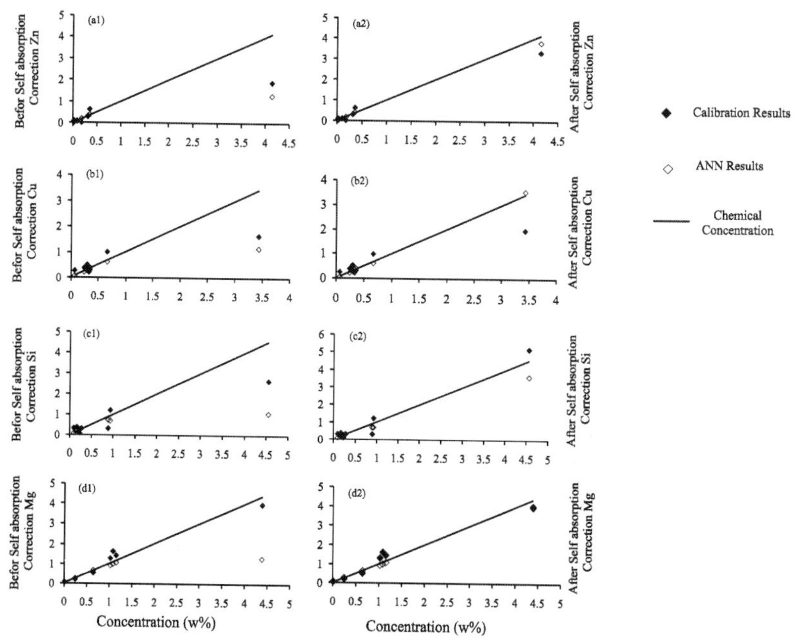

Figure 3. Comparison between the results of ANN and calibration methods with the chemical concentrations (1) without self-absorption correction and (2) with self-absorption correction for the elements of (a) Zn 334.40 nm, (b) Cu 324.40 nm, (c) Si 288.16 nm, and (d) Mg 285.21 nm.

By utilizing the above simple equation, Rezaei et al. [23] corrected the aluminum intensities and then, they predicted the known concentrations in the standard samples with two calibration curves and artificial neural network (ANN) to compare the accuracies of these methods. They used laser-induced breakdown spectroscopy (LIBS) technique for concentration predictions of six elements: Mn, Si, Cu, Fe, Zn, and Mg in seven Al samples. Then, the calibration curve and ANN techniques acquired by six samples are applied for prediction of the elements concentration of the seventh standard sample. In this experiment, a Q-switched Nd:YAG laser at 1,064 nm with a repetition rate of 10 Hz, diagonal output beam of 2.3 mm, and pulse width of 8 ns is focoused on samples. Laser pulse operating at TEM00 mode is adjusted for 50 mJ. The spectra are recorded with an ICCD with exposure time of 1 s and gate width of 5 μs. A comparison between two prediction methods of ANN and calibration curve with their real concentrations in standard samples for four elements of Zn, Cu, Mg, and Si is shown in Figure 3. As it can clearly be seen, at high concentrations, a considerable deviation from real data appeared in Figures 3(a1–d1) in the cases before correcting the self-absorption effect, while when taking into account the self-absorption effect, ANN prediction improves very much in Figures 3(a2–d2). As it is expected, the predictions of the farthest right points on Figure 3, i.e. before considering that self-absorption is not very reliable. Since, these data are a bit greater than the concentration ranges of the training analysis. Results indicate that after self-absorption correction, and at high concentrations, except for Si, the ANN method illustrates more accurate results with a lower relative error compared to the calibration curve method. These results express that the ANN approach is better than the traditional calibration technique for concentration prediction after self-absorption correction, so that the predicted intensities with ANN are nearer to the real emission spectra. The reason for this fact is that ANN obeys a nonlinear behavior at training stage. Moreover, the used constraint on ANN at the training step causes considerable improvement in its accuracy, while the calibration curve follows a linear function which induces higher errors in its predictions.

4.3. Curve of growth

Curve of growth (COG) method makes a relation between the emission intensity and the optical depth. First, this technique was applied for light sources of resonance vapor lamps [24] and flames [25, 26]. Then, Gornushkin et al. [27] applied a COG method for laser-induced plasma spectroscopy. Recently, Aragon and Aguilera represented several effects of different parameters such as variations of optical depth [11], plasma inhomogeneity [10], and delay time [12] on evolutions of COG curves. They fitted the theoretical COG equations to the experimental results and then, extracted plasma parameters, such as number density of neutral emitting atoms and damping constant. Moreover, they utilized the COG curves for estimation of the magnitude of self-absorption parameter and for the evaluation of the concentration at which transition from thin to thick plasma happens. They proposed that the integrated intensity of a spectral line ($W.m^{-2}.sr^{-1}$) in an optically thick plasma can be calculated from [11]:

$$I = I_P(v_0)\int_0^\infty (1 - e^{-\tau(v)})dv, \tag{19}$$

where, v_0 is the central frequency (Hz) of the spectral line, $I_P(v_0)$ is the Planck blackbody distribution ($W.m^{-2}.Sr^{-1}.Hz^{-1}$), and $\tau(v)$ is the optical depth which in a homogeneous plasma in LTE condition can be expressed as:

$$\tau(v) = k'(v)l = \frac{e^2}{4\varepsilon_0 mc} Nf_{ij}l \frac{g_i e^{-\frac{E_i}{K_B T}}}{Z(T)} \times \left(1 - e^{-\frac{E_j - E_i}{K_B T}}\right) L(v), \tag{20}$$

$k'(v)$ is the effective absorption coefficient (m^{-1}), which includes the contribution of induced emission and absorption, l is the plasma length (m), f_{ij} is the transition oscillator strength (dimensionless), and $L(v)$ is the normalized Voigt line profile (Hz^{-1}) as follows:

$$L(\lambda) = \frac{2(\ln 2/\pi)^{1/2}}{\Delta v_D} \frac{a}{\pi} \int_{-\infty}^{+\infty} \frac{e^{-t^2} dt}{(y-t)^2 + a^2}. \tag{21}$$

In Eq. (21), the dimensionless parameters of y and a are defined as

$$a = \frac{(\Delta v_N + \Delta v_L)}{\Delta v_D}\sqrt{\ln 2} \approx \frac{\Delta v_L}{\Delta v_D}\sqrt{\ln 2}. \tag{22}$$

$$y = \frac{2(v - v_0)}{\Delta v_D}\sqrt{\ln 2}, \tag{23}$$

where, Δv_D Δv_N and Δv_L are are the Doppler, natural and Lorentzian line widths (Hz), respectively. It should be noted that the optical depth of a spectral line depends on the multiplication of $Nf_{ij}l$. The relation between the line intensity I and $Nf_{ij}l$ is expressed by Eq. (19) and is called a curve of growth equation. The main problem in utilizing the LIBS technique for analysis is the complex relation between number densities of emitting species N and the concentration in the sample x. In this study, it is assumed that the matrix effects are negligible, so that N is proportional to x by the following equation:

$$N = N'\frac{x}{100}. \tag{24}$$

In the above equation, N indicates the number densities of emitting elements (m^{-3}) in the plasma for the sample which contains 100% concentration. By inserting N in Eq. (24) to Eq. (20), the optical depth can be calculated versus wavelength λ(m) as:

$$\tau(\lambda) = 10^{-2}k_t N'xlL(\lambda). \tag{25}$$

The coefficient k_t, which is dependent to the transition parameters, can be calculated by determination of the plasma temperature as

$$k_t = \frac{e^2\lambda_0^2}{4\varepsilon_0 mc^2}f_{ij}\frac{g_i e^{-\frac{E_i}{K_B T}}}{Z(T)}\left(1 - e^{-\frac{E_j - E_i}{K_B T}}\right). \tag{26}$$

In the above equation, λ_0 is the central wavelength (m), c is the light speed in vacuum (m.s^{-1}), e is the electron charge (C), g_i is degeneracy of lower level, m is electron mass (kg), f_{ij} is the transition oscillator strength (dimensionless), K_B is Boltzmann constant (J.K^{-1}), $Z(T)$ is partition function (dimensionless), ε_0 is is the permittivity of free space (F.m^{-1}), E_i and E_j are energies of lower and higher levels, respectively. By applying Eq. (26) for the optical depth, the spectral line intensity can be obtained versus the concentration of the emitting element in the sample as below:

$$I(x) = I_p(\lambda_0)\int_0^\infty \left(1 - e^{-AxL(\lambda)}\right)d\lambda. \tag{27}$$

A parameter in Eq. (27) is the plasma perpendicular radiating area (m^2), which is defined by

$$A = 10^{-2}k_tN'l. \tag{28}$$

The asymptotic behavior of the COG in low and high concentrations can be obtained by

$$I_{low} \cong I_p Ax. \tag{29}$$

and

$$I_{high} \cong I_p\left[2A(\Delta\lambda_L)\right]^{1/2}x^{1/2}. \tag{30}$$

In double logarithmic scale, by considering the entire concentration range, a linear equation can be depicted. The intersection point of the asymptotes shows the following abscissa:

$$x_{int} = \frac{2(\Delta\lambda_L)}{A}. \tag{31}$$

This point determines the transition limit from thin to thick plasma and starting of the saturation of the spectral line. For instance, the measured spectral intensities of two neutral Fe lines are depicted versus concentration for the Fe-Ni samples. Here, the experiment is performed by focusing of a Q-switched Nd:YAG laser with 1064 nm wavelength and 4.5 ns line width at 20 Hz repetition rate on Fe-Ni alloy samples in atmospheric air. The pulse energy is adjusted for 100 mJ by utilizing of an optical attenuator. Moreover, the plasma emission is collected with temporal resolution, using a delay time of 5 μs from the laser pulse and a gate width of 1 μs. For comparison, the theoretical results are shown in this figure, too. These curves are illustrated in linear and double-logarithmic scales.

As it is seen in Figure 4 the atomic line at 375.82 nm, which contains a higher value of the k_t coefficient rather than the spectral line at 379.50 nm, exposes to a more intense self-absorption in the LIP experiment, which appears as a nonlinear COG curve at lower concentrations. Then, this group developed COG curves by introducing the CSigma graph [2], which comprises different lines of various elements in similar ionization states for LIBS technique. The method is based on the Saha, Boltzmann, and radiative transfer equations for plasmas in LTE condition. Cσ graphs rely on the evaluation of cross-section of a line, i.e. σ_l for each of the experimental

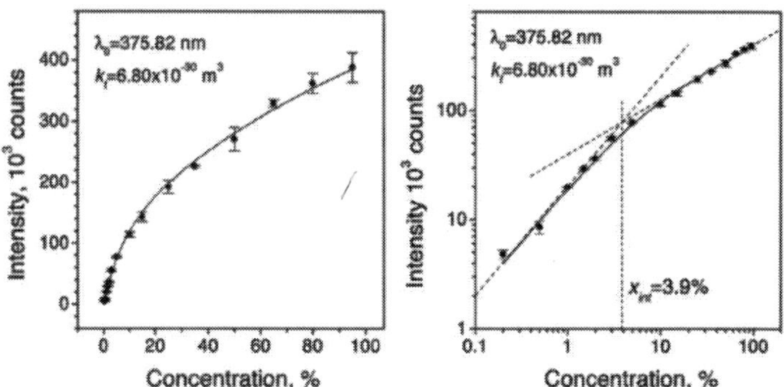

Figure 4. Theoretical and experimental COG for Fe I spectral line on a linear form in left and a double-logarithmic scale on right side. The magnitudes of k_i and x_{int} are indicated in this figure.

results, by knowing the electron density, temperature, and the line atomic data. Then, they fitted the experimental Cσ graphs to calculated curves and four parameters of βA (β is the system instrumental factor equal to the multiplication of the spectral efficiency by the solid angle of detection and A is the transverse area of the plasma region in which the emission is detected), Nl (columnar density), T, N_e could be determined for characterization of LIBS plasma for different ionic and neutral species. The details of the mathematical calculations are expressed in ref. [2].

4.4. Calibration free

Bulajic et al. [14] devised an algorithm for self-absorption correction which was first utilized for three different certified steel NIST samples and for three ternary alloys (Au, Ag, Cu) with known concentrations. Then, it was suggested as a tool for automatic correction of different standardless materials by laser-induced breakdown spectroscopy by using a calibration-free algorithm. The results illustrated that the self-absorption corrected calibration-free method presents reliable conclusions and improves the accuracy by nearly one order of magnitude. The main advantages of applying calibration curve is minimizing the matrix effect, which induces errors in precise evaluation of plasma parameter.

In this work, for each value of SA, 30 different samples are generated. Each synthetical line is appropriately fitted with the analytical software, which yields an estimation value for the parameters $\Delta\lambda_L$ and $\Delta\lambda_G$. In Figure 5, statistical results of $\Delta\lambda_L$ for different values of SA are reported, which explains how, for self-absorbed spectral lines, the experimental Lorentzian width deviates from the 'real' magnitude. Furthermore, by begining from the measured Lorentzian width, it is not feasible to obtain the true value because of the distortion of Voigt profile and the dispersion of the profile of the calculated line widths. It is found more reliable methods for acquiring the true spectral Lorentzian width, by starting from the total line width (i.e. Gaussian plus Lorentzian), which is proved to depend on the SA parameter based on the following expression (see Demtroder [28]):

$$\Delta\lambda_{obs} = \frac{\Delta\lambda_{true}}{\sqrt{SA}} \approx \frac{\sqrt{\Delta\lambda_L^2 + \Delta\lambda_G^2}}{\sqrt{SA}}. \tag{32}$$

Actually, a very good agreement is seen between the values of the total broadening quantities calculated according to the fit of the simulated self-absorbed profiles and those calculated by Eq. (32), when the magnitude of $\Delta\lambda_{true}$ is known. Hence, by assumption of knowing the self-absorption coefficient SA, the total true width can easily be found by utilizing Eq. (32), and then, the contribution of Doppler broadening $\Delta\lambda_G$ and instrumental broadening $\Delta\lambda_L$ will be obtained.

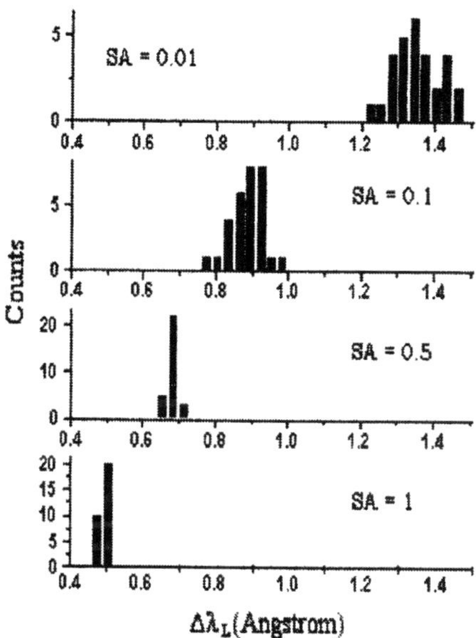

Figure 5. Statistical analysis of Lorentzian width magnitudes attained by fitting with simulated lines with $\Delta\lambda_L$ and $\Delta\lambda_G$ fixed to 0.5 and 0.1 A, respectively, for spectral line of Cu at 324.7 nm, and SA are 1, 0.5, 0.1, and 0.01 from the bottom.

Moreover, Figure 6 illustrates the simulation result of the copper line profile at 324.7 nm at different values of self-absorption for comparing with precious alloy sample containing 40% of Cu. Here, the simulation is performed by curve of growth method as mentioned before. In this figure, the 'flat-top' shape, which is a representation of the self-absorption effect, is clearly seen. Since the peak height reduction can't be well recognized as self-absorption effect, an algorithm is developed for automatic evaluation of self-absorption phenomenon. The performance of self-absorption correction is presented into the CF-LIBS procedure with a recursive algorithm, which is shown by a diagram in mentioned reference.

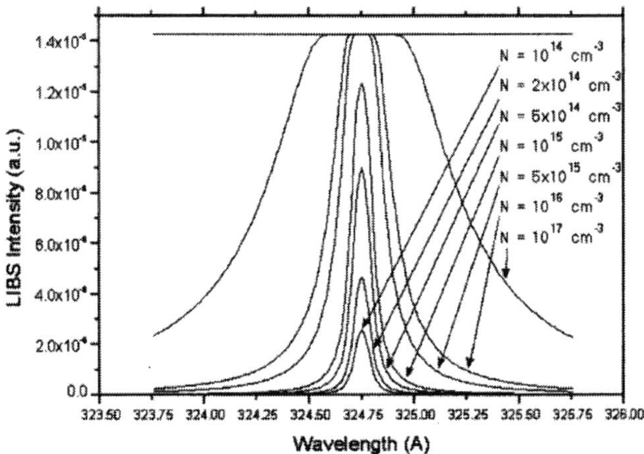

Figure 6. Simulation of Cu spectral line profile at 324.7 nm at different degrees of self-absorption.

4.5. Internal reference method

Sun and Yu [17] introduced a simplified procedure for correction of self-absorption by calibration-free laser-induced breakdown spectroscopy technique (CF-IBS). They utilized an internal reference line for each species. Then, they made a comparison between this spectral line with the other line intensity from the same species of the reference line to evaluate the self-absorption magnitude of the other spectral lines.

They started their method, i.e. internal standard self-absorption correction (IRSAC) by assumption of the plasma being in LTE condition, they estimated the measured integral line intensity as below:

$$I_\lambda^{ij} = f_\lambda^b F C_s A_{ij} \frac{g_i}{Z_s(T)} e^{-E_i/k_B T}, \tag{33}$$

where, A_{ij}, K_B, C_s, g_i, and λ are the transition probability, Boltzmann constant, the total densities of the emitting species s, the degeneracy of upper level i and wavelength of the transition, respectively. f_λ^b is self-absorption coefficient (the same SA) at wavelength λ, which has a magnitude between 0 and 1. F is a constant which includes the optical efficiency of the collection system, the total density of plasma and its volume. $Z_s(T)$ is the partition function of the analyzed spectral line. Here, the transition parameters of g_i, A_{ij} and E_i are inserted from spectral databases, and the magnitudes of F, C_s and T are extracted from the experimental data. According to the calibration-free method, the concentration of all the elements in the sample can be calculated by

$$C_s = \frac{1}{F} U_s(T) e^{q_s}. \tag{34}$$

where, $q_s = \ln \frac{C_s F}{U_s(T)}$. The self-absorption coefficient can be estimated by considering the ratio of the other emission intensities to an internal reference line for each species as

$$\frac{f_\lambda^b}{f_{\lambda_R}^b} = \frac{I_\lambda^{ij}}{I_{\lambda_R}^{mn}} \frac{A_{mn} g_m}{A_{ij} g_i} e^{-E_m - E_i / k_B T}, \tag{35}$$

here, $I_{\lambda_R}^{mn}$, λ_R, and $f_{\lambda_R}^b$ are spectral line intensity, wavelength, and self-absorption coefficient, respectively, of the mentioned reference line. A_{mn}, E_m, and g_m are the transition parameters related to the atomic levels m and n. It is assumed that the internal reference line has negligible self-absorption so that $f_{\lambda_R}^b \approx 1$. Hence, the self-absorption coefficient of other spectral lines are calculated by the following equation:

$$f_\lambda^b = \frac{I_\lambda^{ij} A_{mn} g_m}{I_{\lambda_R}^{mn} A_{ij} g_i} e^{-E_m - E_i / k_B T}. \tag{36}$$

Finally, the corrected line intensities without any self-absorption can be evaluated from the signal ratio of the measured spectral line intensity to the self-absorption coefficient as

$$\hat{I}_\lambda^{ij} = \frac{I_\lambda^{ij}}{f_\lambda^b} = \frac{I_{\lambda_R}^{mn} A_{ij} g_i}{A_{mn} g_m} e^{-E_m - E_i / k_B T}. \tag{37}$$

By utilizing Eq. (37), the self-absorption correction for each spectral line can be estimated. When these corrections are inserted to every point in the Boltzmann plot, the scattering of the points from each species around the best fit line will be decreased, and then more accurate quantitative results will be attained.

The outline of the IRSAC model is shown in this reference. Based on Eq. (36), for estimation of self-absorption coefficients, the plasma temperature must be known, which is initially calculated from Boltzmann plot curve without requiring any correction. The corrected line intensities by the previous temperature are utilized for evaluation of a new temperature. Since the magnitude of self-absorption straightly depends on the plasma temperature, the optimal self-absorption coefficients can be calculated by an iterative procedure up to attain a convergence in the correlation coefficients on the Boltzmann plot. After convergence of the correlation coefficients, the corrected points will be placed approximately on parallel linear Boltzmann plot and the calculated temperatures from two successive iterations will alter a little. For more illustrations, a schematic of the Boltzmann plot for neutral atoms and singly ionized element is shown in Figure 7 before correction by the basic CF-LIBS method and IRSAC model.

In this experiment, a Q-switched Nd:YAG laser with pulse duration of 10 ns, laser energy of 200 mJ, wavelength of 1064 nm, and repetition rate of 1–15 Hz is focused on an aluminum alloy sample. For this sample, the acquisition delay time and integration gate width are adjusted for 2.5 μs and 1 ms, respectively.

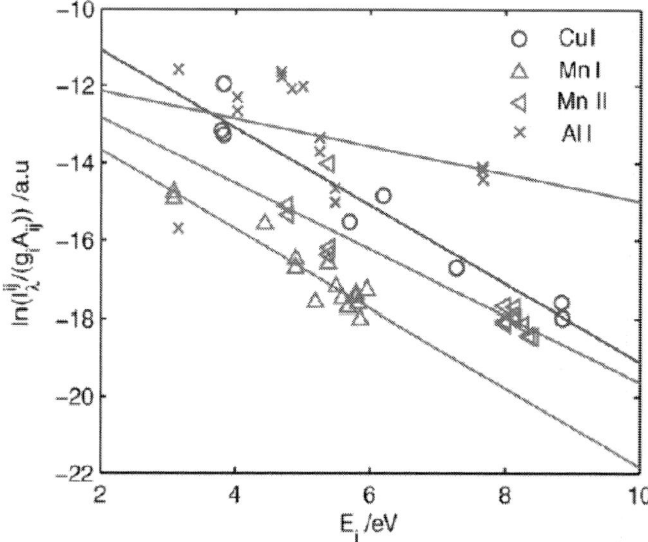

Figure 7. Initial Boltzmann plot which is depicted from spectral intensities of the alloys of aluminium samples.

4.6. Duplicating mirror

Moon et al. [8] duplicated the emission from a plasma by putting a spherical mirror at twice the distance of its focal length from the plasma. They evaluated the existence of optically thick plasma conditions by a very quick check-up. They acquired two line profiles (with and without applying the mirror) for determination of the amount of self-absorption in order to correct the spectral lines.

By taking into account the theoretical consideration, they evaluated the emission from a spatially homogeneous plasma distribution by the presence and absence of a mirror as below:

$$B_{\lambda,1} = B_\lambda^{bb}(1 - \exp[-k_\lambda l]). \tag{38}$$

$$B_{\lambda,2} = B_{\lambda,1} + GB_{\lambda,1}\exp(-k_\lambda l) = B_{\lambda,1}\left[1 + G\exp(-k_\lambda l)\right]. \tag{39}$$

In the above equation, the indexes 1 and 2 refer to the measurements with and without inserting the mirror, respectively. B_λ^{bb} indicates the spectral radiance of black body emission by the Wien or the Planck laws. In this calculation, G comprises both reflection and absorption losses of the mirror and in addition, solid angles produce imperfect matching. G can be computed by taking the ratio of signal intensity of continuous radiation, where $k_\lambda = 0$, i.e. at the line wings with negligible absorption as

$$R^c = \frac{B_{\lambda,2}^c}{B_{\lambda,1}^c} = 1 + G. \tag{40}$$

Furthermore, the ratio of peak intensities with and without considering mirror can be expressed as

$$R_\lambda = \frac{B_{\lambda,2}}{B_{\lambda,1}} = 1 + G\exp(-k_\lambda l). \tag{41}$$

The optical depth variations can also be evaluated by knowing the parameters of R^c and R_λ as:

$$\tau = k_\lambda l = \ln\frac{\left[R^c - 1\right]}{\left[R_\lambda - 1\right]}. \tag{42}$$

Finally, the correction factor $K_{\lambda,corr}$ (which is the inverse of the self-absorption coefficient SA) can be calculated experimentally versus the ratios of R^C and R^λ as:

$$K_{\lambda,corr} = \ln\frac{\left[(R^c - 1)/(R_\lambda - 1)\right]}{1 - \dfrac{(R_\lambda - 1)}{(R^c - 1)}}. \tag{43}$$

Furthermore, the duplication factor can be calculated from the following equation as

$$D_\lambda(\lambda) = \frac{B_\lambda(\lambda, 2fn_a l) - B_\lambda(\lambda, fn_a l)}{B_\lambda(\lambda, fn_a l)}, \tag{44}$$

where, $D_\lambda(\lambda)$ illustrates the relative growth in spectral line intensity or integral absorption created by doubling the value of $(fn_a l)$ with two asymptotic magnitudes of 1 (at low optical depths) and 0.415 (at high optical depths). Here, f, n_a and l refer to oscillation strength, number density, and plasma length, respectively. In this work, the plasma emission is produced by irradiation of a Nd:YAG laser with 90 ± 5 mJ pulse energy, 1064 nm laser wavelength, 6 ns pulse duration, and 1 Hz repetition frequency on the sample surface. The ICCD is adjusted for a delay time of 1 μs. Figure 8 expresses the results for analysis of Cu spectral line at 510.55 nm. Figures 8(a) and 8(b) show the evolution of R_λ and $K_{\lambda,corr}$ as a function of wavelength. As it is seen in Figure 8(a), the peak of spectral line saturates faster rather than that in line wings by doubling the plasma length. For obtaining the exact value of the correction factor $K_{\lambda,corr}$, the ratio of R_λ is estimated pixel by pixel. Furthermore, Figure 9 illustrates the evolution of calculated self-absorption correction factor $K_{\lambda,\,corr}$ as a function of R_λ and R^c. As it is seen in this figure, for the experimental R^c data, and by approaching R_λ to unity, the spectral lines are severely self-absorbed by including high values for correction factor $K_{\lambda,\,corr}$.

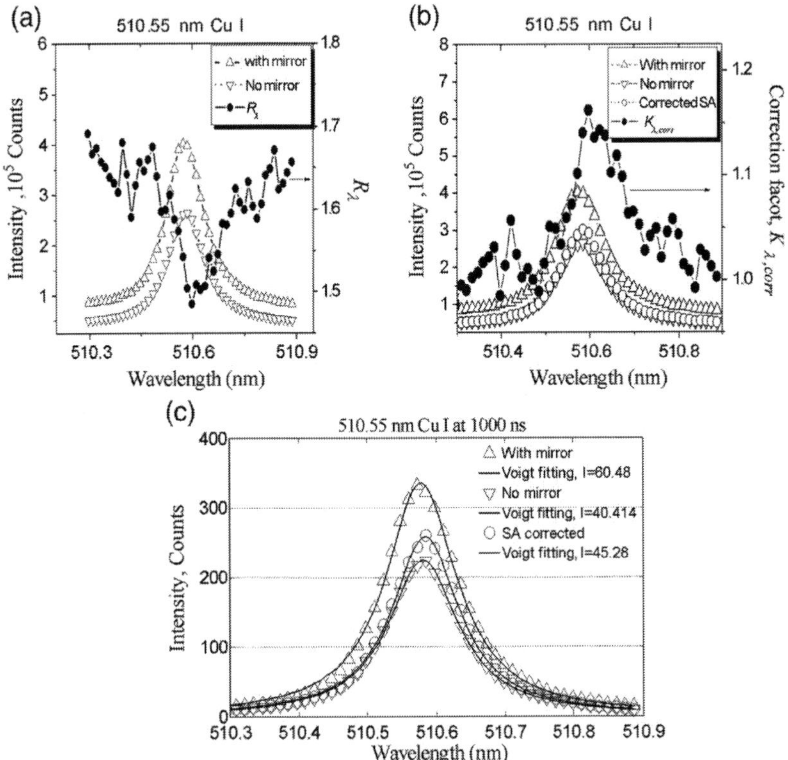

Figure 8. (a) Calculated data of R_λ versus wavelength, (b) evolution of correction factor $K_{\lambda,corr}$ versus line profiles corrected by self-absorption effect and wavelength. All curves are related to the spectral line Cu I at 510.55 nm, and at 1.0 μs delay time.

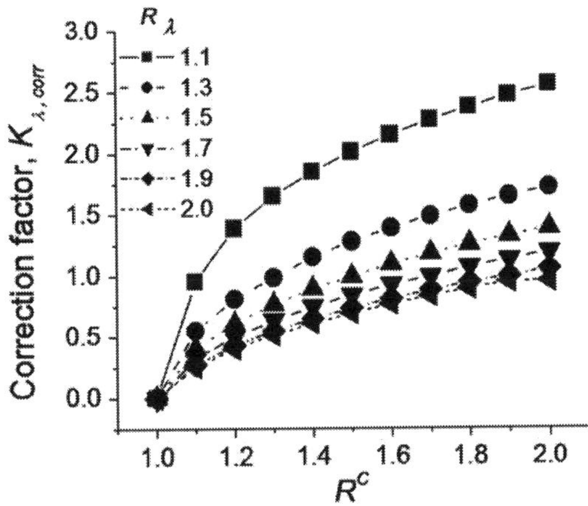

Figure 9. The dependence of the calculated correction factor $K_{\lambda,corr}$ to R_C, for different amounts of R_λ, or different levels of self-absorption.

4.7. Three lines method

Rezaei and Tavassoli [18] introduced the three lines method for studying optically thick plasma in local thermodynamic equilibrium by LIBS method without needing any spectral correction. They performed a LIBS experiment on an aluminum target in air atmosphere by utilizing the two techniques of spectroscopy and shadowgraphy, as well as by applying a theoretical approach.

In this study, plasma parameters were accurately determined by obtaining the plasma length, electron density, and intensities of three spectral lines from experiments. The model explains that instead of utilizing two spectral lines in thin plasmas, three lines are needed in thick plasmas for accurate evaluation of plasma temperature.

The thick plasma emission per unit volume, per unit time, and per unit frequency can be evaluated by replacing the multiplication of self-absorption coefficient by thin plasma intensity as:

$$I_{thick} = SA \times I_{thin} = \frac{(1 - e^{-kl})}{(1 - e^{-\frac{hc}{\lambda_0}\frac{1}{k_BT}})} \frac{C8\pi}{\lambda_0^2} e^{-\frac{(E_u - E_l)}{k_BT}} \ (SI). \tag{45}$$

The parameters inserted in the above equation were introduced in previously mentioned equations. Here, Z is the partition function, which is computed by two and three level methods [29]. C is instrumental function which includes the efficiency and the solid angle of the detection system [30]. Here, k (m⁻¹) is the absorption coefficient which contains both contribution of the stimulated emission of upper level and the absorption of lower level in the SI units such as in the following equation:

$$k = \frac{g_u A_{ul} N_{Al} \lambda_0^2}{8\pi Z} e^{-\frac{E_l}{k_B T}} (1 - e^{-\frac{h\nu_0}{k_B T}}) L(\nu, \nu_0, \gamma_{ul}). \tag{46}$$

$L(\nu, \nu_0, \gamma_{ul})$ is the Lorentzian line profile because of Stark broadening mechanism. γ_{ul} is the decay rate with a straight dependence to the line width as below [31]:

$$\gamma_{ul} = 2\pi(c / \lambda_0^2)\Delta\lambda_{stark}. \tag{47}$$

$\Delta\lambda_{stark}$ is full-width at half-maximum (FWHM) of the spectral line, which is produced due to Stark effect and is defined as follows [32]:

$$\Delta\lambda_{stark} = \frac{2\omega n_e}{n_{ref}}, \tag{48}$$

here, ω and n_e are electron impact parameter and electron number density, respectively. n_{ref} is the reference electron density (here, 10^{16} cm^{-3}) in which ω is calculated.

As mentioned in Eq. (45), for the spatially integrated plasma emission, the ratio of selected spectral lines can be written as below:

$$\frac{I_{thick\_1}}{I_{thick\_2}} - \frac{(1-e^{-k_1 l})}{(1-e^{-k_2 l})} \times \frac{(1-e^{-\frac{hc}{\lambda_{02}}\frac{1}{k_B T}})}{(1-e^{-\frac{hc}{\lambda_{01}}\frac{1}{k_B T}})} \times \frac{\lambda_{02}^2}{\lambda_{01}^2} \times \frac{e^{\frac{1}{k_B T}(E_{l1}-E_{u1})}}{e^{\frac{1}{k_B T}(E_{l2}-E_{u2})}} = 0. \tag{49}$$

$$\frac{I_{thick\_2}}{I_{thick\_3}} - \frac{(1-e^{-k_2 l})}{(1-e^{-k_3 l})} \times \frac{(1-e^{-\frac{hc}{\lambda_{03}}\frac{1}{k_B T}})}{(1-e^{-\frac{hc}{\lambda_{02}}\frac{1}{k_B T}})} \times \frac{\lambda_{03}^2}{\lambda_{02}^2} \times \frac{e^{\frac{1}{k_B T}(E_{l2}-E_{u2})}}{e^{\frac{1}{k_B T}(E_{l3}-E_{u3})}} = 0. \tag{50}$$

It should be noted that both of the above equations satisfy at particular T and N_{Al}. Consequently, the cross of two equations (i.e. left-hand side of equations) with the contour of zero (i.e. right-hand side of equations) is the answer. Therefore, by depicting the contour plot of the above equations and crossing them, the passive parameters of T and N_{Al} is gotten at a particular delay time and at specific laser energy. Then, by inserting these data (answer) in one of the above equations, the instrumental function C will be obtained.

4.8. Line ratio analysis

Bredice et al. [33] utilized the theoretical treatment companioned by experimental results for estimation of the amount of self-absorption in single and collinear double pulse configuration. They used the two-line ratio analysis of the same species of manganese element in Fe–Mn alloys to characterize the homogenous plasma parameters. Moreover, they calculated the self-absorption coefficient by considering the line ratios in different conditions: two lines reveal weak or no self-absorption, two lines experience strong self-absorption, two lines belong to the same multiplet, and two lines are general cases as well. Their results are summarized as the following:

1.   Limit case: two negligible self-absorbed lines:

When two lines are weakly self-absorbed, i.e. $(SA)_1 = (SA)_2 = 1$ (or $\kappa_{1,2}(\lambda_0)l << 1$), the ratio of the two spectral intensities from the similar species can be expressed as

$$\frac{(N_p)_2}{(N_p)_1} \approx \frac{(A_{ki}g_k e^{-\frac{E_k}{k_B T}})_2}{(A_{ki}g_k e^{-\frac{E_k}{k_B T}})_1}. \qquad \kappa_{1,2}(\lambda_0)l << 1 \qquad (51)$$

It should be noted that in this calculation, N_p indicates the integral intensity of the selected spectral line and also, other parameters are introduced before.

2.   Limit case: severely self-absorbed spectral lines:

When two spectral lines are exposed to strong self-absorption, the self-absorption coefficients can be calculated as:

$$SA \approx \frac{1}{\kappa(\lambda_0)l}. \qquad (52)$$

Therefore, the spectral ratios of spectral emissions of the same species, both exposed to strong self-absorption, can be calculated by

$$\frac{(N_p)_2}{(N_p)_1} \approx \left[\frac{(A_{ki}g_k e^{-\frac{E_k}{k_B T}})_2}{(A_{ki}g_k e^{-\frac{E_k}{k_B T}})_1}\right] \cdot \left[\frac{\left(\frac{\lambda_0^4 A_{ki}g_k e^{-\frac{E_i}{k_B T}}}{\Delta\lambda_0}\right)_1}{\left(\frac{\lambda_0^4 A_{ki}g_k e^{-\frac{E_i}{k_B T}}}{\Delta\lambda_0}\right)_2}\right]^\beta \qquad \kappa_{1,2}(\lambda_0)l >> 1 \qquad (53)$$

3. Limit case: two lines related to the same multiplet

In two spectral lines belonging to the same multiplet, the atomic energies of E_i and E_k are similar and it is obtain an intensity ratio that is constant with plasma temperature variation. In this regime, by considering the above situations, the intensity ratio can be located between the two extremes as

$$\frac{(N_p)_2}{(N_p)_1} = \left[\frac{(A_{ki}g_k)_2}{(A_{ki}g_k)_1}\right]^{(1-\beta)}\left[\frac{(\lambda_0^4)_1}{(\lambda_0^4)_2}\right]^{\beta} \qquad \kappa_{1,2}(\lambda_0)l \gg 1, \tag{54}$$

$$\frac{(N_p)_2}{(N_p)_1} \approx \frac{(A_{ki}g_k)_2}{(A_{ki}g_k)_1} \qquad \kappa_{1,2}(\lambda_0)l \ll 1 \tag{55}$$

in the case of high and low self-absorption, respectively.

4. General case

For two arbitrary spectral lines, the spectral intensities ratios can be evaluated by

$$\frac{(N_p)_2}{(N_p)_1} = \frac{(A_{ki}g_k e^{-\frac{E_k}{k_BT}})_2}{(A_{ki}g_k e^{-\frac{E_k}{k_BT}})_1}\left(\frac{k(\lambda_0)_1}{k(\lambda_0)_2}\frac{(1-e^{-k(\lambda_0)_2 l})}{(1-e^{-k(\lambda_0)_1 l})}\right)^{\beta}$$

$$= a\left[\frac{b(1-e^{-k(\lambda_0)_2 l})}{(1-e^{bk(\lambda_0)_2 l})}\right]^{\beta}. \tag{56}$$

$$a = \frac{(A_{ki}g_k e^{-\frac{E_k}{k_BT}})_2}{(A_{ki}g_k e^{-\frac{E_k}{k_BT}})_1},$$

$$b = \frac{(\lambda_0^4 A_{ki}g_k e^{-\frac{E_i}{k_BT}})_1}{(\lambda_0^4 A_{ki}g_k e^{-\frac{E_i}{k_BT}})_2}\left(\frac{(\Delta\lambda_0)_2}{(\Delta\lambda_0)_1}\right). \tag{57}$$

In order to find numerically $k(\lambda_0)_2 l$, the intensity ratio of two spectral lines $\frac{(N_p)_2}{(N_p)_1}$, the line broadening ratio $\frac{(\Delta\lambda_0)_2}{(\Delta\lambda_0)_1}$, and plasma temperature T must be determined from the experiment.

By knowing the $k(\lambda_0)l$ for a spectral line, the self-absorption coefficient SA can be carefully calculated.

For instance, in order to show the application of the mentioned theoretical equations, Figure 10 is depicted. In this experiment, Nd-YAG laser pulses at 1064 nm with 7 and 12 ns FWHM and various laser energies are irradiated on Fe-Mn alloys. The experiment is performed at different places in single and double pulse configurations. Figure 10 illustrates the signal ratios of the spectral lines of Mn II at 293.3 to Mn II at 294.9 nm (lines related to the same multiplet) and ratio of Mn II at 293.3 to spectral line of Mn II at 348.3 nm (lines belong to different multiplets) as a function of delay time. Temporal evolution of intensity ratio is plotted as a function of the acquisition delay time in two situations of single and double pulse measurements. In double pulse measurements, the delay time is considered after the arrival of the second pulse on the sample. Since the first two lines, i.e. Mn II lines at 293.3 and 294.9 nm, belong to the similar multiplet, the theoretically mentioned equation is applicable. As shown in the theory, if two limit cases of low or high self-absorption are satisfied, the intensity ratio of these two lines will be particularly independent of the plasma variation.

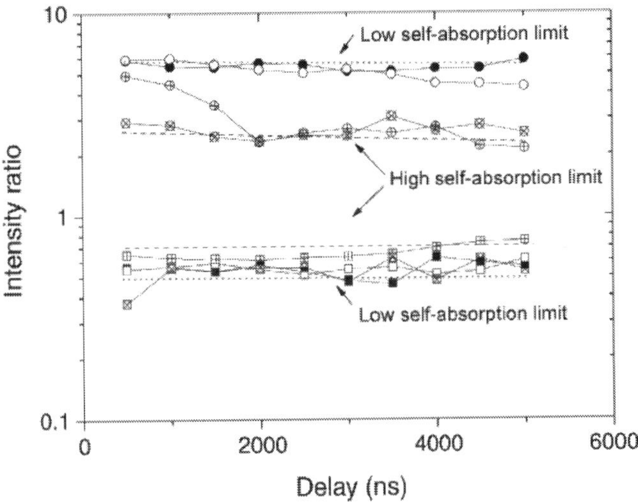

Figure 10. Temporal evolution of intensity ratios of of two spectral lines of Mn II at 293.3 and Mn II at 294.9 nm (full squares, laser energy of 60 mJ, open squares, 120 mJ and crossed squares, 200 mJ laser energy, diagonally crossed squares, 60 + 60 mJ double pulse irradiation) and for Mn II at 293.3/Mn II at 348.3 nm lines (full circles, 60 mJ laser energy, open circles, 120 mJ laser energy, crossed circles, 200 mJ laser energy, diagonally crossed circles, 60 + 60 mJ double pulse measurements).

It should be noted that the dotted line shows the limit of low self-absorption and dashed line indicates the high self-absorption limits. Furthermore, in both lines, y axis is in logarithmic scale. Cristoforetti and Tognoni [19] calculated the concentration ratio of different elements by assumption of holding a homogeneity condition in plasma without needing any self-absorption correction. Furthermore, by obtaining the columnar densities, they computed the plasma temperature and the number densities of different plasma species.

In this work, first, by numerical calculation of the optical depth $k(\lambda_0)l$, the SA parameter is computed by exploiting Eq. (16), as shown in Figure 11. After that, the columnar density n_{il} can be easily extracted by rewriting the equation of optical depth as

$$\left(n_i l\right)_{17} = 1770 \frac{\Delta\lambda_0}{f\lambda_0^2} k(\lambda_0)l,$$

(58)

here, $\Delta\lambda_0$ is the expected line width in the case of thin plasma, $(n_{il})_{17}$ is explained in 10^{17} cm^{-2} units, and λ_0 and $\Delta\lambda_0$ are written in Angstrom units. By knowing the columnar density, the inverse of plasma temperature can be calculated from the slope of the line obtained by fitting the points in the Saha–Boltzmann plot.

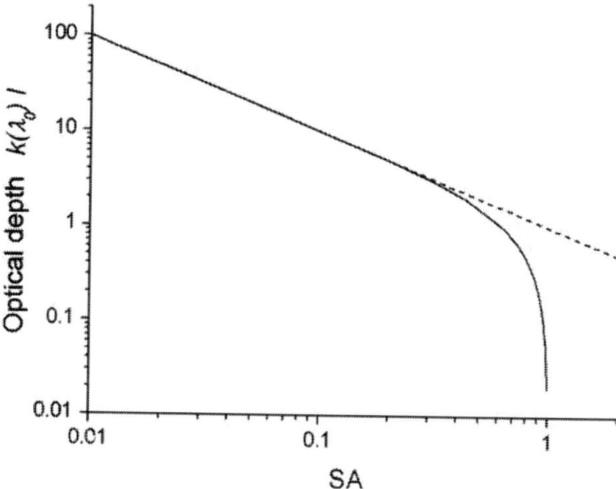

Figure 11. Relation between optical depth and the SA coefficient.

5. Inhomogeneous plasma

Different schematics of nonhomogeneous plasmas including 2, 5, N, and 250 sections in cylindrical and spherical shapes are investigated for thick plasma analysis by considering self-absorption correction as in the figures below:

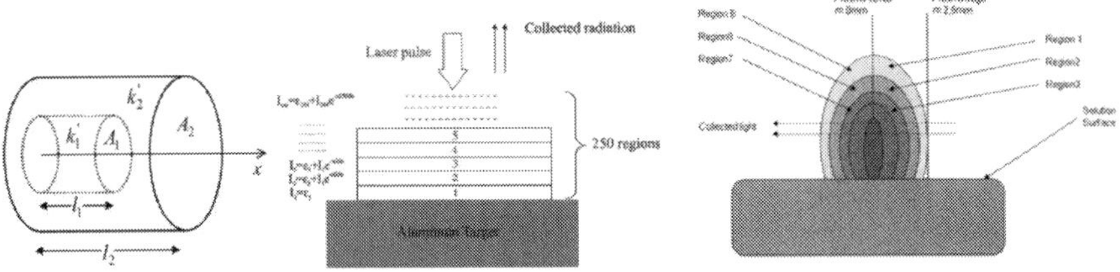

Figure 12. Different schematics of nonhomogeneous thick plasmas [10, 13, 34].

The methods of all of above schematics are approximately similar, so that all of them utilize the emission of internal layers plus attenuation from outer regions. Therefore, the calculations related to 250 layers as well as another numerical model including N layer are explained in the following section.

Rezaei et al. [34] studied the spectral emissions of an aluminum sample located in argon and helium noble gases at 1 atm pressure, by applying a numerical calculation. They computed the plasma parameters by coupling the thermal model of laser ablation, hydrodynamic of plasma expansion, and Saha–Eggert equations. In that model, the spectral emissions were constructed from the superposition of some strong lines of aluminum and several strong lines of ambient gases, which were superimposed on a continuous radiation composed of bremsstrahlung and recombination emissions. Moreover, they calculated the spectral emissions in two cases of thin and thick plasmas by considering the self-absorption influence.

In this work, the plasma is supposed to be consisted of 250 layers with 60-μm thickness (as shown in Figure 12). Each section of this plasma is characterized by specific plasma parameters, such as temperature, electron density, mass densities, and number densities of plasma species as a function of delay time. The plasma radiations of different layers are collected in a parallel direction to the laser pulse. The emission of the first layer quite above the sample surface due to its own radiation can be expressed by:

$$I_{(1)} = \varepsilon_{spec(1)} = \sum_{j=1}^{N_s} N_u^{(1)} A_{ul}^j h\nu_j L_j^{(1)}(\nu,\nu_j,\gamma_{ul}^j).\tag{59}$$

The contribution of the second layer toward the optical collecting system, which comprises both of its radiation and attenuation of the first layer is calculated from:

$$I_{(2)} = \sum_{j=1}^{N_s} N_j^{(2)} A_{ul}^j h\nu_j L_j^{(2)}(\nu,\nu_j,\gamma_{ul}^j) + \varepsilon_{spec(1)} e^{-k_{(2)}dx} = \varepsilon_{spec(2)} + I_{(1)} e^{-k_{(2)}dx}.\tag{60}$$

Consequently, the spectral intensity of the nth layer due to the whole sequential absorption is defined as:

$$I_{(n)} = \sum_{j=1}^{N_s} N_j^{(n)} A_{ul}^j h\nu_j L_j^{(n)} (\nu, \nu_j, \gamma_{ul}^j) + I_{(n-1)} e^{-k_{(n)}dx} = \varepsilon_{spec(n)} + I_{(n-1)} e^{-k_{(n)}dx}. \tag{61}$$

According to the two-level system, the absorption coefficient is expressed by inserting the contributions of the absorption and induced emission between the levels of u and l as

$$k_{(n)} = \frac{h\nu_j}{c}(B_{ul}N_l - B_{lu}N_u)L^{(n)}{}_j(\nu,\nu_j,\gamma_{ul}^j)(\text{SI}), \tag{62}$$

Here, B_{ul} and B_{lu} are Einstein coefficients related to the absorption and induced emission (m³J s), respectively. Finally, the self-absorbed spectral intensity is the emission of last layer (i.e 250th layer) which includes the entire regions shown below for attenuation as follows:

$$I_{selfabsorb} = \varepsilon_{(250)} + I_{(249)} e^{-k_{(250)}dx}. \tag{63}$$

For comparison between thick plasma emissions with a non-self-absorbed spectral line in a thin plasma condition, the summation over all strong line radiations, including the contribution of the whole layers, can be considered as

$$I_{nonselfabsorb}(\nu,\nu_j) = \sum_{k=1}^{k_s} \sum_{j=1}^{Ns} \varepsilon_{spec}^j(\nu,\nu_j), \tag{64}$$

where, k_s and N_s are the number of plasma layers and the number of strong lines, respectively. For instance, the effect of pulse laser energy on Al spectral lines at 394.40 and 396.15 nm is shown in Figure 13 in the logarithmic scale. Here, a Gaussian-shaped laser pulse, with a wavelength of 266 nm, FWHM of 10 ns, under different laser irradiances is focused on aluminum sample. As it is seen by increasing laser intensities, the self-absorption coefficient grows. The magnitudes of self-absorption coefficients in two ambient gases of Ar and He are inserted in this reference.

Furthermore, Ben Ahmed and Cowpe proposed a nonhomogeneous plasma with five layers as shown in Figure 12, and, then, they calculated the total observed intensity by taking into account the i and j regions as:

$$I = \sum_i \left(\frac{I_i}{\tau_i}(1 - exp(-\tau_i))exp(\sum_{j>i}(-\tau_j)) \right). \tag{65}$$

Lazic et al. [16] considered a cylinder with length L including homogenous density and temperature, which is divided into N similar thin layers with length δL surrounded by another

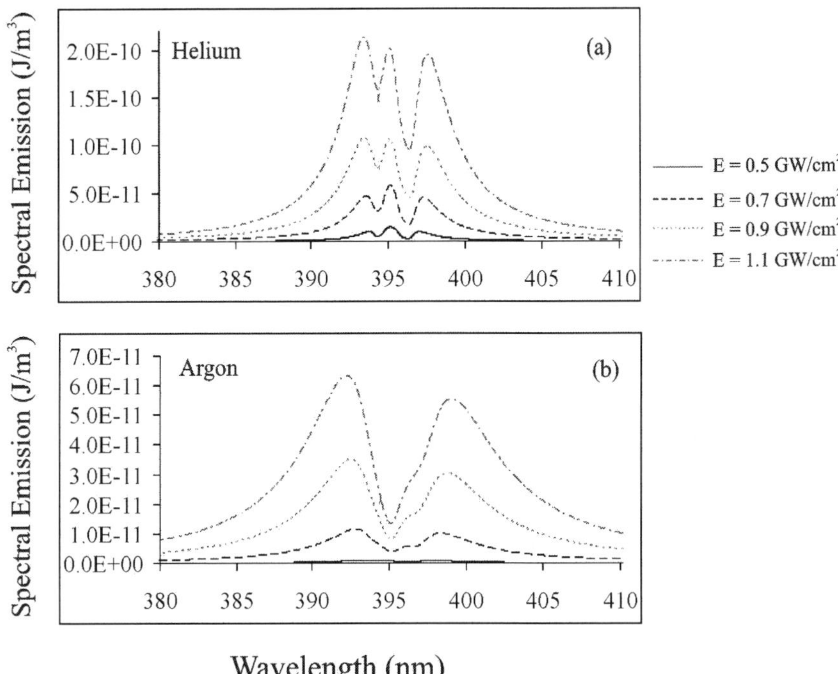

Figure 13. Self-reversal evolutions of the spectral lines of 394.40 and 396.15 nm, at different laser energies of 0.5, 0.7, 0.9, and 1.1 GW/cm² in (a) helium and (b) argon gas.

outer thin layer. Similar to the above-mentioned method, by considering successive absorption and using the seri summation result, they expressed the total line intensity escaping from plasma as

$$I_\alpha^{ki} = FC_\alpha \frac{f_\alpha^{ki}(T)}{U_\alpha(T)} \frac{1-e^{-\beta^{ik}C_\alpha^i L}}{L/\delta L\left(1-e^{-\beta^{ik}C_\alpha^i \delta L}\right)}.$$

(66)

where, C_α is the total species concentration and C_α^i is the concentration of species at lower level i which is related to the C_α by Boltzmann equation. f_α^{ki} comprises the all parameters correspond to ki transition. $U_\alpha(T)$ is partition function and F is a constant attributed to the experimental condition. Then, by assumption of existence of a thin colder plasma surrounding the cylindrical plasma, the overall emission is obtained as below:

$$I_\alpha^{ki} = F_1 C_\alpha \frac{f_\alpha^{ki}(T)}{U_\alpha(T)} \times \frac{1 - e^{-\frac{\beta^{ik} H_\alpha^i C_\alpha L}{U_\alpha(T)}}}{L/\delta L \left(1 - e^{-\frac{\beta^{ik} H_\alpha^i C_\alpha \delta L}{U_\alpha(T)}} \right)} + F_2 C_\alpha \frac{f_\alpha^{ki}(T)}{U_\alpha(T)}.$$

(67)

where, $H_\alpha^i = g_i e^{-\frac{E_i}{K_B T}}$. F_1 and F_2 are constants related to optically thick and thin plasma, respectively, and are straightly dependent on the plasma geometry.

By combination of the above equations, a relation is obtained which connect the raw measured element concentration C_m to effective element concentration C_E as follow:

$$C_m = \frac{a_1}{a_2} \left(1 - e^{-a_2 C_E} \right) + a_3 C_E.$$

(68)

The first term in the above equation is attributed to the optically thick plasma and a_i are coefficients that can be determined by fitting to the experimental results. Figure 14 illustrates the measured spectral emission of Cu at 327.39 nm versus certified concentration for different selected samples. In this experiment, a third harmonic Nd:YAG pulsed laser beam with energies of 6–10 mJ is focused on soil and sediment samples. The acquisition delay time is adjusted for 300 ns and the gate width is 1000 ns for all examined samples.

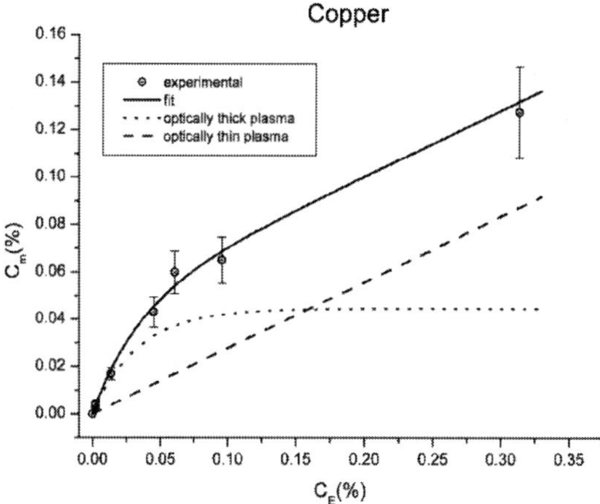

Figure 14. Measured LIBS emission for Cu spectral line at 327.39 nm as a function of certified concentrations.

Some research groups used self-absorbed spectral lines for charactrizing the plasma properties by proposing appropriate models without needing to any self-absorption correction as mentioned in Section (7. IV). In addition, Gornushkin et al. [35] introduced a semiempirical model for an optically thick inhomogeneous plasma in LTE condition. In this model, the input parameters are the ratio of atomic species and plasma pressure or the number of plasma elements, which were all measured from the experiments. Some functions are introduced for calculation of plasma temperature and its size variation. The outputs of this model are time and space evolutions of species number densities, variations of optical depth and spectral line profiles, as well as the resulting intensity of spectral emission close to the transition of strong nonresonance atomic line. The main application of this model is the prediction of electron density, temperature, and the mechanism of line broadening.

According to this model, the relation of the observed spectral radiance I_v(erg.s^{-1}.cm^{-2}.sr^{-1}.Hz^{-1}) with absorption coefficient $\kappa(v, x, T)$ and distribution of volume emission coefficient $\varepsilon_v(v, x, T)$ in optically thick plasma can be expressed by radiation transfer equation as

$$I_v = \int_{-x_0}^{x_0} \varepsilon_v(v,x,T)e^{-\tau(v,x,T)}dx. \tag{69}$$

$\tau(v, x, T)$ is the plasma optical thickness which is expressed by:

$$\tau(v,x,T) = \int_{-x}^{x_0} \kappa(v,x,T)dx. \tag{70}$$

x_0 and $-x_0$ refer to the plasma dimension from edge to edge along the line of sight. The magnitudes of volume emission $\varepsilon_v(v, x, T)$ and the absorption $\kappa(v, x, T)$ coefficients in an inhomogenous plasma are not constant and vary with the position along the line of sight. Bartels and Zwicker [4, 5] proved that in the absence of stimulated transitions, Eq. (69) can be expressed as the multiplication of the below three terms as:

$$I_v = A(T_m).M(E_l, E_u).Y\left[\tau(v,x,T),p\right]. \tag{71}$$

The first term, $A(T_m)$ is assumed as a source function which is related only to T_m and is the exact explanation for Wien's law:

$$A = \frac{2hv^3}{c^2}\exp(-\frac{hv}{kT_m}). \tag{72}$$

It should be noted that Omenetto et al. [36] expressed that the source function in Wien's law can be applied for plasmas in which stimulated transitions are not important and the condition of $hv/kT_m \gg 1$ holds, while Planck's law is utilized when stimulated transitions are taken into account.

The second term in Eq. (71) considers, to some extent, the influences of plasma heterogeneity on radiation and for spectral line with naturally and van der Waals broadening equals to:

$$M = \sqrt{\frac{E_l}{E_u}}. \qquad if \qquad \frac{kT_m}{E_l} << 1. \tag{73}$$

For singly ionized atoms and for neutral lines with Stark broadened, M can be expressed as

$$M = \sqrt{\frac{E_l + 0.5x_0}{E_u + 0.5x_0}}. \qquad if \qquad \frac{kT_m}{E_l + 0.5x_0} << 1 \tag{74}$$

Here, E_l and E_u are the lower and upper levels excitation energies, respectively and x_0 is the ionization energy. Furthermore, the above mentioned conditional statements in Eqs. (84) and (85) are not satisfied for resonance neutral lines, but they are logical assumptions for ionic spectral lines as well as for nonresonance neutral lines.

The third term in Eq. (71) considers the influences of heterogeneous mixing and optical thickness. The function $Y[\tau(v, x, T), p]$ is approximated by the below equation:

$$Y[\tau(v,x,T),p] = e^{-\frac{\tau(v,x,T)}{2}} [\frac{\tau(v,x,T)}{2}(1-p)$$
$$+ p\sinh\left(\frac{\tau(v,x,T)}{2}\right) \tag{75}$$
$$+ \frac{1}{\sqrt{p}}\sinh\left(\frac{\tau(v,x,T)}{2}\sqrt{p}\right)].$$

Here, the parameter p is explained as

$$p = \frac{6}{\pi}arctg\frac{M^2}{\sqrt{1+M^2}}. \tag{76}$$

Therefore, by substitution of above equations into Eq. (71), spectral emission intensity for optically thick inhomogenous plasma I_v can be obtained. In this model, the spatial distribution of plasma temperature is approximated by a second-order polynomial function with a maximum value in the plasma center.

Figure 15 represents the calculated emission distributions for different initial number of atoms and various Si/N atom ratios obtained at several distinct delay times (from 1 to 4 a.u.). In this figure, it is seen that at low densities (10^{16}), the plasma continuous emission (curves at t_1, t_2), monotonically reduces by time passing and growing the plasma. In addition, at high densities (10^{17}), a considerable value of early plasma continuum is absorbed within the plasma plume, which causes less emission radiation at time t_1 compared to times t_2 and t_3. This results in the LIP to approach a black body emitter at initial delay times.

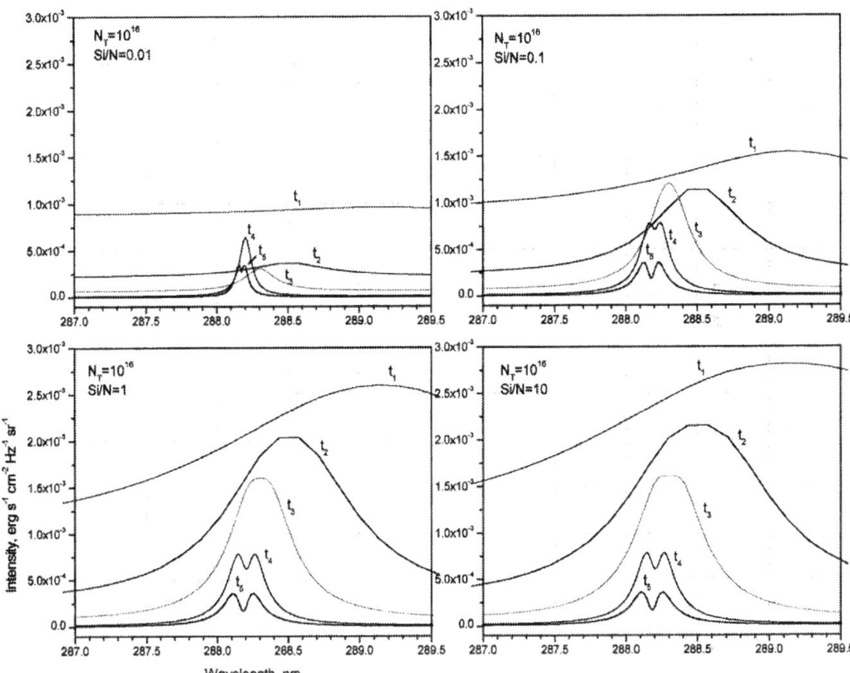

Figure 15. Calculated radiation profiles for LIP in initial times, including different proportions of 10^{16} N and Si atoms: from Si/N = 0.01 to Si/N =10. Times t_1 to t_5 refers to the magnitudes of 1, 1.5, 2, 3, and 4 in relative units.

Moreover, it must be mentioned that different research groups [37–41] discussed about spectrum analysis and extraction of plasma parameters in symmetric and asymmetric self-reversal line shapes produced by different sources that can be studied by readers for completeness of information.

6. Conclusion

In this chapter, a brief description of the different methods of analyzing the thick laser-produced plasmas was represented by considering the theoretical and experimental techniques. In some works, the self-absorption was corrected and then, the plasma parameters was extracted, while in others, some proposed models were explained and thereafter, plasma features was obtained straightly without needing any correction.

Finally, it should be noted that different parameters such as laser features (its wavelength [42], double or single pulse scheme [33, 43], energy [34], and pulse duration [42]), ambient gas condition (the nature and its pressure [42, 44]) and measurement device circumstance (delay time [8, 34], gate width, and exposure time), and the sample characteristics (metal and biological cases) could affect the amount of self-absorption as well. The details of these effects can be studied in the related references.

Author details

Fatemeh Rezaei*

Address all correspondence to: fatemehrezaei@kntu.ac.ir

Department of Physics, K. N. Toosi University of Technology, Shariati, Tehran, Iran

References

[1] Ovsyannikov, A., *Plasma Diagnostics*. 2000: Cambridge Int Science Publishing.

[2] Aragón, C. and J. Aguilera, *CSigma graphs: A new approach for plasma characterization in laser-induced breakdown spectroscopy*. Journal of Quantitative Spectroscopy and Radiative Transfer, 2014. 149: p. 90–102.

[3] Cowan, R.D. and G.H. Dieke, *Self-absorption of spectrum lines*. Reviews of Modern Physics, 1948. 20(2): p. 418.

[4] Zwicker, H., *Evaluation of plasma parameters in optically thick plasmas*. Plasma Diagnostics, 1968.

[5] Bartels, H., *Z. Phys 125 597 Bartels H 1949*. Z. Phys, 1949. 126: p. 108.

[6] Amamou, H., et al., *Correction of self-absorption spectral line and ratios of transition probabilities for homogeneous and LTE plasma*. Journal of Quantitative Spectroscopy and Radiative Transfer, 2002. 75(6): p. 747–763.

[7] Amamou, H., et al., *Correction of the self-absorption for reversed spectral lines: application to two resonance lines of neutral aluminium*. Journal of Quantitative Spectroscopy and Radiative Transfer, 2003. 77(4): p. 365–372.

[8] Moon, H.-Y., et al., *On the usefulness of a duplicating mirror to evaluate self-absorption effects in laser induced breakdown spectroscopy*. Spectrochimica Acta Part B: Atomic Spectroscopy, 2009. 64(7): p. 702–713.

[9] Aguilera, J. and C. Aragón, *Characterization of laser-induced plasmas by emission spectroscopy with curve-of-growth measurements. Part I: Temporal evolution of plasma parameters and self-absorption.* Spectrochimica Acta Part B: Atomic Spectroscopy, 2008. 63(7): p. 784–792.

[10] Aguilera, J., J. Bengoechea, and C. Aragón, *Curves of growth of spectral lines emitted by a laser-induced plasma: influence of the temporal evolution and spatial inhomogeneity of the plasma.* Spectrochimica Acta Part B: Atomic Spectroscopy, 2003. 58(2): p. 221–237.

[11] Aragon, C., J. Bengoechea, and J. Aguilera, *Influence of the optical depth on spectral line emission from laser-induced plasmas.* Spectrochimica Acta Part B: Atomic Spectroscopy, 2001. 56(6): p. 619–628.

[12] Aragón, C., F. Peñalba, and J. Aguilera, *Curves of growth of neutral atom and ion lines emitted by a laser induced plasma.* Spectrochimica Acta Part B: Atomic Spectroscopy, 2005. 60(7): p. 879–887.

[13] Ben Ahmed, J. and J. Cowpe, *Experimental and theoretical investigation of a self-absorbed spectral line emitted from laser-induced plasmas.* Applied optics, 2010. 49(18): p. 3607–3612.

[14] Bulajic, D., et al., *A procedure for correcting self-absorption in calibration free-laser induced breakdown spectroscopy.* Spectrochimica Acta Part B: Atomic Spectroscopy, 2002. 57(2): p. 339–353.

[15] El Sherbini, A., et al., *Evaluation of self-absorption coefficients of aluminum emission lines in laser-induced breakdown spectroscopy measurements.* Spectrochimica Acta Part B: Atomic Spectroscopy, 2005. 60(12): p. 1573–1579.

[16] Lazic, V., et al., *Self-absorption model in quantitative laser induced breakdown spectroscopy measurements on soils and sediments.* Spectrochimica Acta Part B: Atomic Spectroscopy, 2001. 56(6): p. 807–820.

[17] Sun, L. and H. Yu, *Correction of self-absorption effect in calibration-free laser-induced breakdown spectroscopy by an internal reference method.* Talanta, 2009. 79(2): p. 388–395.

[18] Rezaei, F. and S.H. Tavassoli, *A new method for calculation of thick plasma parameters by combination of laser spectroscopy and shadowgraphy techniques.* Journal of Analytical Atomic Spectrometry, 2014. 29(12): p. 2371–2378.

[19] Cristoforetti, G. and E. Tognoni, *Calculation of elemental columnar density from self-absorbed lines in laser-induced breakdown spectroscopy: A resource for quantitative analysis.* Spectrochimica Acta Part B: Atomic Spectroscopy, 2013. 79: p. 63–71.

[20] Hermann, J., C. Boulmer-Leborgne, and D. Hong, *Diagnostics of the early phase of an ultraviolet laser induced plasma by spectral line analysis considering self-absorption.* Journal of applied physics, 1998. 83: p. 691–696.

[21] D'Angelo, C.A., D.M.D. Pace, and G. Bertuccelli, *Semiempirical model for analysis of inhomogeneous optically thick laser-induced plasmas.* Spectrochimica Acta Part B: Atomic Spectroscopy, 2009. 64(10): p. 999–1008.

[22] O'neill, R., *Algorithm AS 47: function minimization using a simplex procedure.* Applied Statistics, 1971: p. 338–345.

[23] Rezaei, F., P. Karimi, and S. Tavassoli, *Effect of self-absorption correction on LIBS measurements by calibration curve and artificial neural network.* Applied Physics B, 2014. 114(4): p. 591–600.

[24] Mitchell, A. and M.W. Zemansky, *Resonance Radiation.* 1934: Cambridge.

[25] Hinnov, E., *A method of determining optical cross sections.* JOSA, 1957. 47(2): p. 151–155.

[26] Alkemade, C.T.J., et al., *Th. Metal Vapours in Flames.* 1982, Pergamon Press: Oxford.

[27] Gornushkin, I., et al., *Curve of growth methodology applied to laser-induced plasma emission spectroscopy.* Spectrochimica Acta Part B: Atomic Spectroscopy, 1999. 54(3): p. 491–503.

[28] Demtröder, W., *Laser Spectroscopy: Basic Concepts and Instrumentation.* 2013: Springer Science & Business Media.

[29] D'Ammando, G., et al., *Computation of thermodynamic plasma properties: a simplified approach.* Spectrochimica Acta Part B: Atomic Spectroscopy, 2010. 65(8): p. 603–615.

[30] Aguilera, J. and C. Aragón, *Characterization of laser-induced plasmas by emission spectroscopy with curve-of-growth measurements. Part II: Effect of the focusing distance and the pulse energy.* Spectrochimica Acta Part B: Atomic Spectroscopy, 2008. 63(7): p. 793–799.

[31] Silfvast, W.T., *Laser Fundamentals.* 2004: Cambridge University Press.

[32] Griem, H.R., *Plasma Spectroscopy.* New York: McGraw-Hill, 1964, 1964. 1.

[33] Bredice, F., et al., *Evaluation of self-absorption of manganese emission lines in Laser Induced Breakdown Spectroscopy measurements.* Spectrochimica Acta Part B: Atomic Spectroscopy, 2006. 61(12): p. 1294–1303.

[34] Rezaei, F., P. Karimi, and S. Tavassoli, *Estimation of self-absorption effect on aluminum emission in the presence of different noble gases: comparison between thin and thick plasma emission.* Applied Optics, 2013. 52(21): p. 5088–5096.

[35] Gornushkin, I., et al., *Modeling an inhomogeneous optically thick laser induced plasma: a simplified theoretical approach.* Spectrochimica Acta Part B: Atomic Spectroscopy, 2001. 56(9): p. 1769–1785.

[36] Omenetto, N., J. Winefordner, and C.T.J. Alkemade, *An expression for the atomic fluorescence and thermal-emission intensity under conditions of near saturation and arbitrary*

self-absorption. Spectrochimica Acta Part B: Atomic Spectroscopy, 1975. 30(9): p. 335–341.

[37] Fishman, I., G. Il'in, and M.K. Salakhov, *Spectroscopic diagnostics of a strongly inhomogeneous optically thick plasma. Part 1. The formation of asymmetric self-reversed emission and absorption lines: determination of electron impact half-width and electron concentration.* Spectrochimica Acta Part B: Atomic Spectroscopy, 1995. 50(9): p. 947–959.

[38] Fishman, I., G. Il'in, and M.K. Salakhov, *Temperature determination of an optical thick plasma from self-reversed spectral lines.* Journal of Physics D: Applied Physics, 1987. 20(6): p. 728.

[39] Fishman, I., G. Il'in, and M.K. Salakhov, *Spectroscopic diagnostics of a strongly inhomogeneous optically thick plasma. Part 2. Determination of atom concentration and variations of different physical values in the plasma cross-section using asymmetric self-reversed emission and absorption lines.* Spectrochimica Acta Part B: Atomic Spectroscopy, 1995. 50(10): p. 1165–1178.

[40] Karabourniotis, D., *Plasma temperature determination from the maximum intensity of a symmetric self-reversed line.* Journal of Physics D: Applied Physics, 1983. 16(7): p. 1267.

[41] Ilin, G., M.K. Salakhov, and I. Fishman, *Analysis of asymmetric self-reversal of spectral lines for the case of lateral observation of a spectrum.* Zhurnal Prikladnoi Spektroskopii, 1976. 24: p. 201–207.

[42] Lee, Y.-I., et al., *Influence of atmosphere and irradiation wavelength on copper plasma emission induced by excimer and Q-switched Nd: YAG laser ablation.* Applied Spectroscopy, 1997. 51(7): p. 959–964.

[43] Gautier, C., et al., *Main parameters influencing the double-pulse laser-induced breakdown spectroscopy in the collinear beam geometry.* Spectrochimica Acta Part B: Atomic Spectroscopy, 2005. 60(6): p. 792–804.

[44] Rezaei, F. and S.H. Tavassoli, *Quantitative analysis of aluminum samples in He ambient gas at different pressures in a thick LIBS plasma.* Applied Physics B, 2015: p. 1–9.

Chapter 14

Industrial Applications of Laser-Induced Breakdown Spectroscopy

Yoshihiro Deguchi and Zhenzhen Wang

Additional information is available at the end of the chapter

http://dx.doi.org/10.5772/61915

Abstract

Laser-induced breakdown spectroscopy (LIBS) is an analytical detection technique based on atomic emission spectroscopy to measure elemental composition. With the development of lasers and detection systems, applications of LIBS encompass a broad range, including physics, engineering, space missions, environment, etc. due to the unique features of little or no sample preparation, noncontact, fast response, and multielemental analysis. The fundamental and application have been extensively studied to improve LIBS technique. This chapter largely targets the engineering fields, especially practical applications. Laser-induced breakdown spectroscopy will be discussed in this chapter including its fundamentals, industrial applications, and challenges.

Keywords: Laser-induced breakdown spectroscopy (LIBS), Industrial applications, Challenge

1. Introduction

Laser-induced plasma was introduced as a spectroscopic emission source only two years after the invention of laser. LIBS is the acronym of "laser-induced breakdown spectroscopy", and it is also called LIPS (laser-induced plasma spectroscopy), laser spark spectroscopy (LSS), and other related names [1]. LIBS is an analytical detection technique based on atomic emission spectroscopy to measure elemental composition. A laser pulse focuses in or on a sample, which can be gas, liquid, aerosol, or solid, to form the micro-plasma. The spectra emitted are used to determine the elemental constituents of measured samples.

1.1. Theory

The initiation, formation, and decay of the plasma are complex physical and chemical processes. In the LIBS process, a laser beam is focused on a small area of the sample. When the laser energy exceeds the breakdown threshold, the plasma with high temperature and high density is produced in the portion. The core of plasma is firstly produced by the absorption of the incident laser energy, such as multiphoton ionization. The creation of the plasma core induces the rapid growth of plasma through the absorption of the laser light by electrons and the electron impact ionization process in it. After the termination of the laser pulse, the plasma continues expanding because of its high temperature and pressure gradients compared with the ambient conditions. At the same time, the recombination of electrons and ions proceeds due to the collision process and the temperature decreases gradually compared with that in the plasma generation process. Emission signals arise in the plasma cooling period [2]. In the plasma, the ions, atoms, and molecules distributed in the different levels transit from the high energy level to the low energy level, emitting the strong emission spectra. The emission intensity from the atomized species provides the elemental compositions of the materials. The light corresponding to a unique wavelength of each element is emitted from excited atoms in plasma, as shown in Figure 1.

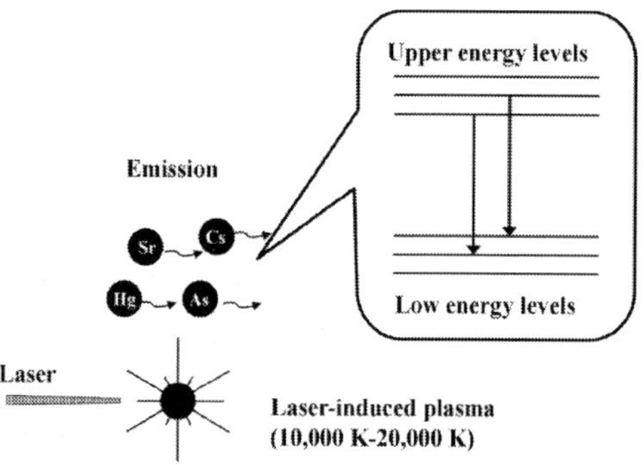

Figure 1. Laser-induced plasma processes

A calibration of the LIBS signal is necessary for quantitative analysis. Despite the fact that the LIBS processes involved are complex, the emission intensity from the atomized species can be described by the following equation with the assumption of uniform plasma temperature [2]:

$$I_i = n_i K_{i,j} g_{i,j} \exp\left(-\frac{E_{i,j}}{kT}\right) \tag{1}$$

where I_i is the emission intensity of species i, n_i is the concentration of species i, $K_{i,j}$ is a variable that includes the Einstein A coefficient from the upper energy level j, $g_{i,j}$ is the statistical weight of species i at the upper energy level j, $E_{i,j}$ is the upper-level energy of species i, k is the Boltzmann constant, and T is the plasma temperature. Equation (1) is applicable under the conditions of local thermodynamic equilibrium (LTE). In Eq. (1), there are several factors that affect the emission intensity I_i, including plasma temperature, plasma nonuniformity, matrix effects, etc. It is very difficult to solve the LIBS process theoretically because it contains laser–material interactions, rapid temperature changes over 10,000 K in a nanosecond or picosecond timescale, and plasma cooling phenomena including the recombination process of ions, electrons, and neutrals. Therefore, the appropriate correction factors must be contained in $K_{i,j}$ to obtain the quantitative results.

The factors that affect the target signal in LIBS process generally include the background noise, stability of plasma, nonuniformity of plasma, matrix effects, dirt on measurement windows, etc. The main background noise of LIBS is the continuum emission from plasma itself. Atomic emissions appear after a certain time delay, which means that LIBS signals appear during the plasma cooling process. Therefore, it is important to choose the appropriate delay time and gate width to reduce the effect of the background noise. The fluctuation of plasma signal is an intrinsic characteristic in LIBS. Use of the intensity ratio and plasma temperature correction can reduce the fluctuation to some extent. The laser-induced plasma has its structure in it. In the plasma generation process, the plasma structure depends largely on the laser density pattern and measured material conditions. The LIBS signal depends on the measurement area across the plasma. In this sense, the correction method also includes the effects of plasma nonuniformity. Matrix effects are the combined effects of all components other than the measured species. The changes of these components may cause the alteration of LIBS signal intensity even if the number density of the measured species is the same in the measurement period. Matrix effects are usually corrected by the experimental calibrations. The attenuation of LIBS signals by the contamination of measurement windows is automatically neglected because the signal intensity ratio is usually used for the elemental composition analyses. Considering the measurement stability and soundness of windows, however, the cleanliness of measurement windows should be maintained, especially in the practical applications.

1.2. Geometric arrangement and measurement species

The LIBS apparatus fundamentally consists of laser, measured materials, lens, spectrometer, ICCD camera, and related devices. A typical geometric arrangement of LIBS is shown in Figure 2. Lasers such as a pulsed Nd:YAG laser are used as a light source. The output laser beam is focused onto the measurement area using the focal lens to induce the plasma. The plasma emission is focused onto the optical fiber. Emission signals are finally detected by the combination of a spectrometer, an ICCD camera, and auxiliary equipment. According to the measured materials of solid, liquid, and gas phases, different measurement chamber or platform can be employed. It is worth noting that the reflection of a laser light from the windows must be considered carefully. Its reflection often results in the damages of optics due to the high-energy laser light. The reflection from plasma is sometimes tricky and malicious for LIBS systems. Damages of optics by the reflection from plasma cause troubles in some cases, especially in the analyses of liquids.

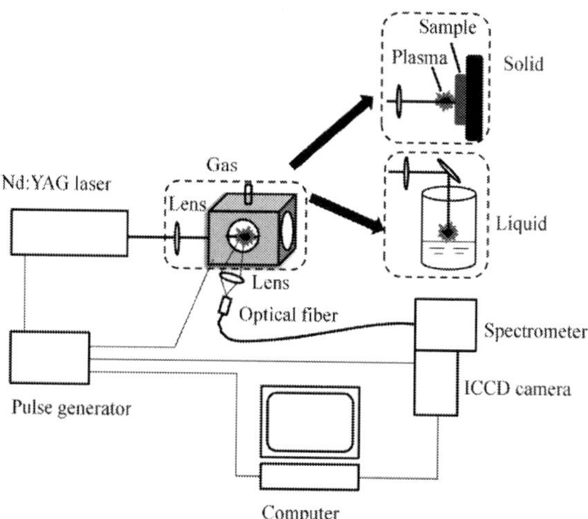

Figure 2. Typical geometric arrangement of LIBS

The basic components of LIBS system are similar but the component specifications are tailored to the particular applications. These specifications include the equipment types, physical parameters, and the technical specifications belonging to operational performance.

2. Fundamentals

Since it is the basis of successful application of LIBS technology, the fundamental study is of great importance. With the development of LIBS, the mechanism of laser–material interaction, plasma generation, plasma–environment interaction, self-absorption effect, signal enhancement, and some other fundamental research have been studied extensively to promote LIBS technique [3-7]. Various quantitative analytical methods have also been studied and improved, such as traditional calibration method, internal calibration method, calibration-free method, partial least squares (PLS) method, etc.

2.1. Plasma and its models

Plasma is a local assembly of atoms, ions, and free electrons, overall electrically neutral. Plasma is characterized by a variety of parameters. The degree of ionization is the most basic parameter. The ratio of electrons to other species is less than 10% in the plasma, called weakly ionized plasma. On the other hand, highly ionized plasma may have atoms stripped of many of their electrons, resulting in very high electron to atom or ion ratios. The plasma produced in LIBS typically belongs to the category of weakly ionized plasma. The goal of LIBS technique is to create the optically thin plasma, which is in local thermodynamic equilibrium and whose elemental composition is the same as that of the sample.

The LIBS plasma features inhomogeneities that can lead to spatial differentiation. This fact is important in choosing the temporal window in order to accumulate spectroscopic data. The spatial and temporal evolution of LIBS plasma from a steel target was monitored using time of flight and shadowgraph techniques [8]. Two regions in the plume were observed, one characterized by air and continuum emissions produced by shock wave ionization, and the other one by emissions from ablated material. The sufficiently high laser fluence and acquisition delay time are necessary to assure the homogeneity for the analytical applications. The homogeneity of LIBS plasma was investigated using the curve of growth method employing five Fe(I) lines [9]. In that formalism, the line shapes as a function of temperature and concentration were modeled. The agreement between modeled and experimental line shapes implied that the Stark effect was the dominant broadening mechanism in the plasma. The temperatures obtained from neutral and ion spectral lines were studied [10]. The different temperatures studied can be obtained from Boltzmann and Saha plots. The difference was explained by the spatial variation of the plasma temperature and densities leading to a difference in spatial locus for populations in the upper levels of transitions for neutrals and ions.

Plasma models are becoming more comprehensive and detailed. A radiative model of LIBS plasma expanding into a vacuum was validated by the experiments [11]. The inverse problem was specifically addressed, which means finding the initial conditions by the comparison of the calculated synthetic spectra and the experimentally measured ones. The composition of the material was effectively deduced from the calculated spectra. The plasma was considered to be characterized by a single temperature and electron density. The combination of the original modeling work on laser-evaporated plasma plume expansion into a vacuum and ablation leading to vaporization and particle formation was studied [12]. The interaction of a nanosecond pulse with a copper target was modeled in vacuum. Some of the parameters were studied including the melting and evaporation of the target, the plume expansion and plasma formation, the ionization degree and density profiles of neutral; once-ionized; and doubly ionized copper and electrons, and the resultant plasma shielding.

2.2. LIBS detection ability

Most fundamental studies focused on signal enhancement to improve the detection limit. The measured results showed that the spatial confinement and fast discharge would be able to enhance the signal from several times to dozens of times, while dual pulse is able to enhance signal 100–1000 times [13-15]. Besides signal enhancement, there were also some other studies worth mentioning. The self-absorption in laser-induced plasma was studied. The results suggested that the self-absorption effect could be alleviated by the selection of suitable atomic line, operating at higher pulse energy and detecting with longer delay [16]. The pressure effect on the plasma emission from fundamental 0.1 to 40 MPa in bulk seawater was investigated [17]. The time-resolved LIBS emission results demonstrated that plasma emission is weakly dependent on the ambient pressure during the early stage of plasma and the pressure has a significant influence on the plasma form during plasma evaluation at a later stage of plasma.

The detection ability of trace species using LIBS has been improved with the development of LIBS devices. The utilization of short pulse laser for plasma generation has been extensively studied [18,19]. Short pulse irradiation allowed for a specificity of excitation that could yield LIBS signals more tightly correlated to particular chemical species and showed significantly lower background emission. A new method to control the LIBS plasma generation process is necessary for the enhancement of detection limit, i.e., low pressure and short pulse LIBS [20-22]. Because of the pressure, volatility, and quenching effects of liquid, the plasma lifetime of liquid sample is shorter compared with that of solid and gas phases. Meanwhile, sputtering of liquid sample by LIBS plasma often raises the problem of the measurement windows. The sensitivity, stability, and repeatability of LIBS signal are much lower, leading to the increasing difficulty of its analyses. Numerous papers have reported LIBS measurement of different forms of liquid phase materials including the solidification, liquid bulk, liquid surface, and others [23-28], which shows different detection features and detection limit.

2.3. Quantitative analysis

The ultimate goal of LIBS technique is to provide a quantitative analysis with high precision and accuracy. Usually, a quantitative analysis begins with determining the response of a system for a given concentration or mass of the analyte of interest, which usually takes the form of a calibration curve. The calibration is usually strongly dependent on the analysis conditions, such as the stability of the laser pulse energy, the sample and sampling procedure, the physical and chemical properties of the sample, etc. The dependence of elemental signals of LIBS on the plasma temperature attributes to a very complex process in plasma. Several studies have reported the LTE condition of plasma in several types of plasmas [29]. The plasma temperature is a very important factor for the quantification of the LIBS measurement. There are several calibration methods to analyze the measured species quantitatively, including the traditional calibration method, internal calibration method, calibration-free method, etc. [30,31].

As for the simple samples, the emission intensity of the measured species is linear with the species content under the ideal condition. The traditional calibration model is relatively simple and convenient. However, the influences of matrix effect and element interference are not considered in the model. The accuracy becomes worse when the complex samples are measured or the experimental parameters fluctuate. The internal calibration method is a commonly used spectral analysis model with strict conditions. The elements with the features of high content, low detection limit, and good stability are mainly selected as the internal calibration elements. Usually, the compositions of the calibration sample and measured sample are not entirely consistent. When the measured samples contain various elements, the accuracy will be affected due to the matrix effect.

A new procedure is proposed for calibration-free quantitative elemental analysis of materials using LIBS technique. The method based on an algorithm developed and patented overcomes the matrix effects. The precise and accurate quantitative results on elemental composition of materials can be acquired without the use of calibration curves. Some applications of the method have been illustrated, e.g., the quantitative analysis of the composition of metallic alloys [32]. This model of CF-LIBS is applicable under the conditions of LTE and optically thin, as well as the assumed conditions without the element interference and self-absorption. Research recently focused on the correction for self-absorption. Multivariate analysis (MVA) is an effective mathematical and statistical approach for LIBS data analysis, since it can utilize much quantitative information from the complex LIBS spectra. Partial least squares (PLS) is such an MVA method and has shown great potential for LIBS quantitative measurement. The model utilizes the multiline spectral information of the measured element and characterizes the signal fluctuations due to the variation of plasma characteristic parameters, such as plasma temperature, electron number density, and total number density, for signal uncertainty reduction [33,34]. LIBS can be used to provide the quantitative analysis of a variety of samples in the laboratory and in the field. However, each application has some unique characteristics that must be dealt with in order to optimize performance. In the real applications of LIBS, the procedures for obtaining quantitative results reproducibly will be developed.

A much deeper understanding of LIBS fundamental physics is the key to overcome the bottlenecks for wide applications of LIBS, such as the relatively low measurement repeatability due to the plasma property and morphology fluctuations, the relatively low accuracy suffered from matrix effects, etc. The plasma generation and evolution processes are complicated processes. Much more work is still required to improve the qualitative and quantitative analyses, as well as the applications of LIBS technique.

3. Applications

LIBS has attracted great attention in various industries as a qualitative and quantitative analytical detection technique due to its noncontact, fast-response, and multidimensional features. With the development of laser and detection systems, LIBS has been applied in various fields, including combustion, metallurgy, food, human, the Mars project, and so on. Especially, the advantages of this method are more significant in the areas of combustion, metallurgy, and harsh environments. Many applications have successfully demonstrated the monitoring of plant control factors using LIBS. LIBS has been actively applied to commercial plants such as iron and steel making, thermal power, waste disposal, and so on. Environmental monitoring and safety applications are also the growing fields for LIBS. Applications of LIBS have covered all industry fields, including analyses from food, plant to space missions, which will be discussed in detail in the next section. In these cases, ruggedness and reliability become important requirements.

3.1. Applications for plants

LIBS, with the features of excellent temporal and spatial resolutions, appears to be a very promising analysis method in steel industry where element distribution measurements of materials at all stages of production provide the information of material quality and production process. By the continuous monitoring of element distribution, the raw materials with narrow composition tolerances can be available ahead of further processing. LIBS measurement of geological materials on the conveyor belts was studied and discussed preliminarily [35,36]. A multispectral line calibration method was proposed for the quantitative analysis of elemental compositions. Its feasibility and superiority over a single-wavelength determination have been confirmed by comparison with the traditional chemical analysis of the copper content in the ore. Two iron ore samples were employed to complete the mineralogical classification using a combination of LIBS and principal components analysis (PCA)/principal components regression (PCR) [37,38]. The combined method of LIBS and PCR was applied to determine the elemental compositions of a series of run-of-mine iron ore samples, which exhibited the potential for in situ determination of ore composition. The calibration models of LIBS have also been studied and discussed in the measurement of ores. The different data-driven multivariate statistical predictive algorithms, such as Principal Components Regression (PCR), Partial Least Squares Regression (PLSR), Multi-Block Partial Least Squares (MB-PLS), and Serial Partial Least Squares Regression (S-PLSR), were compared for the quantitative analysis in iron ore measured using LIBS to improve the performance of the quantitative measurements [39,40]. The on-line measurement system of LIBS has been discussed for the real applications. Figure 3 shows a LIBS system for on-line measurement using extractive sampling. For example, an analytical instrument based on LIBS technique was developed to operate on-line in the harsh environment of iron-ore pelletizing plants. The detection system was successful for the measurements of Si, Ca, Mg, Al, and graphitic C contents in different iron ore slurries prior to filtration and pelletizing [41]. A method for automated quantitative analysis of ores was developed using a commercial LIBS instrument fitted with a developed computer-controlled auto-sampler [42]. The preparation and analysis time for each sample was less than 5 min. The similar method was suitable for a range of ores and minerals.

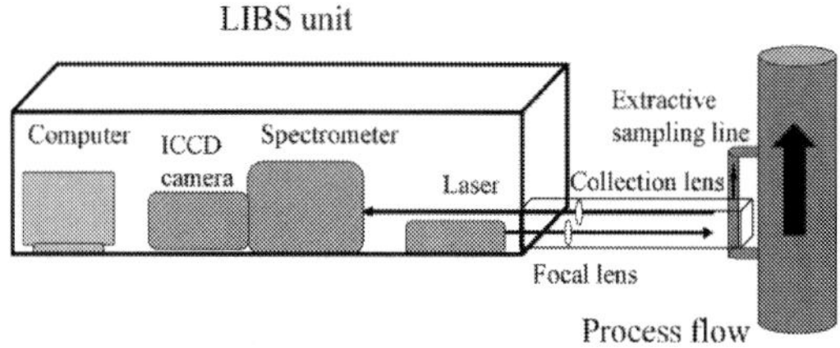

Figure 3. A LIBS system for on-line measurement using extractive sampling

Operating characteristics of coal-fired boilers are heavily influenced by factors such as the differences in fuel properties and combustion conditions. In order to achieve the optimal operation of multiple coal-fired boilers, it is necessary to accurately understand the coal quality and the state of combustion, and to adjust the control parameters. LIBS technique has been widely applied to analyze the compositional characterization of coal [43-45]. A nonlinearized multivariate dominant-factor-based partial least-squares (PLS) model was applied to coal elemental concentration measurement using LIBS [46]. Unburned carbon in fly ash is an important factor to estimate the combustion efficiency of boiler. Fly ash and bottom ash resulting from the coal combustion in a coal-fired power plant were analyzed using binders. Once the experimental conditions and features are optimized, application of LIBS may be a promising technique for combustion process control even in on-line mode [47,48]. Software-controlled LIBS systems including LIBS apparatus and sampling equipment have been designed for possible application to power plants for on-line quality analysis of pulverized coal and unburned carbon in fly ash [49,50], which shows the capability of reliable and real-time measurement using LIBS. LIBS has been applied for detection of unburned carbon in fly ash, char, and pulverized coal without any sample preparation. Figure 4 presents the examples of LIBS spectral lines obtained from fly ash. The calibration difficulty of aerosol sample was surpassed by using the correction factors for quantitative measurement. This automated LIBS apparatus was applied in a boiler-control system of a power plant with the objective of achieving optimal and stable combustion [51,52], which enabled real-time measurement of unburned carbon in fly ash, as shown in Figure 5. The boiler control in the real power plant was demonstrated to achieve an optimized operation without time consumption.

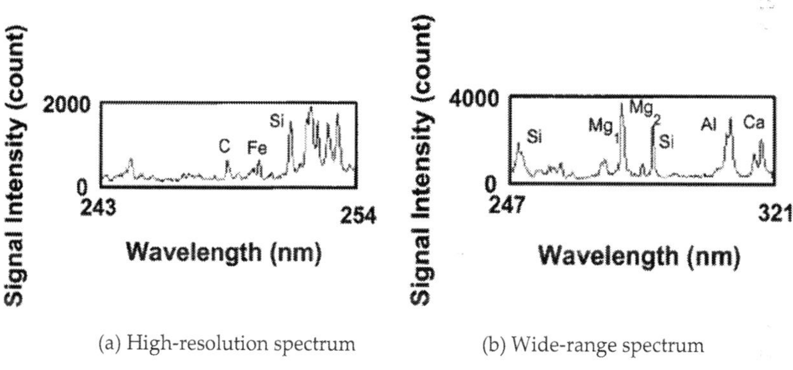

(a) High-resolution spectrum   (b) Wide-range spectrum

Figure 4. Fly ash LIBS spectra [52]

The safe and rational utilization is very important for nuclear power application. The radioactive contamination is a serious problem for the environment and human health. The radioactive materials released from the nuclear power plant are one of the main sources. Simultaneously, nuclear weapons testing fallout, some industry waste discharge, and radioactive substances employed for research also contribute to the issue [53,54]. The atmosphere, water and soil are polluted by these released radioactive materials. There are several serious pollutions to environment and human not only in the surrounding area of the nuclear power plant but also in the outlying regions [55-57]. LIBS has been investigated as a potential analytical tool to improve operations and safeguards for electrorefiners such as those used in processing spent nuclear fuel [58]. Detection of uranium and other nuclear materials is very important for nuclear safeguards and security. The spatial and temporal evolutions of uranium species in laser-produced plasmas were investigated. A set of optimal operating conditions was determined based on the experimental results, which is important for obtaining the

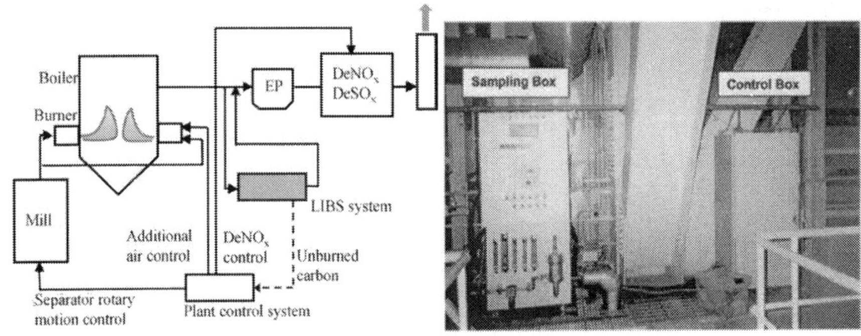

(a) Optimal boiler control system   (b) Photograph of the measurement apparatus

Figure 5. Unburned-carbon measurement in thermal power plant [52]

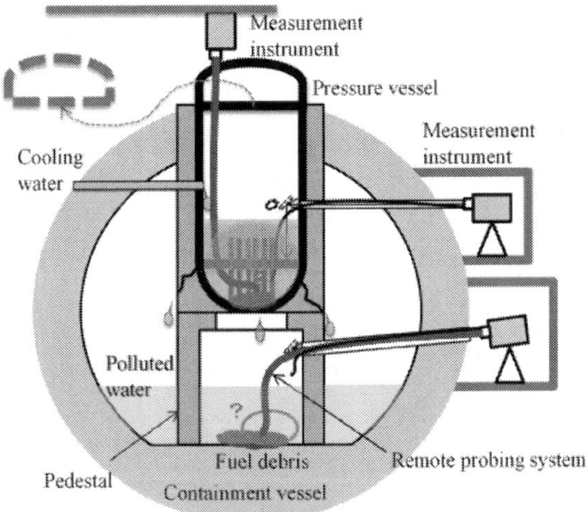

Figure 6. Image of the inside of the post-accident Fukushima Daiichi nuclear power plant and the inspection technique using optical fiber [60]

optimal spectral intensity from samples containing very small amounts of uranium [59]. LIBS can be applied to monitor radioactive elements, which is of utmost importance in case of leakage of radioactive materials from a nuclear power plant [60,61]. Figure 6 shows a schematic of the imaging observation using the imaging fiber, which was equipped for additional electric delivery. A transportable fiber coupled LIBS instrument was developed, which is feasible for the material analysis of underwater debris under a high-radiation field.

3.2. Applications for food, humans, and archaeology

As for LIBS applications to food, composition and contamination measurements of flours of wheat, barley, etc., have been demonstrated. The feasibility of quantifying trace elements in powdered food samples by spatially resolved LIBS has been demonstrated under a reduced argon atmosphere. The selection of the location in the plasma is crucial for obtaining the best signal-to-background ratio of analytical signal and to avoid background continuum and line broadening. The operating parameters affected the plasma property were optimized and used for further analysis of trace elements in starch-based food samples. Spatially resolved LIBS has been shown to be an accurate technique for determining trace elements of ppm level in starch-based food samples directly with an acceptable precision without any tedious digestion and dilution procedure [62]. A procedure for the analysis of K, P, Mg, and Ca in crop plant samples using a commercially available LIBS spectrometer was also developed. Real plant samples employed as the calibration standards were analyzed by ICP-OES or AAS after microwave digestion. A satisfactory agreement between LIBS and AAS/ICP-OES results was achieved [63]. The trace and ultra-trace element detection and qualitative analysis in fresh vegetables have been demonstrated using LIBS technique [64]. For a typical root vegetable such as potato, spectral analysis of the plasma emission reveals more than 400 lines emitted by 27 elements and 2 molecules, C_2 and CN. Many elements such as Mg, Al, Cu, Cr, K, Mn, Rb, Cd, and Pb have been measured by LIBS, as shown in Figure 7. These results demonstrate the potential of an interesting tool for botanical and agricultural studies as well for food quality/safety and environment pollution assessment and control.

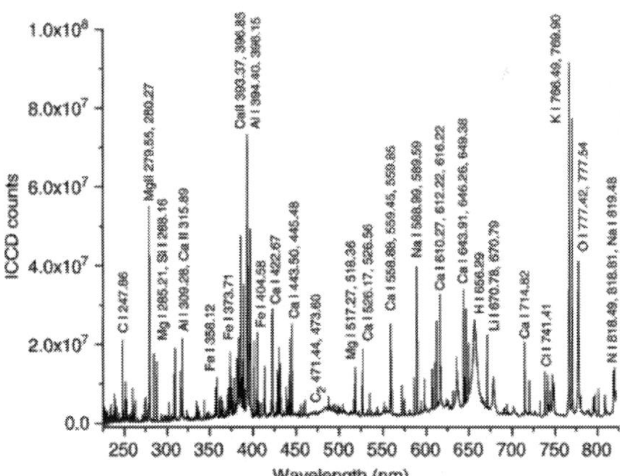

Figure 7. Typical LIBS spectrum of a fresh potato [64]

The application of LIBS to the analysis of important minerals and the accumulation of potentially toxic elements in calcified tissue has been reported [65], which exemplified for quantitative detection and mapping of Al, Pb, and Sr in representative samples, including teeth

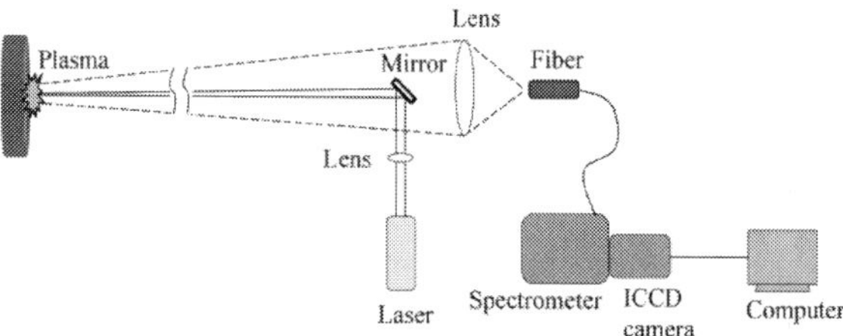

Figure 8. LIBS schematic diagram for measurements from a distance

and bones. In order to identify and quantify the major and trace elements in the tissues, one- and two-dimensional profiles and maps were generated. The state of the tooth has also been diagnosed using prominent constituent transitions in laser-excited tooth [66]. The spectroscopic observations in conjunction with discriminate analysis showed that calcium attached to the hydroxyapatite structure of the tooth was affected severely at the infected part of the tooth. It is possible to distinguish the healthy and caries infected tooth using emission spectroscopy and ICCD imaging of the expanding plasma.

Advancement in LIBS technique has led to its increased use in the fields of conservation, art history, and archaeology [67,68]. A prototype LIBS system was used to determine the elemental composition of multilayer structures in a metal jug from the mid-twentieth century [69]. The piece was highly deteriorated due to environmental damage. The LIBS technique was used as part of a historical investigation that required the determination of the material employed. The jug was selectively sampled at different points on the surface using the stereoscope. By sampling at different points, the surface composition was determined. Furthermore, the presence of two layers of Pb and Cu and their thicknesses were determined through in-depth analysis.

3.3. Applications for space and underwater explorations

One of the more exotic and exciting applications of LIBS instrumentation is for space missions to planet surfaces. Current technological developments of lasers, spectrometers and detectors have made the use of LIBS for space exploration feasible. Figure 8 shows the LIBS schematic diagram for measurements from a distance. LIBS technique greatly increases the scientific return from new missions by providing extensive data relating to planetary geology, which is one of the main goals of space exploration. Meanwhile, the geologic analysis can provide some information of a planet's history, e.g., whether earlier conditions were favorable for life. Several studies have addressed the feasibility of LIBS for space exploration [70-72]. The feasibility of LIBS for stand-off analysis of geological samples under Martian atmospheric conditions has been demonstrated. Under Martian conditions, the analyzed signals appear to be somewhat enhanced compared to that recorded at atmospheric pressure due to the increased ablation.

It is a big challenge to apply LIBS to ocean in situ detection, which has been studied with the development of LIBS. In order to apply LIBS to in situ elemental analysis in the deep ocean, the multielemental analysis of high-pressure aqueous solutions has been studied. The affected factors, such as pressure, laser energy, and so on, have been discussed [73,74]. The potential of LIBS for the underwater chemical characterization of archaeological materials has been also demonstrated [75,76], which involves the delivery of a focused laser pulse toward the distant target through the aqueous media and then the transmission of the light emitted by the laser-induced plasma back to the detection system. Figure 9 shows the LIBS spectra corresponding to different submersed materials obtained in laboratory. The performance of the remote LIBS system was evaluated in a measurement campaign in the Mediterranean Sea. The pictures taken during the on-site trials using LIBS are illustrated in Figure 10. The seashell as the biomineralization product records the growth development and the ocean ecosystem evolution. Therefore, the seashell has been studied as a representative for marine research [77]. LIBS-Raman combination was introduced to obtain the compositional distribution of scallop shell on the surface and also in the shallow layers, which suggested that the micro-chemical diagnostics of LIBS-Raman was a potential way to construct a 3D analysis for the shell research. There are also other techniques that combine LIBS and other methods for the 3D surface analysis.

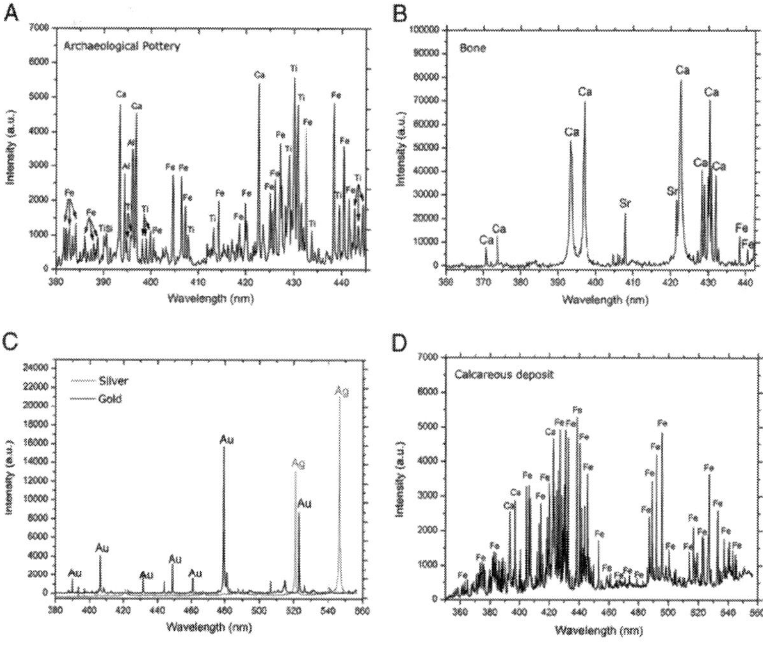

(a) Archaeological pottery (b) Bone sample (c) Precious metals (d) Calcareous deposit

Figure 9. Underwater LIBS spectra of different submersed materials [75]

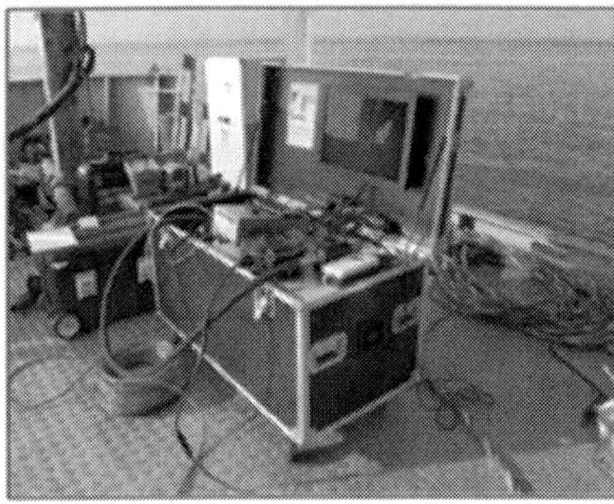

(a) Remote LIBS instrument on the research vessel

(b) Diver working at a 30 m depth

Figure 10. Pictures taken during the on-site trials on the Mediterranean Sea [75]

3.4. Other applications

Applications of LIBS have covered all industry fields, including analyses from food, plant, to space missions. Apart from the applications mentioned above, there are a variety of other applications of LIBS. In engine applications, LIBS has been used to measure the fuel–air ratios in combustion. If the fuel composition does not change, the fuel–air ratio can be inferred from the elemental analysis of unburned and burned gases. It is useful to know that the equivalence ratio can be inferred from burned gas measurement because the elemental composition does not change during reactions [78]. LIBS can be also used for the elemental analysis of particles, such as soot, which contain not only carbon but also metallic elements [79,80]. Tighter environmental regulations recently have focused on global limit of harmful substances released from industry, traffic, and domestic waste. Due to the sensitive and fast analysis features, LIBS has the capability to be used as a continuous-emission monitor for toxic metals, such as Be, Cd, Cr, Hg, and Pb. The sampling methodology and signal processing have been improved [81-83]. The utilization of LIBS technique has been extensively studied in different phase conditions, i.e., solid, liquid, and gas materials, which show different laser-induced plasma processes. The wide pressure dependence and various atmospheric compositions have been studied to understand the LIBS phenomena [84]. One of the challenging targets of LIBS is the enhancement of detection limit of gas phase materials. A new method to control the LIBS plasma generation process has been proposed for the enhancement of detection limit, i.e., low pressure and short pulse LIBS.

LIBS is a promising technique for in situ elemental analysis. The advancement of portable LIBS systems becomes a key technique. There have been a growing number of applications using LIBS in life science, medical fields, and so on. A new mobile instrument for LIBS analysis was developed, which is based on double-pulse LIBS and a calibration-free LIBS technique. Some applications have been presented including the analysis of cultural heritage, environmental diagnostics, and metallurgy [85]. Laser-induced breakdown spectroscopy and Raman spectroscopy have several features that make a combined instrument for remote analysis. These two techniques are very useful and feasible as the combination of elemental compositions from LIBS and molecular vibrational information from Raman spectroscopy strongly complement each other. Remote LIBS and Raman spectroscopy spectra were taken together on a number of mineral samples [86,87]. Figure 11 shows the Raman spectra, the combined Raman and LIBS spectra, and the LIBS spectra of calcite ($CaCO_3$) in air in the 534–699 nm wavelength range. The Raman lines in Figure 11(b) are marked with the letter "R." On the other hand, an approach to further increase the sensitivity of LIBS for the determination of traces is the combination of LIBS and laser-induced fluorescence, which has been studied. The combination of LIBS and LIF allows linking the multispecies capability of LIBS for a broad range of analytes with the high sensitivity of LIF for individual selected species [88].

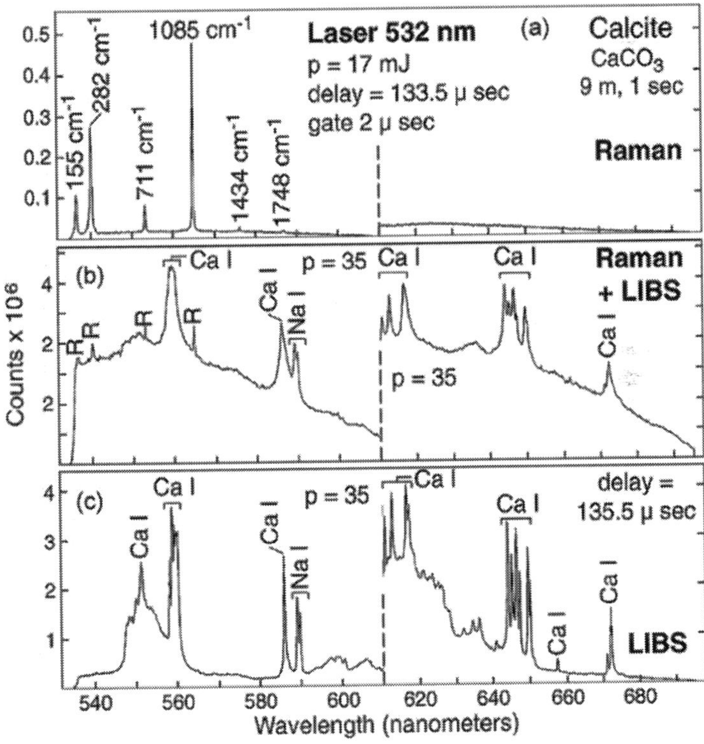

(a) Raman spectra (b) Combined Raman and LIBS spectra (c) LIBS spectra

Figure 11. Combined remote Raman and LIBS spectra of calcite in the 534–699 nm range [86]

4. Challenges

LIBS features various merits of the noncontact, fast response, and multidimensional detection and has been widely studied and applied in different fields as the qualitative and quantitative analytical detection technique. However, one of the major drawbacks of LIBS is the difficulty of quantitative analysis. There are numerous correction methods for LIBS to achieve the quantitative information, which are usually application-dependent. On the other hand, it has become increasingly important to monitor factors in plant conditions in order to improve the operation of industrial plants. Furthermore the long-term continuous use of LIBS system is a considerable factor for industrial applications. As a consequence, the improvement of measurement accuracy, quantitative analysis, and real-time measurement is very necessary for the operation of the overall plants using LIBS system.

4.1. Accuracy and durability

The laser-induced plasma processes are different from the phase samples in various applications. The measurement methods and parameters should be determined according to the specific conditions. There are several important factors that need to be considered when obtaining quantitative information using LIBS. Choosing the appropriate experimental parameters, therefore, is important to make the theoretical treatment applicable for quantitative measurements. On the other hand, data processing and modeling play an important role in LIBS for the analytical results of the measured spectra. An ideal data processing method should be based on a deep understanding of plasma physics and should be capable of minimizing the noise effects, compensating for the signal fluctuations, and reducing the matrix effects. There have been several calibration methods such as the Boltzmann plot method using many emission lines to increase the correction precision. The calibration methods should be developed to realize the quantitative analyses with the precision and accuracy of a measurement. As for the on-line application in the industry, the system simplicity and real-time measurement capability are also significant factors to be considered. The methods for quantitative analyses should be workable and satisfactory for practical applications.

The real advantage of LIBS technique is that the results are delivered continuously and in real time compared with periodic sampling and standard analytical methods with the time consumption. Consequently, LIBS gives a more representative reading of the state of the process, particularly when rapid perturbations occur, and allows process optimization and quality improvement. Current research aims to develop the commercial equipment for continuous industrial applications. However, in these applications the long-term stability and durability of LIBS devices, especially lasers, is one of the challenges. LIBS employs pulsed lasers and their lifetime often limits the plant applications, especially the long-term continuous use for plant monitoring and control. In a harsh environment, actually, all the devices should be paid attention to, including lasers.

4.2. Instrument development

The applications of LIBS technique have recently been proposed in the fields of materials science, industrial process control, environmental protection, cultural heritage conservation, etc. All of these applications would benefit from a mobile instrument. Therefore, the availability of affordable commercial instrumentation, the standardization of measurement procedures, and the calibration standards are required for reproducible and reliable quantitative LIBS analysis in situ.

The focus of a laser beam of LIBS, for instance, is one of the most important factors to be considered when applying LIBS to industrial processes with the change of a target profile. 3D profile information of the object is required for the positioning of a focused laser beam. The noncontact-type profile measurement systems, in general, can be divided into three categories, including a measurement machine integrated with a triangulation laser probe [89], a measuring machine integrated with a laser line projector and one/two CCD cameras [90-92], and a measurement machine integrated with a structured fringes projector and two CCD cameras [93]. To digitize small complex objects with dimensions smaller than about 30 mm, using a measurement machine integrated with a triangulation laser probe is a good strategy due to its small spot size. In general, phase shifting algorithm is applied to calculate the phase map and the 3D profile of an object using the structured fringes projection system [94]. If a 3D profile measurement system can be integrated with a LIBS system, the measured 3D profile information of the object can be used for the real-time positioning of a focused laser beam in a LIBS system.

It has become increasingly important to monitor factors in plant conditions in order to improve the operation of industrial plants. As a consequence, improved on-line monitoring techniques for plant control factors are necessary to enhance the capability of maintaining the overall plant operation. The associated monitoring and control techniques are necessary for the continued operational improvement. Emphasis is placed mainly on instrument development for applications as well as fundamental scientific investigations.

Author details

Yoshihiro Deguchi[1*] and Zhenzhen Wang[2]

*Address all correspondence to: ydeguchi@tokushima-u.ac.jp

1 Graduate School of Advanced Technology and Science, The University of Tokushima, Tokushima, Japan

2 State Key Laboratory of Multiphase Flow in Power Engineering, Xi'an Jiaotong University, Xi'an, China

References

[1] Miziolek AW, Palleschi V, Schechter I. *Laser Induced Breakdown Spectroscopy*. Cambridge University Press: Cambridge, United Kingdom; 2008. p. 1-4.

[2] Deguchi Y. *Industrial Applications of Laser Diagnostics*. CRS Press, Taylor & Francis: New York, USA; 2011. p. 107-113.

[3] Guo LB, Li CM, Hu W, Zhou YS, Zhang BY, Cai ZX, Zeng XY, Lu YF. Plasma confinement by hemispherical cavity in laser-induced breakdown spectroscopy. *Appl Physics Lett*. 2011; 98: 131501-1-131501-3. doi:10.1063/1.3573807

[4] Lui SL, Cheung NH. Resonance-enhanced laser induced plasma spectroscopy for sensitive elemental analysis: Elucidation of enhancement mechanisms. *Appl Physics Lett*. 2002; 81(27): 5114-5116. DOI: 10.1063/1.1532774

[5] Lui SL, Cheung NH. Resonance-enhanced laser induced plasma spectroscopy: Ambient gas effects. *Spectrochim Acta Part B-Atomic Spectroscopy*. 2003; 58: 1613-1623. DOI: 10.1016/S0584-8547(03)00139-3

[6] Hou ZY, Wang Z, Liu JM, Ni WD, Li Z. Signal quality improvement using cylindrical confinement for laser induced breakdown spectroscopy. *Optics Express*. 2013; 21(13): 15974-15979. DOI:10.1364/OE.21.015974

[7] Li XF, Zhou WD, Cui ZF. Temperature and electron density of soil plasma generated by LA-FPDPS. *Front Physics*. 2012; 7(6): 721-727. DOI 10.1007/s11467-012-0254-z

[8] Corsi M, Cristoforetti G, Hildalgo M, Iriarte D, Legnaioli S, Palleschi V, Salvetti A, Tognoni E. Temporal and spatial evolution of a laser-induced plasma from a steel target. *Appl Spectro*. 2003; 57(6): 715-721. DOI: 10.1366/000370203322005436

[9] Aguilera JA, Bengoechea J, Aragón C. Curves of growth of spectral lines emitted by a laser-induced plasma: influence of the temporal evolution and spatial inhomogeneity of the plasma. *Spectrochim Acta Part B-Atomic Spectroscopy*. 2003; 58: 221-237. DOI: 10.1016/S0584-8547(02)00258-6

[10] Aguilera JA, Aragón C. Characterization of a laser-induced plasma by spatially resolved spectroscopy of neutral atom and ion emissions. Comparison of local and spatially integrated measurements. *Spectrochim Acta Part B-Atomic Spectroscopy*. 2004; 59: 1861-1876. DOI:10.1016/j.sab.2004.08.003

[11] Gornushkin IB, Kazakov AY, Omenetto N, Smith BW, Winefordner JD. Experimental verification of a radiative model of laser-induced plasma expanding into vacuum. *Spectrochim Acta Part B-Atomic Spectroscopy*. 2005; 60: 215-230. DOI:10.1016/j.sab.2004.11.009

[12]   Bogaerts A, Chen Z, Gijbels R, Vertes A. Laser ablation for analytical sampling: what can we learn from modeling? *Spectrochim Acta Part B-Atomic Spectroscopy.* 2003; 58: 1867-1893. DOI:10.1016/j.sab.2003.08.004

[13]   Sun DX, Su MG, Dong CZ, Wen GH. A comparative study of the laser induced breakdown spectroscopy in single- and collinear double-pulse laser geometry. *Plasma Sci Technol.* 2014; 16(4): 374-379. DOI: 10.1088/1009-0630/16/4/13

[14]   Zhou WD, Su XJ, Qian HG, Li KX, Li XF, Yu YL, Ren ZJ. Discharge character and optical emission in a laser ablation nanosecond discharge enhanced silicon plasma. *J Analy Atomic Spectrometry.* 2013; 28: 702-710. DOI: 10.1039/c3ja30355a

[15]   Li KX, Zhou WD, Shen QM, Ren ZJ, Peng BJ. Laser ablation assisted spark induced breakdown spectroscopy on soil sample. *J Analy Atomic Spectrometry.* 2010; 25: 1475-1481.DOI: 10.1039/b922187e

[16]   Sun LX, Yu HB. Correction of self-absorption effect in calibration-free laser-induced breakdown spectroscopy by an internal reference method. *Talanta.* 2009; 79: 388-395. DOI:10.1016/j.talanta.2009.03.066

[17]   Hou HM, Tian Y, Li Y, Zheng RE. Study of pressure effects on laser induced plasma in bulk seawater. *J Analy Atomic Spectrometry.* 2014; 29: 169-175. DOI: 10.1039/c3ja50244a

[18]   Megan RLH, Joseph M, Robert O, Jennifer B, Yamac D, Caroline M, James BS. Ultrafast laser-based spectroscopy and sensing: applications in LIBS, CARS, and THz spectroscopy. *Sensors.* 2010; 10: 4342-4372. DOI: 10.3390/s100504342

[19]   Eland KL, Stratis DN, Lai T, Berg MA, Goode SR, Angel SM. Some comparisons of LIBS measurement using nanosecond and picosecond laser pulses. *Appl Spectroscopy.* 2001; 55(3): 279-285. DOI: 10.1366/0003702011951894

[20]   Wang ZZ, Deguchi Y, Kuwahara M, Yan JJ, Liu JP. Enhancement of laser-induced breakdown spectroscopy (LIBS) detection limit using a low-pressure and short-pulse laser-induced plasma process. *Appl Spectroscopy.* 2013; 67(11): 1242-1251. DOI: 10.1366/13-07131

[21]   Wang ZZ, Deguchi Y, Kuwahara M, Zhang XB, Yan JJ, Liu JP. Sensitive measurement of trace mercury using low pressure laser-induced plasma. *Jpn J Appl Physics.* 2013; 52: 11NC05-1-11NCO5-6. DOI: 10.7567/JJAP.52.11NC05

[22]   Zhang XB, Deguchi Y, Wang ZZ, Yan JJ, Liu JP. Detection characteristics of iodine by low pressure and short pulse laser-induced breakdown spectroscopy (LIBS). *J Analy Atomic Spectrometry.* 2014; 29: 1082-1089. DOI: 10.1039/c4ja00044g

[23]   Yaroshchyk P, Morrison RJS, Body D, Chadwick BL. Quantitative determination of wear metals in engine oils using LIBS: The use of paper substrates and a comparison between single- and double-pulse LIBS. *Spectrochim Acta Part B-Atomic Spectroscopy.* 2005; 60: 1482-1485. DOI: 10.1016/j.sab.2005.09.002

[24] Chen ZJ, Li HK, Liu M, Li RH. Fast and sensitive trace metal analysis in aqueous solutions by laser-induced breakdown spectroscopy using wood slice substrates. *Spectrochim Acta Part B-Atomic Spectroscopy*. 2008; 63: 64-68. DOI: 10.1016/j.sab.2007.11.010

[25] Sakka T, Yamagata H, Oguchi H, Fukami K, Ogata YH. Emission spectroscopy of laser ablation plume: Composition analysis of a target in water. *Appl Surf Sci*. 2009; 255: 9576-9580. DOI: 10.1016/j.apsusc.2009.04.086

[26] Sakka T, Masai S, Fukami K, Ogata YH. Spectral profile of atomic emission lines and effects of pulse duration on laser ablation in liquid. *Spectrochim Acta Part B-Atomic Spectroscopy*. 2009; 64: 981-985. DOI: 10.1016/j.sab.2009.07.018

[27] St-Onge L, Kwong E, Sabsabi M, Vadas EB. Rapid analysis of liquid formulations containing sodium chloride using laser-induced breakdown spectroscopy. *J Pharmaceut Biomed Analy*. 2004; 36: 277-284. DOI: 10.1016/j.jpba.2004.06.004

[28] Kumar A, Yueh FY, Miller T, Singh JP. Detection of trace elements in liquids by laser-induced breakdown spectroscopy with a Meinhard nebulizer. *Appl Optics*. 2003; 42(30): 6040-6046. DOI: 10.1364/AO.42.006040

[29] Zhao YH, Horlick G. A spectral study of charge transfer and Penning processes for Cu, Zn, Ag, and Cd in a glow discharge. *Spectrochim Acta Part B-Atomic Spectroscopy*. 2006; 61: 660-673. DOI: 10.1016/j.sab.2006.05.010

[30] Noll R. *Laser-Induced Breakdown Spectroscopy: Fundamentals and Applications*. Springer Berlin Heidelberg, 2012. p. 206-220. DOI: 10.1007/978-3-642-20668-9

[31] Cremers DA, Radziemski LJ. *Handbook of Laser-Induced Breakdown Spectroscopy*. John Wiley & Sons Ltd, England. 2006. p. 201-203.

[32] Ciucci A, Corsi M, Palleschi V, Rastelli S, Salvetti A, Tognoni E. New procedure for quantitative elemental analysis by laser-induced plasma spectroscopy. *Appl Spectroscopy*. 1999; 53(8): 960-964(5). DOI: 10.1366/0003702991947612

[33] Li XW, Wang Z, Lui SL, Fu YT, Li Z, Liu JM, Ni WD. A partial least squares based spectrum normalization method for uncertainty reduction for laser-induced breakdown spectroscopy measurements. *Spectrochim Acta Part B-Atomic Spectroscopy*. 2013; 88: 180-185. DOI: 10.1016/j.sab.2013.07.005

[34] Wang Z, Feng J, Li LZ, Ni WD, Li Z. A non-linearized PLS model based on multivariate dominant factor for laser-induced breakdown spectroscopy measurements. *J Analy Atomic Spectrometry*. 2011; 26: 2175-2182. DOI: 10.1039/c1ja10113g

[35] Lorenzen CJ, Carlhoff C, Hahn U, Jogwich M. Applications of laser-induced emission spectral analysis for industrial process and quality control. *J Analy Atomic Spectrometry*. 1992; 7: 1029-1035. DOI: 10.1039/ja9920701029

[36] Wan X, Wang P. Remote quantitative analysis of minerals based on multispectral line-calibrated laser-induced breakdown spectroscopy (LIBS). *Appl Spectroscopy*. 2014; 68(10): 1132-1136. DOI: 10.1366/13-07203

[37] Death DL, Cunningham AP, Pollard LJ. Multi-element and mineralogical analysis of mineral ores using laser induced breakdown spectroscopy and chemometric analysis. *Spectrochim Acta Part B-Atomic Spectroscopy*. 2009; 64: 1048-1058. DOI: 10.1016/j.sab.2009.07.017

[38] Death DL, Cunningham AP, Pollard LJ. Multi-element analysis of iron ore pellets by laser-induced breakdown spectroscopy and principal components regression. *Spectrochim Acta Part B-Atomic Spectroscopy*. 2008; 63: 763-769. DOI: 10.1016/j.sab.2008.04.014

[39] Yaroshchyk P, Death DL, Spencer SJ. Quantitative measurements of loss on ignition in iron ore using laser-induced breakdown spectroscopy and partial least squares regression analysis. *Appl Spectroscopy*. 2010; 64(12): 1335-1341. DOI: 10.1366/000370210793561600

[40] Yaroshchyk P, Death DL, Spencer SJ. Comparison of principal components regression, partial least squares regression, multi-block partial least squares regression, and serial partial least squares regression algorithms for the analysis of Fe in iron ore using LIBS. *J Analy Atomic Spectrometry*. 2012; 27: 92-98. DOI: 10.1039/c1ja10164a

[41] Barrette L, Turmel S. On-line iron-ore slurry monitoring for real-time process control of pellet making processes using laser-induced breakdown spectroscopy: graphitic vs. total carbon detection. *Spectrochim Acta Part B-Atomic Spectroscopy*. 2001; 56: 715-723. DOI: 10.1016/S0584-8547(01)00227-0

[42] Rosenwasser S, Asimellis G, Bromley B, Hazlett R, Martin J, Pearce T, Zigler A. Development of a method for automated quantitative analysis of ores using LIBS. *Spectrochim Acta Part B-Atomic Spectroscopy*. 2001; 56: 707-714. DOI: 10.1016/S0584-8547(01)00191-4

[43] Mateo MP, Nicolas G, Yañez A. Characterization of inorganic species in coal by laser-induced breakdown spectroscopy using UV and IR radiations. *Appl Surf Sci*. 2007; 254: 868-872. DOI: 10.1016/j.apsusc.2007.08.043

[44] Yuan TB, Wang Z, Li LZ, Hou ZY, Li Z, Ni WD. Quantitative carbon measurement in anthracite using laser-induced breakdown spectroscopy with binder. *Appl Optics*. 2012; 51(7): B22-B29. DOI: 10.1364/AO.51.000B22

[45] Ctvrtnickova T, Mateo MP, Yañez A, Nicolas G. Characterization of coal fly ash components by laser-induced breakdown spectroscopy. *Spectrochim Acta Part B-Atomic Spectroscopy*. 2009; 64: 1093-1097. DOI: 10.1016/j.sab.2009.07.032

[46] Feng J, Wang Z, Li L, Li Z, Ni WD. A nonlinearized multivariate dominant factor-based partial least squares (PLS) model for coal analysis by using laser-induced

breakdown spectroscopy. *Appl Spectroscopy.* 2013; 67(3): 291-300. DOI: 10.1366/11-06393

[47] Ctvrtnickova T, Mateo MP, Yañez A, Nicolas G. Laser induced breakdown spectroscopy application for ash characterisation for a coal fired power plant. *Spectrochim Acta Part B-Atomic Spectroscopy.* 2010; 65: 734-737. DOI: 10.1016/j.sab.2010.04.020

[48] Blevins LG, Shaddix CR, Sickafoose SM, Walsh PM. Laser-induced breakdown spectroscopy at high temperatures in industrial boilers and furnaces. *Appl Optics.* 2003; 42(30): 6107-6118. DOI: 10.1364/AO.42.006107

[49] Yin W, Zhang L, Dong L, Ma WG, Jia ST. Design of a laser-induced breakdown spectroscopy system for on-line quality analysis of pulverized coal in power plants. *Appl Spectroscopy.* 2009; 63(8): 865-872. DOI: 10.1366/000370209788964458

[50] Zhang L, Ma W, Dong L, Yang XJ, Hu ZY, Li ZX, Zhang YZ, Wang L, Yin WB, Jia ST. Development of an apparatus for on-line analysis of unburned carbon in fly ash using laser-induced breakdown spectroscopy (LIBS). *Appl Spectroscopy.* 2011; 65(7): 790-796. DOI: 10.1366/10-06213

[51] Noda M, Deguchi Y, Iwasaki S, Yoshikawa N. Detection of carbon content in a high-temperature and high-pressure environment using laser-induced breakdown spectroscopy. *Spectrochim Acta Part B-Atomic Spectroscopy.* 2002; 57: 701-709. DOI: 10.1016/S0584-8547(01)00403-7

[52] Kurihara M, Ikeda K, Izawa Y, Deguchi Y, Tarui H. Optimal boiler control through real-time monitoring of unburned carbon in fly ash by laser-induced breakdown spectroscopy. *Appl Optics.* 2003; 42(30): 6159-6165. DOI: 10.1364/AO.42.006159

[53] Papp Z, Dezső Z, Daróczy S. Significant radioactive contamination of soil around a coal-fired thermal power plant. *J Environ Radioactivity.* 2009; 59: 191-205. DOI: 10.1016/S0265-931X(01)00071-6

[54] Brim H, Venkateswaran A, Kostandarithes HM, Fredrickson JK, Daly MJ. Engineering deinococcus geothermalis for bioremediation of high-temperature radioactive waste environments. *Appl Environ Microbiol.* 2003; 69(8): 4575-4582. DOI: 10.1128/AEM.69.8.4575-4582.2003

[55] Sato I, Kudo H, Tsuda S. Removal efficiency of water purifier and adsorbent for iodine, cesium, strontium, barium and zirconium in drinking water. *J Toxicologic Sci.* 2011; 36(6): 859-834.

[56] Kim CK, Byun JI, Chae JS, Choi HY, Choi SW, Kim DJ, Kim YJ, Lee DM, Park WJ, Yim SA, Yun JY. Radiological impact in Korea following the Fukushima nuclear accident. *J Environ Radioactivity.* 2012; 111: 70-82.DOI 10.1016/j.jenvrad.2011.10.018

[57] Manolopoulou M, Vagena E, Stoulos S, Ioannidou A, Papastefanou C. Radioiodine and radiocesium in Thessaloniki, Northern Greece due to the Fukushima nuclear accident. *J Environ Radioactivity.* 2011; 102: 796-797. DOI: 10.1016/j.jenvrad.2011.04.010

[58] Hanson C, Phongikaroon S, Scott JR. Temperature effect on laser-induced break-down spectroscopy spectra of molten and solid salts. *Spectrochim Acta Part B-Atomic Spectroscopy*. 2014; 97: 79-85. DOI: 10.1016/j.sab.2014.04.012

[59] LaHaye NL, Harilal SS, Diwakar PK, Hassanein A. Persistence of uranium emission in laser-produced plasmas. *J Appl Physics*. 2014; 115: 163301-1-163301-8. DOI: 10.1063/1.4873120

[60] Saeki M, Iwanade A, Ito C, Wakaida I, Thornton B, Sakka T, Ohba H. Development of a fiber-coupled laser-induced breakdown spectroscopy instrument for analysis of underwater debris in a nuclear reactor core. *J Nuclear Sci Technol*. 2014; 51(7-8): 930-938.DOI: 10.1080/00223131.2014.917996

[61] Gong YD, Choi DW, Han BY, Yoo JH, Han SH, Lee YH. Remote quantitative analysis of cerium through a shielding window by stand-off laser-induced breakdown spectroscopy. *J Nuclear Mater*. 2014; 453: 8-15. DOI: 10.1016/j.jnucmat.2014.06.022

[62] Cho HH, Kim YJ, Jo YS, Kitagawa K, Arai N, Lee YI. Application of laser-induced breakdown spectrometry for direct determination of trace elements in starch-based flours. *J Analy Atomic Spectrometry*. 2001; 16: 622-627. DOI: 10.1039/b100754h

[63] Pouzar M, Černohorský T, Průšová M, Prokopčáková P, Krejčová A. LIBS analysis of crop plants. *J Analy Atomic Spectrometry*. 2009; 24: 953-957. DOI: 10.1039/b903593a

[64] Juvé V, Portelli R, Boueri M, Baudelet M, Yu J. Space-resolved analysis of trace elements in fresh vegetables using ultraviolet nanosecond laser-induced breakdown spectroscopy. *Spectrochim Acta Part B-Atomic Spectroscopy*. 2008; 63: 1047-1053. DOI: 10.1016/j.sab.2008.08.009

[65] Samek O, Beddows DCS, Telle HH, Kaiser J, Liška M, Cáceres JO, Gonzáles Ureña A. Quantitative laser-induced breakdown spectroscopy analysis of calcified tissue samples. *Spectrochim Acta Part B-Atomic Spectroscopy*. 2001; 56, 865-875. DOI: 10.1016/S0584-8547(01)00198-7

[66] Thareja RK, Sharma AK, Shukla S. Spectroscopic investigations of carious tooth decay. *Med Engin Physics*. 2008; 30, 1143-1148. DOI: 10.1016/j.medengphy.2008.02.005

[67] Nevin A, Spoto G, Anglos D. Laser spectroscopies for elemental and molecular analysis in art and archaeology. *Appl Physics A: Mater Sci Process*. 2012; 106: 339-361. DOI: 10.1007/s00339-011-6699-z

[68] Giakoumaki A, Melessanaki K, Anglos D. Laser-induced breakdown spectroscopy (LIBS) in archaeological science—applications and prospects. *Analy Bioanaly Chem*. 2007; 387: 749-760. DOI : 10.1007/s00216-006-0908-1

[69] Osorio LM, Cabrera LVP, García MAA, Reyes TF, Ravelo I. Portable LIBS system for determining the composition of multilayer structures on objects of cultural value. *J Physics: Conference Series*. 2011; 274: 1-6. DOI: 10.1088/1742-6596/274/1/012093

[70] McCanta MC, Dobosh PA, Dyar MD, Newsom HE. Testing the veracity of LIBS analyses on Mars using the LIBSSIM program. *Planetary Space Sci.* 2013; 81: 48-54. DOI: 10.1016/j.pss.2013.03.004

[71] Cho Y, Sugita S, Kameda S, Miura YN, Ishibashi K, Ohno S, Kamata S, Arai T, Morota T, Namiki N, Matsui T. High-precision potassium measurements using laser-induced breakdown spectroscopy under high vacuum conditions for in situ K-Ar dating of planetary surfaces. *Spectrochim Acta Part B-Atomic Spectroscopy.* 2015; 106: 28-35. DOI: 10.1016/j.sab.2015.02.002

[72] Knight AK, Scherbarth NL, Cremers DA, Ferris MJ. Characterization of laser-induced breakdown spectroscopy (LIBS) for application to space exploration. *Appl Spectroscopy.* 2000; 54(3): 331-340. DOI: 10.1366/0003702001949591

[73] Hou HM, Tian T, Li Y, Zheng RE. Study of pressure effects on laser induced plasma in bulk seawater. *J Analy Atomic Spectrometry.* 2014; 29: 169-175. DOI: 10.1039/c3ja50244a

[74] Marion Lawrence-Snyder, Jonathan P. Scaffidi, William F. Pearman, Christopher M. Gordon, S. Michael Angel. Issues in deep ocean collinear double-pulse laser induced breakdown spectroscopy: Dependence of emission intensity and inter-pulse delay on solution pressure. *Spectrochim Acta Part B-Atomic Spectroscopy.* 2014; 99: 172-178. DOI: 10.1016/j.sab.2014.06.008

[75] Guirado S, Fortes FJ, Lazic V, Laserna JJ. Chemical analysis of archeological materials in submarine environments using laser-induced breakdown spectroscopy. On-site trials in the Mediterranean Sea. *Spectrochim Acta Part B-Atomic Spectroscopy.* 2012; 74-75: 137-143. DOI: 10.1016/j.sab.2012.06.032

[76] Fortes FJ, Guirado S, Metzinger A, Laserna JJ. A study of underwater stand-offlaser-induced breakdown spectroscopy for chemical analysis of objects in the deep ocean. *J Analy Atomic Spectrometry.* 2015; 30: 1050-1056. DOI: 10.1039/c4ja00489b

[77] Lu Y, Li YD, Li Y, Wang YF, Wang S, Bao ZM, Zheng RE. Micro spatial analysis of seashell surface using laser-induced breakdown spectroscopy and Raman spectroscopy. *Spectrochim Acta Part B-Atomic Spectroscopy.* 2015; 110: 63-69. DOI: 10.1016/j.sab.2015.05.012

[78] Ferioli F, Buckley SG, Puzinauskas PV. Real-time measurement of equivalence ratio using laser-induced breakdown spectroscopy. *Int J Engine Res.* 2006; 7: 447-457. DOI: 10.1243/14680874JER01206

[79] Lombaert K, Morel S, Le Moyne L, Adam P, Tardieu de Maleissye J, Amouroux J. Nondestructive analysis of metallic elements in diesel soot collected on filter: Benefits of laser induced breakdown spectroscopy. *Plasma Chem Plasma Process.* 2004; 24(1): 41-56. DOI: 10.1023/B:PCPP.0000004881.17458.0d

[80] Strauss N, Fricke-Begemann C, Noll R. Size-resolved analysis of fine and ultrafine particulate matter by laser-induced breakdown spectroscopy. *J Analy Atomic Spectrometry*. 2010; 25: 867-874. DOI: 10.1039/b927493f

[81] Zhang HS, Yueh FY, Singh JP. Laser-induced breakdown spectrometry as a multimetal continuous-emission monitor. *Appl Optics*. 1999; 38(9): 1459-1466. DOI: 10.1364/AO.38.001459

[82] Buckley SG, Johnsen HA, Hencken KR, Hahn DW. Implementation of laser-induced breakdown spectroscopy as a continuous emissions monitor for toxic metals. *Waste Manag*. 2000; 20: 455-462. DOI: 10.1016/S0956-053X(00)00011-8

[83] Vander Wal RL, Ticich TM, West JR, Householder PA. Trace metal detection by laser-induced breakdown spectroscopy. *Appl Spectroscopy*. 1999; 53(10): 1226-1236.

[84] Effenberger AJ, Scott JR. Effect of atmospheric conditions on LIBS spectra. *Sensors*. 2010; 10: 4907-4925. DOI: 10.3390/s100504907

[85] Bertolini A, Carelli G, Francesconi F, Francesconi M, Marchesini L, Marsili P, Sorrentino F, Cristoforetti G, Legnaioli S, Palleschi V, Pardini L, Salvetti A. Modì: a new mobile instrument for in situ double-pulse LIBS analysis. *Analy Bioanaly Chem*. 2006; 385: 240-247. DOI: 10.1007/s00216-006-0413-6

[86] Sharma SK, Misra AK, Lucey PG, Lentz RCF. A combined remote Raman and LIBS instrument for characterizing minerals with 532 nm laser excitation. *Spectrochim Acta Part A-Mol Biomol Spectroscopy*. 2009; 73: 468-476. DOI: 10.1016/j.saa.2008.08.005

[87] Wiens RC, Sharma SK, Thompson J, Misra A, Lucey PG. Joint analyses by laser-induced breakdown spectroscopy (LIBS) and Raman spectroscopy at stand-off distances. *Spectrochim Acta Part A-Mol Biomol Spectroscopy*. 2005; 61: 2324-2334. DOI: 10.1016/j.saa.2005.02.031

[88] Hilbk-Kortenbruck F, Noll R, Wintjens P, Falk H, Becker C. Analysis of heavy metals in soils using laser-induced breakdown spectrometry combined with laser-induced fluorescence. *Spectrochim Acta Part B-Atomic Spectroscopy*. 2001; 56: 933-945. DOI: 10.1016/S0584-8547(01)00213-0

[89] Shiou FJ, Chen HH, Tsai CH. Development of a non-contact complete 3D surface measurement system for small complex objects using a triangulation laser probe. *J Chinese Instit Engin*. 2010; 33(2): 141-151. DOI: 10.1080/02533839.2010.9671606

[90] Fan KC, Tsai TH. Optimal shape error analysis of the matching image for a freeform surface. *Robotics Compu Integr Manufact*. 2001; 17: 215-222. DOI: 10.1016/S0736-5845(00)00029-6

[91] Nguyen HC, Lee BR. Laser-vision-based quality inspection system for small-bead laser welding. *Int J Precision Engin Manufact*. 2014; 15(3): 415-423.DOI: 10.1007/s12541-014-0352-7

[92] Shiou FJ, Deng YW. Simultaneous flatness and surface roughness measurement of a plastic sheet using a fan-shaped laser beam scanning system. *Key Engin Mater*. 2008; 381-382: 233-236. DOI: 10.4028/www.scientific.net/KEM.381-382.233

[93] GOM GmbH. User's manual of ATOS 4.7, GOM GmbH, Braunschweig, Germany; 2001.

[94] Chang M, Lin, KH. Non-contact scanning measurement utilizing a spacing mapping method. *Optics and Lasers in Engin*. 1998; 30(6): 503-512. DOI: 10.1016/S0143-8166(98)00053-0

Plasmas for Environmental Technology

Electrical Discharge in Water Treatment Technology for Micropollutant Decomposition

Patrick Vanraes, Anton Y. Nikiforov and Christophe Leys

Additional information is available at the end of the chapter

http://dx.doi.org/10.5772/61830

Abstract

Hazardous micropollutants are increasingly detected worldwide in wastewater treatment plant effluent. As this indicates, their removal is insufficient by means of conventional modern water treatment techniques. In the search for a cost-effective solution, advanced oxidation processes have recently gained more attention since they are the most effective available techniques to decompose biorecalcitrant organics. As a main drawback, however, their energy costs are high up to now, preventing their implementation on large scale. For the specific case of water treatment by means of electrical discharge, further optimization is a complex task due to the wide variety in reactor design and materials, discharge types, and operational parameters. In this chapter, an extended overview is given on plasma reactor types, based on their design and materials. Influence of design and materials on energy efficiency is investigated, as well as the influence of operational parameters. The collected data can be used for the optimization of existing reactor types and for development of novel reactors.

Keywords: electrode configuration, electrohydraulic discharge, energy yield, organic degradation efficiency, dielectric barrier discharge

1. Introduction

In this introductory section, the current status and limitations of both conventional and advanced water treatment systems are explained in detail, with a focus on their performance on micropollutant removal. This context is necessary to understand the role that plasma technology can play in the future challenges of water purification, which is the main topic of this chapter. Further, it opens a clear perspective on research that still needs to be done in order to turn the plasma treatment of water into a mature technology.

With ongoing improvement of chemical analytical methods, a wide spectrum of compounds and their transformation products are increasingly detected in water bodies and sewage sludge. Many of these compounds occur in low concentrations in the range of microgram to nanogram per liter. They are therefore called micropollutants. Among these are food additives, industrial chemicals, pesticides, pharmaceuticals, and personal care products. Even with such low concentrations, various environmental effects have been observed. Continuous release of antibiotics in the environment leads, for example, to increasing resistance of microorganisms [1]. Chronic exposure of aquatic life to endocrine disruptive compounds causes feminization, masculinization, and immunomodulating effects in fish and frogs, which has tremendous effects on the ecosystem [2, 3]. Additionally, there is growing concern for the direct acute and chronic effects of micropollutants on human health and safety [4–8]. Although more insight is gained on the impact of individual micropollutants, their synergistic, additive, and antagonistic effects are still vastly unknown. Some countries or regions have adopted regulations for a small number of compounds. Nevertheless, effluent limitation guidelines and standards do not exist for most micropollutants.

The primary source of many micropollutants in aquatic systems is the effluent of conventional wastewater treatment plants. An example of such treatment plant is schematically shown in Figure 1. The treatment process occurs in sequential steps:

- Primary treatment consists of a series of mechanical and/or physicochemical treatment steps, which remove solids, oil, and fat. This step is usually common to all wastewater treatment plants [9]. During pretreatment, large debris is eliminated, as it can damage or clog pumps and sewage lines in later steps. Next, abrasive particles are withdrawn, such as grit, sand, oil, and fat. As a third step, a primary clarifier removes the last part of solid material before the wastewater is send to the secondary treatment unit. With this solid material, as much as possible biodegradable material is removed as well in order to significantly reduce oxygen demand and therefore costs in the later biological treatment unit. Overall, primary treatment removes almost half of the suspended solids in raw wastewater [10]. Sometimes, pH adjustment for neutralization is added as a last step of primary treatment to make the effluent suitable for secondary treatment [11].

- Secondary treatment usually consists of biological conversion of dissolved and colloidal organic compounds into stabilized, low energy compounds, by means of a diversified group of microorganisms in presence of oxygen. Additionally, new cells of biomass are generated during this biological treatment step. The mixture of microorganisms with inorganic and organic particles contained in the suspended solids is referred to as activated sludge. Secondary treatment can differ substantially between wastewater treatment plants, as it depends on the nature of the activated sludge [9].

- Tertiary treatment, also called water polishing, consists of any additional treatment that is carried out to make the effluent quality more suitable for discharge in the receiving environment. Chlorination is the most common method of wastewater disinfection due to its low costs. Nevertheless, it can lead to formation of chlorinated organic compounds that may be carcinogenic or harmful to the environment. Therefore, also ultraviolet disinfection is increasingly finding application in wastewater treatment plants [12]. Additional pollutant removal can be achieved with sand filtration, biological nutrient removal, adsorption on activated carbon, membrane processes, and advanced oxidation processes [9, 12]. However, many conventional wastewater treatment plants have no or limited tertiary treatment, due to the high costs associated with most of these methods.

As should be noted, organic pollution that enters a wastewater treatment plant has three outlets: effluent water, excess sludge, and exhaust gas in the form of CO_2 or volatile organic compounds. Excess sludge disposal is complex since it contains a high concentration of harmful substances and only a small part of it consists of solid matter. Sludge treatment is required to reduce the water and hazardous organic content and to render the processed solids suitable for reuse or final disposal. Therefore, sludge is processed in sequential steps of thickening, digestion, and dewatering before disposal [10].

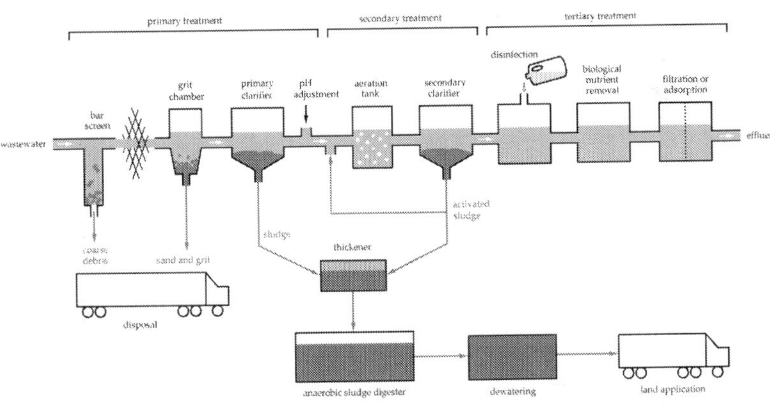

Figure 1. Simplified scheme of a conventional wastewater treatment plant, adapted from Flores Alsina and Benedetti et al. [10, 13].

Up to now, conventional treatment plants cannot sufficiently remove micropollutants, as shown by many studies worldwide [14]. Several options are available to improve the elimination of these contaminants, including:

- Prevention by application of products without micropollutants or only with micropollutants which are easily removed

- Reassessment and optimization of current treatment processes

- Pretreatment of hospital and industrial effluents

- End-of-pipe upgrading of wastewater treatment plants

Preventive measures will always be limited by the increased demand for industrial, pharmaceutical, and personal care products. Moreover, highly stable micropollutants such as the pesticide atrazine are detected even several years after the discontinuation of their use. Hence, more effective water treatment processes are required. Existing processes in water treatment plants can be optimized by increased sludge ages and hydraulic retention times in conjunction with nutrient removal stages and the varying redox conditions associated with them. Temperature and pH control can also enhance micropollutant removal. According to the reviews by Jones et al. and Liu et al., this is expected to be the most economically feasible approach to increase overall water treatment plant performance [15, 16]. While such measures indeed lead to more effective removal of micropollution in general, they often have limited or negligible effect on several specific persistent micropollutants, as shown in the extended review by Luo et al. [14].

The removal difference among different compounds can be partly ascribed to micropollutant properties. For example, Henry's law constant as a measure of compound volatility gives an indication how easily a contaminant will be removed by aeration and convection. In activated sludge processes, the solid–water distribution coefficient K_d has been proposed as a relative accurate indicator of sorption behavior [17, 18]. It is defined as the partition of a compound between the sludge and the water phase and takes into account both hydrophobicity and acidity of the molecule. The biodegradation of micropollutants is more complex to predict, but compound structure is an important indicator. Highly branched side chains and sulfate, halogen, or electron withdrawing functional groups generally make a contaminant less biodegradable. Also, saturated or polycyclic compounds show high resistance to biodegradation [19, 20].

In order to remove such persistent compounds, an additional secondary or tertiary treatment step for wastewater is required. As a cost-effective alternative, the treatment of hospital and industrial effluent can solve the problem at its source. For these purposes, several advanced treatment techniques have been proposed in recent years. Each removal option has its own limitations and benefits in removing trace contaminants. A complete review on this topic was conducted by Luo et al. [14]. Advanced biological treatment techniques are activated sludge, membrane bioreactors, and attached growth technology. As in the case of conventional biological treatment, these methods are commonly unable to remove polar persistent micropollutants. Coagulation–flocculation processes yield ineffective elimination of most micropollutants. In contrast, electrochemical separation treatment methods such as electrocoagulation and internal microelectrolysis give very good results, as discussed by Sirés and Brillas [21]. Activated carbon, nanofiltration, and reverse osmosis are generally also very effective for removal of trace contaminants. However, they are associated with high energy costs and high financial costs. Moreover, all these separation methods have the additional problem of toxic concentrated residue disposal, as they only remove the compounds without further decomposition into less toxic by-products. Advanced oxidation techniques, on the other hand, are able to oxidize toxic compounds to smaller molecules, ideally with full mineralization to CO_2 and H_2O [22]. Nevertheless, their weakest point is as well their high energy demand. Additionally, they can be hazardous to the environment, by their unwanted production of CO_2, long-living oxidants, and potentially toxic oxidation products. Therefore, research needs to focus on optimizing existing systems in order to overcome these problems.

Optimal water treatment schemes will eventually be decided upon, achieving effluent limitations set by national or international environmental regulations at a reasonable cost. In their extended reviews, Jones et al. and Liu et al. conclude that it seems unpractical for activated carbon, nanofiltration, reverse osmosis, and advanced oxidation techniques to be widely used in conventional wastewater treatment [15, 16]. However, with more and more shortages of drinking water all over the world, the recycling of wastewater treatment plant effluents as a drinking water source seems just a question of time. For such purpose, these technologies may be advantageous due to their high removal efficiency. Moreover, with the recent developments in energy harvesting in wastewater treatment [23, 24], one needs to consider the possibility to make wastewater treatment plants self-sustainable in power consumption, even with additional implementation of advanced treatment. As mentioned above, the treatment of hospital and industrial effluent can be a cost-effective approach as well. This last approach has gained a lot of attention over the past several years. For an overview of research done in this field for the case of hospitals, the reader is referred to the reviews of Verlicchi et al. [25, 26].

Comninellis et al. propose in their perspective article a strategy for wastewater treatment, depending on the water's total organic content, biodegradability, toxicity, and other physico-chemical requirements, such as transparency [27]. According to this strategy, the use of cost-effective biological treatment is only advised when total organic content is high enough. If such wastewater is not biodegradable, advanced oxidation can be used as pretreatment step, to enhance biodegradability, and reduce toxicity. One needs to consider, however, that presence of oxidant scavengers can sabotage the oxidation efficiency. Therefore, the used oxidation technology needs to be geared toward the wastewater under treatment. In the case that total organic content is low, wastewater possesses little metabolic value for the microorganisms. Then, advanced oxidation technology that effectively mineralizes the targeted pollution can be applied as a one-step complete treatment method. Alternatively, separation treatment can be applied prior to advanced oxidation, where pollutants are transferred from the liquid to another phase and subsequently posttreated.

Following this line of thought, advanced oxidation techniques take a promising place in the quest for micropollutant removal, as they appear the most effective methods for the decomposition of biorecalcitrant organics. However, in this time of a growing energy crisis and concerns over global warming, removal efficiency should not be the only objective. Sustainable development on the whole must to be considered. Therefore, the main objective of research on advanced oxidation technology should be optimization in terms of energy cost and effluent toxicity as well as its compatibility with biological treatment.

Examples of advanced oxidation processes are ozonation, hydrogen peroxide addition, chlorination, Fenton process, UV irradiation, radiolysis, microwave treatment, subcritical wet air oxidation, electrochemical oxidation, homogeneous and heterogeneous catalytic oxidation, ultrasonication, and combinations thereof, such as peroxonation, photocatalysis, and electro-Fenton process.

One type of oxidation method is typically insufficient for micropollutant removal, while a combination of oxidation methods with each other or with other advanced treatment techniques leads to significant improvement up to complete removal, as concluded in many reviews, such as for antibiotics [28], for pharmaceuticals [29], for UV-based processes [30], and for the general case [31, 32]. Subcritical wet air oxidation is not feasible for micropollutant removal [33].

Oturan and Aaron conclude from their review that chemical methods such as Fenton's process and peroxonation perform worse than photochemical, sonochemical, and electrochemical advanced oxidation techniques [32]. For the latter three technologies, they further conclude the following:

- Among the photochemical methods, photo-Fenton process and heterogeneous photocatalysis possessed in most cases a better efficiency than H_2O_2 and O_3 photolysis. Moreover, the solar photo-Fenton process and solar photocatalysis reduce energy consumption even further. The heterogeneous catalyst TiO_2 has many advantages, as it is chemically highly stable, biologically inert, very easy to produce, inexpensive, and possessing an energy gap comparable to that of solar photons.

- The sonochemical combination of Fenton's process with ultrasonication has effective results at lab scale, but application at the industrial level in real time is needed to demonstrate its economic and commercial feasibility.

- Electrochemical methods have the advantage of minimizing or eliminating the use of chemical reagent. Anode oxidation and electro-Fenton have very good performance when a boron-doped diamond anode is used. They can be combined with other advanced oxidation methods for further improvement of efficiency. Examples of such combinations are photoelectro-Fenton, solar photoelectro-Fenton, sonoelectro-Fenton, and peroxyelectrocoagulation. One needs to consider, however, that boron-doped diamond anodes are very expensive [21].

In conclusion, a synergetic combination of multiple oxidants and oxidation mechanisms is recommended for efficient micropollutant decomposition. Therefore, water treatment by means of plasma discharge takes an interesting and promising place among the advanced oxidation techniques, as it can generate a wide spectrum of oxidative species and processes in proximity of the solution under treatment, including shock waves, pyrolysis, and UV radiation. The hydroxyl radical OH is often named as the most important oxidant, due to its high standard oxidation potential of 2.85 V and its unselective nature in organic decomposition. Further, plasma in contact with liquid can generate significant amounts of O_3 and H_2O_2. These two oxidants are frequently used in other advanced oxidation methods and lead together to the peroxone process. Other important reactive oxygen plasma species include the oxygen radical O, the hydroperoxyl radical HO_2, and the superoxide anion O_2^-. When electrical discharge occurs in air, also reactive nitrogen species such as the nitrogen radical, the nitric oxide radical NO, and the peroxynitrite anion $ONOO^-$ will play an important role. Oxidants can either enter the liquid phase through transfer from the gas phase or be formed directly in the liquid phase at the plasma–water interface by interaction of plasma species with water or dissolved molecules. Radical hydrogen H, a powerful reducing agent, is directly formed in aqueous phase by the electron collision with water molecules. Plasma discharge also leads to aqueous electrons, which even have a stronger reduction potential. A more detailed overview of oxidative plasma species and processes without nitrogen-containing oxidants was conduct-ed by Joshi and Thagard [34]. Recent insights on the chemistry of plasma-generated aqueous peroxynitrite and its importance in water treatment are given by Brisset and Hnatiuc and Lukes et al. [35, 36].

The presence of different oxidants reduces the selectivity of an oxidation method. Direct oxidation of organics by ozone is, for example, very selective. This is illustrated with the reaction rate constants listed by Jin et al., Sudhakaran and Amy, and Von Gunten [37–39], which range from 10^{-5} to 3.8×10^7 $M^{-1}s^{-1}$. Hydroxyl, on the other hand, is considered unselective in the decomposition of organics, with reaction rate constants from 2.2×10^7 to 1.8×10^{10} M^{-1} s^{-1}, as listed in the same 3 references and [40]. Hence, organic oxidation by means of plasma is expected to be rather unselective as well. Nevertheless, the chemistry behind plasma treatment is very complex due to the interactions between the various reactive species in the gaseous phase, in the aqueous phase and at their interface. Moreover, this chemistry is strongly dependent on the used electrode configuration and material, discharge regime, applied voltage waveform, water properties, and feed gas. Therefore, the optimization of plasma reactors for water treatment in terms of energy costs and effluent toxicity is a complicated task, which still requires more research effort and insight. On the positive side, plasma-based water treatment has already shown itself as a versatile technology, which can find application in the treatment of biological, organic, and inorganic contamination, after sufficient optimization has been reached. As an additional advantage, its flexible design allows it to be easily combined with other advanced treatment techniques. Such combinations can lead to interesting synergetic effects and further optimization.

As pointed out already by Sillanpää et al. [41], no information has been reported in scientific literature on the treatment cost of advanced oxidation processes. To our knowledge, plasma technology has only been compared with other water treatment methods in the review of Sirés and Brillas [21]. According to their review, the main drawback of plasma technology would be its high energy requirement. However, the authors did not validate this claim with quantitative data. Comprehensive quantitative assessment is needed to compare different techniques better from both economic and technical points of view, as Luo et al. conclude in their review [14]. For the specific case of plasma technology for water treatment, only 3 reviews have extensively compared different reactors in their energy efficiency for organic decompo-sition. Malik was the first to do this for dye degradation [42]. Recently, Bruggeman and Locke and Jiang et al. have made a comparison for mostly phenolic compounds [43, 44]. Their findings will be discussed in next sections. Nevertheless, the main comparative parameter in these reviews is energy yield G_{50} for 50% pollutant removal, expressed in g/kWh, which is only used in literature on plasma treatment. Moreover, G_{50} is strongly dependent on the initial pollutant concentration. It is therefore not fit for comparison with other advanced treatment techniques.

In this chapter, several plasma technologies are discussed that can be applied in water treatment. First, different approaches for reactor classification will be analyzed in detail. In Section 3, an overview will be given of different plasma reactor types, where classification is based on reactor design and reactor materials. The influence of multiple working parameters on reactor energy efficiency is discussed in Section 4. Section 5 gives a summary with future prospects and concluding remarks on the research that is still required in this field.

2. Approaches for reactor classification

Plasma reactors for water treatment can be classified in many ways, depending on several criteria. Such classifications have already been made in a few reviews. One popular approach starts from 2 or 3 main plasma–water phase distributions and subdivides reactors further based on their electrode configuration, as in Bruggeman and Leys and Locke et al. [45, 46]. Here, we will adapt this approach with 6 plasma–water phase distributions, where electrical discharge is generated

1. directly in the water bulk,

2. directly in the water bulk with externally applied bubbles (bubble discharge reactors),

3. in gas phase over water bulk or film,

4. in gas phase with water drops or mist,

5. as a combination of the previous types, or

6. not in direct contact with the solution under treatment.

Accordingly, these reactor types are called (1) electrohydraulic discharge, (2) bubble discharge, (3) gas phase discharge, (4) spray discharge, (5) hybrid, and (6) remote discharge reactors. The basic idea behind this classification is that the total plasma–water interface surface is an important, determining parameter for a reactor's energy efficiency. A larger interface surface is expected to cause higher pollutant degradation efficiency, in agreement with the reviews of Malik, Jiang et al., and Bruggeman and Locke [42–44]. Interface surface can be enlarged by generating plasma in bubbles, by spraying the solution through the active plasma zone, and by making the solution flow as a thin film along the discharge. It can be further enlarged by extending the plasma volume, which has led to many possible choices of electrode configurations and geometries. Additionally, dielectric barriers and porous layers are often introduced in the setup to avoid unwanted energy losses to Joule heating of water and spark formation, while enhancing the local electric field for easier breakdown. In Section 3, this approach for reactor classification will be studied in more detail.

Another popular approach classifies reactors based on the used discharge regime and applied voltage waveform. The following discharge regimes should be considered:

a. Corona and streamer discharge

b. Glow discharge

c. Dielectric barrier discharge (DBD)

d. Arc discharge

Microdischarge and Townsend discharge are not included in this list, as they are not used for water treatment plasma reactors to our knowledge, except in ozone generation. Spark discharge is to be understood as a transient form of arc discharge and therefore falls in the fourth category. All four discharge regimes can be formed in the gas phase or the liquid phase, although we only found one very short and recent report on submerged DBD discharge without bubbles for water disinfection [47]. In the gas phase, corona discharge needs lowest power input and arc discharge the highest. Underwater discharge requires additional energy for plasma onset by cavitation, but it has the advantage of a large plasma–liquid contact surface. Locke et al. and Jiang et al. suggested in their reviews to apply low-energy plasma such as corona and glow discharge for the treatment of water with low contaminant concentration. On the other hand, high-energetic arc discharge might be more effective for high pollutant concentration [44, 46].

Depending on the voltage waveform used, the following types of discharge can be distinguished:

i.   DC discharge

ii.   AC (low frequency) discharge

iii.   Radio frequency discharge

iv.   Microwave discharge

v.   Monopolar pulsed discharge

vi.   Bipolar pulsed discharge

Variations are possible, such as periodically interrupted AC discharge to avoid excessive heating of the plasma gas [48]. Multiple voltage-related parameters can influence reactor energy efficiency, such as voltage amplitude and polarity, sinusoidal or pulse frequency, and pulse rise time and width. Their influence will be discussed in more detail in Section 4. By combining different voltage waveforms with the different discharge regimes in either the liquid or the gas phase, a list of discharge types can be obtained. Malik, for example, identified 27 distinct reactor types in his literature study, using this approach [42]. Locke et al. made a classification into 7 types [46]:

- Pulsed corona and corona-like electrohydraulic discharge

- Pulsed spark electrohydraulic discharge

- Pulsed arc electrohydraulic discharge

- Pulsed power electrohydraulic discharge

- Gas phase glow discharge

- Gas phase pulsed corona discharge

- Hybrid gas–liquid electrical discharge

This classification partly overlaps with the 6 types of plasma–water phase distribution mentioned above, where bubble discharge is included in hybrid gas–liquid discharge. Jiang et al. add 4 more types [44]:

- DC pulseless corona electrohydraulic discharge

- Dielectric barrier discharge

- Gas phase gliding arc discharge

- DC arc discharge torch

While the first 3 types generate plasma in or in contact with the liquid, the last type introduces the water under treatment directly in the torch, where it subsequently gets vaporized and becomes the plasma forming gas. It can therefore be classified as a subtype of spray discharge reactors. For more summarized information on the basic chemistry and physics of the discharge types in this list, the reader is referred to [43, 46].

The list illustrates how the 6 reactor types based on plasma–water phase distribution can be further split into subtypes. This has been done in a more extensive way by Locke and Shih, who identified more than 30 subtypes during their comparative study for the reactor energy efficiency of H_2O_2 production [49]. While this method has proven to be useful for their purpose, it disregards the influence of reactor design, materials, and working parameters. As Locke and Shih point out, studying this influence is a challenging task, but it will reveal valuable information on reactor optimization.

A small number of reviews have been made on plasma reactor energy efficiency for organic decomposition, which mostly focused on the influence of plasma–water phase distribution, discharge type, and a few working parameters. Nevertheless, no comprehensive overview is found in literature up to now on reactor design and reactor materials. The next section will deal with this topic. While the section is not meant to be fully comprehensive, we want to give a broader overview of reactor types reported in literature than has been done in prior reviews. Such overview is not only useful to get easier and faster understanding of reactor operation and development but also serves as source of inspiration for future reactor designs. Moreover, it reveals the close relationship between electrode configuration and discharge type. Both popular and more exotic, unique designs will be discussed, with extra focus on several general working parameters in Section 4. This review also includes plasma reactors that have been used for inorganic removal or biological treatment, as these reactors are also valuable candidates for aqueous organic decomposition.

3. Overview of reactor types

3.1. Electrohydraulic discharge reactors

Electrohydraulic discharge reactors have been studied for many years due to their importance in electrical transmission processes and their potential for water treatment. From theoretical point of view, they are attractive for water treatment due to the relatively high ratio of plasma–water contact surface to plasma volume and proximity of plasma to the water surface. Moreover, they generate shock waves that can aid in organic decomposition. Nevertheless, electrohydraulic discharge reactors are usually less efficient than other reactor types for water decontamination [42, 50]. This is likely due to the additional input energy required for cavitation, i.e., gas phase formation during discharge onset.

Most commonly reported types of electrohydraulic discharge are pulsed arc and pulsed corona discharge. To our knowledge, all arc electrohydraulic discharge reactors reported in literature have pulsed input power. Mostly, a rod-to-rod electrode configuration is used (Figure 2a), but there are also reports on a reactor with a grounded L-shaped stationary electrode and a vibrating rod electrode (Figure 2b) [51, 52]. As learned from personal communication with Dr. Naum Parkansky, the vibrating electrode has the purpose to facilitate electrical breakdown and mix the treated solution. Apart from organic degradation, pulsed arc electrohydraulic discharge is also gaining more recent attention for biological treatment [53, 54]. Often, refractory metal such as tantalum, titanium, tungsten, or a corresponding alloy is selected as electrode material, as it needs to be sufficiently resistant to corrosion and shock waves. Tungsten has proven to be less corrosive than titanium and titanium alloy [53]. Particles that eroded from titanium electrodes enhanced methylene blue decomposition during aging in one study. The authors explained this enhancement with titanium peroxide formation from interaction of H_2O_2 with the particle surface [55]. In contrast, particles that eroded from low carbon steel electrodes diminished the decomposition of the same pollutant, possibly through the catalytic decomposition of H_2O_2 and scavenging of OH radicals [52]. While an increasing energy efficiency is reported for the decomposition of 2,4,6-trinitrotoluene [56] and methyl-tert-butyl ether [57] when the electrode gap is reduced, the opposite effect has been observed for atrazine degradation in another study [58]. Further investigation is needed to understand this apparent contradiction. In a small comparative study, Hoang et al. found the energy efficiency of 4-chlorophenol decomposition of this type of reactor to be one order of magnitude lower than degradation efficiency with UV, UV/H_2O_2, and O_3 systems [59].

Submerged pin-to-pin electrode configuration, as in Figure 2c, is uncommon in scientific literature. Such reactor has been used with high frequency bipolar pulsed power with reduced voltage and low pulse energy by Potocký et al. [61], to lower temperature loading of the electrodes and to clean both electrode tips continuously from any possible adjacent products. With addition of a high inductance in series with the discharge, transition from glow type to arc type discharge can be suppressed [62]. For sufficiently large interelectrode distance and low voltage amplitude, unbridged nonarc discharge was observed. For closer electrodes and higher voltage, bridged arc discharge was obtained, which, according to the authors, was initiated with spark formation at both electrodes [63].

A pin-to-plate electrode configuration (Figure 2d) is often used in electrohydraulic discharge reactors, either for DC glow [64, 65] or pulsed corona [66–68] discharge with positive polarity. The pin curvature radius determines the local electric field strength and is therefore an important parameter that influences discharge initiation [69]. For pulsed corona discharge, the anode pin material has been reported to lead to catalytic effects for organic decomposition.

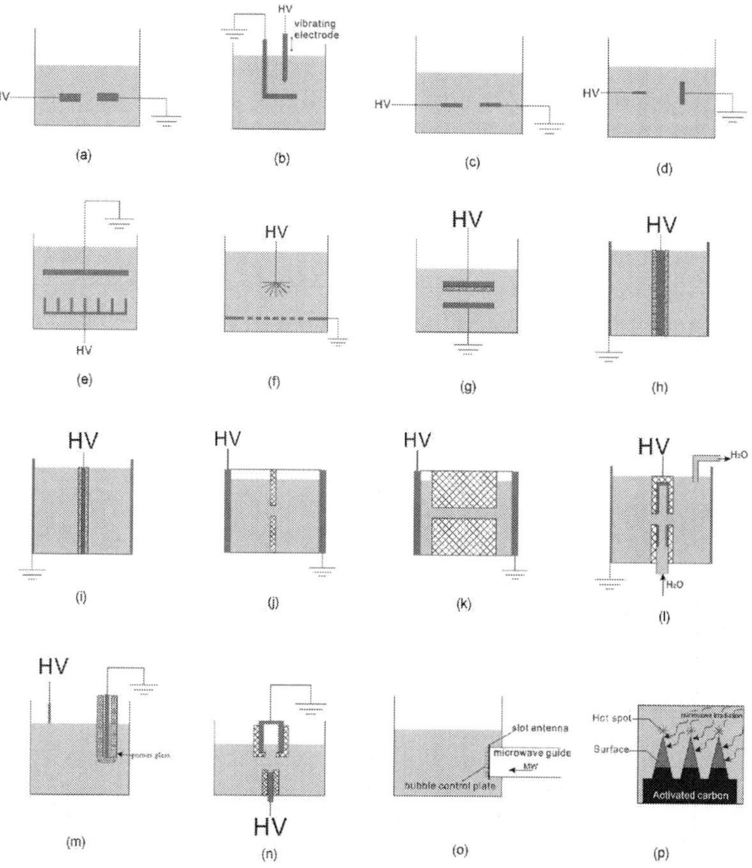

Figure 2. Types of electrohydraulic discharge reactors: (a) pulsed arc, (b) pulsed arc with vibrating electrode, (c) pin-to-pin, (d) pin-to-plate, (e) multi-pin-to-plate, (f) brush-to-plate, (g) plate-to-plate with porous ceramic coating, (h) coaxial rod-to-cylinder with ceramic coating on rod, (i) coaxial wire-to-cylinder with ceramic coating on wire, (j) diaphragm discharge, (k) capillary discharge, (l) coaxial diaphragm discharge reactor from Šunka et al. [60] with perforations in tubular electrode covered by polyethylene layer, (m) contact glow discharge electrolysis, (n) RF-discharge in cavitation bubble on electrode, (o) microwave bubble plasma from waveguide with antenna slot, and (p) "hot spot" plasma formation on activated carbon surface under influence of microwave irradiation.

Platinum enhances pollutant degradation as compared to a NiCr, but only in combination with certain electrolytes. In the case of ferrous salts, this is due to reduction of Fe^{3+} to Fe^{2+} by erosion particles from the platinum electrode [66, 67]. Erosion of tungsten, on the other hand, causes catalysis of oxidation by plasma-generated H_2O_2 [70].

To achieve a higher plasma volume, the pin electrode can be replaced by a multipin electrode [71], a brush electrode [72], or a plate electrode coated with a thin ceramic layer [73], respectively, shown in Figures 2e–g. The concentration of predischarge in the pores of the ceramic layer enhances the electric field strength on the electrode surface [68]. Due to inhomogeneities such as entrapped microbubbles inside the ceramic layer, the electric field can be locally even higher. As a result, a large number of streamers can be generated with lower input voltage as compared to uncoated electrode systems. Moreover, a ceramic coating does not only facilitate an upscale of the system but also serves as a support for a suitable catalyst of the plasma chemical reactions. Also, certain ceramic materials can enhance organic decomposition by catalytic effects [74]. Usually, positive high-voltage pulses are applied to the coated electrode. In some cases, negative high voltage is used to avoid arc formation, as this can damage the coating [73]. However, bipolar pulses are advised since monopolar pulses cause a polarized charge buildup on the ceramic, which can quench the electrical discharge [75].

To enlarge the plasma volume even further, coaxial geometry has been used, where either a coated rod [75] or wire [68, 74] high-voltage electrode is located at the symmetry axis of the grounded cylindrical electrode (Figures 2h–i). Analog to the curvature of pin electrodes, the diameter of the inner electrode is an important system parameter.

When a submerged anode and cathode are separated from each other with a perforated dielectric barrier, electrohydraulic discharge will occur through cavitation at the perforation. For larger ratio of perforation diameter to thickness, this type of discharge is called diaphragm discharge (Figure 2j), while for lower values, the term capillary discharge is used (Figure 2k). DC glow discharge [76, 77] and pulsed corona discharge [78] are commonly used for this reactor, but also AC power input is possible. The strongly inhomogeneous electric field during plasma onset has a similar structure to the one in pin-to-plate geometry. Therefore, similar plasma features can be expected. Energy efficiency for dye and phenol degradation with a single diaphragm is the same as in the case of a pin-to-plate electrode system [46]. Also, similarities with contact glow discharge electrolysis are reported [77]. As an important difference, diaphragm discharge is not in direct contact with the electrodes, which prevents electrode erosion [76]. Sunka et al. developed a coaxial reactor where a polyethylene covered tubular anode with perforations was placed inside a cylindrical cathode (Figure 2l) [60]. The generated plasma was reported to be similar as well.

Another common type of electrohydraulic discharge is contact glow discharge electrolysis. As depicted in Figure 2m, a pointed anode is placed with its tip in the water surface. It is separated from the submerged cathode by means of a sintered glass barrier. In such reactor, glow discharge is generated at the anode tip in a vapor layer surrounded by water. Plasma volume can be increased by increasing the anode number. Stainless steel performed better as anode material than platinum for Acid Orange 7 decoloration [79].

Electrohydraulic discharge can also be generated with RF or microwave power, but such reactor types are less common. Figure 2n shows a reactor where plasma is generated in a cavitation bubble on the tip of an RF electrode [80]. Producing cavitation bubbles by means of microwave power is more complicated. Therefore, a slot antenna can be placed in between the liquid and a microwave guide, as illustrated in Figure 2o. The electric field intensity can be enhanced by installing a quartz plate with holes, a so-called bubble control plate, on the slot antenna. Ishijima et al. reported an increase of methylene blue decomposition efficiency with a factor of 20 after installation of the bubble control plate and tripling the amount of slot antennas [81]. Another way to produce underwater plasma with microwave power is by adding a microwave-absorbing material with high surface area, such as activated carbon, to the solution under treatment. Under influence of microwave irradiation, delocalized π-electrons on the activated carbon surface gain enough energy to jump out of the surface and generate confined plasmas (Figure 2p), also called hot spots, which are known to increase organic decomposition efficiency [82].

3.2. Bubble discharge reactors

Since electrohydraulic discharge generally has low energy efficiency due to the difficulty of initiating discharge directly in the water phase, a lot of attention has gone to enhancing efficiency by discharge formation in externally applied bubbles. Bubbling has the additional advantage of mixing the solution. Moreover, discharge initiation in the gas phase minimizes electrode erosion, which lengthens the lifetime of the system. Bubbling gas through the discharge region greatly increases radical density in the plasma, as for example observed by Sun et al. [69] for O_2 and Ar bubbles. Obviously, feed gas plays a determining role. Yasuoka et al. measured highest efficiency for Ar bubbles, in which plasma spread extensively along the inner surface, while lowest efficiency was obtained with He plasma, which has smallest plasma–water contact surface [83]. The importance of the working gas in plasma reactors in general will be further discussed in Section 4.

A common method is to pump gas upward through a nozzle anode, located underneath a grounded electrode [69], as shown in Figure 3a. Often, the nozzle electrode is placed inside a dielectric tube up to its tip, to avoid any energy leakage toward the water. Alternatively, a pin anode is sometimes placed inside a dielectric nozzle which transports the feed gas (Figure 3b) [86]. The pin tip can be placed below or above the nozzle extremity. Many variations can be encountered in literature, such as a pin anode inside a perforation in a dielectric plate (Figure 3c) [87] and different nozzle orientations (Figures 3d–e) [88, 89]. All choices in nozzle or perforation material, shape, dimensions, and orientation determine the bubble shape during formation and its position after detachment, which significantly influences the electric field in the interelectrode region and therefore the plasma characteristics. This complicates comparison of different reactors. Another option is to place the high-voltage electrode underneath the perforated dielectric plate, as shown by Yasuoka et al., Sato et al., and Yamatake et al. [83, 90, 91] (Figures 3f–h), where the ring-shaped grounded electrode is located around the bubble. Also, here, electrode geometry and position influence the electric field. Yasuoka et al. found their single hole reactor (Figure 3h) to be more energy efficient than advanced oxidation with photochemical persulfate, photocatalyst heteropoly acid, photodegradation, and ultrasonic cavitation for the decomposition of 2 surface active compounds [83]. By increasing the number of nozzles or holes, energy efficiency can be enhanced. In a study by Sato et al. [90], a reactor with a single hole (Figure 3f) was compared to a reactor with 9 holes (Figure 3g). Discharge power deposited per hole was lower in the reactor with 9 holes, which seemed to minimize self-quenching of OH radicals, resulting in higher efficiency. Following this line of thought, a multibubble system as in Figure 3i with a high-voltage mesh in the gas phase attached to a

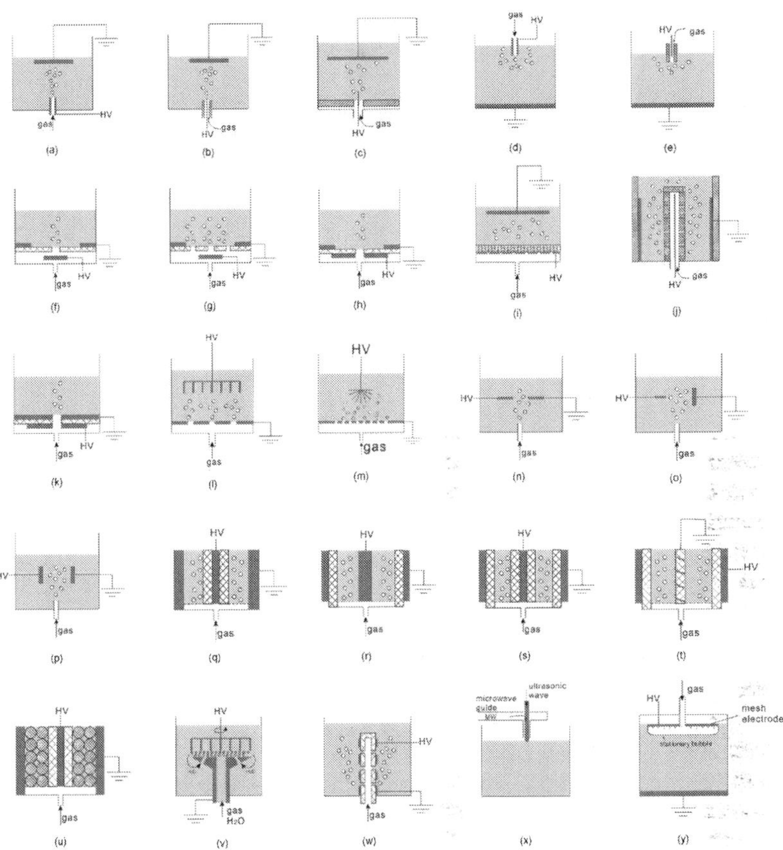

Figure 3. Types of bubble discharge reactors: (a) upward nozzle electrode-to-plate, (b) upward nozzle containing needle electrode-to-plate, (c) hole containing needle electrode-to-plate, (d) downward nozzle electrode-to-plate, (e) downward nozzle containing needle electrode-to-plate, (f–h) hole above electrode and inside circular ground electrode, (i) multibubble discharge on porous ceramic with high-voltage mesh, (j) multibubble discharge on porous ceramic tube surrounding a high-voltage wire electrode, (k) bubble discharge on hole in grounded electrode, (l–m) bubbles rising toward high-voltage electrode, (n–p) bubbles rising in between electrodes, (q) DBD reactor with inner barrier, (r) DBD reactor with outer barrier, (s) DBD reactor with double barrier, (t) DBD reactor with spiral electrode from Aoki et al. [84], (u) glass bead packed-bed DBD reactor, (v) coaxial arc discharge reactor with rotating multipin electrode from Johnson et al. [85], (w) multielectrode slipping surface discharge reactor, (x) microwave discharge in ultrasonic cavitation bubble, and (y) stationary bubble under high-voltage mesh.

porous ceramic seems a promising alternative [92]. Similarly, bubbling gas through a porous ceramic tube containing a high-voltage wire electrode [93] (Figure 3j) gives a large plasma–water contact surface as well. Increasing gas flow rate had no influence in this reactor on decomposition efficiency of phenol, while energy efficiency was enhanced for Acid Orange II removal.

Another situation is found when bubbles are formed on the grounded electrode, as shown by Yamatake et al. and Nikiforov [91, 94] (Figure 3k). Reactor from Figure 3h has more aggressive reaction with water and improved durability of electrode in comparison to the reactor from Figure 3k, as concluded by Yamatake et al. [91]. Nevertheless, one needs to take into account the applied voltage type. While for most of the reactors above positive pulsed corona discharge is used, less commonly AC voltage [94] or positive DC voltage [89, 91] is applied, leading to significantly different phenomena. As Yamatake has shown, plasma can be stably generated in the reactor from Figure 3h even without gas flow since O_2 gas is generated from electrolysis by application of positive DC voltage, while this is not the case for the reactor from Figure 3k. The microdischarge channel makes plasma more stable than that in the study of Kurahashi et al. [89], where the same phenomenon has been studied. Therefore, the system of Figure 3k requires higher gas flow due to the short lifetime of the oxygen radical, while in the reactor of Figure 3h decomposition did not depend on flow rate due to direct reaction with water.

Gas can also be bubbled from the ground electrode toward a multipin or brush high-voltage electrode, as shown by Chen et al. and Wang et al. [72, 95] (Figures 3l–m), or in between 2 sideways positioned electrodes, as shown by Lee et al., Miichi et al., and Vanraes et al. [96–98] (Figures 3n–p). The discharge systems are often very similar to the ones described in Section 3.1 without bubbles. For a plate-to-plate configuration, pulsed streamer discharge mostly occurred inside bubbles adjacent to the electrodes [97].

A last common type of bubble discharge reactor has a coaxial DBD geometry. The cylindrical dielectric barrier can either be placed at the inner [99] or at the outer electrode [91], or at both [91] (Figures 3q–s). For a single barrier, a reticulate electrode can be placed in contact with the water in order to enhance the local electric field [99]. With double barrier, unwanted erosion of the electrodes can be avoided. In the study of Yamatake et al. [91], a double-barrier reactor was found to have higher decomposition efficiency of acetic acid in comparison to a single barrier reactor with similar dimensions, which is most likely caused by a difference in energy density. Monopolar pulsed, bipolar pulsed, and AC high-voltage are most commonly applied to the inner electrode. In the study of Aoki et al. [84], however, the reactor from Figure 3t was used to generate RF glowlike plasma in bubbles. The grounded spiral electrode impeded the motion of the bubbles in the small gap between the electrodes, which increases the probability of discharge in the bubbles. Nevertheless, most of the input energy seems to be dissipated as heat in this system, resulting in low energy efficiency. Bubble movement can also be impeded by adding obstacles in the water bulk. For the single barrier reactor of [100] (Figure 3u), the addition of spherical glass beads significantly enhanced the energy efficiency of indigo carmine decomposition. Porous ceramic sphere gave worse performance than glass beads, but better one than inert conductive fragments. As an interesting research question, it is still unclear whether bubbles in contact with the electrodes in these DBD systems give rise to a better efficiency than freely rising bubbles or not.

A more exotic and patented reactor design is investigated by Johnson et al. [85] (Figure 3v). In this reactor, the high-voltage electrode is a pin array that can rotate at speeds up to 2500 rpm. Oxygen is pumped through the stationary electrode and nebulized to form a bubble mist between the electrodes. DC voltage is applied to the system, generating arc discharge. Rotating the electrode distributes erosion particles evenly on the stationary electrode, preventing pitting and unwanted changes in the relative distance between the electrodes. Moreover, it reduces mass transfer limitations that are apparent in pin-to-plate reactors. Additionally, it lowers the inception voltage by abating the effective distance between the pin electrodes and the stationary electrode. Energy efficiency for methyl tert-butyl ether decomposition increased with an increase in spin rate. The system was found to be more energy efficient than many corona-based technologies but still requires further optimization.

Anpilov et al. developed the multielectrode slipping surface spark discharge system depicted in Figure 3w [101]. A series of cylindrical electrodes are mounted on the outside of a dielectric tube. One of the extreme electrodes is grounded, and high-voltage pulses are applied to the other extreme electrode. All other electrodes are on floating potential. To increase system efficiency, the outer surface of the electrodes is coated with a thin insulating layer. Gas is pumped into the electrode gaps through drilled holes in the dielectric tube. When a high-voltage pulse is applied, plasma discharge occurs initially in the first interelectrode gap adjoining the high-voltage electrode. The breakdown of this gap quickly transports the high-voltage potential to the next electrode, which leads to the breakdown of the second gap. The process repeats itself until the grounded electrode has been reached. This way, the discharge load on each electrode can be kept low, enhancing erosion resistance and increasing the system lifetime. The effectiveness of the system has been demonstrated for disinfection of biological wastewater, methane conversion, and aqueous organic waste decomposition [102].

Apart from bubble cavitation by electrical discharge, the classical gas pumping method, and gas formation by electrolysis, bubbles can also be generated with ultrasonic cavitation. Since microwave power cannot easily generate cavitation bubbles, a combination of ultrasound and microwave discharge can be an attractive method. An example is given by Horikoshi et al. [103] (Figure 3x). Stationary bubbles, however, are not commonly used in bubble discharge reactors, unless for diagnostic purposes. An example is given by Yamabe et al. [104], with the reactor of Figure 3y. This reactor was used to investigate plasma formation and propagation along the gas–water interface, which is a common feature for bubble discharge reactors and which makes it, therefore, different in nature than most gas phase discharge reactors.

3.3. Gas phase discharge reactors

Electrical discharge in the gas phase is usually more energy efficient for organic degradation than discharge in the liquid phase [42, 43, 50]. In this section, we will distinguish 4 subgroups of gas phase discharge reactors: corona and glow discharge over a horizontal water surface (Figure 4), DBD over a horizontal water surface (Figure 5), falling water film reactors (Figure 6), and arc discharge over a water surface (Figure 7).

3.3.1. Corona and glow discharge over water surface

The most standard version of a discharge over water surface has a pin-to-water configuration with a grounded water electrode, as depicted in Figure 4a. The type of discharge produced in this reactor, corona, glow, or transient glow-to-spark, depends on the applied voltage, pin curvature, interelectrode distance, and voltage polarity [105, 106]. For the application of water treatment, both positive and negative DC and monopolar pulsed voltage have been reported. AC input power is less common but has been used as well [107]. Plasma volume can be increased by replacing the high-voltage pin electrode with a multipin [108], a brush [72], or a horizontal wire [109] (Figures 4b–d). In the study of Miyazaki et al. [108] with a multipin electrode, the energy efficiency for a certain amount of phenol decomposition was independent of the type of discharge, the voltage amplitude, the polarity of the applied voltage, and the amount of pin electrodes.

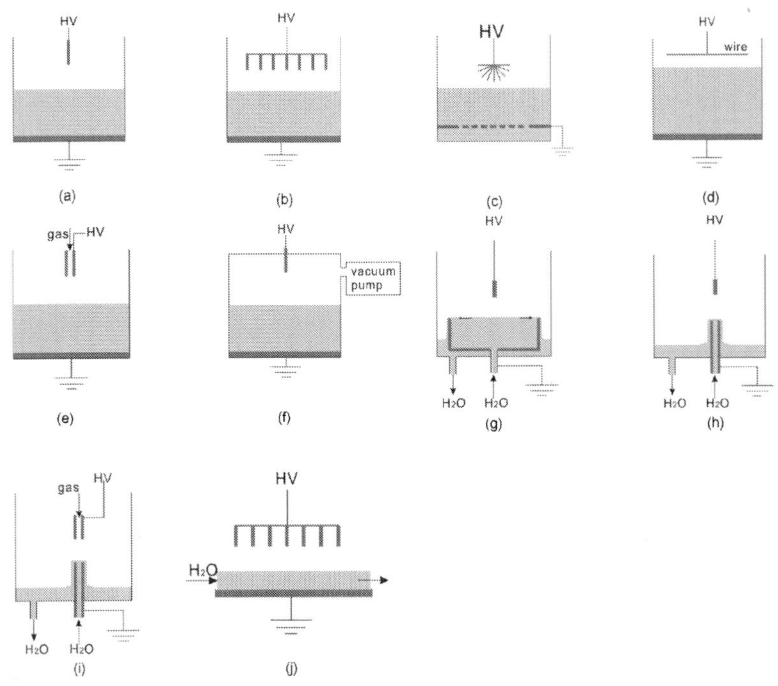

Figure 4. Types of reactors with corona or glow discharge over water surface: (a) pin-to-water, (b) multi-pin-to-water, (c) brush-to-water, (d) wire-to-water, (e) nozzle electrode-to-water, (f) low-pressure pin-to-water glow discharge reactor, (g) pin above radial water flow, (h) pin above flowing liquid electrode, (i) miniature microjet above flowing liquid electrode, and (j) multipin above water flow.

According to Dors et al., atmospheric pressure glow discharge in air produces gaseous nitrogen oxides, leading to formation of undesirable aqueous nitrates and nitrites, while DC positive corona produces ozone in air without any traces of nitrogen oxides [110]. The energy efficiency of phenol oxidation in their system, depicted in Figure 4e, was comparable to the results obtained in pulsed corona discharge systems. Sharma et al., however, compared their low-pressure negative DC glow discharge reactor (Figure 4f) with bench scale data of atmospheric pressure corona discharge and concluded that the power cost for pentachlorophenol decomposition was lower for their system [111]. Additionally, the operating cost of their reactor was found to be comparable with power cost of UV-based advanced oxidation technologies. Yet it is unclear how feasible such low-pressure system is for applications with large water volume or continuous water flow.

It is important to note that dimensions and movement of the water phase in this type of reactors can influence energy efficiency significantly. In the system of Sharma et al., for example, stirring rate increased the rate of pentachlorophenol removal [111]. Water movement also plays an important role in radial flow reactors (Figure 4g). In the reactors from Jamróz et al. [112] with small sized flowing liquid cathode (Figures 4h–i), degradation efficiency depended strongly on the water flow rate, while the Ar flow rate from the miniature flow Ar microjet (Figure 4i) gave negligible effect on methyl red decomposition. The importance of water and gas flow rates will be further discussed in Section 4.3. As mentioned above, making the solution flow as a thin film along the discharge is another way to enhance the oxidation process. Promising results from a pilot-scale system with negative pulsed corona from multiple carbon fiber cathodes above a flowing water film (Figure 4j) have been published by Even-Ezra et al., Gerrity et al., and Mizrahi and Litaor [113–115]. The system was similar or more efficient than a pilot-scale UV/H_2O_2 advanced oxidation process and achieved similar energy efficiency to those reported in the literature for other advanced oxidation processes [113, 114]. Moreover, the plasma pilot system with additional ozone injection was more cost-effective than three other commercialized advanced oxidation systems (O_3/H_2O_2, O_3/UV, and $O_3/H_2O_2/UV$) [115]. However, the total capital costs and reliability of large-scale gas discharge reactors for water treatment are still relatively unclear.

3.3.2. DBD over water surface

DBD over horizontal water surface is most commonly powered with AC voltage, but occasionally pulsed high voltage has been used as well. Often, glass is used as dielectric barrier, especially quartz glass, while Al_2O_3 ceramic barriers have also been reported less frequently. Interestingly, reactors with DBD over water often have energy efficiency for organic decomposition that increases with input power [116–118]. The most standard reactor design for DBD in the gas phase over a water surface is shown in Figure 5a. In the study of Hu et al. [116], energy efficiency in such reactor was found to increase by decreasing the distance between the dielectric barrier electrode and the water surface. This can be explained with a decrease in plasma volume and thus in unused plasma reactions far from the water surface. A water batch can also be placed in between two dielectric barriers to avoid erosion of one of the electrodes, as in the study of Hijosa-Valsero et al. [119] (Figure 5b). A more exotic way of bringing DBD over a water surface is shown in Figure 5c, where water is kept on floating potential. However, this device has not been applied for organic degradation and is instead used for biomedical applications [120]. In another less common design, the water can be located above the dielectric barrier, with an uncovered high-voltage electrode positioned above the water surface, as in the wire-to-water reactor of Marotta et al. [121] (Figure 5d). In this reactor, the energy efficiency of phenol decomposition was 3.2 times higher with stainless steel wires as compared to Ni/Cr wires. However, it is unclear whether this effect should be attributed to the larger diameter of the Ni/Cr wires or their possible inhibiting effect on ozone or other oxidants. The material of the ground electrode in contact with water in many DBD reactors is, on the other hand, clearly important. In the study of Lesage et al. [122], the use of a stainless steel substrate resulted in a better decomposition efficiency in comparison to the use of brass. This is explained with corrosion of the brass substrate under influence of nitrate, leading to formation of aqueous nitrite, which scavenges OH radicals and thus inhibits the degradation process. Stainless steel, as a more inert metal, does not have this effect.

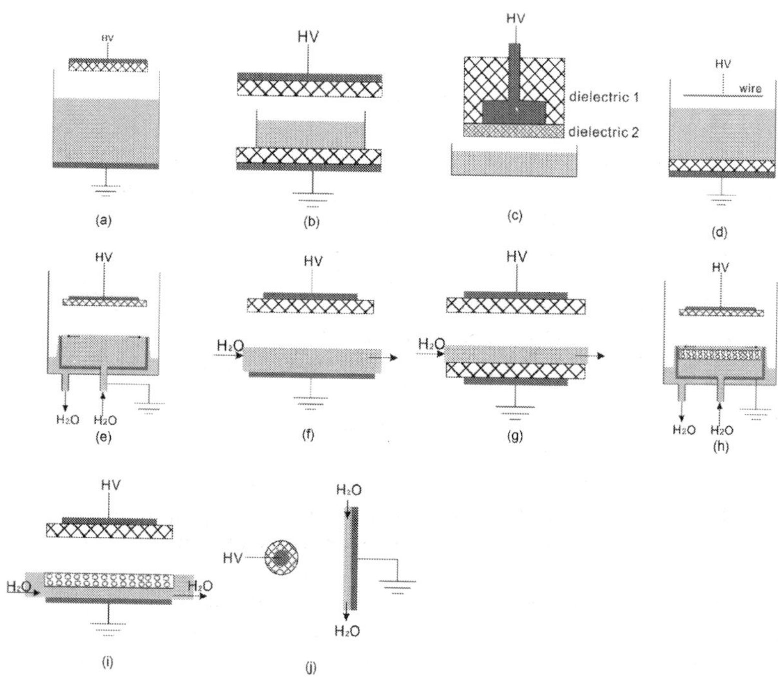

Figure 5. Types of reactors with DBD over water surface: (a) DBD-to-water, (b) water batch in between DBD, (c) DBD above floating electrode, (d) wire-to-water DBD, (e) DBD above radial water flow, (f) DBD above water flow, (g) water flow in between DBD, (h) DBD over radial water flow on porous ceramic, (i) DBD over water flow on porous ceramic, and (j) DBD rod to falling water film.

Also for DBD-based systems, the movement of water influences the degradation efficiency. Reactors with radial flow [117] or flowing water films [118, 123] are investigated in literature for organic decomposition (Figures 5e–g). Both situations have also been reported with incorporation of a porous ceramic segment in the zone between electrodes (Figures 5h–i) [124, 125]. This porous segment serves as guide for the flowing water. It allows the water to remain undisturbed by the electrical discharge due to hydrophilic force. This enables a reduction of the discharge gap and subsequently an increase in the intensity, stability, homogeneity, and efficiency of the discharge. Under such a configuration, a transition from filamentary mode to semi-homogeneous mode of the plasma discharge can be realized [125]. Moreover, such ceramic can be effectively used as substrate for photocatalysts, where both substrate and catalyst remain unchanged after use [124].

An exotic reactor type where a rod high-voltage electrode with dielectric cover is placed next to a falling water film has been investigated by Lesage et al. [122, 123] (Figure 5j). The reactor was found to be significantly more efficient than a gliding arc (see Section 3.3.4) over the same falling water film, partly due to less corrosion of the brass substrate in contact with the water.

3.3.3. Coaxial reactors with falling water film

A relatively common falling water film reactor that does not use DBD is shown in Figures 6a–b and is often referred to with the term wetted-wall reactor. Either a rod [126] or more commonly a wire [127–130] high-voltage electrode is placed along the axis of a grounded cylindrical electrode. The falling water film flows along the inner wall of the cylinder electrode where it comes in contact with streamer or corona discharge. Mostly, positive pulsed power is applied on the inner electrode, but also negative DC [127] has been reported. Usually, corona discharge is formed in such reactors for all voltage waveforms, while spark discharge is undesired due to excessive energy dissipation to Joule heating. To prevent spark formation, it is necessary to have the entire inner wall area covered by the water flow [130]. The choice of gas flow direction is important, as concluded by Faungnawakij et al. [131] for negative DC corona. Experiments showed a downward airflow to be more effective than upward airflow for acetaldehyde degradation. In the study of Sano et al. [127], the energy efficiency of a wetted-wall reactor with negative DC voltage applied to the inner wire was calculated to be 3 to 4 times higher than in a wire-to-water corona reactor (Figure 4d) over flowing water with negative DC voltage. For the same reactor, energy efficiency was found to be highest for conditions of a smooth water surface, i.e., for a minimal water flow rate where the flow entirely covers the anode inner wall and for an optimal current, which does not disturb the flow by strong ion wind. There is an optimal wall radius, where decomposition efficiency is maximal. For higher radius, many plasma-generated short-lived radicals cannot reach the water film in time, while for smaller radius, the plasma–water contact surface decreases [127]. Sato et al. compared four kinds of coaxial reactors with falling water film and positive pulsed power. They found phenol most energy efficiently removed with the configuration of Figure 6b [128]. Sealing such reactor seems beneficial for energy efficiency, due to better confinement of the produced ozone [129].

Most falling water film reactors generate plasma by DBD, either with AC or monopolar pulsed high voltage. Pulsed DBD in coaxial configuration using O_2 is considered as one of the most efficient electrical discharge systems evaluated because of the large surface area and small electrode distance [50]. Several configurations are possible, but in the most common design, the water film flows over the surface of an inner stainless steel rod electrode placed inside a glass cylindrical vessel which acts as dielectric barrier. The outer electrode can be a metal mesh or a metal painted layer which is located around the vessel. Four versions of this reactor design

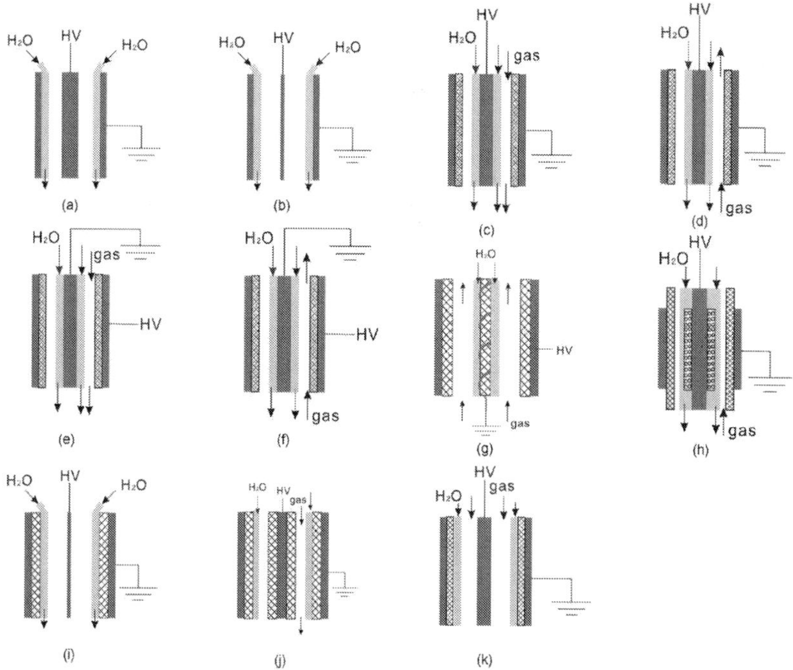

Figure 6. Types of coaxial reactors with falling water film: (a) wetted-wall reactor with rod electrode, (b) wetted-wall reactor with wire electrode, (c–f) 4 variations of falling water film DBD reactor with outer barrier, (g) falling water film DBD reactor with inner spiral electrode, (h) falling water film on glass fiber fabric in DBD reactor with outer barrier, (i) wetted-wall DBD reactor with wire electrode, (j) wetted-wall DBD reactor with double barrier, and (k) configuration of coaxial whirlpool reactor of [132].

are found in literature, where the inner electrode is either grounded or connected to the high voltage and with an upward or downward gas flow (Figures 6c–f). With the reactor of Figure 6f, energy efficiencies of micropollutant decomposition are about one order of magnitude higher than for a water batch in between a DBD reactor (Figure 5b) [119]. A slightly different configuration was used by Ognier et al. [133], where a tungsten wire that was rolled around a dielectric rod served as inner grounded electrode (Figure 6g). In this study, volatile aqueous compounds were treated. The more volatile the compound was, as expressed with the Henry's law constant, the more efficiently it was removed. Therefore, degradation processes of pollutants in the gas phase should be considered in plasma reactors, depending on the volatility of the compound. In the study of Bubnov et al. [134], the inner electrode of a coaxial DBD reactor was covered with a 1-mm-thick porous hydrophilic glass-fiber fabric (Figure 6h). This fabric allows a more homogeneous water flow and higher water retention time. Moreover, it can function as substrate for catalysts, such as Cu and Ni compounds, which enhanced the decomposition efficiency in the research of Bubnov et al. [134].

Less frequently, the water film is chosen to flow along the inner wall of the dielectric barrier. Morimoto et al., for example, investigated the effect of placing a dielectric barrier inside the wetted-wall reactor of Figure 6b, as shown in Figure 6i. Addition of the barrier allows to decrease the interelectrode gap without formation of spark discharge, which is expected to increase energy efficiency. With the application of positive nanosecond pulsed high voltage on the inner wire, the treatment efficiency of the DBD system was found to be, surprisingly, less energy efficient for indigo carmine decomposition as compared to the normal wetted-wall reactor. Another type of wetted-wall DBD reactor with falling water film is investigated by Rong et al. [135], for a reactor with double dielectric barrier (Figure 6j). A more exotic type of DBD reactor with modified water-gas mixing is reported by Chen et al. [132]. The system is powered with high frequency bipolar tailored voltage pulses. It has a configuration as shown in Figure 6k, but water and air are introduced in the reactor with high flow rate of 5 L/min and 100 to 200 L/min, respectively, causing a whirlpool. For this reactor, gas flow rate was shown to have negligible effect on decomposition efficiency of methyl orange.

3.3.4. Arc discharge over water surface

Gliding arc discharge above a water surface (Figure 7a) is a popular approach for water treatment with plasma. In this reactor type, two diverging electrodes are placed above a water solution. An electric arc forms at the shortest electrode gap and glides along the electrode's axis under influence of a gas flow directed toward the water surface. The arc length increases on moving and its temperature decreases, turning the arc from thermal plasma into quenched plasma while breaking into a plume. A new arc then forms at the narrowest gap and the cycle continues. Unfortunately, many publications on this type of reactor are unclear about the use of AC or DC voltage, but AC power is definitely a common choice. Important research in this field has focused on enlarging the plasma treated water surface with adjustments in design. One possibility is to use a couple of controlled electrodes in between the electrode gap to facilitate breakdown, increasing current intensity and allowing a larger interelectrode distance [136]. Another option is to use three main electrodes supplied by two power sources, as proposed by Burlica et al. [137] (Figure 7b). Both approaches have shown to increase reactor efficiency. Gliding arc discharge can also be used for the treatment of falling water films, as shown by Lesage et al. [123] (Figure 7c). Arc discharge with an active water electrode is less commonly researched (Figure 7d). According to Janca et al. [138], the energy efficiency of such system was found to be strongly dependent on the type of discharge produced, such as arc or gliding arc. For more detailed information on water treatment by means of gliding arc, the reader is referred to the review by Brisset et al. [136].

3.4. Spray discharge reactors

3.4.1. Low-energy spray discharge reactors

Although spray discharge reactors have received more attention in recent years due to their high reported energy efficiencies [42–44], still more research is required to characterize and optimize them. One of the most common spray reactors has a wire-to-cylinder geometry

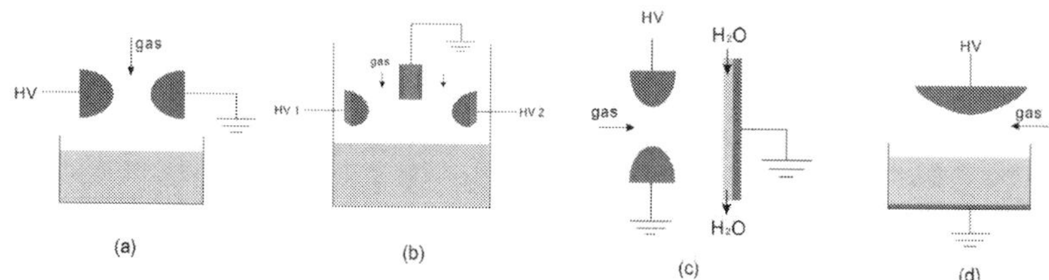

Figure 7. Types of gliding arc discharge reactors over water surface: (a) standard configuration, (b) configuration with extra main electrode, (c) gliding arc discharge to falling water film, and (d) gliding arc discharge with active water electrode.

(Figure 8a), similar to wetted-wall reactors. This reactor type is always operating with positive pulsed corona or streamer discharge, according to our literature review. In the study of Kobayashi et al. [139], different spraying nozzles were used to investigate the influence of water location on the energy efficiency of indigo carmine decomposition. As the results showed, spraying water as droplets into the discharge area is more effective than making it flow as a water film on the inner reactor wall. Moreover, droplets that were sprayed close to the reactor wall underwent 1.5 times faster decolorization than droplets near the wire electrode. Energy efficiency was found to be independent of droplet size for same water flow rate [140].

Sugai et al. adjusted this reactor by addition of packed-bed of pellets (Figure 8b) or fluorocarbon wires (Figure 8c) in order to increase the droplet retention time in the discharge space [141]. The packed pellets were hollow polyethylene balls with 14 holes per ball to increase discharge. The fluorocarbon wires were woven as insulation grids in the outer cylindrical electrode. Addition of the pellets decreased the energy efficiency significantly, due to the narrowing of the discharge space. The fluorocarbon wires, however, kept the discharge space unaltered and increase the energy efficiency with 2–10%.

In the study of An et al. [142], a similar electrode configuration was used as in Figure 8a, but with tooth wheels assembled on the inner electrode (Figure 8d). Here, droplets were not created by spraying, but due to condensation of steam under influence of up-flowing air. Alternatively, positive pulsed corona can also be generated in a spray reactor around wire anodes in parallel with two grounded plate cathodes [143], as depicted in Figure 8e. In another approach, water is sprayed in between rod electrodes that are each surrounded with a dielectric barrier (Figure 8f). Monopolar pulsed voltage of both polarities is reported in literature [144, 145]. In the study of Wang et al. [146], grounded water was sprayed from dielectric nozzles in proximity of a high-voltage dielectric barrier plate electrode (Figure 8g). When AC high voltage was applied to the plate, the by electrostatic induction generated electrostatic force pulled the water droplets to the glass dielectric layer. The energy efficiency of indigo carmine decomposition depended on both voltage amplitude and air gap distance, for which optimal values were found. Another interesting reactor type is based on the electrospray process to simultaneously generate and treat water spray under influence of high voltage, as shown in Figure

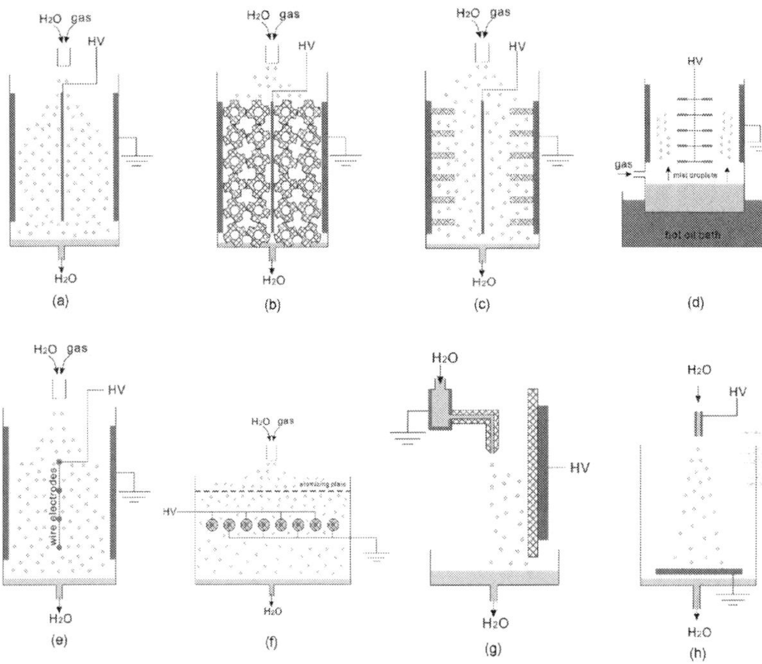

Figure 8. Types of low-energy spray discharge reactors: (a–c) wire-to-cylinder corona or streamer reactor without or with packed-bed of pellets or fluorocarbon wires for increased droplet retention, (d) mist droplets in tooth wheel-to-cylinder corona reactor, (e) multi-wire-to-plate corona reactor, (f) spray through DBD rods, (g) spray from grounded electrode to DBD plate, and (h) electrospray reactor.

8h. In the study of Elsawah et al. [147], water is treated that way with positive pulsed corona electrospray.

3.4.2. Spray arc reactors

Water spray can also be introduced in gliding arc discharge for treatment, as depicted in Figure 9a. This method is found to be more energy efficient than gliding arc discharge over a water surface as discussed in Section 3.3.4 [136, 137]. The decomposition of 4-chlorophenol in such spray reactor became more energy efficient with increasing gas–water mixing rate [148]. Efficiency is higher with electrodes from stainless steel than for aluminum or brass electrodes. Also here, extending the plasma volume by the use of a controlled electrode couple or an extra main electrode enhances energy efficiency (Figure 9b) [137]. The process can also be optimized by introducing the water with a flat spraying nozzle perpendicular to the gas flow to improve the contact with the plasma (Figure 9c) [149]. For more detailed information on water treatment by means of gliding arc spray reactors, the reader is referred to the review by Brisset et al. [136]. Water can be treated as well with a DC plasma torch, where it is usually directly introduced into the torch as plasma forming gas. A discussion on this treatment method is provided by Brisset et al. [44].

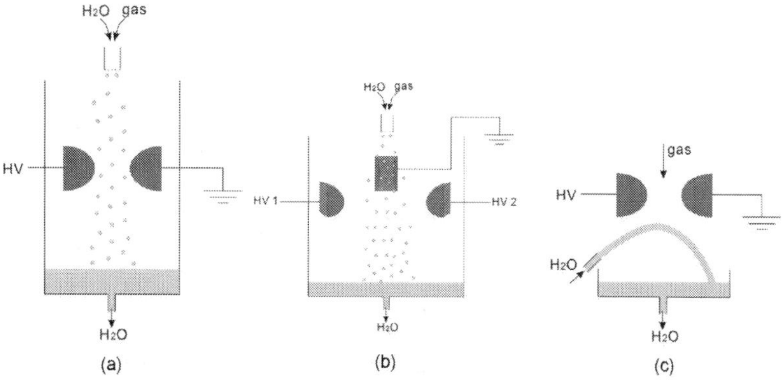

Figure 9. Types of spray gliding arc reactors: (a) standard configuration, (b) configuration with extra main electrode, and (c) water jet under angle through gliding arc discharge.

3.5. Hybrid reactors

Electrohydraulic discharge and gas phase discharge can simultaneously be generated when a high-voltage electrode is placed in the water phase with a grounded electrode above the water surface in the gas phase. Mostly, positive pulsed corona is generated in these systems reported in literature. Several electrode configurations are possible, such as an underwater pin to plate in gas [150] (Figure 10a) and an underwater pin to multipin in gas [151] (Figure 10b). In some reactors, a second high-voltage electrode is placed in the gas phase, either powered by the same high-voltage source [151] (Figure 10c) or by a second one [152] (Figure 10d). Also here, energy efficiency can be enhanced by discharge formation in externally applied bubbles, leading to hybrid reactors that combine bubble discharge with gas phase discharge. One example is given by Ren et al. [153], where bubbles are formed on high-voltage nozzle electrodes located underneath a plate electrode in the gas phase (Figure 10e). Hybrid reactors with discharge in both water and gas phase are sometimes proposed for the treatment of gaseous and aqueous pollutants simultaneously, as in the case of volatile pollutants [44, 46]. Their energy efficiency for organic decomposition is moderately higher than the one of hydraulic discharge reactors [42, 43].

Another type of hybrid reactor combines the treatment of a falling water film and droplets. This situation naturally occurs by spraying water from a shower nozzle at an angle. In the study of Kobayashi et al. [139], a wetted-wall hybrid reactor with inner wire anode was used, where water solutions from the falling film and from the droplets were collected separately (Figure 10f). Indigo carmine was decomposed 0.57 times faster in the droplets than in the water

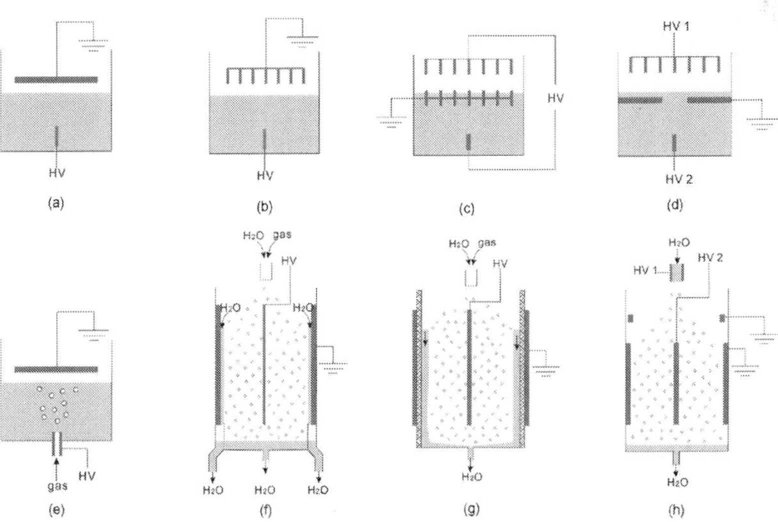

Figure 10. Types of hybrid reactors: (a, b) underwater high-voltage pin to plate or multipin in gas, (c) reactor from Lukes et al. [151], (d) reactor from Lukes et al. [152], (e) bubble discharge on underwater nozzle electrode to plate in gas, (f) wetted-wall spray corona reactor with inner wire electrode, (g) wetted-wall spray DBD reactor with outer barrier, and (h) electrospray through wire-to-mesh corona reactor.

film, as could be expected from the higher energy efficiency of spray discharge reactors as compared to falling water film reactors. In the study of Nakagawa et al. [154], the water film was formed on the dielectric barrier which separated the inner rod anode and the grounded mesh surrounding the barrier (Figure 10g). For low water flow rate, the energy efficiency of rhodamine B decomposition was similar for 3 of such reactors with different inner diameter and barrier thickness. For higher flow rates, however, one reactor performed significantly better. This reactor had larger inner diameter and equal barrier thickness compared to one of the other reactors.

A special case is found when droplets from an electrospray are treated a second time with plasma discharge. In the study of Njatawidjaja et al. [155], electrostatically atomized droplets passed through pulsed corona discharge in between a wire-to-mesh electrode configuration (Figure 10h). In both parts of the reactor, positive polarity performed better than negative one for the decomposition of Chicago sky blue dye. In the electrospray, positive DC voltage produced a larger number of finer droplets with a wider spray angle than in the case of negative voltage.

3.6. Remote discharge reactors

The concept of remote discharge reactors for water treatment is not new. An early example is ozonation, where ozone is generated by means of plasma discharge in gas phase and subsequently transported toward the solution under treatment. More recently, electrical discharge reactors have been developed where remotely generated plasma gas is bubbled through the solution. As main difference, the plasma gas does not only contain ozone, but also other reactive species, such as H_2O_2 and OH. In the study of Tang et al. [156], this is accomplished by using humid air as feed gas of a DBD gas phase reactor (Figure 11a). With corona discharge in dry air, positive and negative ions can be generated in addition to ozone for water treatment, as in the air ionization device reported by Wohlers et al. [157]. Yamatake et al. compared the bubble discharge reactor from Figure 3k with ground electrode in contact with water to a reactor where plasma is generated separately from the water and subsequently bubbled through the solution (Figure 11b) [158]. Both reactors had identical electrode configuration, used oxygen as feed gas, and were powered by the same positive DC voltage source. Acetic acid was not decomposed in the remote discharge reactor, while decomposition was significant in the bubble discharge reactor. This difference was explained with the production of oxygen radicals, which acted as main oxidant during bubble discharge but had too short lifetime to reach the water in the remote discharge reactor.

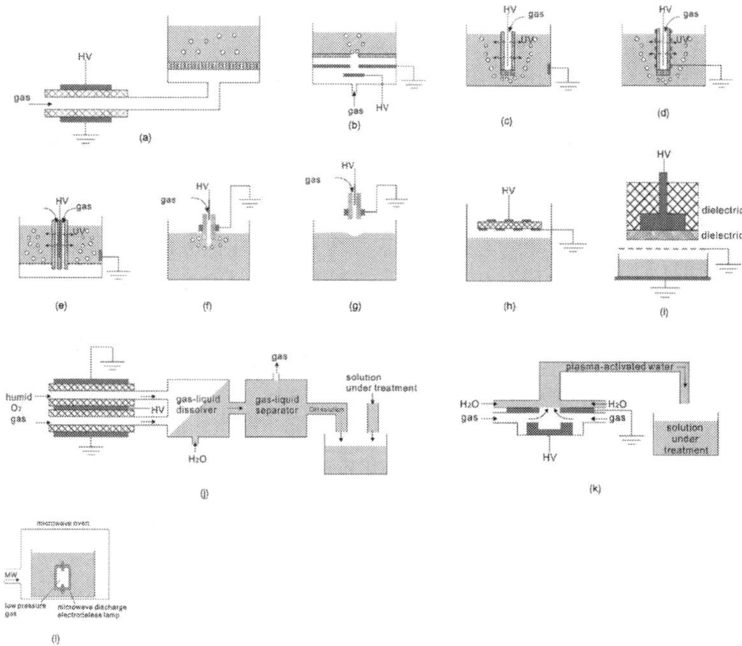

Figure 11. Types of remote discharge reactors: (a, b) plasma gas bubbling reactors, (c–e) plasma gas bubbling reactors with UV irradiation through quartz barrier, (f) underwater DBD plasma jet, (g) DBD plasma jet over water surface, (h) metal strips on quartz disc for DBD above water surface, (i) setup from Dobrynin et al. [159] with removable grounded mesh, (j) reactor from Zhang et al. [160] for production of hydroxyl radical solution, (k) gliding arc reactor from Kim et al. [161] for plasma-activated water production, and (l) reactor with tungsten-triggered microwave discharge electrodeless lamp.

Plasma gas bubbling can be combined with UV irradiation by generating plasma inside a quartz tube submerged in the solution. Several variations are reported in literature, where the high-voltage electrode inside the tube is a screw [162], a spiral wire [163], a rod covered by dielectric barrier [164], or a glass tube filled with NaCl solution [165]. A metal mesh surrounding the tube [162] or a metal rod in contact with the water [165] serves as ground electrode (Figures 11c–d). DBD plasma is usually generated inside the tube with AC power. The working gas is pumped in the tube and subsequently bubbled through the solution by means of a gas diffuser or a series of air distribution needles (Figure 11e) [164]. Comparison with other oxidation techniques has indicated that such reactors may be competitive technology to other plasma systems such as the hybrid reactor by Nakagawa et al. [154] (Figure 10g) and to photocatalytic oxidation [165, 166].

A system closely related to these plasma gas bubbling reactors is a submerged DBD plasma jet for water treatment. In the study of Foster et al. [167], a nanosecond pulsed DBD plasma jet is investigated for oxidation of aqueous organic pollutants (Figure 11f). Its decomposition efficiency was higher than the one reported in literature for glow discharge and pulsed corona discharge [167]. The energy efficiency of methylene blue decomposition significantly dropped with increasing treatment time. As the results suggest, using multiple plasma jets powered in parallel can improve the process significantly. The plasma jet can also be placed above the water surface (Figure 11g), as is done for the production of plasma-activated water by Ma et al. [168]. Nevertheless, application of non-thermal plasma jets for water treatment is relatively uncommon for now.

A few systems are reported in literature where plasma is generated above a water surface without direct contact with the solution. In the study of Olszewski et al. [169], copper strips adhered to both sides of a quartz disc were used as electrode configuration for methyl orange decomposition with AC DBD plasma (Figure 11h). In the study of Dobrynin et al. [159], inactivation of spores was investigated by means of the DBD electrode system of Figure 5c above a removable mesh for reference experiments (Figure 11i). Experiments with and without the mesh were compared to reveal the role of UV irradiation.

Recently, a new approach is gaining popularity in which water treated with plasma, often termed plasma-activated water, is added to the solution under treatment. Plasma-activated water contains several long-living oxidants that are able to inactivate biological organisms. According to Zhang et al., long-living aqueous hydroxyl radicals were produced in their setup by mixing DBD-treated humid O_2 gas with water (Figure 11j) [160]. This hydroxyl radical solution was sprayed in a sea enclosure to effectively inactivate red tide organisms. Plasma-activated water can be produced by means of any of the reactors mentioned above, but gliding arc discharge is most commonly used. Figure 11k depicts the gliding arc reactor used by Kim et al. [161] for plasma-activated water generation utilizing a vortex flow with two circular disk electrodes. Introducing the produced water–air mixture in the solution under treatment through microbubble generators significantly enhanced the process. Up to now, plasma-activated water is used for disinfection of water, while its effects on organic contamination are still largely unknown.

In principle, plasma technology is also frequently used in water treatment technology solely as UV source in plasma lamps or excimer lamps. An upcoming new technology is the microwave discharge electrodeless lamp, which self-ignites under influence of microwave power. As an example, Figure 111 schematically shows a tungsten-triggered microwave discharge electrodeless lamp proposed by Horikoshi et al. [170] for low microwave power levels. The tungsten wire was embedded in a synthetic quartz tube attached to the lamp system to act as a trigger. In that manner, the high microwave power usually required for self-ignition in aqueous medium is avoided.

4. Influence of working parameters on energy efficiency

In addition to reactor design and materials, there are several other factors that influence reactor energy efficiency, which need to be considered for reactor optimization. Roughly, these factors can be split into two groups: working parameters determining reactor operation and solution parameters. Since solution parameters, such as water temperature, pH, conductivity, and water matrix, are hard to control, especially for large volume of influent water, they are not discussed in this chapter. For a discussion on their influence, the reader is referred to [44]. In contrast, operational parameters such as applied voltage characteristics, working gas, and flow rates of gas and solution are easier to adjust. Therefore, they deserve additional attention for further reactor optimization. In this section, we will shortly review the importance of voltage waveform, working gas, and gas and water flow rates. Their influence on reactor efficiency for organic decomposition will be illustrated with examples from literature.

For a given reactor, energy efficiency depends on voltage-related parameters. In the pulsed bubble discharge reactor of [92] and the positive pulsed streamer discharge in wetted-wall reactor of [129], energy efficiency increased for rising voltage amplitude. In a negative pulsed DBD falling water film reactor [171] and a positive pulsed corona electrospray reactor [147], however, increasing voltage amplitude reduced the energy efficiency. In the gas phase DBD reactor of [169], the interruption period of pulse-modulated AC voltage had no effect on methyl orange degradation, while lowering its duty cycle from 100% to 25% increased energy efficiency 2.11 times. The authors explained the latter effect with additional dye degradation during plasma off time under influence of long living reactive species such as O_3 and H_2O_2. Sinusoidal voltage frequency is an important parameter, as it can lead to different plasma phenomena, which explains the distinction of AC, radio frequency, and microwave discharge. Nevertheless, very limited information is available in literature on the dependence of the energy efficiency of the sinusoidal frequency for a given reactor design. In the study of Lesage et al. [122], no change in energy efficiency was observed for 4-chlorobenzoic acid decomposition with voltage frequency increase from 500 Hz to 2000 Hz for AC powered DBD over moving water film. Increasing AC frequency from 1.5 kHz to 15.6 kHz kept the energy efficiency of a coaxial falling water film DBD reactor in the same order of magnitude as well, in spite of the additional heating that resulted from the higher frequency [172]. In the case of pulsed discharge, pulse properties such as rise time and width are expected to be important. For pulsed positive corona discharge in humid O_2/N_2 atmosphere, the energy efficiency of radical and excited species production increases with decreasing pulse width [173]. This is in agreement with the higher efficiency of indigo carmine decomposition for shorter pulse width observed by Sugai et al. [174] for a positive pulsed streamer discharge spray reactor. A faster pulse rise rate generated thicker streamers and a higher energy efficiency in the same reactor [175]. Positive pulsed arc electrohydraulic discharge was reported to have increased the energy efficiency of sulfadimethoxine when pulse duration was brought back from 100 µs to 20 µs [51]. Remarkably, also pulse frequency can influence reactor efficiency. In a bubble reactor with positive pulsed corona, the energy efficiency of 2,4-DCP degradation increased with increasing pulse frequency [176]. In contrast, the energy efficiency of indigo carmine reduction dropped for increasing pulse frequency in a positive pulsed streamer wetted-wall reactor [129], in a negative pulsed DBD spray reactor [144], and in a positive pulsed DBD hybrid film and spray reactor [154]. In a negative pulsed DBD falling water film reactor [177] and a positive pulsed corona spray reactor [143], no influence of the frequency was observed. In bipolar pulsed electrohydraulic reactors, breakdown voltage decreases with frequency [62, 96]. Reversing voltage polarity can also cause significant changes in plasma properties and thus treatment efficiency. In the study of Lee et al. [96], positive polarity of pulsed corona in electrohydraulic discharge greatly enhanced the energy efficiency of methyl orange decomposition in pin-to-plate electrode configuration in comparison with negative polarity. The authors explain this observation with the space charge effect, which causes positive corona streamers to be faster and longer, hence increasing radical production and plasma–water contact surface. In the study of Yasuoka et al. [178], higher efficiency was observed as well for positive polarity in a DC bubble discharge reactor. For phenol decomposition in gas phase discharge reactors with a 20% O_2 and 80% N_2 atmosphere, negative DC was found to be more efficient than positive one by Sano et al. [179], while according to Miyazaki et al. [108], no significant difference was seen for both polarities with pulsed power. In a pulsed DBD falling water film reactor, better results were obtained with negative polarity as compared to positive one [171]. In the bubble discharge reactor of Figure 3h with O_2, He, Ar, or Ne bubbles, positive DC voltage performed better than negative one for the decomposition of interfacial active agents [83]. For negative polarity, electrolysis occurred with formation of hydrogen and oxygen gases. For positive polarity, the interfacial anion agents were more concentrated at the bubble surface due to electrostatic attraction. Positive plasma species collided with the water surface, where a cathode drop is formed. These unknown species likely enhanced the decomposition of the agents.

The working gas determines many plasma features for a given input voltage waveform, such as breakdown voltage, electron density and temperature, plasma homogeneity and intensity, generated reactive species, etc. Air is the most frequently used working gas for water treatment plasma reactors due to its wide availability. Pure oxygen gas, however, is often found to give more efficient organic degradation, while nitrogen gas leads to lower efficiency [42, 44, 50]. This can be partly explained with formation of OH radicals and O_3 in oxygen. In N_2 containing gases, however, toxic aqueous nitric products are generated, which decrease solution pH and act as scavengers of oxidants such as OH radicals [50]. Noble gases like helium and argon are sometimes used, especially in bubble discharge reactors. Often, argon leads to the faster decomposition of phenols but performs worse than oxygen for other compounds [50]. In a wetted-wall reactor (Figure 6b) with argon, streamer discharge has been observed which had slightly better phenol decomposition energy efficiency than corona in oxygen [128]. Noble gases, however, are expected to be less economically feasible for use on larger scale due to their high price. Interestingly, treatment efficiency of electrohydraulic discharge reactors can also be altered by bubbling different gases through the solution under treatment, where bubbles are kept away from the discharge zone. Sahni and Locke observed a decrease in nitroform anion decomposition by pulsed underwater corona when the solution was oxygenated with O_2 gas or deoxygenated with N_2, He, or Ar gas as compared to discharge without prior bubbling, but the authors could not explain this effect [180]. Also, water content of the working gas should be taken into account, as it influences formation of important oxidants such as OH and H_2O_2.

Gas flow rate is another factor that needs to be taken into consideration. In three bubble discharge reactors, increasing bubble flow rate enhanced the decomposition process [158, 181, 182], while according to Reddy et al. [183] no significant effect was observed. In the hybrid reactor with bubbles from Figure 10e, oxidation rate first increased and then reached a stationary value with rising gas flow rate [153]. In the remote discharge reactor with bubbling from Chen et al. [184], increasing gas flow also enhanced decomposition. For gas phase discharge reactors, the influence of gas flow rate seems less pronounced. In a positive pulsed corona-like discharge over water, phenol decomposition was slower with increasing oxygen flow and slightly dropped with increasing airflow, while argon flow rate had no influence [185, 186]. In a pulsed DBD falling water film reactor, the effect of the oxygen flow rate on methylene blue degradation was not significant [171].

In reactors with moving solution, the water flow rate often influences the decomposition process. In coaxial falling water film reactors with corona discharge, phenol degradation rate was unchanged with faster water flow for positive pulsed voltage [128], while it dropped in case of negative DC, which was attributed to higher roughness of the water surface [127]. In a pulsed DBD falling water film reactor, the energy efficiency of methylene blue degradation decreased with increasing water flow rate [171]. In contrast, 4-chlorobenzoic acid was decomposed faster with increasing water flow for AC gliding arc discharge over falling water film [123]. Measurements in spray discharge reactors indicate the existence of an optimal water flow rate. In a positive pulsed streamer spray reactor, the energy efficiency of indigo carmine decomposition initially increased and then stabilized with increasing water flow rate due to saturation of aqueous ozone [140]. Moreover, the efficiency was independent of droplet size. In a similar corona reactor, the energy efficiency of oxalic acid decomposition first increased with rising water flow rate, reached a maximal value and subsequently dropped again [187]. The optimal water flow rate increased with applied pulse frequency. Rising water flow rate decreased breakdown voltage in a positive pulsed DBD spray reactor [145]. For the hybrid positive pulsed DBD spray and falling water film reactor of Figure 10g, the energy efficiency of rhodamine B decomposition enhanced with increasing water flow [154].

5. Summary and concluding remarks

Removal of hazardous micropollutants is often insufficient by means of modern conventional wastewater treatment plants. Preventive measures and optimization of conventional biological treatment are suggested as most cost-effective solutions. Nevertheless, preventive measures are limited by increasing demand, while the optimization of conventional techniques often has negligible effect on many persistent micropollutants. Therefore, advanced treatment techniques such as electrochemical separation, activated carbon, nanofiltration, and reverse osmosis have recently received more attention for their effective removal of micropollutants. These techniques, however, are associated with high costs and the additional problem of hazardous concentrate disposal. Advanced oxidation techniques are a promising alternative, as they are the most effective available methods to decompose biorecalcitrant organics. As a main drawback, their energy costs are high up to now, preventing their implementation on large scale. Alternatively, their application can be limited to the treatment of important micropollutant sources, such as hospital and industrial effluent.

Among the advanced oxidation techniques, water treatment by means of electrical discharge takes an interesting place since it is able to generate a wide spectrum of oxidative species, leading to a low selectivity of the degradation process. Further, the optimization of this technology is complex due to the wide variety in reactor design and materials, discharge types, and operational parameters. In this chapter, plasma reactors are comprehensively classified based on their design and materials, in contrast to other reviews where focus lies more on applied voltage and discharge type. Six main reactor types are distinguished. In electrohydraulic discharge reactors and bubble discharge reactors, plasma is generated directly in the liquid bulk, respectively, without and with external application of bubbles. In gas phase discharge reactors and spray discharge reactors, plasma is generated in het gas phase, respectively, over a water bulk or film and in contact with water drops or mist. Reactors that use a combination of these types simultaneously are classified as hybrid reactors. In the last type of reactors, referred to as remote discharge reactors, plasma is not generated in direct contact with the solution under treatment.

Most commonly, electrohydraulic discharge reactors use pulsed arc or positively pulsed corona discharge, where electrode material can have substantial influence on organic decomposition after plasma contact due to formation of erosion particles in water. For the case of arc discharge, energy efficiency is reported to be dependent on interelectrode distance. For pulsed corona, high-voltage pin curvature radius is an important parameter as well. The corona plasma volume can be enlarged by replacing the high-voltage pin with a multipin electrode or a high-voltage electrode covered with a thin porous ceramic layer. Diaphragm and capillary discharge are expected to have similar plasma features to corona discharge in pin-to-plate electrode configuration, which is in agreement with the similar energy efficiency. Contact glow discharge electrolysis is another common type of electrohydraulic discharge, where electrode material also plays an important role. In more exotic types of electrohydraulic discharge reactors, plasma formation is preceded by cavitation under application of RF or microwave power.

Adding external bubbles in the water bulk has the advantages of easier plasma onset, immediate mixing of the solution, minimizing electrode erosion and increasing radical density. Most commonly, bubbles are generated by pumping gas through a dielectric or an electrode, which is shaped as a nozzle, perforated plate, or porous ceramic. Choices in bubble gas, nozzle or perforation material, shape, dimensions, and orientation strongly influence plasma properties, which complicates reactor comparison. Often, the high-voltage electrode is positioned directly in contact with the bubble or in de gas phase in contact with the bubble. In that case, energy efficiency can be enhanced by increasing the number of nozzles or holes. Alternatively, bubbles can be positioned in between submerged electrodes. Some common bubble discharge reactors use a coaxial DBD configuration, where the gas is bubbled in axial direction. Energy efficiency in these systems can be increased by using a double barrier or adding glass beads in the water bulk. More exotic types of bubble discharge reactors have been reported in literature, with promising first results.

Corona and glow discharge in gas phase over grounded water bulk or film is mostly generated with pulsed power. Negative pulsed corona over flowing water film in a pilot system has been shown to have better or comparable energy efficiency than other advanced oxidation processes. Based on one report, energy efficiency of organic decomposition seems to be independent of the type of discharge, voltage amplitude, polarity of the applied voltage, and amount of pin electrodes. In contrast, DBD over water bulk or film has an energy efficiency that is reported to increase with increasing voltage amplitude and decreasing interelectrode distance. Interelectrode distance can be decreased substantially by adding a porous ceramic segment at the water surface. Movement of the water phase by stirring or by making it flow as a film along the discharge increases energy efficiency substantially, making water flow rate an important parameter for optimization. Based on this principle, coaxial reactors with falling water film are gaining more popularity. Such reactors can be further optimized by adjusting gas flow direction and electrode and barrier dimensions. A last common gas phase discharge reactor uses arc discharge over the water bulk or film. Larger arc-treated water surface and energy efficiency can be reached by using a couple of controlled electrodes or a second high-voltage electrode.

Spray discharge reactors have received more attention in recent years due to their high reported energy efficiencies. In the case of positive pulsed corona, treatment is more efficient for droplets near the inner reactor wall, while droplet size has no influence on energy efficiency. Such reactors can be further optimized by adding fluorocarbon wires along the inner reactor wall for larger droplet retention time. Other spray discharge reactors treat droplets with DBD, electrospray or arc discharge. In the case of gliding arc, the process can be optimized by addition of a couple of controlled electrodes or a second high-voltage electrode and by introducing the water perpendicular to the gas flow to improve contact with plasma. DC plasma torches can also be used for water treatment by injecting the solution into the torch as plasma forming gas.

One type of hybrid reactor is designed by placing a high-voltage electrode in the water phase and a ground electrode or second high-voltage electrode in the gas phase above the water surface, without or with the addition of bubbles. Their energy efficiency for organic decomposition is moderately higher than the one of hydraulic discharge reactors. Another hybrid reactor type naturally occurs by spraying water from a shower nozzle at an angle, causing a falling water film in combination with spray. In such case, organic decomposition in droplets is more efficient than in the water film. A more exotic hybrid reactor that deserves more attention consists of a spray discharge reactor where droplets are formed by electrospray with additional plasma treatment afterwards.

Remote discharge reactors can be encountered in many configurations. In the most standard design, plasma gas is remotely generated and sequentially bubbled through the solution, as in the well-known example of ozonation. Plasma gas bubbling can be combined with UV irradiation by generating plasma inside a quartz tube submerged in the solution. Research has indicated that such reactors may be competitive technology to systems with direct plasma treatment. Plasma can also be used solely for UV irradiation, as in plasma lamps, excimer lamps and microwave discharge electrodeless lamps. Recently, the generation and application of plasma-activated water for water disinfection is gaining popularity, while its effects on organic contamination are still largely unknown.

Additionally, the importance of voltage waveform, working gas and flow rates of gas, and water for further optimization are shortly reviewed in this chapter. Energy efficiency has different dependency for different reactor types on voltage parameters, such as voltage type, amplitude, polarity, sinusoidal frequency, pulse rise time, pulse duration, and pulse frequency. For example, positive polarity causes higher efficiency of electrohydraulic and bubble discharge reactors in all considered cases, explained with the space charge effect. On the other hand, negative polarity gives better performance in gas phase discharge reactors. The choice of working gas can significantly alter plasma chemistry and therefore treatment efficiency and by-product formation. While atmospheric air is often chosen due to its wide availability, oxygen generally enhances the process. Argon often performs better for phenol degradation but is expected to be less economically feasible for use on large scale. Increasing gas flow rate typically enhances decomposition in bubble discharge and remote discharge reactors, whereas its effect is less pronounced in gas phase discharge reactors. Also, adjustment of water flow can significantly increase energy efficiency. Measurements in spray discharge reactors, for instance, indicate the existence of an optimal water flow rate. While dependency of a few of these parameters on energy efficiency seems adequately established for a limited number of reactor types, a deeper literature study and more experimental investigation are required for additional confirmation and a better understanding in most of the cases. Due to the wide variety in plasma reactors and the distinct, unique features of every reactor, researchers are motivated to report new results in this field, including clear descriptions of reactor design and materials.

One needs to keep a few additional influences in mind when interpreting energy efficiencies reported in literature. First of all, solution parameters such as water temperature, pH, conductivity, and water matrix can have significant effect on plasma chemistry and reactions in the water bulk. Often, deionized water at room temperature is used as solvent in plasma reactors. In other cases, however, deviations in energy efficiency can be caused by a difference in solution parameters. Also, it should be remarked that micropollutant measurement is generally performed a certain time after plasma treatment with analytical chemistry methods, such as gas chromatography–mass spectrometry and liquid chromatography–mass spectrometry. During this time, postreactions with long-living oxidants can occur, decomposing the micropollutant to a greater extent. This effect is usually neglected by researchers, complicating accurate comparison between reported energy efficiencies. On the positive side, this aging effect can be beneficial in applications where sufficiently long hydraulic retention time is possible after plasma treatment. Last but not least, most of the reported electrical discharge reactors only treat small water volumes in batch mode without or with recirculation of water. As a result, the determined energy efficiency can be significantly different for the same reactor type in single-pass mode, where water is flowing through the system only once. The latter case is more representative for industrial application and thus deserves more attention.

In this chapter, only improvement of the plasma process in terms of reactor design, materials, and working parameters is discussed. Further optimization can be achieved by combining plasma discharge with other advanced treatment methods, such as adsorption, Fenton's reagent, photocatalysis, and ultrasonication. Such combinations have been reported before but require additional attention and further exploration. As should be noted, plasma technology can also be used for synthesis, pretreatment, regeneration, and posttreatment of materials and matter involved in water treatment processes, such as nanotubes, membranes, activated carbon, excess sludge, and organic concentrate from filtration. In-line application of these methods during the water treatment process needs to be considered as a possible alternative to direct water treatment with advanced oxidation processes since energy demand and overall costs can be pressed significantly this way. However, more experimental investigation and thorough cost analysis is necessary to confirm this claim.

Future application of plasma discharge for water treatment will largely depend on its effectiveness and energy efficiency as compared to other advanced oxidation processes and treatment methods in general, but additional criteria need to be taken into consideration as well. Sustainability, ease of operation, capital costs, and costs related to maintenance, gas input, and additional energy for pumps will also determine whether a system will be adopted on a large scale. Moreover, an extensive study of generated oxidation by-products and long-living oxidants in treated water is necessary to assure that overall toxicity is consistently and sufficiently decreased after plasma treatment. Up to now, reports on these topics are largely lacking in literature or limited to only a few specific cases.

Author details

Patrick Vanraes*, Anton Y. Nikiforov and Christophe Leys

*Address all correspondence to: Patrick.Vanraes@UGent.be

Department of Applied Physics, Research Unit Plasma Technology, Ghent University, Ghent, Belgium

References

[1] L. Rizzo, *et al.*, "Urban wastewater treatment plants as hotspots for antibiotic resistant bacteria and genes spread into the environment: a review," *Science of the Total Environment,* vol. 447, pp. 345–360, 2013.

[2] T. Hayes, *et al.*, "Herbicides: feminization of male frogs in the wild," *Nature,* vol. 419, pp. 895–896, 2002.

[3] S. Milla, *et al.*, "The effects of estrogenic and androgenic endocrine disruptors on the immune system of fish: a review," *Ecotoxicology,* vol. 20, pp. 305–319, 2011.

[4] I. R. Falconer, *et al.*, "Endocrine-disrupting compounds: a review of their challenge to sustainable and safe water supply and water reuse," *Environmental Toxicology,* vol. 21, pp. 181–191, 2006.

[5] R. McKinlay, *et al.*, "Endocrine disrupting pesticides: implications for risk assessment," *Environment International,* vol. 34, pp. 168–183, 2008.

[6] W. Mnif, *et al.*, "Effect of endocrine disruptor pesticides: a review," *International Journal of Environmental Research and Public Health,* vol. 8, pp. 2265–2303, 2011.

[7] K. Fent, *et al.*, "Ecotoxicology of human pharmaceuticals," *Aquatic Toxicology,* vol. 76, pp. 122–159, 2006.

[8] A. Pruden, *et al.*, "Antibiotic resistance genes as emerging contaminants: studies in northern Colorado," *Environmental Science & Technology,* vol. 40, pp. 7445–7450, 2006.

[9] A. L. Batt, *et al.*, "Comparison of the occurrence of antibiotics in four full-scale wastewater treatment plants with varying designs and operations," *Chemosphere,* vol. 68, pp. 428–435, 2007.

[10] X. Flores Alsina, "Conceptual design of wastewater treatment plants using multiple objectives," Doctor PhD, Universitat de Girona, Girona, 2008.

[11] Y.-P. Wang and R. Smith, "Design of distributed effluent treatment systems," *Chemical Engineering Science,* vol. 49, pp. 3127–3145, 1994.

[12] I. Michael, *et al.*, "Urban wastewater treatment plants as hotspots for the release of antibiotics in the environment: a review," *Water Research,* vol. 47, pp. 957–995, 2013.

[13] L. Benedetti, *et al.*, "Multi-criteria analysis of wastewater treatment plant design and control scenarios under uncertainty," *Environmental Modelling & Software,* vol. 25, pp. 616–621, 2010.

[14] Y. Luo, *et al.*, "A review on the occurrence of micropollutants in the aquatic environment and their fate and removal during wastewater treatment," *Science of the Total Environment,* vol. 473–474, pp. 619–641, 2014.

[15] O. A. Jones, *et al.*, "Questioning the excessive use of advanced treatment to remove organic micropollutants from wastewater," *Environmental Science & Technology*, vol. 41, pp. 5085–5089, 2007.

[16] Z.-h. Liu, *et al.*, "Removal mechanisms for endocrine disrupting compounds (EDCs) in wastewater treatment—physical means, biodegradation, and chemical advanced oxidation: a review," *Science of the Total Environment*, vol. 407, pp. 731–748, 2009.

[17] A. Joss, *et al.*, "Removal of pharmaceuticals and fragrances in biological wastewater treatment," *Water Research*, vol. 39, pp. 3139–3152, 2005.

[18] T. A. Ternes, *et al.*, "Peer reviewed: scrutinizing pharmaceuticals and personal care products in wastewater treatment," *Environmental Science & Technology*, vol. 38, pp. 392A-399A, 2004.

[19] O. H. Jones, *et al.*, "Human pharmaceuticals in wastewater treatment processes," *Critical Reviews in Environmental Science and Technology*, vol. 35, pp. 401–427, 2005.

[20] N. Tadkaew, *et al.*, "Removal of trace organics by MBR treatment: the role of molecular properties," *Water Research*, vol. 45, pp. 2439–2451, 2011.

[21] I. Sirés and E. Brillas, "Remediation of water pollution caused by pharmaceutical residues based on electrochemical separation and degradation technologies: a review," *Environment International*, vol. 40, pp. 212–229, 2012.

[22] J. M. Poyatos, *et al.*, "Advanced oxidation processes for wastewater treatment: state of the art," *Water, Air, and Soil Pollution*, vol. 205, pp. 187–204, 2010.

[23] O. Lefebvre, *et al.*, "Microbial fuel cells for energy self-sufficient domestic wastewater treatment—a review and discussion from energetic consideration," *Applied Microbiology and Biotechnology*, vol. 89, pp. 259–270, 2011.

[24] A. Helal, *et al.*, "Feasibility study for self-sustained wastewater treatment plants—using biogas CHP fuel cell, micro-turbine, PV and wind turbine systems," 2013.

[25] P. Verlicchi, *et al.*, "What have we learned from worldwide experiences on the management and treatment of hospital effluent?—an overview and a discussion on perspectives," *Science of the Total Environment*, vol. 514, pp. 467–491, 2015.

[26] P. Verlicchi, *et al.*, "Hospital effluents as a source of emerging pollutants: an overview of micropollutants and sustainable treatment options," *Journal of Hydrology*, vol. 389, pp. 416–428, 2010.

[27] C. Comninellis, *et al.*, "Advanced oxidation processes for water treatment: advances and trends for R&D," *Journal of Chemical Technology and Biotechnology*, vol. 83, pp. 769–776, 2008.

[28] V. Homem and L. Santos, "Degradation and removal methods of antibiotics from aqueous matrices—a review," *Journal of Environmental Management*, vol. 92, pp. 2304–2347, 2011.

[29] J. Rivera-Utrilla, *et al.*, "Pharmaceuticals as emerging contaminants and their removal from water. A review," *Chemosphere*, vol. 93, pp. 1268–1287, 2013.

[30] W. Yang, *et al.*, "Treatment of organic micropollutants in water and wastewater by UV-based processes: a literature review," *Critical Reviews in Environmental Science and Technology*, vol. 44, pp. 1443–1476, 2014.

[31] H. R. Ghatak, "Advanced oxidation processes for the treatment of biorecalcitrant organics in wastewater," *Critical Reviews in Environmental Science and Technology*, vol. 44, pp. 1167–1219, 2014.

[32] M. A. Oturan and J.-J. Aaron, "Advanced oxidation processes in water/wastewater treatment: principles and applications. A review," *Critical Reviews in Environmental Science and Technology*, vol. 44, pp. 2577–2641, 2014.

[33] M. Klavarioti, *et al.*, "Removal of residual pharmaceuticals from aqueous systems by advanced oxidation processes," *Environment International*, vol. 35, pp. 402–417, 2009.

[34] R. P. Joshi and S. M. Thagard, "Streamer-like electrical discharges in water: part II. Environmental applications," *Plasma Chemistry and Plasma Processing*, vol. 33, pp. 17–49, 2013.

[35] J.-L. Brisset and E. Hnatiuc, "Peroxynitrite: a re-examination of the chemical properties of non-thermal discharges burning in air over aqueous solutions," *Plasma Chemistry and Plasma Processing*, vol. 32, pp. 655–674, 2012.

[36] P. Lukes, *et al.*, "Aqueous-phase chemistry and bactericidal effects from an air discharge plasma in contact with water: evidence for the formation of peroxynitrite through a pseudo-second-order post-discharge reaction of H2O$_2$ and HNO$_2$," *Plasma Sources Science and Technology*, vol. 23, p. 015019, 2014.

[37] X. Jin, *et al.*, "Reaction kinetics of selected micropollutants in ozonation and advanced oxidation processes," *Water Research*, vol. 46, pp. 6519–6530, 2012.

[38] S. Sudhakaran and G. L. Amy, "QSAR models for oxidation of organic micropollutants in water based on ozone and hydroxyl radical rate constants and their chemical classification," *Water Research*, vol. 47, pp. 1111–1122, 2013.

[39] U. Von Gunten, "Ozonation of drinking water: Part I. Oxidation kinetics and product formation," *Water Research*, vol. 37, pp. 1443–1467, 2003.

[40] W. R. Haag and C. D. Yao, "Rate constants for reaction of hydroxyl radicals with several drinking water contaminants," *Environmental Science & Technology*, vol. 26, pp. 1005–1013, 1992.

[41] M. E. Sillanpää, *et al.*, "Degradation of chelating agents in aqueous solution using advanced oxidation process (AOP)," *Chemosphere*, vol. 83, pp. 1443–1460, 2011.

[42] M. A. Malik, "Water purification by plasmas: which reactors are most energy efficient?" *Plasma Chemistry and Plasma Processing*, vol. 30, pp. 21–31, 2010.

[43] P. J. Bruggeman and B. R. Locke, "Assessment of potential applications of plasma with liquid water," in *Low Temperature Plasma Technology—Methods and Applications*, P. K. Chu and L. XinPei, Eds., Boca Raton, FL: CRC Press, 2014, pp. 367–399.

[44] B. Jiang, *et al.*, "Review on electrical discharge plasma technology for wastewater remediation," *Chemical Engineering Journal*, vol. 236, pp. 348–368, 2014.

[45] P. Bruggeman and C. Leys, "Non-thermal plasmas in and in contact with liquids," *Journal of Physics D: Applied Physics*, vol. 42, p. 053001, 2009.

[46] B. Locke, *et al.*, "Electrohydraulic discharge and nonthermal plasma for water treatment," *Industrial & Engineering Chemistry Research*, vol. 45, pp. 882–905, 2006.

[47] E.-J. Lee, *et al.*, "Application of underwater dielectric barrier discharge as a washing system to inactivate *Salmonella typhimurium* on perillar leaves," in *Plasma Sciences (ICOPS), 2015 IEEE International Conference on*, 2015, pp. 1–1.

[48] P. Vanraes, *et al.*, "Removal of atrazine in water by combination of activated carbon and dielectric barrier discharge," *Journal of Hazardous Materials*, vol. 299, pp. 647–655, 2015.

[49] B. R. Locke and K.-Y. Shih, "Review of the methods to form hydrogen peroxide in electrical discharge plasma with liquid water," *Plasma Sources Science and Technology*, vol. 20, p. 034006, 2011.

[50] M. Hijosa-Valsero, *et al.*, "Decontamination of waterborne chemical pollutants by using atmospheric pressure nonthermal plasma: a review," *Environmental Technology Reviews*, vol. 3, pp. 71–91, 2014.

[51] N. Parkansky, *et al.*, "Submerged arc breakdown of sulfadimethoxine (SDM) in aqueous solutions," *Plasma Chemistry and Plasma Processing*, vol. 28, pp. 583–592, 2008.

[52] N. Parkansky, *et al.*, "Submerged arc breakdown of methylene blue in aqueous solutions," *Plasma Chemistry and Plasma Processing*, vol. 32, pp. 933–947, 2012.

[53] Q.-X. Lin, *et al.*, "The research on pulsed arc electrohydraulic discharge with discharge electrode and its application to removal of bacteria," *IEEE Transactions on Plasma Science*, vol. 43, pp. 1029–1039, 2015.

[54] L. Zhu, *et al.*, "The research on the pulsed arc electrohydraulic discharge and its application in treatment of the ballast water," *Journal of Electrostatics*, vol. 71, pp. 728–733, 2013.

[55] N. Parkansky, *et al.*, "Decomposition of dissolved methylene blue in water using a submerged arc between titanium electrodes," *Plasma Chemistry and Plasma Processing*, vol. 33, pp. 907–919, 2013.

[56] P. Lang, *et al.*, "Oxidative degradation of 2,4,6-trinitrotoluene by ozone in an electrohydraulic discharge reactor," *Environmental Science & Technology*, vol. 32, pp. 3142–3148, 1998.

[57] D. M. Angeloni, *et al.*, "Removal of methyl-tert-butyl ether from water by a pulsed arc electrohydraulic discharge system," *Japanese Journal of Applied Physics*, vol. 45, pp. 8290–8293, 2006.

[58] N. Karpel Vel Leitner, *et al.*, "Generation of active entities by the pulsed arc electro-hydraulic discharge system and application to removal of atrazine," *Water Research*, vol. 39, pp. 4705–4714, 2005.

[59] L. V. Hoang and B. Legube, "Degradation of 4-chlorophenol by pulsed arc discharge in water-estimation of the energy consumption," *IEEE Transactions on Dielectrics and Electrical Insulation*, vol. 16, pp. 1604–1608, 2009.

[60] P. Šunka, *et al.*, "Potential applications of pulse electrical discharges in water," *Acta Physica Slovaca*, vol. 54, pp. 135–145, 2004.

[61] Š. Potocký, *et al.*, "Needle electrode erosion in water plasma discharge," *Thin Solid Films*, vol. 518, pp. 918–923, 2009.

[62] P. Baroch, *et al.*, "Generation of plasmas in water: utilization of a high-frequency, low-voltage bipolar pulse power supply with impedance control," *Plasma Sources Science and Technology*, vol. 20, p. 034017, 2011.

[63] T. Takeda, *et al.*, "Morphology of high-frequency electrohydraulic discharge for liquid-solution plasmas," *IEEE Transactions on Plasma Science*, vol. 36, pp. 1158–1159, 2008.

[64] Q. Lu, *et al.*, "Degradation of 2,4-dichlorophenol by using glow discharge electrolysis," *Journal of Hazardous Materials*, vol. 136, pp. 526–531, 2006.

[65] Q. Lu, *et al.*, "Glow discharge induced hydroxyl radical degradation of 2-naphthylamine," *Plasma Science and Technology*, vol. 7, p. 2856, 2005.

[66] S. Mededovic, *et al.*, "Aqueous-phase mineralization of S-triazine using pulsed electrical discharge," *IJPEST*, vol. 1, pp. 82–90, 2007.

[67] S. Mededovic and B. R. Locke, "Side-chain degradation of atrazine by pulsed electrical discharge in water," *Industrial & Engineering Chemistry Research*, vol. 46, pp. 2702–2709, 2007.

[68] P. Sunka, *et al.*, "Generation of chemically active species by electrical discharges in water," *Plasma Sources Science and Technology*, vol. 8, p. 258, 1999.

[69] B. Sun, *et al.*, "Optical study of active species produced by a pulsed streamer corona discharge in water," *Journal of Electrostatics*, vol. 39, pp. 189–202, 1997.

[70] P. Lukes, *et al.*, "The catalytic role of tungsten electrode material in the plasmachemical activity of a pulsed corona discharge in water," *Plasma Sources Science and Technology*, vol. 20, p. 034011, 2011.

[71] H. Zheng, *et al.*, "p-Nitrophenol Enhanced Degradation in High-Voltage Pulsed Corona Discharges Combined with Ozone System," *Plasma Chemistry and Plasma Processing,* vol. 33, pp. 1053–1062, 2013.

[72] C. W. Chen, *et al.*, "Ultrasound-assisted plasma: a novel technique for inactivation of aquatic microorganisms," *Environmental Science & Technology,* vol. 43, pp. 4493–4497, 2009.

[73] M. M. Sein, *et al.*, "Studies on a non-thermal pulsed corona plasma between two parallel-plate electrodes in water," *Journal of Physics D: Applied Physics,* vol. 45, p. 225203, 2012.

[74] P. Lukes, *et al.*, "Effect of ceramic composition on pulse discharge induced processes in water using ceramic-coated wire to cylinder electrode system," *Czechoslovak Journal of Physics,* vol. 52, pp. D800-D806, 2002.

[75] P. Lukes, *et al.*, "The role of surface chemistry at ceramic/electrolyte interfaces in the generation of pulsed corona discharges in water using porous ceramic-coated rod electrodes," *Plasma Processes and Polymers,* vol. 6, pp. 719–728, 2009.

[76] Y. J. Liu and X. Z. Jiang, "Phenol degradation by a nonpulsed diaphragm glow discharge in an aqueous solution," *Environmental Science & Technology,* vol. 39, pp. 8512–8517, 2005.

[77] L. Wang, "4-Chlorophenol degradation and hydrogen peroxide formation induced by DC diaphragm glow discharge in an aqueous solution," *Plasma Chemistry and Plasma Processing,* vol. 29, pp. 241–250, 2009.

[78] V. Goncharuk, *et al.*, "The diaphragm discharge and its use for water treatment," *Journal of Water Chemistry and Technology,* vol. 30, pp. 261–268, 2008.

[79] X. Jin, *et al.*, "An improved multi-anode contact glow discharge electrolysis reactor for dye discoloration," *Electrochimica Acta,* vol. 59, pp. 474–478, 2012.

[80] T. Maehara, *et al.*, "Influence of conductivity on the generation of a radio frequency plasma surrounded by bubbles in water," *Plasma Sources Science and Technology,* vol. 20, p. 034016, 2011.

[81] T. Ishijima, *et al.*, "Efficient production of microwave bubble plasma in water for plasma processing in liquid," *Plasma Sources Science and Technology,* vol. 19, p. 015010, 2010.

[82] N. Remya and J.-G. Lin, "Current status of microwave application in wastewater treatment—a review," *Chemical Engineering Journal,* vol. 166, pp. 797–813, 2011.

[83] K. Yasuoka, *et al.*, "An energy-efficient process for decomposing perfluorooctanoic and perfluorooctane sulfonic acids using dc plasmas generated within gas bubbles," *Plasma Sources Science and Technology,* vol. 20, p. 034009, 2011.

[84] H. Aoki, *et al.*, "Plasma generation inside externally supplied Ar bubbles in water," *Plasma Sources Science and Technology,* vol. 17, p. 025006, 2008.

[85] D. C. Johnson, *et al.*, "Treatment of methyl tert-butyl ether contaminated water using a dense medium plasma reactor: a mechanistic and kinetic investigation," *Environmental Science & Technology,* vol. 37, pp. 4804–4810, 2003.

[86] W. Bian, *et al.*, "Enhanced degradation of p-chlorophenol in a novel pulsed high voltage discharge reactor," *Journal of Hazardous Materials,* vol. 162, pp. 906–912, 2009.

[87] Y. Zhang, *et al.*, "Methyl orange degradation by pulsed discharge in the presence of activated carbon fibers," *Chemical Engineering Journal,* vol. 159, pp. 47–52, 2010.

[88] N. Lu, *et al.*, "Electrical characteristics of pulsed corona discharge plasmas in chitosan solution," *Plasma Science and Technology,* vol. 16, pp. 128–133, 2014.

[89] M. Kurahashi, *et al.*, "Radical formation due to discharge inside bubble in liquid," *Journal of Electrostatics,* vol. 42, pp. 93–105, 1997.

[90] K. Sato, *et al.*, "Water treatment with pulsed discharges generated inside bubbles," *Electrical Engineering in Japan,* vol. 170, pp. 1–7, 2010.

[91] A. Yamatake, *et al.*, "Water purification by atmospheric DC/pulsed plasmas inside bubbles in water," *International Journal of Environmental Science and Technology,* vol. 1, pp. 91–95, 2007.

[92] J. Li, *et al.*, "Degradation of phenol in water using a gas–liquid phase pulsed discharge plasma reactor," *Thin Solid Films,* vol. 515, pp. 4283–4288, 2007.

[93] N. Lu, *et al.*, "Treatment of dye wastewater by using a hybrid gas/liquid pulsed discharge plasma reactor," *Plasma Science and Technology,* vol. 14, pp. 162–166, 2012.

[94] A. Y. Nikiforov, "An application of AC glow discharge stabilized by fast air flow for water treatment," *IEEE Transactions on Plasma Science,* vol. 37, pp. 872–876, 2009.

[95] H. Wang, *et al.*, "Enhanced generation of oxidative species and phenol degradation in a discharge plasma system coupled with TiO_2 photocatalysis," *Applied Catalysis B: Environmental,* vol. 83, pp. 72–77, 2008.

[96] H. Lee, *et al.*, "The effect of liquid phase plasma for photocatalytic degradation of bromothymol blue," *Science of Advanced Materials,* vol. 6, pp. 1627–1631, 2014.

[97] T. Miichi, *et al.*, "Generation of radicals using discharge inside bubbles in water for water treatment," *Ozone: Science & Engineering,* vol. 24, pp. 471–477, 2002.

[98] P. Vanraes, *et al.*, "Electrical and spectroscopic characterization of underwater plasma discharge inside rising gas bubbles," *Journal of Physics D: Applied Physics,* vol. 45, p. 245206, 2012.

[99]   R. Zhang, *et al.*, "Plasma induced degradation of indigo carmine by bipolar pulsed dielectric barrier discharge (DBD) in the water–air mixture," *Journal of Environmental Sciences (China),* vol. 16, pp. 808–812, 2004.

[100] R.-b. Zhang, *et al.*, "Enhancement of the plasma chemistry process in a three-phase discharge reactor," *Plasma Sources Science and Technology,* vol. 14, pp. 308–313, 2005.

[101] A. Anpilov, *et al.*, "Electric discharge in water as a source of UV radiation, ozone and hydrogen peroxide," *Journal of Physics D: Applied Physics,* vol. 34, p. 993, 2001.

[102] E. M. Barkhudarov, *et al.*, "Multispark discharge in water as a method of environmental sustainability problems solution," *Journal of Atomic and Molecular Physics,* vol. 2013, pp. 1–12, 2013.

[103] S. Horikoshi, *et al.*, "A novel liquid plasma AOP device integrating microwaves and ultrasounds and its evaluation in defluorinating perfluorooctanoic acid in aqueous media," *Ultrasonics Sonochemistry,* vol. 18, pp. 938–942, 2011.

[104] C. Yamabe, *et al.*, "Water treatment using discharge on the surface of a bubble in water," *Plasma Processes and Polymers,* vol. 2, pp. 246–251, 2005.

[105] P. Bruggeman, *et al.*, "Water surface deformation in strong electrical fields and its influence on electrical breakdown in a metal pin–water electrode system," *Journal of Physics D: Applied Physics,* vol. 40, p. 4779, 2007.

[106] P. Bruggeman, *et al.*, "Influence of the water surface on the glow-to-spark transition in a metal-pin-to-water electrode system," *Plasma Sources Science and Technology,* vol. 17, p. 045014, 2008.

[107] L. O. d. B. Benetoli, *et al.*, "Effect of temperature on methylene blue decolorization in aqueous medium in electrical discharge plasma reactor," *Journal of the Brazilian Chemical Society,* vol. 22, pp. 1669–1678, 2011.

[108] Y. Miyazaki, *et al.*, "Pulsed discharge purification of water containing nondegradable hazardous substances," *Electrical Engineering in Japan,* vol. 174, pp. 1–8, 2011.

[109] M. Magureanu, *et al.*, "Pulsed corona discharge for degradation of methylene blue in water," *Plasma Chemistry and Plasma Processing,* vol. 33, pp. 51–64, 2013.

[110] M. Dors, *et al.*, "Phenol oxidation in aqueous solution by gas phase corona discharge," *Journal of Advanced Oxidation Technologies,* vol. 9, pp. 139–143, 2006.

[111] A. Sharma, *et al.*, "Destruction of pentachlorophenol using glow discharge plasma process," *Environmental Science & Technology,* vol. 34, pp. 2267–2272, 2000.

[112] P. Jamróz, *et al.*, "Atmospheric pressure glow discharges generated in contact with flowing liquid cathode: production of active species and application in wastewater purification processes," *Plasma Chemistry and Plasma Processing,* vol. 34, pp. 25–37, 2014.

[113] I. Even-Ezra, *et al.*, "Application of a novel plasma-based advanced oxidation process for efficient and cost-effective destruction of refractory organics in tertiary effluents and contaminated groundwater," *Desalination and Water Treatment,* vol. 11, pp. 236–244, 2009.

[114] D. Gerrity, *et al.*, "An evaluation of a pilot-scale nonthermal plasma advanced oxidation process for trace organic compound degradation," *Water Research,* vol. 44, pp. 493–504, 2010.

[115] A. Mizrahi and I. M. Litaor, "The kinetics of the removal of organic pollutants from drinking water by a novel plasma-based advanced oxidation technology," *Desalination and Water Treatment,* vol. 52, pp. 5264–5275, 2013.

[116] Y. Hu, *et al.*, "Application of dielectric barrier discharge plasma for degradation and pathways of dimethoate in aqueous solution," *Separation and Purification Technology,* vol. 120, pp. 191–197, 2013.

[117] S. P. Li, *et al.*, "Degradation of nitenpyram pesticide in aqueous solution by low-temperature plasma," *Environmental Technology,* vol. 34, pp. 1609–1616, 2013.

[118] J. Feng, *et al.*, "Degradation of aqueous 3,4-dichloroaniline by a novel dielectric barrier discharge plasma reactor," *Environmental Science and Pollution Research,* vol. 22, pp. 4447–4459, 2015.

[119] M. Hijosa-Valsero, *et al.*, "Removal of priority pollutants from water by means of dielectric barrier discharge atmospheric plasma," *Journal of Hazardous Materials,* vol. 262, pp. 664–673, 2013.

[120] G. Fridman, *et al.*, "Floating electrode dielectric barrier discharge plasma in air promoting apoptotic behavior in melanoma skin cancer cell lines," *Plasma Chemistry and Plasma Processing,* vol. 27, pp. 163–176, 2007.

[121] E. Marotta, *et al.*, "Advanced oxidation process for degradation of aqueous phenol in a dielectric barrier discharge reactor," *Plasma Processes and Polymers,* vol. 8, pp. 867–875, 2011.

[122] O. Lesage, *et al.*, "Degradation of 4-chlorobenzoïc acid in a thin falling film dielectric barrier discharge reactor," *Industrial & Engineering Chemistry Research,* vol. 53, pp. 10387–10396, 2014.

[123] O. Lesage, *et al.*, "Treatment of 4-chlorobenzoic acid by plasma-based advanced oxidation processes," *Chemical Engineering and Processing: Process Intensification,* vol. 72, pp. 82–89, 2013.

[124] S. Li, *et al.*, "Degradation of thiamethoxam in water by the synergy effect between the plasma discharge and the TiO$_2$ photocatalysis," *Desalination and Water Treatment,* vol. 53, pp. 3018–3025, 2014.

[125] P. Baroch, *et al.*, "Special type of plasma dielectric barrier discharge reactor for direct ozonization of water and degradation of organic pollution," *Journal of Physics D: Applied Physics,* vol. 41, p. 085207, 2008.

[126] S. Rong and Y. Sun, "Wetted-wall corona discharge induced degradation of sulfadiazine antibiotics in aqueous solution," *Journal of Chemical Technology & Biotechnology,* vol. 89, pp. 1351–1359, 2013.

[127] N. Sano, *et al.*, "Decomposition of phenol in water by a cylindrical wetted-wall reactor using direct contact of gas corona discharge," *Industrial & Engineering Chemistry Research,* vol. 42, pp. 5423–5428, 2003.

[128] M. Sato, *et al.*, "Decomposition of phenol in water using water surface plasma in wetted-wall reactor," *International Journal of Plasma Environmental Science & Technology,* vol. 1, pp. 71–75, 2007.

[129] T. Yano, *et al.*, "Decolorization of indigo carmine solution using nanosecond pulsed power," *Dielectrics and Electrical Insulation, IEEE Transactions on,* vol. 16, pp. 1081–1087, 2009.

[130] M. Morimoto, *et al.*, "Indigo carmine solution treatment by nanosecond pulsed power with a dielectric barrier electrode," *Dielectrics and Electrical Insulation, IEEE Transactions on,* vol. 22, pp. 1872–1878, 2015.

[131] K. Faungnawakij, *et al.*, "Removal of acetaldehyde in air using a wetted-wall corona discharge reactor," *Chemical Engineering Journal,* vol. 103, pp. 115–122, 2004.

[132] Y. Chen, *et al.*, "A discharge reactor with water-gas mixing for methyl orange removal," *International Journal of Plasma Environmental Science and Technology,* 2008.

[133] S. Ognier, *et al.*, "Analysis of mechanisms at the plasma–liquid interface in a gas–liquid discharge reactor used for treatment of polluted water," *Plasma Chemistry and Plasma Processing,* vol. 29, pp. 261–273, 2009.

[134] A. G. Bubnov, *et al.*, "Plasma-catalytic decomposition of phenols in atmospheric pressure dielectric barrier discharge," *Plasma Chemistry and Plasma Processing,* vol. 26, pp. 19–30, 2006.

[135] S.-P. Rong, *et al.*, "Degradation of sulfadiazine antibiotics by water falling film dielectric barrier discharge," *Chinese Chemical Letters,* vol. 25, pp. 187–192, 2014.

[136] J.-L. Brisset, *et al.*, "Chemical reactivity of discharges and temporal post-discharges in plasma treatment of aqueous media: examples of gliding discharge treated solutions," *Industrial & Engineering Chemistry Research,* vol. 47, pp. 5761–5781, 2008.

[137] R. Burlica, *et al.*, "Organic dye removal from aqueous solution by glidarc discharges," *Journal of Electrostatics,* vol. 62, pp. 309–321, 2004.

[138] J. Janca, *et al.*, "Investigation of the chemical action of the gliding and "point" arcs between the metallic electrode and aqueous solution," *Plasma Chemistry and Plasma Processing*, vol. 19, pp. 53–67, 1999.

[139] T. Kobayashi, *et al.*, "The effect of spraying of water droplets and location of water droplets on the water treatment by pulsed discharge in air," *IEEE Transactions on Plasma Science*, vol. 38, pp. 2675–2680, 2010.

[140] T. Sugai, *et al.*, "The effect of flow rate and size of water droplets on the water treatment by pulsed discharge in air," *IEEE Transactions on Plasma Science*, vol. PP, p. 1, 2015.

[141] T. Sugai, *et al.*, "Improvement of efficiency for decomposition of organic compound in water using pulsed streamer discharge in air with water droplets by increasing of residence time," in *Pulsed Power Conference, 2009. PPC'09. IEEE*, 2009, pp. 1056–1060.

[142] G. An, *et al.*, "Degradation of phenol in mists by a non-thermal plasma reactor," *Chemosphere*, vol. 84, pp. 1296–1300, 2011.

[143] S. Preis, *et al.*, "Formation of nitrates in aqueous solutions treated with pulsed corona discharge: the impact of organic pollutants," *Ozone: Science & Engineering*, vol. 36, pp. 94–99, 2014.

[144] S. Jiang, *et al.*, "Investigation of pulsed dielectric barrier discharge system on water treatment by liquid droplets in air," *Dielectrics and Electrical Insulation, IEEE Transactions on*, vol. 22, pp. 1866–1871, 2015.

[145] J. Kornev, *et al.*, "Generation of active oxidant species by pulsed dielectric barrier discharge in water-air mixtures," *Ozone: Science & Engineering*, vol. 28, pp. 207–215, 2006.

[146] Z. Wang, *et al.*, "Plasma decoloration of dye using dielectric barrier discharges with earthed spraying water electrodes," *Journal of Electrostatics*, vol. 66, pp. 476–481, 2008.

[147] M. Elsawah, *et al.*, "Corona discharge with electrospraying system for phenol removal from water," *IEEE Transactions on Plasma Science*, vol. 40, pp. 29–34, 2012.

[148] C. M. Du, *et al.*, "Degradation of 4-chlorophenol using a gas–liquid gliding arc discharge plasma reactor," *Plasma Chemistry and Plasma Processing*, vol. 27, pp. 635–646, 2007.

[149] D. Moussa, *et al.*, "Plasma-chemical destruction of trilaurylamine issued from nuclear laboratories of reprocessing plants," *Industrial & Engineering Chemistry Research*, vol. 45, pp. 30–33, 2006.

[150] D. R. Grymonpré, *et al.*, "Hybrid gas–liquid electrical discharge reactors for organic compound degradation," *Industrial & Engineering Chemistry Research*, vol. 43, pp. 1975–1989, 2004.

[151] P. Lukes, *et al.*, "Hydrogen peroxide and ozone formation in hybrid gas–liquid electrical discharge reactors," *Industry Applications, IEEE Transactions on,* vol. 40, pp. 60–67, 2004.

[152] P. Lukes, *et al.*, "Generation of ozone by pulsed corona discharge over water surface in hybrid gas–liquid electrical discharge reactor," *Journal of Physics D: Applied Physics,* vol. 38, pp. 409–416, 2005.

[153] H. Ren, *et al.*, "Oxidation of ammonium sulfite by a multi-needle-to-plate gas phase pulsed corona discharge reactor," *Journal of Physics: Conference Series,* vol. 418, p. 012128, 2013.

[154] Y. Nakagawa, *et al.*, "Decolorization of rhodamine B in water by pulsed high-voltage gas discharge," *Japanese Journal of Applied Physics,* vol. 42, pp. 1422–1428, 2003.

[155] E. Njatawidjaja, *et al.*, "Decoloration of electrostatically atomized organic dye by the pulsed streamer corona discharge," *Journal of Electrostatics,* vol. 63, pp. 353–359, 2005.

[156] Q. Tang, *et al.*, "Gas phase dielectric barrier discharge induced reactive species degradation of 2,4-dinitrophenol," *Chemical Engineering Journal,* vol. 153, pp. 94–100, 2009.

[157] J. Wohlers, *et al.*, "Application of an air ionization device using an atmospheric pressure corona discharge process for water purification," *Water, Air, and Soil Pollution,* vol. 196, pp. 101–113, 2008.

[158] A. Yamatake, *et al.*, "Water treatment by fast oxygen radical flow with DC-driven microhollow cathode discharge," *IEEE Transactions on Plasma Science,* vol. 34, pp. 1375–1381, 2006.

[159] D. Dobrynin, *et al.*, "Cold plasma inactivation of Bacillus cereus and Bacillus anthracis (anthrax) spores," *IEEE Transactions on Plasma Science,* vol. 38, pp. 1878–1884, 2010.

[160] Z. Zhang, *et al.*, "Killing of red tide organisms in sea enclosure using hydroxyl radical-based micro-gap discharge," *IEEE Transactions on Plasma Science,* vol. 34, pp. 2618–2623, 2006.

[161] H. S. Kim, *et al.*, "Use of plasma gliding arc discharges on the inactivation of *E. coli* in water," *Separation and Purification Technology,* vol. 120, pp. 423–428, 2013.

[162] L. Jie, *et al.*, "Degradation of organic compounds by active species sprayed in a dielectric barrier corona discharge system," *Plasma Science and Technology,* vol. 11, p. 211, 2009.

[163] T. C. Wang, *et al.*, "Multi-tube parallel surface discharge plasma reactor for wastewater treatment," *Separation and Purification Technology,* vol. 100, pp. 9–14, 2012.

[164] D. Zhu, *et al.*, "Wire–cylinder dielectric barrier discharge induced degradation of aqueous atrazine," *Chemosphere,* vol. 117, pp. 506–514, 2014.

[165] M. Tichonovas, *et al.*, "Degradation of various textile dyes as wastewater pollutants under dielectric barrier discharge plasma treatment," *Chemical Engineering Journal,* vol. 229, pp. 9–19, 2013.

[166] Y. S. Mok, *et al.*, "Application of dielectric barrier discharge reactor immersed in wastewater to the oxidative degradation of organic contaminant," *Plasma Chemistry and Plasma Processing,* vol. 27, pp. 51–64, 2007.

[167] J. E. Foster, *et al.*, "A comparative study of the time-resolved decomposition of methylene blue dye under the action of a nanosecond repetitively pulsed DBD plasma jet using liquid chromatography and spectrophotometry," *IEEE Transactions on Plasma Science,* vol. 41, pp. 503–512, 2013.

[168] R. Ma, *et al.*, "Non-thermal plasma-activated water inactivation of food-borne pathogen on fresh produce," *Journal of Hazardous Materials,* vol. 300, pp. 643–651, 2015.

[169] P. Olszewski, *et al.*, "Optimizing the electrical excitation of an atmospheric pressure plasma advanced oxidation process," *Journal of Hazardous Materials,* vol. 279, pp. 60–66, 2014.

[170] S. Horikoshi, *et al.*, "Microwave discharge electrodeless lamps (MDEL). III. A novel tungsten-triggered MDEL device emitting VUV and UVC radiation for use in wastewater treatment," *Photochemical & Photobiological Sciences,* vol. 7, p. 303, 2008.

[171] M. Magureanu, *et al.*, "Decomposition of methylene blue in water using a dielectric barrier discharge: optimization of the operating parameters," *Journal of Applied Physics,* vol. 104, p. 103306, 2008.

[172] B. Jaramillo-Sierra, *et al.*, "Phenol degradation in aqueous solution by a gas–liquid phase DBD reactor," *European Physical Journal Applied Physics,* vol. 56, p. 24026, 2011.

[173] R. Ono, *et al.*, "Effect of pulse width on the production of radicals and excited species in a pulsed positive corona discharge," *Journal of Physics D: Applied Physics,* vol. 44, p. 485201, 2011.

[174] T. Sugai, *et al.*, "Investigation of optimum pulse width of applied voltage for water treatment by pulsed streamer discharge in air spraying water droplets," in *Power Modulator and High Voltage Conference (IPMHVC), 2010 IEEE International,* 2010, pp. 213–216.

[175] T. Sugai and Y. Minamitani, "Influence of rise rate of applied voltage for water treatment by pulsed streamer discharge in air-sprayed droplets," *IEEE Transactions on Plasma Science,* vol. 41, pp. 2327–2334, 2013.

[176] J. Zhang, *et al.*, "Degradation of 2,4-dichorophenol by pulsed high voltage discharge in water," *Desalination,* vol. 304, pp. 49–56, 2012.

[177] M. Magureanu, *et al.*, "Degradation of pharmaceutical compound pentoxifylline in water by non-thermal plasma treatment," *Water Research*, vol. 44, pp. 3445–3453, 2010.

[178] K. Yasuoka, *et al.*, "Degradation of perfluoro compounds and F-recovery in water using discharge plasmas generated within gas bubbles," *International Journal of Environmental Science and Technology*, vol. 4, pp. 113–117, 2010.

[179] N. Sano, *et al.*, "Decomposition of organic compounds in water by direct contact of gas corona discharge: influence of discharge conditions," *Industrial & Engineering Chemistry Research*, vol. 41, pp. 5906–5911, 2002.

[180] M. Sahni and B. R. Locke, "Quantification of reductive species produced by high voltage electrical discharges in water," *Plasma Processes and Polymers*, vol. 3, pp. 342–354, 2006.

[181] Y. Jin, *et al.*, "Optimizing decolorization of methylene blue and methyl orange dye by pulsed discharged plasma in water using response surface methodology," *Journal of the Taiwan Institute of Chemical Engineers*, vol. 45, pp. 589–595, 2014.

[182] R. Zhang, *et al.*, "Kinetics of decolorization of azo dye by bipolar pulsed barrier discharge in a three-phase discharge plasma reactor," *Journal of Hazardous Materials*, vol. 142, pp. 105–110, 2007.

[183] P. M. K. Reddy, *et al.*, "Degradation and mineralization of methylene blue by dielectric barrier discharge non-thermal plasma reactor," *Chemical Engineering Journal*, vol. 217, pp. 41–47, 2013.

[184] X. Chen, *et al.*, "Degradation of 4-chlorophenol in a dielectric barrier discharge system," *Separation and Purification Technology*, vol. 120, pp. 102–109, 2013.

[185] M. Sato, *et al.*, "Aqueous phenol decomposition by pulsed discharges on the water surface," *Industry Applications, IEEE Transactions on*, vol. 44, pp. 1397–1402, 2008.

[186] Z. He, *et al.*, "The formation, role and removal of No3-N during corona discharge in air for phenol removal," *Environmental Technology*, vol. 26, pp. 285–292, 2005.

[187] P. Ajo, *et al.*, "Pulsed corona discharge in water treatment: the effect of hydrodynamic conditions on oxidation energy efficiency," *Industrial & Engineering Chemistry Research*, vol. 54, pp. 7452–7458, 2015.

Chapter 16

Study of CO$_2$ Decomposition in Microwave Discharges by Optical Diagnostic Methods

Tiago Silva, Nikolay Britun, Thomas Godfroid and
Rony Snyders

Additional information is available at the end of the chapter

http://dx.doi.org/10.5772/60692

Abstract

The increasing of the carbon dioxide (CO$_2$) release into the atmosphere is undeniably one of the biggest concerns for the twenty-first century. Among the different strategies proposed for reduction of CO$_2$ emission (carbon capture and sequestration (CCS), renewable energies, etc.), low-temperature plasma technology offers an alternative and rather efficient way to convert CO$_2$ into the valuables chemicals (e.g. syngas) which can be stored and used afterwards. Several CO$_2$ decomposition plasma-related approaches have been proposed in the literature, all having a main task: increasing the energy efficiency associated to the decomposition process, while keeping the conversion rate at reasonably high level. This task is especially challenging since many kinetic mechanisms of CO$_2$ decomposition in low-temperature discharges are not yet well-known, such as the vibrational excitation which plays a key role in achieving high decomposition rates. In this chapter our recent research efforts associated with the experimental study of the CO$_2$ decomposition in microwave surfaguide low-temperature discharges are presented. The research was focused on the systematic investigation of the basic plasma parameters. The discharge area of the reactor was characterized by optical emission spectroscopy using the light emitted from spontaneous relaxation of excited species in plasma. The critical parameters such as gas temperature and dissociation rate were evaluated. In addition to this, the post-discharge area was characterized by two-photon absorption laser-induced fluorescence and gas chromatography techniques in order to investigate the exhaust gas composition. All together, the results overviewed in this chapter provide interesting insights into different kinetic mechanisms of CO$_2$-containing discharges, which play an important role in the CO$_2$ decomposition process.

Keywords: CO$_2$ decomposition, microwave discharge, optical emission spectroscopy, two-photon absorption laser-induced fluorescence, gas chromatography

1. Introduction

The combustion of coal, oil, and natural gas supplies approximately 87% of the primary energy used worldwide [1]. These anthropogenic activities are the main sources of CO_2 emission, which is well known for its contribution to the greenhouse effect and planetary heating. Currently, we are burdening the atmosphere of our world with an additional 40 billion tons of CO_2 every year [2]. Furthermore, the rate of these emissions is dramatically increasing as the world population grows and new economies are developed. As recently reported by the National Oceanic and Atmospheric Administration (NOAA), the level of CO_2 accumulated in the atmosphere has reached a symbolic milestone of 400 parts per million (ppm) [3]. This is the highest level of atmospheric CO_2 in human history. In fact, the planet has not experienced these levels of CO_2 since the *Pliocene* period (more than three million years ago). At this point, we are facing a global warming trend which is undoubtedly one of the biggest concerns for the twenty-first century. The scientific evidence concerning numerous drastic effects on the actual earth's physical and biological systems is overwhelming. Serious consequences include the melting of glaciers, loss of biodiversity, change in the agricultural productivity, etc. Unless we find efficient solutions to reduce our reliance on fossil fuels, the planet will continue to warm up in the coming decades and greater will be the impact of climate change.

1.1. The recycling of CO_2

One of the most promising solutions to stabilize the amount of greenhouse pollution in the atmosphere is the so-called *carbon capture and utilization* (CCU) [4]. Such an approach attempts to use CO_2, either directly or after chemical conversion, as part of a reaction chain to produce value-added products. The supply of CO_2 at local level would be guaranteed in the long term by various sources, from small smoke stacks to large coal-fired plants. In this regard, CCU is a very attractive strategy from an economic point of view, considering that in the near future, large amounts of CO_2 will become available as feedstock of nearly zero or even negative cost [5]. In addition, given the possibility of having cheap renewable energy (e.g., wind power during the night period [5]), there is a huge potential in combining the CCU with CO_2-free electricity. This last point is considered to be a key aspect toward a more general goal of resource since it combines the efficient reuse of a waste followed by the reduction of fossil fuel combustion [6]. An overview on different ways to convert CO_2 with the specific reference of introducing renewable energy in the chemical production chain is given in [6, 7]. Among the different CCU routes, the CO_2 hydrogenation is widely discussed nowadays. This strategy is based on the water-gas shift reaction (WGSR):

$$CO + H_2O \Leftrightarrow CO_2 + H_2 \tag{1}$$

to produce syngas (mixture of CO/H_2) which can be further utilized as a building block of methanol, dimethyl ether, liquid hydrocarbons, and other useful compounds via the Fischer-Tropsch process [7]. The intrinsically high energy density of these fuels and their good transportability make them highly desirable. Figure 1 shows an idealized energy cycle that combines renewable energies with CCU valorization through the WGSR. The CO_2 is recovered from an emission source and transformed into liquid fuels, provided that a sustainable production of CO via renewable electricity is obtained. This is an extremely practical case in which the net production of CO_2 would be zero, while the renewable energy could be stored in CO_2-neutral fuels to be integrated in the existing transport infrastructure.

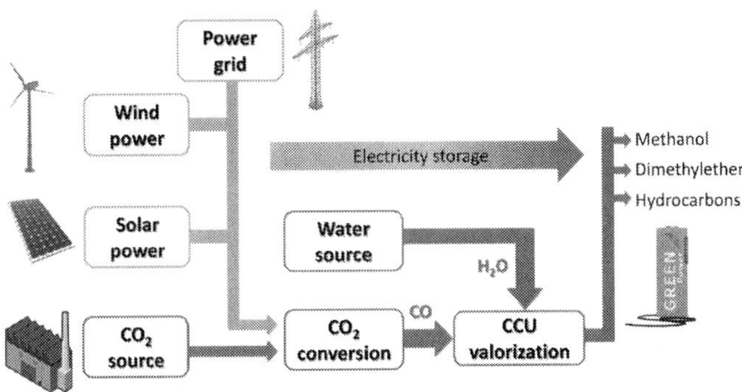

Figure 1. Generic and idealized energy cycle using captured CO_2 and H_2O to yield value-added products via renewable electricity.

Another important strategy that can be used to produce syngas is the dry reforming of CO_2 with methane. This alternative is also particularly attractive since it converts two of the principal gases responsible for the greenhouse effect. In spite of many advantages related to these approaches, there are, however, some critical issues that need to be addressed. In particular, it has to be noted that most of the CCU routes are associated to highly endothermic reactions (e.g., CO_2 conversion in Figure 1). Under classical industrial conditions (e.g., using a typical reactor configuration with packed bed tubes inside a furnace), these reactions are sustained and limited by the rate of heat transfer. This leads to high production costs, which turns the whole conversion chain economically unreasonable [8]. In order to overcome this problem, there is one technology worth investigating: plasma-assisted decomposition. Plasmas provide an ideal environment for CO_2 conversion due to the formation of high energetic charged species that can initiate chemical reactions difficult or impossible to obtain using ordinary thermal mechanisms. In particular, plasma electrons can lead to the formation of vibrationally excited molecules, which are able to dissociate through the vibrational ladder-climbing process [9]. Among different types of plasmas that can be used for CO_2 decomposition, the so-called cold (also known as nonthermal) plasmas are the most promising candidates. These electrical discharges are characterized by nonequilibrium conditions, under which electrons, ions, and neutral species have different translational and – in case of molecules – internal (ro-vibrational) energies. This results in formation of specific nonequilibrium gaseous media in which endothermic processes with increased energy efficiencies and dissociation rates can be achieved [9].

1.2. The CO_2 dissociation in cold plasma

From a practical point of view, all the research involving CO_2 conversion in cold plasmas has one common task: increasing the energy efficiency associated to the decomposition process while keeping the dissociation rate at a reasonably high level. For the sake of comparison between different experiments, the energy efficiency rate (η) of the CO_2 decomposition is usually defined as the ratio of the CO_2 dissociation enthalpy (ΔH) and the injected energy per CO molecule produced in the plasma (E_{CO}):

$$\eta = \Delta H / E_{CO}. \tag{2}$$

Note that ΔH depends on reaction mechanism. The most desirable chemical route to produce CO from CO_2 would be a process or a sequence of processes requiring the smallest amount of energy. In the most direct way, the CO_2 decomposition occurs via:

$$CO_2 \rightarrow CO + O, \quad \Delta H = 5.5 \text{ eV/molecule}. \tag{3}$$

The atomic oxygen created is able then to react either with another oxygen atoms to form O_2 or with vibrationally excited CO_2 (denoted as CO_2^*) according to [9]:

$$O + CO_2^* \rightarrow CO + O_2, \quad \Delta H = 0.3 \text{ eV/molecule}. \tag{4}$$

Combining the above equations, the net reaction for the total dissociation process of one CO_2 molecule yields

$$CO_2 \rightarrow CO + 1/2O_2, \quad \Delta H = 2.9 \text{ eV/molecule}. \tag{5}$$

The energy efficiency rate given by Eq. 2 can be rewritten for the total CO_2 decomposition η (%) according to

$$\eta = \frac{\chi \cdot 2.9}{E_m}, \tag{6}$$

where χ (%) is the CO_2 dissociation rate, while E_m is the specific energy input (in eV/molecule and defined as the ratio of the discharge power over the gas flow rate through the discharge). It is important to stress that in practice, there are many chemical and system inefficiencies (e.g., reverse reactions forming back CO_2) that contribute to a non-efficient decomposition. In this context, the ideal scenario is to find the type of discharge with optimal plasma properties (pressure, flow rate, power, etc.) that minimize these unwanted mechanisms. The most common types of plasma discharges used for CO_2 conversion include the radio frequency (RF) [10], dielectric barrier discharge (DBD) [11–15], gliding arc plasmatron (GAP) [16, 17], glow discharge (GD) [18], and microwave (MW) [9, 19, 20] (see Table 1).

It is interesting to note that among different plasma sources used nowadays, DBD is probably the most widely used in CO_2 conversion research. Although the efficiencies obtained with this source are typically low, the possibility to work at atmospheric pressure under nonequilibrium conditions is very promising. Combined with a catalytic material, these discharges should also improve the selective production of the targeted compounds [15]. In addition, the DBD has a very simple design, which is beneficial for upscaling in industrial applications.

Plasma (f_d)	Power (W)	Pressure (Torr)	E_m (eV/mol)	η(%)	Ref.
RF (13.56 MHz)	~10	0.05–0.3	~39	~ 3	[10]
DBD (0–30 kHz)	~10	760 (1 atm)	~0.2	~ 2	[11]
DBD (6–75 kHz)	80	760 (1 atm)	~2.0	~8	[12]
DBD (1–30 kHz)	45	760 (1 atm)	~5.7	~3.3	[13]
DBD (30–90 kHz)	200	760 (1 atm)	~13.9	~2.0	[14]
DBD (60–130 kHz)	5–80	760 (1 atm)	~0.1-10	<5	[15]
GAP (20 kHz)	~200	760 (1 atm)	~2.0	~30	[16]
GAP (n/a)	n/a	760 (1 atm)	~0.3	~43	[17]
GD (8.1 kHz)	~20	760 (1 atm)	~4.0	~7	[18]
MW (2.45 GHz)	1200	760 (1 atm)	~5.0	~20	[19]
MW (2.45 GHz)	n/a	120	~0.3	~85	[9]
MW (0.91 GHz)	1000	90	~1.9	~35	†

Table 1. List of various plasma sources (excited at frequency f_d) used to decompose CO$_2$ and the corresponding energy efficiencies. † refers to this work

On the other hand, the GAPs have also been receiving a special attention in environmental applications because they can operate at atmospheric pressures while reaching relatively high efficiencies [21]. Nunnally et al. [17] have recently reported a maximum energy efficiency of 43% on CO$_2$ decomposition. This result was attributed to the reverse vortex flow configuration (also known as *tornado* effect) which increases the residence time of the plasma species and leads to a more uniform gas treatment.

Finally, the MW plasma sources are the ones that (until today) were able to provide the highest efficiencies (~85%) for CO$_2$ conversion (see Table 1). As pointed out by Fridman [9], the ability to create a strong nonequilibrium environment in MWs leads to the formation of vibrationally excited CO$_2$ states via vibrational-vibrational (VV) exchange reactions, which favors efficient dissociation. It has to be noted, however, that high energy efficiencies in MWs are usually achieved at reduced pressures (~100–200 Torr), which is not desirable for industrial purposes. Furthermore, MWs operating in the atmospheric regime seem to lead to a clear reduction in the energy efficiency (~20%) as recently reported by Spencer et al. [19]. This is due to the fact that the degree of nonequilibrium tends to decrease at higher pressure in these discharges. Nevertheless, these considerations make the abovementioned discharges particularly attractive to investigate and to further develop toward an industrial implementation.

1.3. The research strategy used

The current study is based on the project P07/34 "Plasma-Surface Interaction" of the Interuniversity Attraction Poles (IAP) action supported by the Belgian Science Policy Office (BELSPO), which (among other subjects) combines experimental and theoretical activities related to CO$_2$ decomposition. The research was focused on the systematic investigation of the basic plasma parameters in microwave discharges aiming at enhancing our understanding on the CO$_2$ decomposition process. The characterization of the discharge has been done through different diagnostic methods, namely, optical emission spectroscopy (OES), two-photon absorption laser-induced fluorescence (TALIF), and gas chromatography

(GC). Due to their nonintrusiveness, these techniques are particularly useful to characterize the CO_2 decomposition in the microwave discharge domain. In this context, we explore the mentioned diagnostic techniques either directly or via development of the additional easy-to-handle tools which, at the same time, may be particularly valuable for maximization of the energy efficiency in real cold plasma discharges targeted to the CO_2 conversion. Among different important parameters investigated, a special attention was given to the measurement of the (i) space-resolved CO_2 dissociation rate, (ii) gas temperature, and (iii) density of products at the end of the reactor exhaust (post-discharge).

2. The experimental part

The plasma investigated in this work was created and sustained through microwave surfaguide-type wave discharges (here abbreviated as MSGDs). This particular kind of microwave plasma source was already proven to be very efficient to produce atomic species in the discharge volume [22]. Working in the pulsed regime, the MSGDs ensure even more efficient molecular dissociation of diatomic or multi-atomic species [23]. The schematic representation of the experimental setup is shown in Figure 2. Two MSGDs excited at different frequencies of 2.45 and 0.915 GHz have been involved in this work. In either case, the discharge was sustained inside a quartz tube (inner radius $R = 7$ mm and length $L = 30$ cm) surrounded by another (polycarbonate) tube for cooling purposes. The mentioned cooling has been performed using an oil flow of 10 °C. The gas mixture was injected from the top of the system and regulated by the electronic mass flow controllers. The pressure range studied was varied from 1 to 90 Torr. The core system was always surrounded by a grounded aluminum grid to prevent a leak of the microwave radiation into the outer space (not shown in Figure 2). The power supply was able to create discrete pulses, typically with lengths in the order of milliseconds to microseconds. A more detailed description of the MSGD can be found in [23]. In respect to the plasma characterization, the OES measurements were performed along the discharge axis, while the TALIF and GC methods were implemented in the post-discharge of the reactor (see Figure 2). The particularities of each diagnostic along with the investigated parameters are described below.

2.1. OES implementation

OES represents a powerful yet nonintrusive characterization technique based on the measurement and analysis of the light emitted from spontaneous relaxation of excited species in plasma. In the current work, the OES was implemented in the 2.45 GHz MSGD system through an Andor Shamrock-750 spectrometer and an Andor iStarDH740-18F-03 ICCD camera with a built-in digital delay generator. In a first approximation (considering an optically thin plasma), the observed line intensity $I(p, q)$ of a particular electronic transition $n(p) \rightarrow n(q)$ is defined as (see, e.g., [24])

$$I(p, q) = h\nu_{pq}(4\pi)^{-1}n(p)A(p, q)L, \tag{7}$$

where $h\nu_{pq}$ is the energy gap between the upper and lower level p and q, $n(p)$ the density of the emitting level, $A(p, q)$ the radiative transition probability, and L the line segment of the plasma along which radiation is collected. A typical emission spectrum recorded at the top of the discharge ($Z = 7$ cm in Figure 2) in a pure CO_2 plasma is shown in Figure 3. As one

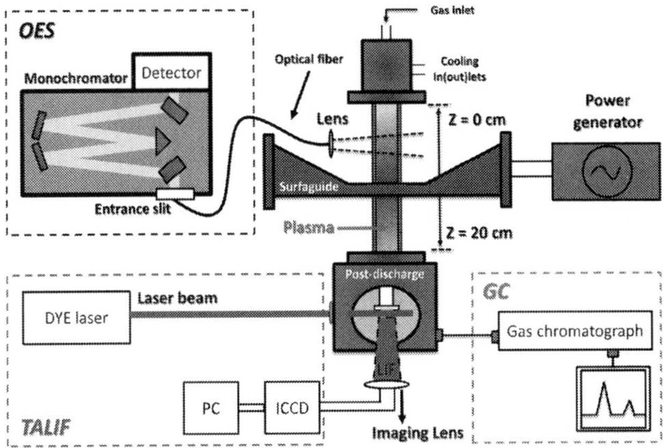

Figure 2. Schematic representation of the MSGD together with the diagnostics techniques used for the plasma characterization.

can see, the Angstrom system ($B^1\Sigma \rightarrow A^1\Pi$) of CO occupies the larger part of the spectrum (~400–700 nm), neighboring the CO third positive system ($b^3\Sigma^+ \rightarrow a^3\Pi_r$) [25]. Other CO bands (e.g., triplet, Herman, and Asundi [25]) were not identified during the measurements, and their contribution is accepted to be negligible for further analysis. The presence of the CO_2^+ ultraviolet doublet ($B^2\Sigma^+ \rightarrow X^2\Pi$) emission at 289 nm and the atomic O peaks in the near-infrared (NIR) part of the spectrum were identified as well. The continuum part of the spectrum induced by CO–O recombination with a maximum at ~450 nm is also visible. It is interesting to note that neither C_2 nor C emission has been observed in this work, which suggests a negligible amount of atomic carbon produced.

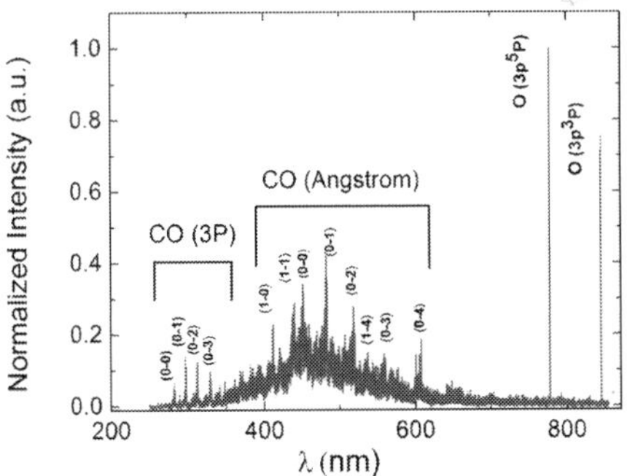

Figure 3. Appearance of the emission spectrum of a pure CO_2 microwave discharge acquired at $Z = 7$ cm. The notation (v'-v'') refers to the vibrational transition, where v'($''$) stands for upper (lower) vibrational energy level. Reproduced with permission from Ref. [20]. Copyright 2014 IOP Publishing.

The parameters investigated by OES in the plasma phase were the (i) rotational temperature (T_{rot}) and (ii) space-resolved CO_2 dissociation rate $\chi(Z)$. OES can be used to determine T_{rot}, provided there is access to a suitable rotational emission band (I_{rot}). In case of Boltzmann equilibrium among the rotational levels, the T_{rot} can be determined via [26]

$$I_{rot} = \frac{C}{Q_R} \frac{1}{\lambda^4} S_J \cdot Exp\left(-\frac{F(J')hc}{k_B T_{rot}} \right) \tag{8}$$

where C is a constant combining all the terms nondependent on the rotational number J, Q_R the rotational state sum, λ the wavelength of the emitted radiation, S_J the J-dependent dimensionless Honl-London factors, $F(J')$ the rotational energy term ($J'(")$ stands for upper (lower) rotational level), k_B the Boltzmann constant, and c the speed of light. Based on Eq. 8, the so-called Boltzmann plot in the coordinates ($F(J')$, $Log(I_{rot}/S_J)$) is often applied to extract the T_{rot}. This parameter is particularly relevant in CO_2 plasmas since it gives a first estimation of the gas temperature T_{gas} [20]. In the current work, the rotational spectra of CO and N_2 have been used for rotational analysis. The Honl-London factors of these molecules can be found in [26].

On the other hand, the estimation of the $\chi(Z)$ cannot be directly obtained from the emission intensity since the number densities of the decomposed species are represented by their non-radiative ground states. The simplest way to overcome this issue by OES is through the so-called *corona* model [27] and the actinometry technique (proposed by Coburn and Chen [28]). Briefly speaking, the corona model assumes that upward transitions (forming excited species) are only due to electron collisions from the ground state, while downward transitions occur via radiative decay. Under these assumptions, the creation-loss balance for the upper level p can be written according to

$$n_g n_e k(g,p) = n(p) \sum_q A(p,q) \tag{9}$$

where n_g is the ground state density, n_e the electron density, $k(g,p)$ the rate coefficient of excitation from the ground level, and $\sum_q A(p,q)$ the sum of all radiative de-excitation processes with origin in the p level. In order to get quantitative information about the prospected n_g, the actinometry technique is applied. In this procedure, a small amount of gas (actinometer) with known concentration n_g^{act} and line intensity $I(i,j)$ with transition $n(i) \rightarrow n(j)$ is added into the discharge. From the ratio of both emission intensities $I(p,q)$ and $I(i,j)$ and combination between Eqs. 7 and 9, the following expression can be deduced:

$$n_g = n_g^{act} \frac{I(p,q)v_{ij}A(i,j)k(g,i)\sum_q A(p,q)}{I(i,j)v_{pq}A(p,q)k(g,p)\sum_j A(i,j)}. \tag{10}$$

In the current research, the actinometry was used to measure the density of CO in the discharge, which leads to the $\chi(Z)$ provided that the initial CO_2 density is known. A small amount (5%) of N_2 was chosen as actinometer due to the proximity of the $N_2(C^3\Pi_u)$ (11.05 eV) and $CO(B^1\Sigma^+)$ (10.78 eV) excited states. Due to this proximity, the populations of these two energetic states should correlate as a result of the electron excitation. The emission lines chosen for the actinometry analysis correspond to the transitions $CO(B^1\Sigma^+)(v'=0) \rightarrow CO(A^1\Pi)(v''=1)$ and $N_2(C^3\Pi_u)(v'=0) \rightarrow N_2(B^3\Pi_g)(v''=2)$, from the Angstrom system of CO and second positive system (SPS) of N_2, respectively. A negligible dissociation rate of

N$_2$ was assumed. This last point is supported by (i) OES measurements in CO$_2$-5%N$_2$ gas mixtures showing a negligible emission of atomic nitrogen [20] and (ii) CO$_2$-N$_2$ modeling data showing a negligible N$_2$ dissociation at low N$_2$ admixtures [29]. The rate coefficients necessary in Eq. 10 were calculated via the Maxwellian distribution (assuming electron temperature T_e ~1 eV as measured in [20]) and the excitation cross sections of CO [30] and N$_2$ [31]. Finally, it is important to mention that Eq. 10 is only valid in CO$_2$ low-pressure discharges with negligible excitation out of metastable levels (e.g., CO($a^3\Pi_r$) and N$_2$($A^3\Sigma_u^+$)). An extended collisional-radiative model would be required in the high-pressure regime, since the excitation out of these intermediate levels is expected to play an important role as well (see, e.g., [32]).

2.2. TALIF implementation

Laser-based plasma diagnostic techniques have largely contributed to understanding plasma kinetics since the 1970s with the development of solid-state lasers [33]. One of these techniques is the TALIF, which is based on the measurement of the excited state fluorescence radiation induced by the excitation of the ground state via resonant absorption of laser photons. The fluorescence signal is proportional to the laser intensity thus allowing detection of the ground state population density. The use of two laser photons is required for most light atoms because the energy gap between the lower and the first electronic excited level exceeds the laser energy produced by conventional dye lasers.

In the current work, the TALIF was applied in the post-discharge region of the 2.45 GHz system through a YAG:Nd pumping laser (Spectra Physics INDI YAG) coupled with a Sirah Cobra Stretch dye laser. The TALIF signal was acquired perpendicularly to the laser beam through an ICCD detector coupled with Nikkor 50 mm lens (see Figure 2). The parameter investigated by TALIF was the oxygen ground state O($^3P_2$) density. The excitation of this level was performed via 2 × 225.586 nm laser photons, while the fluorescence was measured through the radiative decay of the O(^{3}S) level at 844.68 nm. An illustration of the actual fluorescence signal is shown in Figure 4. It is important to stress that the data obtained here by TALIF only yields relative densities. In order to achieve absolute results, the induced fluorescence needs to be calibrated (normally via noble gases [34]). Even though such absolute measurements are particularly promising to improve our understanding on the role played by the oxygen atoms in CO$_2$ discharges, they were not covered by the current study.

2.3. GC implementation

GC offers a sensitive detection of stable reaction products at the exhaust of the plasma reactor. These products are separated inside the gas chromatograph due to their distribution between two non-mixable phases: a stationary phase (solid and/or liquid, filled in the so-called separation column) and the mobile phase (carrier gas, flowing through the column and containing the gaseous sample mixture) [35]. Nowadays, the GC is often applied in many kinds of CO$_2$-containing discharges, including CO$_2$–N$_2$ [29], CO$_2$–H$_2$O [36], CO$_2$–CH$_4$ [37], CO$_2$–Ar [38], etc.

In the current work, the gas chromatograph (Bruker) equipped with a carbon molecular sieve column and a molecular sieve 5A column in series connected with a thermal conductivity detector was implemented in the post-discharge of the 0.91 GHz system. As this plasma

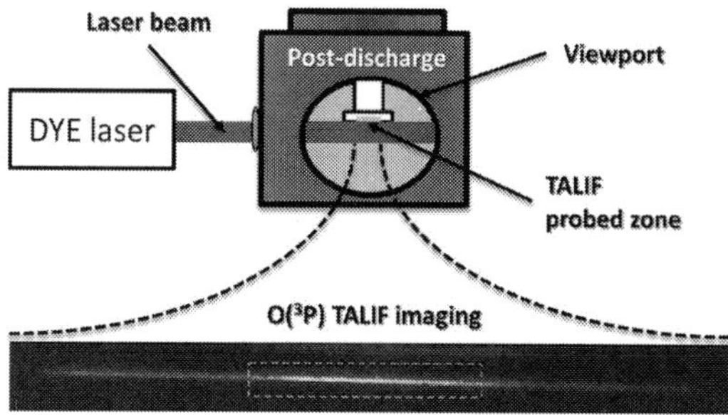

Figure 4. Schematic of the post-discharge region (above), along with the actual O($^3P_2$) image kept by the ICCD detector (below). Dashed rectangle represents the TALIF signal integration area.

source worked in reduced pressure regime, a sampling system was developed between the post-discharge and the gas chromatograph. The low-pressure sample coming from the post-discharge is diluted with neutral gas, in order to reach atmospheric pressure, prior to its injection in the chromatograph. Argon was used as a carrier gas. The CO_2 dissociation rate is calculated by

$$\chi = \frac{A^{off}_{CO_2} - A^{on}_{CO_2}}{A^{off}_{CO_2}}, \tag{11}$$

where $A^{on}_{CO_2}$ ($A^{off}_{CO_2}$) is the chromatograph signal of CO_2 when the plasma is switched on (off). It is important to emphasize that unlike OES, the GC does not provide *in situ* data. However, GC is advantageous over the actinometry method above-described since its applicability is not dependent on the discharge pressure and the accuracy of any population model.

3. The characterization of the discharge area

In this section, we will discuss the results related to investigation of the pulsed CO_2-containing discharges by OES. The energy delivered in the plasma pulse (E_p) during these measurements was varied from 0.8 J to about 1.2 J. The discharge pressure was kept below 5 Torr in this part of work.

3.1. Spatial analysis (OES)

The first insight that one can take regarding the plasma composition is based on the discharge color. For instance, as reported by Timmermans et al. [39], the emission of C_2 molecules in CO_2 discharges (only detected at high pressures due to the significant production of atomic carbon) is represented by a typical green emission. On the contrary, in our case, the light emission from the discharge always showed a strong blue emission due to the presence of CO (see Figure 5). Furthermore, an increase of this light (readily observed by eye), after passing the waveguide (bottom part in this case), has motivated the space-resolved measurements along the discharge tube. Being represented at two different positions in the discharge (top and bottom), the emission spectrum shows certain changes reflecting the molecular decomposition process happening along the gas flow direction (see Figure 5). In addition, one can clearly observe a definite increase of CO Angstrom emission intensity from the top to the bottom of the tube.

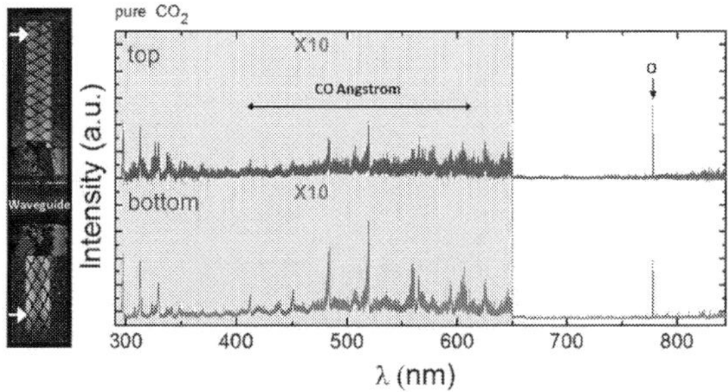

Figure 5. Emission spectra of the CO$_2$ MSGD taken at two different discharge positions (shown by the white arrows). Pressure $\sim$3 Torr.

The abovementioned spatial increase of CO emission is also visible via time-resolved OES imaging (see Figure 6). These measurements provide a density mapping of the excited species in plasma. A filter that passes wavelengths within the 480 nm range was used as indicator of the CO Angstrom emission. Using a plasma on-time of 600 µs, one can clearly see that the CO emission extends axially along the quartz tube with clear predominance at the bottom region of the discharge.

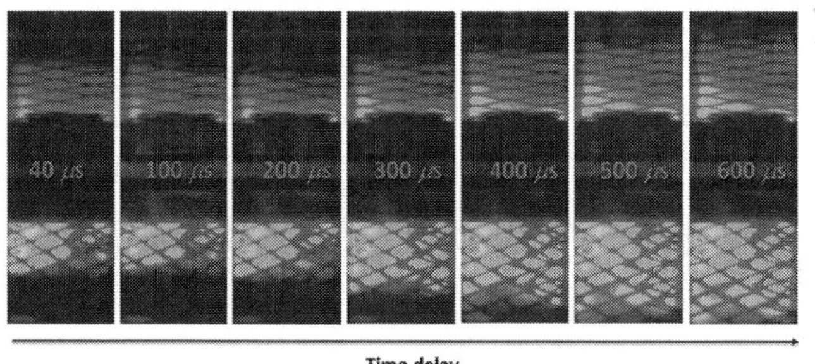

Figure 6. Time-resolved two-dimensional OES imaging of the CO$_2$ MSGD using a 480 nm optical band-pass filter. The iCCD camera was positioned $\sim$15 cm away from the surfaguide. The mean power applied was 300 W with plasma on-time of 600 µs and 50% duty cycle. Each frame is a single-shot picture with a gate step of 20 µs.

In order to clarify the previous observations, the space-resolved temperature measurements were performed. Figure 7 shows T_{gas} calculated via the Boltzmann plot (as described in

Section 2.1) through the CO and N_2 rotational bands. The given error bars correspond to the statistical errors of the Boltzmann fitting in each case. Note that the space interval $Z \approx 8$–13 cm was inaccessible for optical measurements due to the presence of the waveguide. These measurements suggest a linear increase of T_{gas} along the gas flow direction. The results can be interpreted as a trace of the CO_2 decomposition process. Indeed, if the CO_2 molecules in plasma undergo dissociation along the discharge tube, the collision rate increases toward the gas flow direction, which may explain the observed temperature increase. A possible enhancement of the vibrational energy exchange in the plasma phase may also induce vibrational-translational (VT) exchange, which leads to an increase of the gas heating.

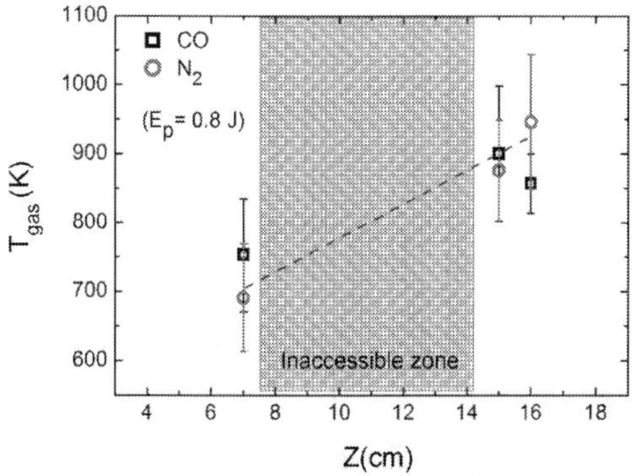

Figure 7. T_{gas} determined based on the N_2 and CO rotational temperatures at different positions of the discharge tube. The $CO_2 + 5\%N_2$ gas mixture is used with pulse (period) of 1.0 (1.5) ms. Adapted with permission from Ref. [20]. Copyright 2014 IOP Publishing.

3.2. CO_2 dissociation rate (actinometry)

The investigation related to the CO_2 dissociation rate obtained by the actinometry method (described in Section 2.1) is presented in this section. The $CO_2 + 5\%N_2$ gas mixture with pulse (period) of 1.0 (1.5) ms was used. The actinometry ratio between CO and N_2 was acquired 0.7 ms after the beginning of the plasma pulse in order to ensure a steady-state plasma regime. As a result of our measurements, a nonuniform $\chi(Z)$ along the gas flow direction was found, as illustrated in Figure 8. The experimental points of $\chi(Z)$ at the top of the discharge are well described by a linear fit (see Figure 8). However, the strong increase (almost four times) between the extremities of the discharge suggests a fast evolution of $\chi(Z)$ in the waveguide vicinity. To fit the obtained $\chi(Z)$ data, the so-called logistic growth has been proposed:

$$\chi(Z) = \chi_0 + \frac{\chi_{max}}{1 + e^{-r(Z - Z_c)}}, \tag{12}$$

where χ_0 (%), Z_c (cm), and χ_{max} (%) are the initial values of the dissociation rate, the middle position of the discharge, and the maximum value of dissociation rate (defined as an average value at the end of the tube). The r (cm^{-1}) is the free fit parameter (see Figure 8). A symmetrical power distribution in the discharge is assumed in this case: $\chi\, (Z = Z_c) \sim$

$\chi_{max}/2$. A derivative of Eq. 12 should correspond to the spatial distribution of the power absorbed in the plasma bulk P_{abs} which is used for CO$_2$ decomposition. Such a distribution can be characterized by the correspondent width (FWHM), i.e., a spatial region where the decomposition is most efficient, as shown in Figure 8, where the red curve represents the derivative of the black one. As one can see, the initial increase of P_{abs} coincides with a small experimental increase of $\chi(Z)$ at the beginning of the discharge ($Z = 4$–7 cm), followed by the fast growth of $\chi(Z)$ in the waveguide region where P_{abs} is supposed to be maximum.

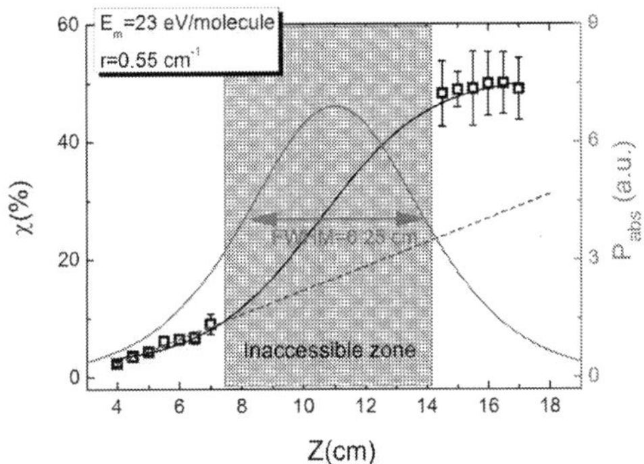

Figure 8. Space-resolved values of the dissociation rate χ (left scale) and the power absorbed in the plasma bulk P_{abs} (right scale) measured at $E_m = 23$ eV/molecule. Reproduced with permission from Ref. [20]. Copyright 2014 IOP Publishing.

Furthermore, it was verified that $\chi(Z)$ increases with E_m as shown in Figure 9. It is interesting to note that our measurements show that $\chi(Z)$ likely reaches a steady state at the bottom of the discharge, whereas a definite linear increase in $\chi(Z)$ is always recognized at its top, i.e., before the gas passes the waveguide. Such differences can be associated with the different chemical processes between these two discharge regions, which should be strongly related to the different power absorption channels. It would be interesting to compare these results with those obtained in an extended discharge tube in order to further investigate this last point.

A linear proportionality between the χ_{max} and E_m is also observed in our case, as shown in Figure 10. A similar behavior of the CO$_2$ decomposition rate which grows linearly when increasing the power and/or decreasing the flow rate is found in [40]. At the same time, based on Eq. 6, η is found to be $\sim$6% at different E_m (see Figure 10).

3.3. Gas temperature analysis (line ratio)

The study of T_{gas} via the CO rotational band was further investigated in order to (i) validate the Boltzmann plot results previously obtained and (ii) search for a simple gas temperature formula based on CO spectral synthesis. Figure 11 shows an example of the so-calculated $CO(B^1\Sigma)(v' = 0) \rightarrow CO(A^1\Pi)(v'' = 1)$ transition based on Eq. 8. The final theoretical spectrum includes the sum of the different rotational branches (P, Q, R corresponding to $\Delta J = J' - J'' = -1, 0, +1$). To reflect the broadening of the spectral lines, a pseudo-Voigt distribution [41], i.e., a combination of a Gaussian and a Lorentzian function, is used:

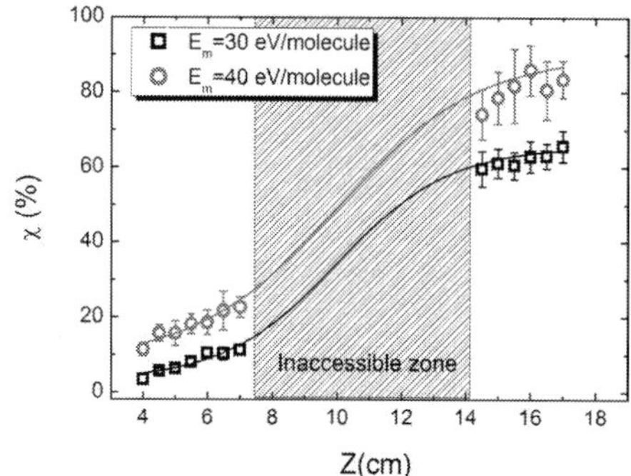

Figure 9. Space-resolved measurements of the CO_2 dissociation rate χ for two different E_m values. The region between about 8 and 14 cm was not optically accessible. Reproduced with permission from Ref. [20]. Copyright 2014 IOP Publishing.

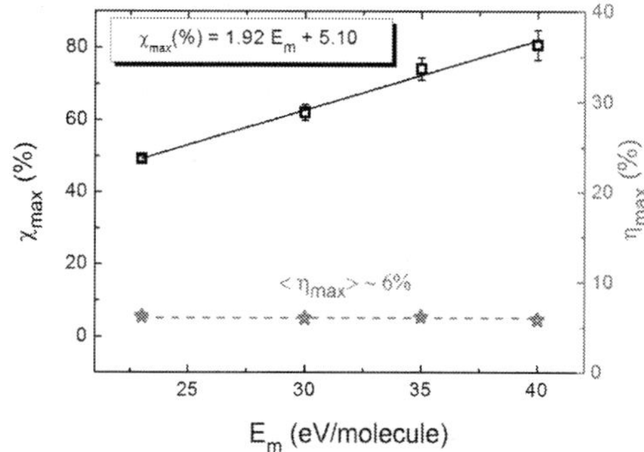

Figure 10. Maximum CO_2 dissociation rate χ_{max} (left scale) along with its energetic efficiency η_{max} (right scale) as a function of E_m. Reproduced with permission from Ref. [20]. Copyright 2014 IOP Publishing.

$$g(p, w_G, w_L) = p \cdot Exp[-4Log(2)(\frac{\lambda - \lambda_0}{w_G})^2] + (1 - p)\frac{w_L}{w_L^2 + 4(\lambda - \lambda_0)^2}, \tag{13}$$

where p is the contribution of the Gaussian function, λ_0 the line central wavelength, and w_G and w_L the full width at half-maximum (FWHM) for the Gaussian and Lorentzian profiles, respectively. The Voigt profile parameters were experimentally determined by measuring the shape of the 435.8 nm Hg line ($^3P_1$–$^3S_1$) from an Ar-Hg lamp (see Figure 11). The Hg line chosen for determination of the monochromator apparatus function is assumed to have (i) a spectral response similar to the one of the CO spectra and (ii) Doppler, Stark, and van der Waals broadening much smaller than the instrumental broadening [42–44]. The rotational peaks located at $P_1 = 481.61$ and $P_2 = 482.48$ nm were chosen in order to build a line-ratio formula. In this procedure, special attention was paid to find isolated peaks with good visibility and good sensitivity to the gas temperature, as discussed in [45]. To avoid bulky calculations, the expression for T_{gas} is then derived, taking into account the contribution of all

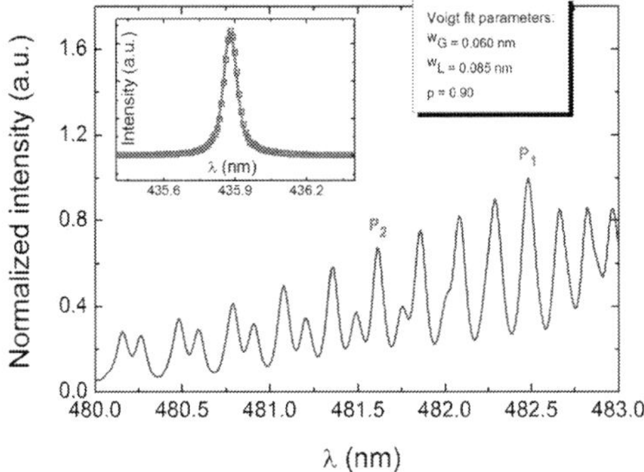

Figure 11. Synthetic spectrum of the molecular transition from the CO Angstrom rotational band calculated at $T_{rot} = 470$ K. Inset: measured Hg (435.8 nm) emission line (squares) and the corresponding fit (solid line) using a pseudo-Voigt function given by Eq. 13. Reproduced with permission from Ref. [41]. Copyright 2014 OAS Publishing.

the P, Q, and R rotational branches based on a semi-analytical approach, namely, when the peak ratio is determined as a function of T_{gas} for each rotational spectrum synthesized. After linearization in the coordinates $(Log(R), 1/T_{gas})$, the final result for the mentioned peaks can be presented in the form:

$$T_{gas}(K) = 343(0.36 + Log(R))^{-1}, \qquad (14)$$

where $R = P_1/P_2$ is the peak ratio. The graphical representation of Eq. 14 between 200 and 10 000 K (typical range of cold plasmas) is given in Figure 12. The relative error of the T_{gas} determination relatively to the accuracy of the line intensity ratio (δT_{gas}) can be determined after differentiating Eq. 14. As a result, we obtain

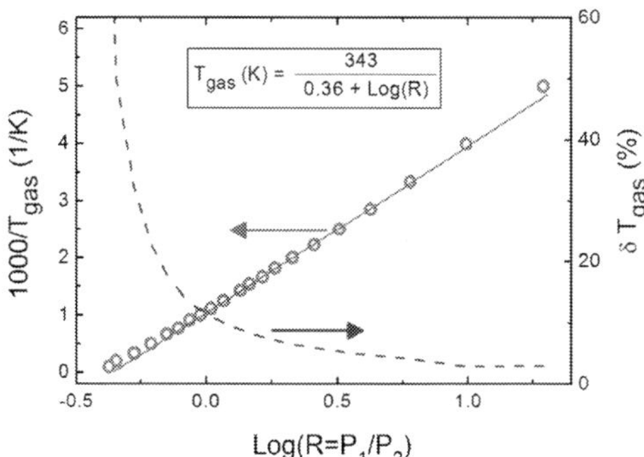

Figure 12. T_{gas} determination chart based on the line ratio between two spectral peaks in the CO Angstrom band (left scale). The red circles and the solid line represent the calculated points and the linear fit, respectively. The dashed line (right scale) indicates the relative error for the gas temperature – δT_{gas}. Reproduced with permission from Ref. [41]. Copyright 2014 OAS Publishing.

$$\delta T_{gas}(K) = \delta R \cdot (0.36 + Log(R))^{-1}. \tag{15}$$

The quantity δT_{gas} is given as a function of $Log(R)$ in Figure 12 where the R relative error $\delta R = 0.05$ is assumed. As we can see, the T_{gas} relative error increases dramatically, when $Log(R) < 0$, exceeding 30% for the temperatures above $\sim$3000 K. This fact limits the applicability of the proposed method at high gas temperatures.

In order to test the proposed method, the time-resolved T_{gas} was measured in pure CO_2 MSGD with a plasma pulse (period) of 1.0(1.5) ms. The proposed line-ratio method is compared with the results obtained by the Boltzmann plot method and direct comparison between the measured and calculated spectra. Figure 13 shows a reasonable agreement in terms of T_{gas} evolution for these different approaches. The uncertainty (about 11% in saturation) of the obtained T_{gas} by the line-ratio method is still in the range of the errors described by Eq. 15 at $T_{gas} \sim 800$ K. These time-resolved measurements also show that, at a fixed E_p, the T_{gas} grows at the beginning of the pulse and saturates after about 0.4 ms. These effects were observed previously for the other gas mixtures in the same type of discharges [46]. First effect corresponds to the gas heating at the beginning of the plasma pulse, whereas the second one reflects the equilibrium between the gas heating and heat dissipation and may be related to VT relaxation processes, as suggested in [46].

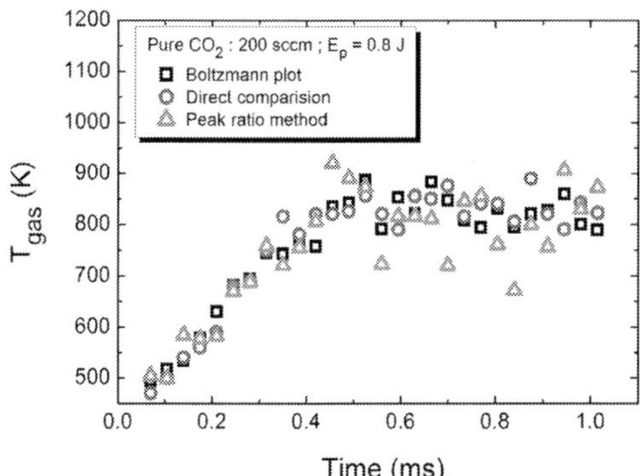

Figure 13. Time evolution of T_{gas} determined based on the CO rotational band using the Boltzmann method (black squares), direct spectral comparison (red circles), and proposed line-ratio method (green triangles). Reproduced with permission from Ref. [41]. Copyright 2014 OAS Publishing.

4. The characterization of the post-discharge area

The research related to the post-discharge area characterization of the microwave reactor is presented in this section. A special attention was devoted to the investigation of parameters that may favor the fine-tuning of CO_2 decomposition. This brings us to the study of the following effects: (i) Ar admixture, (ii) duty cycle, and (iii) discharge pressure. The motivation behind each effect, along with the results obtained, is presented below.

4.1. Ar admixture effect (TALIF)

The influence of Ar atoms in the chemistry of gas discharges containing oxygen species has been extensively studied over the past years (see, e.g., [47, 48]). In this context, it is well known that the Ar metastables (here denoted by Ar^{met}) play a key role as internal energy reservoirs for the production of atomic oxygen species via [48]:

$$\begin{aligned} Ar^{met} + O_2 &\rightarrow O(^3P_2) + O(^3P_2) + Ar \\ &\rightarrow O(^3P_2) + O(^1D) + Ar \\ &\rightarrow O(^3P_2) + O(^1S) + Ar \end{aligned} \qquad (16)$$

Furthermore, due to their long lifetime, the Ar^{met} may also contribute to the chemistry of the afterglow and post-discharges. These considerations have triggered the TALIF study on $O(^3P_2)$ formation as a function of the Ar admixture. The typical results related to the $O(^3P_2)$ density measurements in the Ar–O$_2$ and Ar–CO$_2$ mixtures are shown in Figure 14. Indeed, these measurements strongly suggest that Ar atoms enhance the $O(^3P_2)$ production. At low Ar admixtures, the strong molecular quenching induced by O$_2$/CO$_2$ should lead to a decrease in the Ar^{met} population and less $O(^3P_2)$ produced via Eq. 16. The last observation is supported by theoretical and experimental results obtained in Ar–O$_2$ [47] and Ar–CO$_2$ [49] MSGDs, which show a decrease of Ar^{met} (about three orders of magnitude) in the plasma phase as a consequence of the molecular quenching.

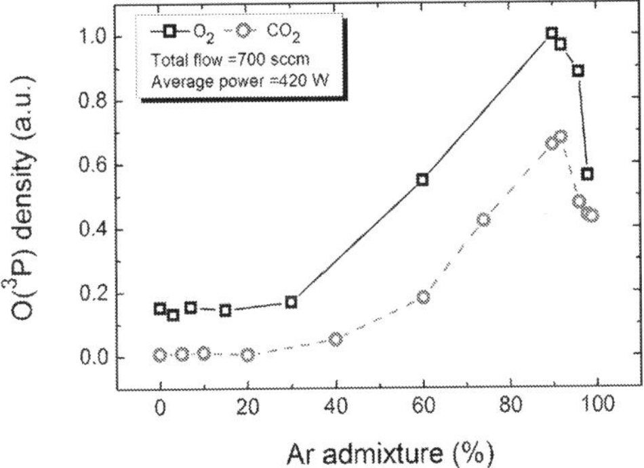

Figure 14. Density of the $O(^3P_2)$ as a function of the Ar admixture in O$_2$ (solid line) and CO$_2$ (dashed line) mixtures. The mean power applied was 400 W with a 0.7 kHz pulsing frequency and 33% duty cycle. The total gas flow in both mixtures was 700 sccm. The maximum Ar admixture used was 99%.

It is interesting to note that as the Ar content increases from ∼90% to 100%, there is a clear decrease of $O(^3P_2)$ as the source of oxygen (CO$_2$/O$_2$) is also decreasing. This leads to the formation of the maximum in $O(^3P_2)$ production around 90% of Ar admixture. Furthermore, it should be emphasized that the electron density is expected to decrease as the Ar admixture decreases since the power used for the production of electron-ion pairs is diverted into vibrational and rotational heating of the molecular gas. As a result of lower electron density, one might also expect a decrease in the contribution of the ladder-climbing processes (e.g., stepwise ionization) out of the Ar^{met}. These considerations (together with low molecular quenching rate at low molecular admixture) may contribute to an increase of Ar^{met} population around 90% of Ar admixture. Further investigations are required in order to validate this last assumption, namely, the measurement of Ar^{met} densities in the post-discharge region of the reactor.

4.2. Duty cycle effect (TALIF)

The pulse discharge operation mode offers an additional feature that may increase the flexibility of the global dynamic plasma response in microwave discharges. In many experiments, the power interruption was usually found to be suitable for reducing gas heating in the discharge by injecting a large amount of power during the plasma on-time while keeping the average power at low values [23]. In case of microwave plasmas, this effect was already proven to be advantageous for the production of atomic nitrogen and oxygen in N_2 and O_2 discharges ([23, 50]).

In this work, the influence of the duty cycle (defined as the ratio between the plasma on-time and its repetition period) on the $O(^3P_2)$ production was investigated by TALIF (see Figure 15). These results clearly show an improvement in the $O(^3P_2)$ production as the duty cycle decreases. The CO_2-20% Ar gas mixture with a pulsing frequency of 0.5 kHz was used during these measurements.

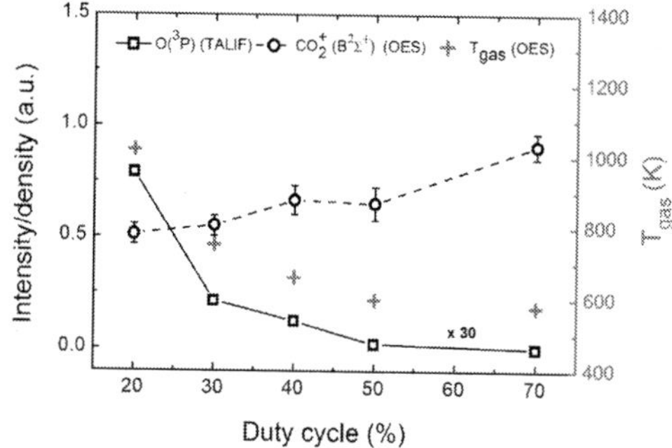

Figure 15. Left scale: Emission of CO_2^+ (dashed line) and density of $O(^3P_2)$ (solid line). The TALIF data was normalized to the maximum value obtained in Figure 14 and further multiplied by a factor of 30. Right scale: Gas temperature (crosses) calculated via Eq. 14. The mean power applied was 250 W with a 0.5 kHz pulsing frequency. The CO_2-20% gas mixture with total flow rate of 700 sccm was used.

It is important to stress that the average power was kept constant. On the other hand, by decreasing the power per pulse as the duty cycle increases, we found a decrease of T_{gas} (calculated via Eq. 14) that follows the $O(^3P_2)$ evolution. Interestingly, the emission of the ultraviolet doublet system of CO_2^+ (measured in the plasma phase by OES) shows an increase with the duty cycle. These observations suggest that low duty cycles favor the dissociative recombination of CO_2, which is responsible for the release of energy in the form of heating into the plasma [23]. There are other factors that may also play an important role in the power modulation effect, namely, the electron parameters and sources of internal energy (e.g., metastable species). For instance, as shown by the theoretical work of Subramonium et al. [51], it can be expected to find higher T_e values at the leading edge of the plasma on-time as the duty cycle decreases. This result supports the data obtained, considering dissociation via electron impact of O$_2$ as the major source of O($^3P_2$).

4.3. Pressure effect (GC)

The pressure effect on the CO$_2$ decomposition was studied by GC (as described in Section 2.3) in this last part of the work. Figure 16 shows the χ (calculated via Eq. 11) and η (calculated via Eq. 6) evolution in the range of 20–90 Torr at different pulsing frequencies. The values obtained at low pressures are in good agreement with the theoretical calculations performed by Bogaerts et al. [52] for microwave plasmas sustained at 20 Torr. As the pressure increases, our experimental data clearly shows an improvement of the CO$_2$ decomposition with $\eta \sim 35\%$ and $\chi \sim 20\%$ at 90 Torr. Figure 16 also suggests that higher η values can be achieved with further increase of pressure. Unfortunately, such pressure range was limited by the current pumping system. As suggested by Fridman [9], these pressure-dependent results can be understood based on the increase of electron-neutral collision frequency in the plasma phase which favors the vibrational excitation of the asymmetric vibrational mode of CO$_2$ and leads to higher conversion rates. A possible decrease of T_e with the increase of discharge pressure might also enhance the vibrational excitation of CO$_2$ since less energy is wasted in electronic excitation. This last observation is supported by T_e measurements recently obtained in CO$_2$–Ar discharges [49]. However, further increase of pressure (eventually reaching the atmospheric range) may lead to higher VT relaxation rates, faster gas heating, and lower efficiencies. In this case, the plasma would act as a heater, which would bring additional problems such as the reverse reactions forming back CO$_2$.

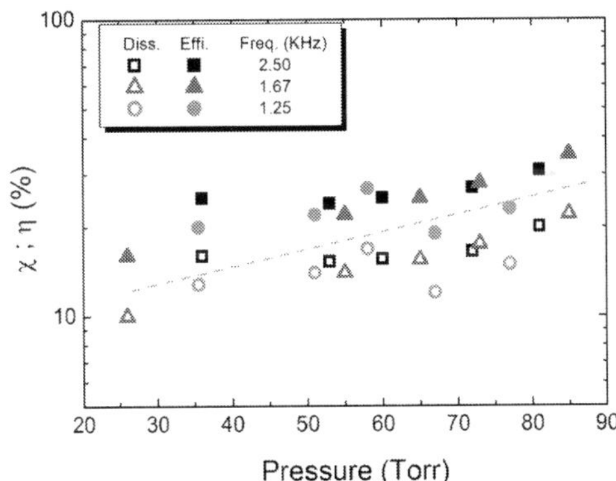

Figure 16. Evolution of the dissociation rate and energy efficiency of CO$_2$ as a function of the discharge pressure at different pulse frequencies. The mean power applied was 2000 W with 50% duty cycle. The flow of CO$_2$ was 15 slm ($E_m \approx 1.9$ eV/molecule).

5. Summary and conclusion

As new solutions to decrease our dependence on fossil fuels are required, the recycling of CO$_2$ into value-added materials may play a critical role in the near future. In this regard, plasma-assisted decomposition offers a practical and efficient way to convert CO$_2$

into valuable products that can be extremely useful in CCU applications. Therefore, and given the necessity to enhance the CO_2 conversion in plasma reactors, there is a constant need to improve the knowledge of CO_2-containing discharges through the development of experimental diagnostic techniques and the improvement of theoretical models. This chapter explores the microwave surfaguide discharge as one potential example of plasma technology that can be used for CO_2 conversion. In order to fully characterize our system and obtain a large set of valuable data, we used three different diagnostics: (i) optical emission spectroscopy, (ii) two-photon absorption laser-induced fluorescence, and (iii) gas chromatography. The first one is particularly useful to study the plasma region of the reactor, while the other two give an important insight on the density of products in the post-discharge region.

As a result of the optical emission spectroscopy measurements (via actinometry) along the discharge tube, a nonuniform distribution of the CO_2 dissociation rate is obtained in the gas propagation direction, which also depends on the specific energy input. This method offers a practical way to construct a CO_2 "dissociation map" through the discharge volume with the spatial resolution depending only on the collimating optics used in the optical emission measurements. With a proper population model to describe the excitation processes, such a mapping should give rather accurate results that may be particularly valuable for maximization of the CO_2 dissociation efficiency in cold plasma discharges.

A special attention has been also given to the measurement of the gas temperature, which is a fundamental parameter to characterize the dissipation of energy through gas heating. In this regard, a simple gas temperature formula based on the line ratio between two rotational peaks of the CO Angstrom rotational emission band was developed. This method is extremely useful if a quick gas temperature estimation without the spectral synthesis is required. It allows for T_{gas} determination, assuming the presence of CO emission from plasma discharge. The applicability of the proposed line-ratio formula is, however, limited by the temperature of about 3000 K, above which the measurement error is rather high ($>$30%). The validity of the proposed approach was additionally approved by the well-known Boltzmann plot, as well as by direct comparison with simulated spectra.

Concerning the post-discharge characterization, the research presented here was mainly focused on the study of parameters that may favor the fine-tuning of CO_2 decomposition. As a result of these measurements, the presence of Ar atoms was confirmed to be beneficial for the formation of the ground state atomic oxygen species. This effect was attributed to the high population density of Ar metastable species in the plasma phase at higher Ar admixtures. The gas mixture CO_2:Ar = 1:9 was found to be the most efficient proportion to produce the ground state atomic oxygen. On the other hand, the study of the power modulation effect also showed interesting results, namely, the increase of oxygen ground state species at lower duty cycle values. Finally, what is related to the pressure effect, a linear increase of CO_2 decomposition was obtained in the range of 20–90 Torr. The maximum value of energy efficiency obtained was about 35% with a reasonable dissociation rate of 20%. The complete set of results obtained in this work are summarized in Table 2.

Overall, the methods explored/developed in this work offer a valuable set of tools that can be applied in microwave and other plasma sources. In addition, the data presented here may also provide a useful input for modeling of the microwave surfaguide-type discharges operating under CO_2 gas mixtures. As a long-term perspective, the authors believe that

Parameter	Method	Result	Observation
Dissociation rate	Actinometry	↗ along the tube	Logistic growth
Energy efficiency	Actinometry	Constant at 6%	23–40 eV/mol.
Gas temperature	OES	↗ along the tube	Linear growth
Gas temperature	OES	↗ with plasma pulse	Satt. at ~0.4 ms
Atomic oxygen	TALIF	↗ with Ar admixture	Max. at 90% Ar
Atomic oxygen	TALIF	↘ with duty cycles	T_{gas} dependent
Energy efficiency	GC	↗ with pressure	~35% at 90 Torr
Dissociation rate	GC	↗ with pressure	~20% at 90 Torr

Table 2. Summary of the various results obtained in this work and the corresponding methods used for the plasma characterization

microwave plasmas are among the most promising candidates to obtain an efficient CO_2 decomposition. However, further investigations related to understanding of the basic plasma processes are still required in the atmospheric pressure regime. The use of catalytic materials [15] (e.g., to improve selective production of decomposed products) and efficient discharge configurations (e.g., using a reverse vortex flow [17]) may turn out as the next step toward a practical implementation of microwave plasmas targeted for massive CO_2 conversion.

Acknowledgments

This work is supported by the Belgian Government through the "Pole d'Attraction Interuniversitaire" (PAI, P7/34, "Plasma-Surface Interaction", Ψ). The authors do appreciate the valuable comments of Prof. Joost van der Mullen (Universite Libre de Bruxelles, Belgium). N. Britun is a postdoctoral researcher of the Fonds National de la Recherche Scientifique (FNRS), Belgium.

Author details

Tiago Silva[1*], Nikolay Britun[1], Thomas Godfroid[2] and Rony Snyders[1,2]

*Address all correspondence to: tiago.dapontesilva@umons.ac.be

1 Chimie des Interactions Plasma-Surface (ChIPS), Universite de Mons, Mons, Belgium

2 Materia Nova Research Center, Mons, Belgium

References

[1] Shearer C, Bistline J, Inman M, Davis SJ. The effect of natural gas supply on US renewable energy and CO_2 emissions. *Environ. Res. Lett.* 2014; **9**:094008.

[2] Le Quiere C, et al. Global carbon budget. *Earth Syst. Sci. Data Discuss.* 2013; **6**:689–760.

[3] Trends in Atmospheric Carbon Dioxide [Internet]. 2015. Available from: http://www.esrl.noaa.gov/gmd/ccgg/trends/; accessed: 2015-08-15.

[4] Chen W. CO_2 conversion for syngas production in methane catalytic partial oxidation. *J. CO_2 Util.* 2014; **5**:1–9.

[5] Jiang Z, Xiao T, Kuznetsov VL, Edwards PP. Turning carbon dioxide into fuel. *Phil. Trans. R. Soc.* 2010; **368**:3343–3364.

[6] Centi G, Perathoner S. Opportunities and prospects in the chemical recycling of carbon dioxide to fuels. *Catal. Today.* 2009; **148**:191–205.

[7] De Falco M, Iaquaniello G, Centi G, editors. CO_2: A Valuable Source of Carbon. Springer, London/Heidelberg/New York/Dordrecht; 2013. 194p.

[8] Amouroux J, Cavadias S, Doubla A. Carbon dioxide reduction by non-equilibrium electrocatalysis plasma reactor. *IOP Conf. Series: Materials Science and Engineering.* 2011; **19**:012005.

[9] Fridman A. Plasma Chemistry. Cambridge University Press, New York; 2008. 978p.

[10] Spencer LF, Gallimore AD. Efficiency of CO_2 dissociation in a radio-frequency discharge. *Plasma Chem. Plasma Processes.* 2011;**31**:79–89.

[11] Lindon MA, Scime EE. CO_2 dissociation using the Versatile atmospheric dielectric barrier discharge experiment (VADER). *Front. Phys.* 2014; **2**:55

[12] Aerts R, Somers W, Bogaerts A. Carbon dioxide splitting in a dielectric barrier discharge plasma: a combined experimental and computational study. *ChemSusChem.* 2015; **8**:702–716.

[13] Ozkan A, Dufour t, Arnoult G, De Keyzer P, Bogaerts A, Reniers F. CO_2-CH_4 conversion and syngas formation at atmospheric pressure using a multi-electrode dielectric barrier discharge. *J. CO_2 Util.* 2015; **9**:74–81.

[14] Paulussen S, Verheyde B, Tu X, De Bie C, Martens T, Petrovic D, Bogaerts A, Sels B. Conversion of carbon dioxide to value-added chemicals in atmospheric pressure dielectric barrier discharges. *Plasma Sources Sci. Technol.* 2010; **19**:034015.

[15] Brehmer F, Welzel S, van de Sanden MCM, Engeln R. CO and byproduct formation during CO_2 reduction in dielectric barrier discharges. *J. Appl. Phys.* 2014; **116**:123303.

[16] Indarto A, Yang DR, Choi J, Lee H, Song HK. Gliding arc plasma processing of CO_2 conversion. *J. Hazard. Mater.* 2007; **146**:309–315.

[17] Nunnally T, Gutsol K, Rabinovich A, Rabinovich A, Gutsol A, Kemoun A. Dissociation of CO_2 in a low current gliding arc plasmatron. *J. Phys. D: Appl. Phys.* 2001; **44**:274009.

[18] Wang J, Xia G, Huang A, Suib SL, Hayashi Y, Matsumoto H. CO_2 decomposition using glow discharge plasmas. *J. Catal.* 1999; **185**:152–159.

[19] Spencer LF, Gallimore AD. CO_2 dissociation in an atmospheric pressure plasma/catalyst system: a study of efficiency. *Plasma Sources Sci. Technol.* 2013; **22**:015019.

[20] Silva T, Britun N, Godfroid T, Snyders R. Optical characterization of a microwave pulsed discharge used for dissociation of CO$_2$. *Plasma Sources Sci. Technol.* 2014; **39**:025009.

[21] Czernichowski A. Gliding arc: applications to engineering and environment control. *Pure Appl. Chem.* 1994; **66**:1301–1310.

[22] Ferreira CM, Moisan M, editors. Microwave Discharges: Fundamentas and Applications, Plenum Press, New York; 1993. 564p.

[23] Godfroid T, Dauchot JP, Hecq M. Atomic nitrogen source for reactive magnetron sputtering. *Chemical Surf. Coatings Technol.* 2003; **174–175**:1276–1281.

[24] Fujimoto T, Plasma Spectroscopy, Oxford: Clarendon Press; 2004. 304p.

[25] Pearse RWB, Gaydon AG. The Identification of Molecular Spectra. Wiley, New York; 1976. p. 407.

[26] Herzberg G, Molecular Spectra and Molecular Structure Vol 1: Spectra of Diatomic Molecules. Litton Educational Publishing, New York; 1950. 658p.

[27] Boffard JB, Lin CC, DeJoseph Jr. Application of excitation cross sections to optical plasma diagnostics. *J. Phys. D: Appl. Phys.* 2004; **37**;R143–R161

[28] Coburn JW, Chen M. Optical-emission spectroscopy of reactive plasmas – a method for correlating emission intensities to reactive particule density. *J. Appl. Phys.* 1980; **6**:3134–3136.

[29] Heijkers S, Snoeckx R, Kozak T, Silva T, Godfroid T, Britun N, Snyders R, Bogaerts A. CO$_2$ conversion in a microwave plasma reactor in the presence of N$_2$: elucidating the role of vibrational levels. *J. Phys. Chem. C.* 2015; **119** 12815–12828.

[30] SIGLO database, www.lxcat.laplace.univ-tlse.fr; retrieved 5 May 2015.

[31] IST-Lisbon database, www.lxcat.laplace.univ-tlse.fr; retrieved 5 May 2015.

[32] Boffard JB, Jung RO, Lin CC, Wendt AE. Optical emission measurements of electron energy distributions in low-pressure argon inductively coupled plasmas. *Plasma Sources Sci. Technol.* 2010; **19**:065001.

[33] Amorim J, Baravian G, Jolly. J. Laser-induced resonance fluorescence as a diagnostic technique in non-thermal equilibrium plasmas. *J. Phys. D: Appl. Phys.* 2000; **33**:R51–R65.

[34] Es-sebbar Et, Benilan Y, Jolly A, Gazeau M-C. Characterization of an N$_2$ flowing microwave post-discharge by OES spectroscopy and determination of absolute ground-state nitrogen atom densities by TALIF. *J. Phys. D: Appl. Phys.* 2009; **42**:135206.

[35] Grob RL, Barry EF. Modern Practice of Gas Chromatography. John Wiley & Sons Inc., Hoboken, New Jersey; 2004. 1064p.

[36] Chen G, Silva T, Georgieva V, Godfroid T, Britun N, Snyders R, Delplancke-Ogletree MP. Simultaneous dissociation of CO_2 and H_2O to syngas in a surface-wave microwave discharge. *Int. J. Hydrogen Energ.* 2015; **40**:3789–3796.

[37] Pinhao NR, Janeco A, Branco JB. Influence of helium on the conversion of methane and carbon dioxide in a dielectric barrier discharge. *Plasma Chem. Plasma P.* 2014; **31**:427–439.

[38] Otake K, Kinoshita H, Kikuchi T, Suzuki R. CO_2 decomposition using electrochemical process in molten salts. *J. Phys.: Conf. Ser.* 2012; **379**:012038.

[39] Timmermans EAH, Jonkers J, Rodero A, Quintero MC, Sola A, Gamero A, Schram DC and van der Mullen J. The behavior of molecules in microwave-induced plasmas studied by optical emission spectroscopy. 2: Plasmas at reduced pressure. *Spectrochim. Acta Part B.* 1999; **54**:1085–1098.

[40] Rond C, Bultel A, Boubert P, Cheron BG. Spectroscopic measurements of nonequilibrium CO_2 plasma in RF torch. *Chem. Phys.* 2008; **354**:16–26.

[41] Silva T, Britun N, Godfroid T, Snyders R. Simple method for gas temperature determination in CO_2-containing discharges. *Opt. Lett.* 2014; **39**:6146–6149.

[42] Munoz J, Dimitrijevic MS, Yubero C, Calzada MD. Using the van der Waals broadening of spectral atomic lines to measure the gas temperature of an argon-helium microwave plasma at atmospheric pressure. *Spectrochim. Acta, Part B: At. Spectrosc.* 2009; **64**:167–172.

[43] Palomares JM, Hubner S, Carbone EAD, de Vries N, van Veldhuizen EM, Sola A, Gamero A, van der Mullen JJAM. H_β Stark broadening in cold plasmas with low electron densities calibrated with Thomson scattering. *Spectrochim. Acta, Part B: At. Spectrosc.* 2012; **73**:39–47.

[44] Konjevic N. Plasma broadening and shifting of non-hydrogenic spectral lines: present status and applications. *Phys. Rep.* 1999; **316**:339–401.

[45] Kotstka S, Roy S, Lakusta P, Mayer T, Renfro M, Gord J, Branam R. Comparison of line-peak and line-scanning excitation in two-color laser-induced-fluorescence thermometry of OH. *Appl. Opt.* 2009; **48**:6332–6343.

[46] Britun N, Godfroid T, Snyders R. Time-resolved study of pulsed $Ar-N_2$ and $Ar-N_2-H_2$ microwave surfaguide discharges using optical emission spectroscopy. *Plasma Sources Sci.Technol.* 2012; **21**:035007.

[47] Kutasi K, Guerra V, Paulo S. Theoretical insight into $Ar-O_2$ surface-wave microwave discharges. *J. Phys. D: Appl. Phys.* 2010; **43**:175201.

[48] Kitajima T, Nakano T, Makabe T. Increased O (^1D) metastable density in highly Ar-diluted oxygen plasmas. *Appl. Phys. Lett.* 2006; **88**:091501.

[49] Silva T, Britun N, Godfroid T, van der Mullen J, Snyders R. Study of Ar and Ar-CO$_2$ microwave surfaguide discharges by optical spectroscopy. Submitted to *J. Appl. Phys.* 2015.

[50] Rousseau A, Granier A, Gousset G, Leprince P. Microwave discharge in H$_2$: influence of H-atom density on the power balance. *J. Phys. D: Appl. Phys.* 1994; **27**:1412–1422.

[51] Subramonium P, Kushner MJ. Two-dimensional modeling of long-term transients in inductively coupled plasmas using moderate computational parallelism. I. Ar pulsed plasmas. *J. Vac. Sci. Technol. A.* 2001; **20(2)**:313–324.

[52] Bogaerts A, Kozak T, Van Laer K, Shoceckx R. FDCDU15 – carbon dioxide utilisation: plasma-based conversion of CO$_2$: current status and future challenges. *Faraday Discuss.* 2015; DOI: 10.1039/C5FD00053J.

Basic Plasma Phenomena

Chapter 17

Stochastic and Nonlinear Dynamics in Low-Temperature Plasmas

Aldo Figueroa, Raúl Salgado-García, Jannet Rodríguez, Farook Yousif Bashir, Marco A. Rivera and Federico Vázquez

Additional information is available at the end of the chapter

http://dx.doi.org/10.5772/62096

Abstract

Low-temperature (LT) plasmas have a substantial role in diverse scientific areas and modern technologies. Their stochastic and nonlinear dynamics strongly determine the efficiency and effectiveness of LT plasma-based procedures involved in applications such as etching, spectrochemical analysis, deposition of thin films on substrates, and others. Understanding and controlling complex behaviors in LT plasmas have become a serious research problem. Modeling their behavior is also a major problem. However, models based on hydrodynamic equations have proven to be useful in their study. In this chapter, we expose the use of fluid models taking into account relevant kinetic processes to describe out from equilibrium LT plasma behavior. Selected topics on the stability, stochastic, and nonlinear dynamics of LT plasmas are discussed. These include the coexistence of diffusive and wave-like particle transport and delayed feedback control of oscillatory regime with relaxation.

Keywords: Low-temperature plasmas, stochastic dynamics, nonlinear behavior, delayed feedback control

1. Introduction

A plasma is a complex system consisting of mutually interacting particles. Some or all of the following can coexist in a plasma: electrons, photons, positive ions, negative ions, metastables, free radicals, and neutral species. In the laboratory, plasmas can be classified into two categories, namely, high-temperature and low-temperature plasmas. The former is obtained in fusion devices, and the plasma is generally in local thermal equilibrium (LTE), which means that each of the existing species is in thermal equilibrium with each other. Typical

LTE plasma temperatures range from 5000 K to 20000 K. Low-temperature plasmas, produced in well-known electrode configurations, are not in thermal equilibrium (non-LTE) as each of the species has its own temperature. In these systems, electrons are characterized by their higher temperatures than those of other heavier species (ions, atoms, etc.). The thermodynamic state (thermal plasma or nonthermal plasma) of the plasma mainly depends on its pressure p. This is understood if one considers that in a high-pressure plasma, many collisions occur leading to a more frequent interchange of energy between species than in nonthermal plasmas. The pressure determines not only the thermodynamic state of the plasmas but also the distance d between electrodes. In fact, the product pd can be used to distinguish the thermal and nonthermal states. Besides pressure and electrode separation, the working gas, the applied electric (or electromagnetic) field, and the configuration of the containing device also determine the thermodynamic state of the plasma. Among the wide variety of gas discharges, in this chapter, we will discuss nonthermal plasmas. Among them, we find the following: direct current (DC) glow discharges, capacitively coupled radio-frequency (RF) discharges, pulsed glow discharges, atmospheric pressure glow discharges, corona discharges, dielectric barrier discharges, and magnetron discharges. For an overview of the physical description and applications of both thermal and nonthermal plasmas, see Ref. [1]. A DC glow discharge is a nonthermal plasma wherein the main role is played by electrons. Normally, they have high temperatures which lead to inelastic collisions determining the ionization rate in the system. In fact, these collision processes sustain the plasma. The temperature of heavier particles is lower than that of electrons. Their kinetics has importance in applications such as etching, spectrochemical analysis, and deposition of thin films on substrates through processes such as sputtering, ion implantation, deposition, etc. When one or both of the electrodes are insulating materials, positive or negative charges accumulate and get charged causing plasma annihilation. This process is avoided by introducing an alternating voltage to the electrodes giving rise to the so-called capacitively coupled discharges which are obtained in radio frequencies. This kind of discharge is especially used for the spectrochemical analysis of nonconductive materials or deposition of thin films on dielectric electrodes [2]. If the applied alternating voltage is in the form of pulses, a very similar discharge to DC glow discharge is obtained in a short period of time. In this way, higher peak voltages and currents can be applied, and more efficient sputtering, ionization, and excitation processes are obtained. Because of the existence of lapses of time where the discharge burns out, the pulsed discharges do not excessively heat up in such a way that the temperature of the gas is lower than the electron temperature. This is conducive to the nonthermal state of the plasma. Atmospheric pressure glow discharges occur at higher (e.g., atmospheric) pressures. Increasing the operating pressure makes it necessary to reduce the distance between electrodes in such a way that discharge conditions are maintained. In this kind of discharge, one (or two) electrode is made of a dielectric material, and the device must be operated with an alternating voltage. However, under these operating conditions, a stable glow discharge can be obtained when some working gases (e.g., helium) are used. In other cases (nitrogen, oxygen, argon), it takes the form of a filamentary glow discharge, however, in a stable state. Atmospheric pressure plasmas are especially suitable for the deposition of both organic and inorganic coatings [3] and the treatment of industrial and biological surfaces [4]. We end the description of nonthermal plasma discharges by considering the dielectric barrier discharge. The device used to produce this kind of discharge is quite similar to that of the atmospheric pressure glow discharge. The main difference is that the latter is homogeneous along the electrodes, while the dielectric barrier discharge produces filamentary discharge of a very short duration. Note, however, as mentioned earlier, that the atmospheric pressure discharge may also consist of those filamentary discharges for certain gases. In each of the filamentary discharges, the high electron energy can excite the gas-provoking chemical reactions and emission of radiation which are of interest in various applications [5].

In this brief overview, it may be appreciated that nonthermal plasmas have a substantial role in diverse scientific areas and modern technologies. This has attracted the interest of both theorists and experimentalists giving rise to the discovery of very complex behaviors which were unknown until 20 years ago. It has also attracted the attention to nonlinear dynamics of nonthermal plasmas, since it strongly determines the efficiency and effectiveness of nonthermal plasma-based procedures. Because of this, understanding and controlling complex behaviors in nonthermal plasmas have become a serious research problem. We find a great variety of complex dynamics in nonthermal plasmas, namely, deterministic chaos, mixed-mode oscillations, and homoclinic chaos as well as order–chaos transitions, stochastic resonance, coherence resonance, etc. In this chapter, we will describe some of them and the control techniques developed to mitigate or even enhance those kinds of complex behaviors in low-temperature discharges. In Section 2, the basis of fluid modeling of nonthermal plasmas is exposed. The dynamic equations of particle densities and electron energy as well as constitutive expressions for the fluxes can be found in the same section. In Section 3, a brief discussion on stability aspects of nonthermal plasmas is given. In Section 4, the plasma dynamics is analyzed as a stochastic process. The stochastic equations are solved by means of a high-order numerical method. We also discuss the coexistence of diffusive and wave-like particle transport regimes in the stationary state. Finally, Section 5 is devoted to the nonlinear behavior of nonthermal plasmas, and several control methods are described. Then the delayed feedback control (DFC) is applied to the oscillation with relaxation regime under specific values of physical parameters of the plasma, resulting in an efficient control of the oscillations.

2. Fluid modeling of nonthermal plasmas

The importance of nonthermal gas discharges in various industrial and technological applications makes understanding and controlling such plasmas necessary. This requires to model a dissipative system out from thermodynamic equilibrium where several nonlinear kinetic processes take place giving rise to complex behaviors. Among others, nonlinear kinetic processes include electron attachment and ionization, ion–ion recombination, etc. Models describing gas discharges can be classified as fluid methods [6], particle methods, and hybrid methods [7] which are combinations of the former and the latter. Hybrid methods are particularly useful when there exist several time scales in the system. In this kind of model, the fast and energetic electrons are treated as particles while the slow components (heavier species) as coexisting fluids. Fluid models (FMs) are used for describing collective movement of particles, such as wave effects, which cannot be obtained from a kinetic equation of the electron behavior. In fluid models, all the species are considered as fluids, and they make use of continuity equations and the energy balance equation (normally written only for electrons). Some additional equations are necessary in order to complete these sets of equations, namely, the Poisson equation for the self-consistent (self-generated) electric field and constitutive equations relating the dissipative fluxes with the thermodynamic forces. These equations resort to well-known phenomenological laws such as Ohm's law, Fourier law, etc. [8, 9]. To complete the framework, expressions for the transport and rate coefficients are necessary. An often-used assumption for these coefficients is the local field approximation. In this approximation, the coefficients are supposed to depend on the reduced local electric field E / p with p the pressure of the working gas . This assumption has a limited validity since it implies that ionization can occur in the cathode sheath solely. Nevertheless, it is known that bright luminous regions of high charge density in the negative glow can be produced by nonlocal ionization processes [6, 10, 11]. Instead, sometimes the local mean energy approximation is used. In this approximation, expressions for the transport and rate coefficients come from the Boltzmann equation as functions of the electron temperature which is obtained by solving the energy balance equation of electrons.

The continuity equations can be written as

$$\frac{\partial n_j}{\partial t} + \nabla \cdot \vec{J}_j = S_j \tag{1}$$

where n_j denotes the densities of the species in the plasma, with $j = p, e, m, n, g, \ldots$ for positive ions, electrons, metastable molecules, negative ions, working gas, etc., respectively. S_j is the volume gain or loss of particles. The particle fluxes are given by the constitutive equations which, if only drift and diffusion are considered, become [12]

$$\begin{aligned}
\vec{J}_p &= \mu_p n_p \vec{E} - D_p \nabla n_p \\
\vec{J}_n &= -\mu_n n_n \vec{E} - D_n \nabla n_n \\
\vec{J}_e &= -\mu_e n_e \vec{E} - \nabla \left(D_e n_e \right) \\
\vec{J}_m &= -D_m \nabla n_m
\end{aligned} \tag{2}$$

where $\mu_p,\ \mu_n,\ \mu_e$ are the mobilities of ions and electrons; $D_p,\ D_n,\ D_e,\ D_m$ the corresponding diffusion coefficients (see Section 3); and $\vec{E}$ the electric field. Note that D_e is included in the gradient operator, since it depends on the mean electronic energy U_e which is a function of the spatial position. The same must be said for the electronic mobility μ_e. The balance equation for the mean electronic energy is given as

$$\frac{\partial(n_e U_e)}{\partial t} + \nabla \cdot \vec{J}_{U_e} = -\vec{J}_e \cdot \vec{E} - n_e S_{U_e} \tag{3}$$

The work done by the electric field on the electrons and the electronic energy changes due to elastic and inelastic collisions (the term with S_{U_e}). In the same equation, the energy flux is given by

$$\vec{J}_{U_e} = -\mu_e^* n_e U_e \vec{E} - \nabla\left(D_e^* n_e U_e\right) \tag{4}$$

with μ_e^* and D_e^* being functions of the mean electronic energy.

The source terms in the continuity equations depend on several kinetic processes. Electrons are mainly generated by ionization (collisions of electrons with the working gas molecules). Negative ions are generated by the attachment of electrons by the working gas molecules and lost by ion–ion recombination. Metastable molecules are generated by electron collisions with the working gas molecules and lost by superelastic collisions of electrons with metastables. In the last process, associative detachment processes of metastables with the working gas molecules are also of importance. The total electric flux (electric current density) depends on particle fluxes as shown in the following:

$$\vec{j} = q\left(\vec{J}_p - \vec{J}_n - \vec{J}_e\right) \tag{5}$$

which obeys the condition

$$\nabla \cdot \vec{j} = 0 \tag{6}$$

The total electric flux can be used to eliminate $\nabla \cdot \vec{J}_e$ in the continuity equation for the electrons. In this way, the electronic transport coefficients, which depend on the energy U_e, are no longer necessary.

Respecting boundary conditions, particle (electrons, ions) fluxes must be specified at the electrodes of the device, often considered as perfect absorbers of particles [11]. The energy flux (due to electrons) must also be specified at electrodes. In the case of conducting electrodes, the boundary condition for Poisson equation should fix the level of the electric potential (including the imposed potential) giving rise to Dirichlet-type and Neumann-type conditions when they are dielectrics (perpendicular electric field imposed on them).

Much is known on low-temperature plasmas by fluid model-based investigations. Fluid models (FM) have been used to study the fundamental properties and dynamic behavior of several gases: argon, oxygen, hydrogen, and helium in a diversity of configurations (see Section

1) and its results confirmed by experiments. The following is not an exhaustive list. In RF discharges, the FM of the plasma has reported the experimentally observed dependence of the harmonic part of the discharge current on the geometry of the device [11]. Coupled with external circuit equations, FMs have been used to reproduce the commonly observed mixed mode oscillations and chaos in glow discharges [13, 14]. The influence of impurities constituted by small amounts of nitrogen on the dielectric controlled discharge at atmospheric pressure in helium has been investigated by means of FM in one-dimensional models [15]. The temporal nonlinear behavior in atmospheric glow and dielectric barrier discharges such as period multiplication and secondary bifurcations has been studied by using a FM approach [16, 17, 18]. Another interesting problem analyzed with FM is the stability of homogeneous state in the positive column of a glow discharge [12]. The results agree with the observation that a hysteresis occurs at the transition from the H-mode to the T-mode.

3. Fluid modeling and instabilities in low-temperature plasmas

3.1. Ionization instability in the positive column of a glow discharge

The ionization instability occurs in the positive column of a glow discharge. It is a consequence of the nonlinear dependence of the ionization rate on the electron temperature. The main dynamic processes behind the instability are the fluctuations of the ionization rate provoked by fluctuations of the electron temperature. Electrons diffuse and are drifted in the electric field being heated and cooled by collisions with other species in the plasma. In this way, the electron temperature becomes a fluctuating quantity and thus the ionization rate. The self-consistent electric field also contributes to this process. The ionization instability can be described through a properly selected set of dynamic variables and the associated fluid equations [19]. The relevant properties are the electron and positive-ion densities, the ionization rate, the electric field, and the electron temperature. Fluid modeling description is based on i) the equation describing the diffusion, production, and recombination of charged particles Eq. (1); ii) an expression of the ionization rate as a function of the electron temperature and the ionization potential; iii) an equation for the self-consistent electric field in terms of the inhomogeneities of the electron density, namely, Poisson equation; and iv) the energy balance equation for electrons, Eq. (3) [19], describing their heating and cooling by the electric field and collisions respectively. Under specific conditions, the solutions of this set of equations show that a local fluctuation of n_e results in a fluctuation of the temperature T_e on the anode side, and then an ionization perturbation propagates toward the anode with a certain group velocity.

3.2. Transition from a uniform nonthermal RF discharge to an arc

The power density in a uniform RF discharge is limited by instabilities which destroy the homogeneity of the plasma by arcing. Several kinds of instabilities have been identified, the thermal instability and the α to β state transition, being the two most encountered. The thermal instability occurs in a sequence of processes [20] leading to a (positive) fluctuation of the electron density δn_e which increases the electric current density δj_e. It results in an increment of power dissipation $\delta(j_e E)$. This results in an increase in gas temperature δT and leads to a decrease in the gas density δn_g. In turn, the reduced electric field fluctuates positively by $\delta(E/n_g)$ causing an increase in the electron temperature δT_e. Hence, an increment in ionization occurs and a further increasing in the electron density δn_e. It has been assumed that the initial fluctuation of the electron density occurs rapid enough to consider that the electric field remains constant. It must be mentioned that, in fact, the described amplification process can be initiated with a fluctuation of any of the involved physical properties. The discharge is unstable if the rate of heating of the working gas, v_h, is higher than the cooling rate, v_c, divided by a factor, v^*, known as the logarithmic sensitivity, given by [20]:

$$v^* = \frac{d\left(\ln\left(v_i\left(E/n_g\right) - v_a\left(E/n_g\right)\right)\right)}{d\left(\ln\left(E/n_g\right)\right)}$$ (7)

where v_i is the ionization frequency and v_a the attachment frequency. The cooling rate v_c enters the energy balance equation of the working gas which is written as follows:

$$c_p \rho \frac{\partial T_g}{\partial t} = \vec{j}_e \cdot \vec{E} - c_p \rho \left(T_g - T_0\right) v_c$$ (8)

The two terms on the right-hand side describe Joule heating and cooling of the gas through conduction and convection. At this point, it is necessary to introduce the growth rate of thermal fluctuations, Ω, in such a way that $\delta T_g \approx e^{\Omega t}$. Using the two previous equations, it can be proved that the growth rate of thermal fluctuations depends on the heating and cooling frequencies as follows:

$$\Omega = v_h(v^* + 1) - v_c$$ (9)

Clearly, if $\Omega < 0$, the discharge is stable, and if $\Omega > 0$, thermal instability occurs, and the discharge is unstable. Calculated values for v^* are in the range of 5–10 [21]. It has been shown that if there is an increment of 10%–20% in the gas temperature, thermal instability occurs.

The α to β state transition is another known instability of the RF gas discharge. This instability occurs between two modes of sustaining the discharge. In the α mode, the discharge is sustained through volumetric ionization, while in the β mode, sustaining takes place through secondary electron emission. The transition appears when the sheath suffers a breakdown after the electric field reaches a critical value. The instability associated with the α to β state transition takes place in a smaller time scale than that of the thermal instability described above.

3.3. Linear stability of a DC-driven electronegative discharge

Consider the homogeneous and stationary state of an electronegative discharge with axial symmetry [12], where the homogeneity of the system is in the axial direction. The equilibrium condition can be expressed in terms of the averaged source terms as $\bar{P}_e=\bar{P}_n=\bar{P}_m=0$ (these averages are written in terms of the cross section averaged density of the working gas, the details in [22]). Except for equilibrium, the generation of metastables by super-elastic collisions (see Section 2) has practically no influence on the linear properties of the discharge. In order to describe the properties of stability in a cylindrical geometry, the radial and axial dynamics are separated by introducing a new set of fields, N_j, in Eq. (1) (we follow the work of Bruhn et al. [12]). They are defined as

$$n_e(r,z,t) = N_e(z,t)\frac{g(r)}{\bar{g}}$$
$$n_n(r,z,t) = N_n(z,t)\frac{h(r)}{\bar{h}} \tag{10}$$
$$n_m(r,z,t) = N_m(z,t)\frac{l(r)}{\bar{l}}$$

where $g(r)$, $h(r)$, and $l(r)$ are the corresponding radial profile functions [23, 24, 25]. The bar over each one denotes a cross-section average. Then, it is assumed that the current density $\vec{j}$ (Eq. (5)) has the radial electron profile since electrons are the dominant components of the axial electric current. In this way

$$\vec{j} = j\frac{g(r)}{\bar{g}}\vec{e}_z = \frac{I(t)}{\pi r_0^2}\frac{g(r)}{\bar{g}}\vec{e}_z \tag{11}$$

where r_0 is the radius of the discharge and $I(t)$ is the electric current. The same assumption is also made for the mean electron energy $U_e=U_e(z,t)$, while the electric field is written as $\vec{E}(r,z,t)=E(z,t)\vec{e}_z+E(r,t)\vec{e}_r$ since its radial component cannot be neglected. Substituting Eq. (10) into Eq. (1), the cross-section averaged balance equations for the axial part of the particle densities are found

$$\frac{\partial N_m}{\partial t} = D_m\frac{\partial^2 N_m}{\partial z^2} + \bar{S}_m$$
$$\frac{\partial N_n}{\partial t} = D_n\frac{\partial^2 N_n}{\partial z^2} + \mu_n\frac{\partial}{\partial z}(N_n E_z) + \bar{S}_n \tag{12}$$
$$\frac{\partial N_e}{\partial t} = D_e\frac{\partial^2 N_e}{\partial z^2} + (D_e - D_n)\frac{\partial^2 N_n}{\partial z^2} + \bar{S}_e - \mu_e\frac{\partial}{\partial z}(N_e E_z) - (\mu_e + \mu_n)\frac{\partial}{\partial z}(N_n E_z)$$

The coefficients in Eq. (12) have been defined in Section 2. Since the explicit expressions, in their most general form, for the averaged source terms in Eq. (12) shed light on the kinetic reactions involved in the dynamics of species existing in the plasma, they are written here:

$$\bar{S}_m = k_m \bar{n} N_e - k_s c_1 N_e N_m - k_{det} c_2 N_n N_m - v_w^m N_m$$
$$\bar{S}_n = k_{at} \bar{n} N_e - k_{det} c_2 N_n N_m + (v_w^n - \underline{v}_w^n) N_n - k_{rec} N_n (c_3 N_e + c_4 N_n)$$
$$\bar{S}_e = k_{io} \bar{n} N_e - k_{att} \bar{n} N_e + k_{det} c_2 N_n N_m - (v_w^e - \underline{v}_w^e) N_e - \left(1 + \frac{\mu_p}{\mu_n}\right) v_w^n N_n + \left(1 - \frac{D_p}{D_n}\right) \underline{v}_w^n N_n$$

(13)

where $c_1 \ldots c_4$ are the so-called shape factors and v_w^m, v_w^n, $\underline{v}_w^n$, v_w^e, $\underline{v}_w^e$ denote the losses of different species in the walls. The metastable working gas molecules are generated by electron impact at the rate k_m and they are lost by superelastic collisions at k_s, as well as by associative detachment at k_{det}. The negative ions are generated by attachment at k_{att}, and they are lost by ion–ion recombination at k_{rec}, where positive and negative ions recombine to form molecular and atomic oxygen. The main generation process of electrons is ionization at the rate k_{io}. After averaging, the balance equation for the mean electronic energy can be written as

$$\frac{\partial (N_e U_e)}{\partial t} = \frac{\partial^2 \left(D_e^* N_e U_e\right)}{\partial z^2} + \frac{\partial \left(\mu_e^* N_e U_e E_z\right)}{\partial z} + E_z \frac{\partial \left(D_e^* N_e\right)}{\partial z} + \mu_e N_e E_z^2 - N_e \bar{S}_u^e - N_m \bar{S}_u^m$$

(14)

where $\bar{P}_u^e$ and $\bar{P}_u^m$ are averaged loss terms given by

$$\bar{S}_u^e = \frac{\mu_e^*}{\mu_e}\left(1 + \frac{\mu_p}{\mu_n}\right) v_w^n U_e + \frac{\mu_e^*}{\mu_e}\left(\frac{D_p}{D_n} - 1\right) \underline{v}_w^n U_e + c_5 \left(D_p - D_n\right) \frac{D_e}{\mu_e} + c_6 \left(\mu_p + \mu_n\right) \left(\frac{D_e}{\mu_e}\right)^2$$
$$\bar{S}_u^n = P_u\left(U_e\right) + \underline{v}_w^u U_e$$

(15)

The quantities μ_j, μ_j^*, and D_j were introduced in Section 2.

Equations (12) and (13) can be used to study the stability of the discharge by introducing nondimensional variables which measure how much the system is far from equilibrium. The particle density deviation is nondimensionalized through the corresponding particle density at equilibrium, while the electric field and the mean electronic energy are nondimensionalized by the values at equilibrium. If the new set of variables are denoted by n_j^*, with $j = m, n, e, U_e, E$, the continuity equations, Eq. (12), can be written up to first order as

$$T_{jk} \frac{\partial n_k^*}{\partial \tau} = \left(A_{jk} + B_{jk} \frac{\partial}{\partial \xi} + C_{jk} \frac{\partial^2}{\partial \xi^2}\right) n_k^* + P_{jk} \frac{1}{l} \int_0^l n_k^* d\xi$$

(16)

where the coefficients A_{jk}, B_{jk}, C_{jk}, P_{jk}, and T_{jk} are known quantities (see Appendix C in [12]), and ξ and τ are dimensionless variables of position and time, respectively. The perturbations n_k^* are expressed as a linear combination of plane waves as follows:

$$n_k^*(\xi,\tau) = \sum_{N=-\infty}^{\infty} n_k^{(N)} e^{i(k_N \xi - \omega_N \tau)} \tag{17}$$

where k_N is the real number and ω_N is the complex number, and $n_k^{(N)}$ is the amplitude of the k th Fourier component of each specific property. The dispersion relations are obtained from Eq. (16) with the expansion of Eq. (17). The complex roots can each be identified with a potential unstable plasma behavior. The specific value of the root ω_N from the dispersion relation determines the time evolution of the perturbation. In case $Re(i\omega_N)>0$, the perturbation grows in time which implies an instability. In this way, $Re(i\omega_N)$ represents the rate of growth of the perturbation. The characteristic $Im(i\omega_N)$ determines if the wave behavior of perturbations is propagative or dispersive in nature. The calculation of the roots ω_N may be difficult since it may involve high-order algebraic equations. The analysis of time scales of different relaxation processes can reduce the order of them. For instance, characteristic time of electron energy relaxation is in the range 10^{-10}–10^{-8}sec, which is very short when compared with the time required for ionization, attachment, etc. [26]. Thus, the time response of the electron energy density to a perturbation is almost instantaneous compared with the time required for a change that occurs in the electric charge density. This allows to omit the term in the time derivative in Eq. (14), which reduces the order of the dispersion relation. The stationary state is stable if all $Im(\omega_N)$ are non-positive and if the instability boundary is given by one vanishing $Im(\omega_N)$. In Figure 1 (from [12]), it is shown the boundary of the mode $N=1$ for $k>0$.

The only mode which has influence on the coupling of the plasma with the external circuit is the mode $N=0$, which has purely imaginary eigenvalues ω_0 in the considered range of plasma parameters. In Figure 2, the instability curves $\omega_0=0$ of the mode for different values of the external resistor can be seen. Above the curves, all the sets of parameters (I_0, k) represent stable mode $N=0$. The line marked as lower critical point corresponds to the lower critical current in Figure 1. In this way, all the states of the mode $N=0$ inside the instability loop of the mode $N=1$ are stable. The stability of the mode $N=0$ determines the nonlinear properties of the system since only in that case a Hopf bifurcation arises at the critical points of the mode $N=1$ [12].

The linear stability analysis has been used to study the stability of the low-pressure positive column in plasmas of electronegative gases [27, 28]. Testrich et al. [28] considered as relevant components of the plasma: electrons, negative ions (O^-), positive ions (O_2^+), and metastable oxygen molecules. Other excited oxygen states were considered to have negligible concentrations. The kinetics included the following: electron attachment and detachment, oxygen ionization, metastable molecules production, and recombination O^-–O_2^+. In Figure 3, the comparison of theoretical with experimental results at 0.5 Torr can be seen. As mentioned earlier, the theoretical loop (right) defines the stable and unstable discharge regimes. Waves with dimensionless wave number k within the loop are linearly unstable. On the left, the detected frequency spectra depending on the discharge electric current in different investigated operating modes are shown. The instability windows were in good agreement with experimental data.

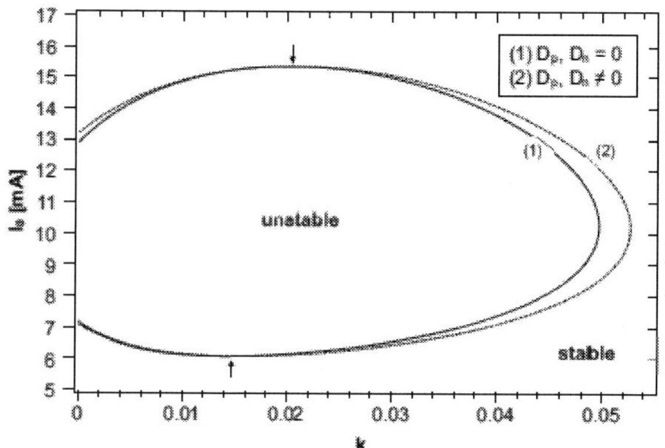

Figure 1. Instability curves for the mode $N=1$ for (1) vanishing diffusion coefficients $D_{p,n}=0$ and (2) $D_{p,n}\neq0$. Critical points are indicated by arrows (from Bruhn et al. [12]).

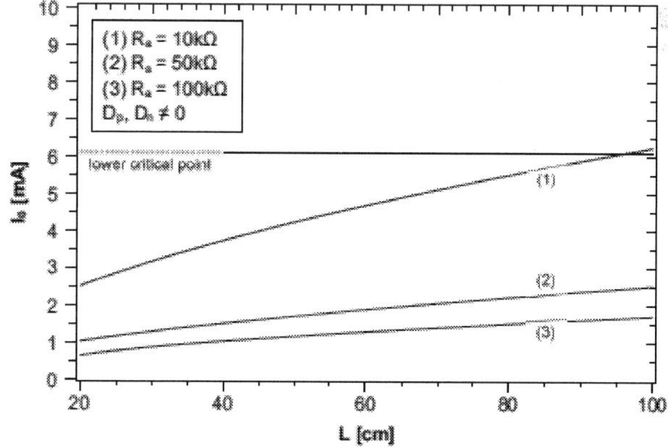

Figure 2. Instability curves for the mode $N=0$, for three different values of the external resistor (from Bruhn et al. [12]).

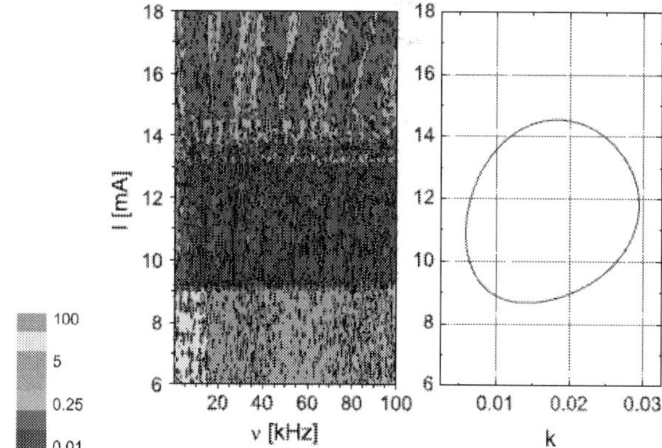

Figure 3. Comparison of experimental results with theoretical investigation on the stability of a DC oxygen discharge. On the left, detected frequency spectra depending on the discharge electric current in different investigated operating modes are shown. The colored plot represents the averaged relative intensity of electric field, 0.1 (blue), 0.25 (cyan), 5 (green), and 100 (orange), in arbitrary units. On the right, the theoretical loop defines the stable and unstable discharge regimes. The experiments were carried out in a pressure range from 0.5 to 0.9 Torr within a discharge current interval from 0.5 to 90 mA. From Testrich et al. [28].

4. Stochastic dynamics in glow discharges

4.1. Plasma dynamics as a stochastic process

It is well known that the random fluctuations of certain observables give information about the transport processes occurring in a given system. This fact is actually a direct consequence of the well-known fluctuation–dissipation theorem which relates the random fluctuations of a given variable to the transport coefficient of the corresponding conjugated thermodynamic variable [29]. It has been used to obtain information on the transport coefficients in glow discharges from the time and space autocorrelation function and their corresponding auto-power spectrum of random fluctuations in plasmas [30]. Moreover, it has been shown experimentally that the random current fluctuations exhibit, under certain conditions, the Markovian property in the fully developed turbulence regime [31].

The above-mentioned works have in common that the random fluctuations are assumed to be homogeneous and isotropic in space. Actually, it is known that glow discharges are highly inhomogeneous, and therefore the fluctuations might vary in space which might be due to nonlocal effects, finite size confinement, instabilities, etc. The study of the influence of inhomogeneities in glow discharges has revealed the presence of two particle transport regimes, namely, a purely diffusive transport regime and a wave-diffusive transport regime [32, 33]. The non-locality, memory effects, and other phenomena inducing inhomogeneities in glow discharges have been introduced in the referred works by considering a space-dependent diffusion coefficient. Within this treatment, it has been found that every type of transport regime induces a characteristic type of decay in the correlation function. Indeed, the purely diffusive regime induces a decaying mode, while the wave-like transport regime induces an oscillatory decay mode. These two types of behavior have actually been observed in fusion plasmas in tokamak and non-tokamak devices ([32] and references therein). It is important to point out that it is possible to transit from one behavior to the other by tuning a parameter value [32]. Moreover, it was also found theoretically that it is actually possible to observe the two transport regimes (diffusive transport and wave-like transport) in a single device at different locations where the plasma density has substantial differences.

4.2. FM model for fluctuations in glow discharges

In this section, we explore the occurrence of the above-described phenomenon, namely, the coexistence of the diffusive transport regime and the wave-like transport regime in glow discharges. For such purposes, we implement numerical experiments with a closely related model to that used in Ref. [32]. The model we consider here corresponds to the case of Eqs. (1)– (3) where stochastic properties are introduced through additive random terms in the constitutive equations as a stochasticity source of the particle fluxes (Eq. (2)). The stochastic terms are intended to model spontaneous rapid variations in the particle fluxes which might be due to some phenomena which are not present in the constitutive equations but which would be relevant for the dynamics of the plasma fluctuations. It is clear that Eq. (2) can be considered as an approximation of a larger set of constitutive equations [9, 32]. Of course, we introduce the stochastic terms as a simple way to model (in a rather heuristic way) such phenomena here. Actually, there is a formal way to do this, but our intention in this section is just to study the phenomena induced by the introduction of those stochastic "corrections." We leave, as a future work, the introduction, in a more rigorous way, of those stochastic terms. For the sake of completeness, we write in the simplest case the stochastic fluid model obtained from the above-mentioned considerations [34]. We consider only two species, namely, positive ions and electrons in an external electric field generated by parallel plates. First, we have the continuity equations given by

$$\frac{\partial n_{\mathrm{p}}}{\partial t} + \nabla \cdot \vec{J}_{\mathrm{p}} = S_{\mathrm{p}}$$

$$\frac{\partial n_{\mathrm{e}}}{\partial t} + \nabla \cdot \vec{J}_{\mathrm{e}} = S_{\mathrm{e}}$$

$$(18)$$

where the source terms are written to include only the ionization and the electron–ion recombination rates.

Next, we introduce the stochastic version of the constitutive equations as

$$\vec{J}_{\mathrm{p}} = \mu_{\mathrm{p}} n_{\mathrm{p}} \vec{E} - D_{\mathrm{p}} \nabla n_{\mathrm{p}} + \hat{f}_{\mathrm{p}}$$

$$\vec{J}_{\mathrm{e}} = -\mu_{\mathrm{e}} n_{\mathrm{e}} \vec{E} - \nabla \left(D_{\mathrm{e}} n_{\mathrm{e}} \right) + \hat{f}_{\mathrm{e}}$$

$$(19)$$

where $\hat{f}_p$ and $\hat{f}_e$ are assumed to be δ -correlated white noise terms with zero mean and Gaussian distribution. Formal aspects of the inclusion of stochastic terms in the constitutive equations can be found in the context of the general equation of the reversible–irreversible coupling, GENERIC [35, 36]. Another approach based on the Markovian method can be seen in Ref. [31]. Within the first formalism, it is shown that fluctuating phenomena which are described by dynamical equations having the GENERIC structure are stationary, Gaussian, and Markovian processes. In general, as it is the case here, with the use of this kind of stochastic terms in Eq. (19), one can incorporate the statistical properties of stationary, Gaussian, and Markovian processes to the mean Eqs. (1)– (3). The set of equations is completed with the Poisson equation describing the self-consistent electric field and the electronic energy conservation equation (Eq. (3)). The former is given by

$$\nabla^2\varphi = \frac{e}{\varepsilon}\left(n_e - n_p\right) \tag{20}$$

where φ is the electric potential function.

We close this subsection with the reduced expressions for the source terms in Eq. (18). In fact, we have

$$S_p = S_e = k_{io}\overline{n}n_e - k_{att}\overline{n}n_e \tag{21}$$

In the last equation, k_{io} and k_{att} are the ionization and recombination $(e-p)$ coefficients, respectively, and $\overline{n}$ the bulk gas density.

4.3. Numerical method of solution of FM model

Our main goal is to study the phenomenon of coexistence of the two regimes of transport previously reported in plasmas [32]. To this end, we numerically simulate the dynamics of Eqs. (18)– (21) in a one-dimensional "box" of length $L = 1$ (see Figure 4) with a noise level of $D = 0.10$. The boundary conditions needed by Eqs. (18)– (20) are specified at each electrode. Accordingly with Figure 4, the grounded anode is at $x = 0$, and the powered cathode is at $x = L$. At the anode, the electron density is taken as zero, since it is assumed here that the recombination rate k_{att} is infinite as it corresponds to a grounded conducting surface. It is also assumed that the ion flux at position $x = 0$ is due only to drift. $\varphi = 0$ is the given value of the plasma potential at $x = 0$. At the cathode, the electronic flux is proportional to the ion flux which is also assumed to be due only to drift.

The numerical method used here is based on the spectral Chebyshev collocation (SCC) method which assumes that an unknown partial differential equation (PDE) solution can be represented by a global, interpolating, Chebyshev partial sum. This form is often convenient, particularly if the solution is required for some other calculation where its rapid and accurate evaluation is important. The global character of spectral methods is beneficial for accuracy.

Once the finite Chebyshev series representing the solution is substituted into the differential equation, the coefficients become determined so that the differential equation is satisfied at certain points within the range under consideration. In this spectral method, the PDE equation is required to be satisfied exactly at the interior points, namely, the Gauss–Lobatto collocation points given by

$$x_i = \cos\left(\frac{i\pi}{N}\right) \tag{22}$$

with $i = 1, \ldots, N-1$.

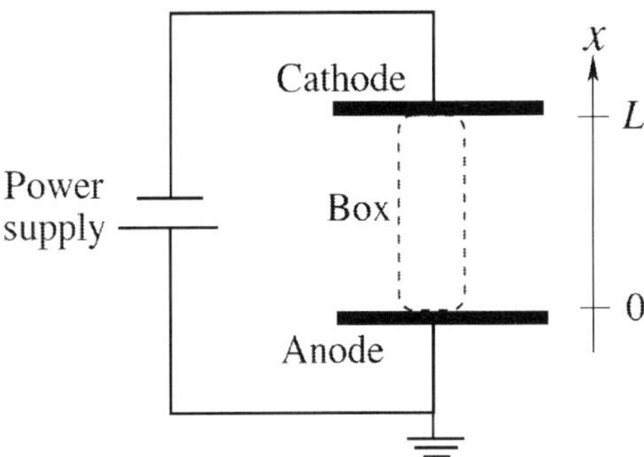

Figure 4. The device with a parallel-plate configuration. The anode and the cathode are conducting surfaces. The anode is grounded and the cathode powered by the shown power supply. The anode is situated at $x = 0$ in the reference frame and the cathode at $x = L$. The figure also shows the narrow box within which the simulations are made with the use of Eqs. (18)–(21). The separation between the plates is L.

In equation (22), N denotes the size of the grid. The number of points is chosen so that, along with the initial or boundary conditions, there are enough equations to find the unknown coefficients. The positions of the points in the range are chosen to make the residual obtained small when the approximate solution is substituted into the differential equation. The range in which the solution is required is assumed finite, and, for convenience, a linear transformation of the independent variable is made to make the new range $(-1, 1)$.

Finally, an additional property of the spectral methods is the easiness with which the accuracy of the computed solution can be estimated. This can be done by simply checking the decrease of the spectral coefficients. There is no need to perform several calculations by modifying the resolution, as is usually done in finite difference and similar methods for estimating the "grid convergence." A further explanation of this spectral method can be found in Refs. [37] and [38]. The spatial derivative terms in Eqs. (18) and (19) were expressed on derivative matrices expanded on Chebyshev polynomials. The matrix-diagonalization method was used to solve

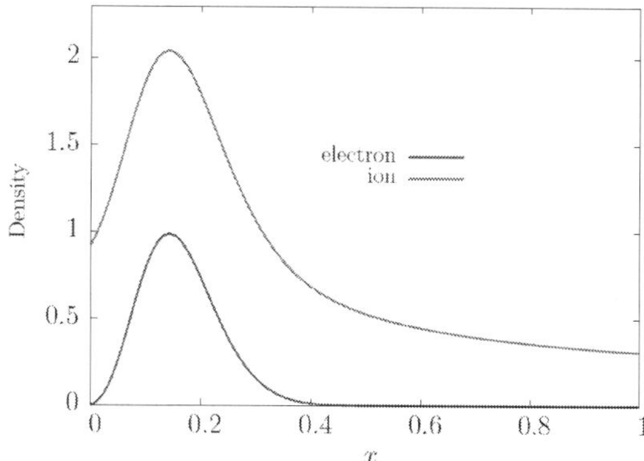

Figure 5. Spatial distribution of electron and ion densities. The anode is at $x = 0$ and the cathode at $x = 1$ (nondimensional position). Both the distributions show maxima at $x = 0.14$. The plasma is not in the neutral condition along the spatial domain.

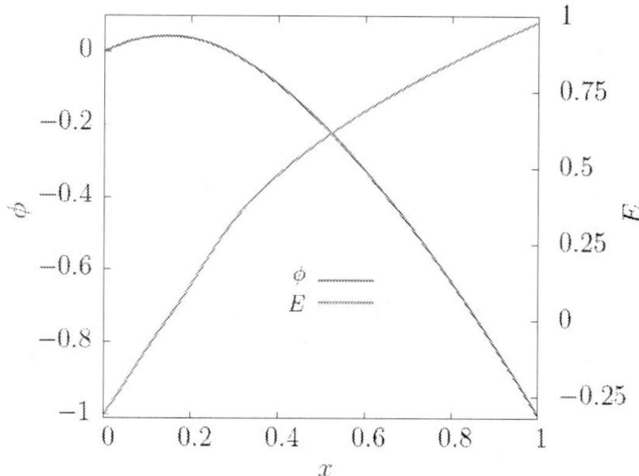

Figure 6. Spatial distribution of the plasma potential and the self-consistent electric field. The electric field has the expected behavior, namely, it decays almost linearly from the cathode, and then it becomes negative in the anode sheath. The distributions seen in Figures 5 and 6 indicate that the negative glow does not exist for the plasma conditions assumed here.

the coefficient equation system in physical space directly. A coordinate transformation was necessary to map the computational interval to $0 < x < 1$.

We then obtain the density of electrons and ions as well as the plasma potential and the self-consistent electric field resulting in the spatial distributions shown in Figures 5 and 6. The simulation in time at three different positions, namely, at $x = 0.14$, $x = 0.5$, and $x = 0.86$, of the positive ion density results in time series representing the density variations in time. Representative numerical time series can be observed in Figure 7a. Each time series is different from each other due to the spatial inhomogeneities of the plasma and the presence of fluctuating phenomena. These behaviors can be qualitatively compared with experimental results that we have obtained in a DC discharge shown in Figure 7b. The measurements were made in a hollow cathode device (90 mm long, 29.5 mm radius) which is depicted in Figure 8. Then we can calculate the autocorrelation function of each time series and compare the resulting autocorrelation functions in the different locations we used to obtain the density variations (Figure 9). The smoothed curves shown in Figure 9 have been averaged over a large number of numerical realizations corresponding to each of the considered positions.

4.4. Discussion

As it can be seen in Figure 5, the electron and ion densities have a maximum at $x = 0.14$, while in the other two points (at $x = 0.5$ and $x = 0.86$) the densities show smaller values. From the graph shown, it can be concluded that the plasma is in a non-neutrality condition in most of the domain [0,1]. In the sheath regions, the electron density is nearly zero. The plasma potential and the self-consistent electric field are shown in Figure 6. The electric field has the expected behavior, namely, it decays almost linearly from the cathode, and then it becomes negative in the anode sheath. The distributions seen in Figures 5 and 6 indicate that the negative glow does not exist for the plasma conditions assumed here.

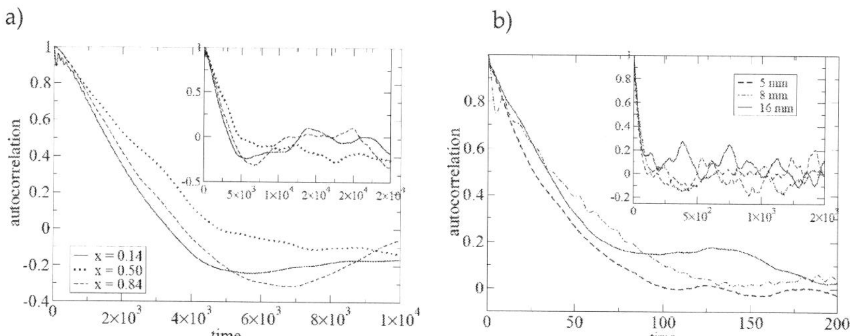

Figure 7. Time series of the positive ion density at three positions in the profile. The series are generated by a) numerical simulation of the time evolution of the density through stochastic version of the model, Eqs. (18)– (21), and b) experimental measuring of the floating potential at each position. The insets show the rapid variations occurring during a longer lapse of simulation in each case. Time is in a.u.

In Figure 9, we observe that the density autocorrelation function has different behaviors at different positions. Though the shapes of the shown curves are very similar, the comparison reveals that in the plasma two different particle transport processes are present, the so-called diffusive and the wave-like transport regimes. These two regimes are characterized by its autocorrelation function as anticipated in Ref. [32]. The autocorrelation function for the pure diffusive transport would be a strictly monotonous decreasing function of the delay time, while the correlation function for the wave-like transport would be an oscillating decaying function. These pure modes are not indeed present in the density variations we have obtained from the numerical simulation we performed, but a transport regime resulting from the coexistence of both of the mentioned regimes is observed. We name this the wave-diffusive regime. The tendency shown by the three correlation functions in Figure 9 reveals that the "less oscillatory" curve (solid line) corresponds to the position at $x = 0.14$ where the electronic density is maximum. At this position, one should indeed expect a "more diffusive" behavior. We note that the oscillating decaying autocorrelation functions correspond to measurements where the plasma has small diffusivity gradients [33], while in the diffusive transport regime where the autocorrelation function has a purely decaying mode, the plasma has high diffusivity gradients.

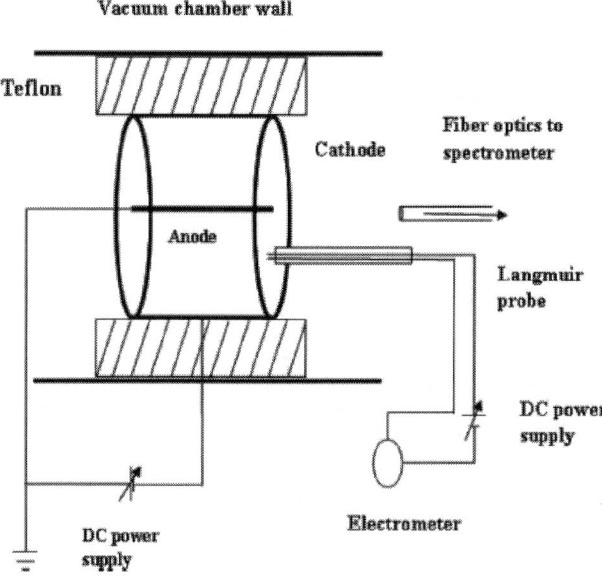

Figure 8. Schematic diagram of the hollow cathode DC discharge device.

Finally, we should mention that this work can be extended in several directions. For example, if we think of the plasma as a system out of equilibrium, we could explore the phenomenon of resonant response reported in Ref. [39]. This phenomenon occurs when the system relaxes to its stationary state in an oscillatory decaying mode. It has been proved that such a frequency actually can be used to stimulate the system in the linear response regime, thus leading to a response in the density fluctuations of the system.

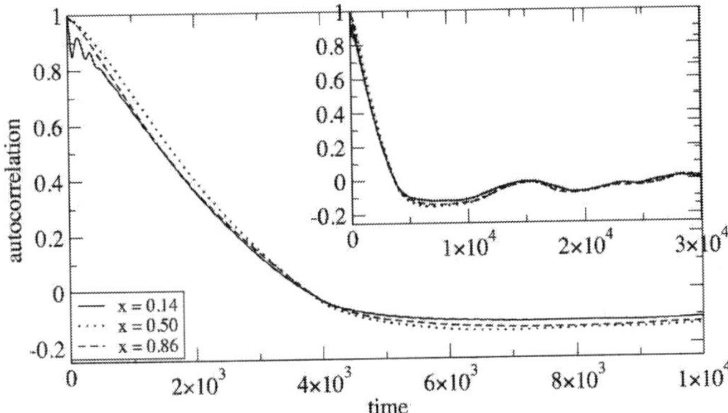

Figure 9. Numerical autocorrelation function for the plasma density fluctuations. The autocorrelation function is calculated from the measurements of the density as a function of time at three different locations in the box. The autocorrelation with a more oscillating decaying mode (solid line) corresponds to measurements near the anode at $x = 0.14$ where the electronic density is maximum. Other measurements (at $x = 0.50$ and $x = 0.86$) result in a less oscillating decaying autocorrelation function. This means that the diffusive transport regime is present in regions where the plasma has a high density of electrons, while the wave-diffusive transport regime is present in regions where the electron density is low. Time is in a.u.

5. Control of nonlinear dynamics in discharge plasmas

5.1. Overview of nonlinear dynamics in discharge plasmas

Among the very interesting nonlinear systems are gaseous plasmas. As mentioned, plasmas are governed by dissipative effects and are far from equilibrium. They exhibit complex or regular behavior depending on the controlling parameters such as discharge current, voltage, and pressure. The transition from one state to another takes place for a change in any one of the parameters [40]. An important feature is the understanding of the complex chaotic behaviors commonly observed in glow discharges. The transition to chaos and constructive effects of noise, such as stochastic resonance, and coherence resonance have been studied in many glow discharge plasma systems [41, 42].

Nonlinear dynamical systems present several routes to chaos such as period doubling, intermittency, quasiperiodicity, etc. [43]. Very important and different transition sequences from periodic to chaotic patterns have also been observed. An example is the so-called alternating periodic chaotic (APC) sequence. Much of the behavior concerning APC sequences is understood in terms of a homoclinic chaotic scenario [44]. Homoclinic orbits are associated with erratic behavior (chaos) in a dynamical system [45]. Homoclinic chaos in discharge tubes dates back to more than twenty years, as well as the chaotic behavior and period doubling in pulsed plasma discharge [46]. Currently, much attention has been given to the generic problem of phase synchronization using these devices. In particular, phase synchronization has been demonstrated under different conditions [47]. Given a chaotic oscillator for which an angle coordinate can be suitably introduced as a state space variable, it is often the case that the phase of this oscillator synchronizes with the phase of an external periodic perturbation, or pacer [48, 49].

Chaos may be undesirable for industrial applications where cycle-to-cycle reproducibility is important, yet for the treatment of cell-containing materials including living tissues, it may offer a novel route to combat some of the major challenges in medicine such as drug resistance. Chaos in low-temperature atmospheric plasmas and its effective control are likely to open up new vistas for medical technologies. In the general framework of nonlinear dynamical systems, a number of strategies have been developed to achieve active control over complex temporal or spatio-temporal behavior [50]. Many of these techniques apply to plasma instabilities. There are some methods of controlling chaos where the main idea is to convert chaotic behavior to periodic behavior by inducing a small perturbation in the system. Depending on the physical mechanism of the specific instability in each case, an appropriate control strategy is chosen out of a variety of different approaches; in particular, discrete feedback, continuous feedback, or spatio-temporal open-loop synchronization [51, 52]. In particular, time-delayed feedback plays a prominent role in controlling chaos; this is one of the successful applications that knowledge acquired in nonlinear science has provided to plasma physics. Pyragas [53] proposed the time-delayed feedback technique, which is based on feedback perturbation in the form of the difference between a delayed output signal and the output signal itself; this is appropriate for laboratory experiments conducted in real time. This method is robust to noise and does not require real-time computer processing to calculate a target unstable periodic orbit (UPO); therefore, it can act on the experimental system continuously over time. The feedback perturbation signal that is applied to the nonlinear system is proportionally adjusted to the difference between the two successive values of an arbitrary dynamic variable. Since plasma is a typical nonlinear dynamical system with a large number of degrees of spatiotemporal freedom, various unexpected phenomena are observed when time-delayed feedback is applied in plasma.

Therefore, investigations into the behavior of nonlinear systems with regard to time delay are currently required. From a theoretical point of view, discharges have been traditionally described taking into account the complex processes involved in the plasma recombination and electric conductivity. Such descriptions require, as mentioned in Section 2, the use of coupled partial differential equations involving spatial and time variables, the transport of momentum and energy of plasma components, the continuity equation, diffusion equation, Poisson equation, etc.

5.2. Time-delayed feedback control

Several algorithms to control chaotic dynamics have been proposed in the last 20 years ([54, 55] and references therein). They include OGY method [56], linear feedback [57, 58], nonfeedback method [59, 60], adaptive control [61, 62], backsteeping method, and sliding mode control. Delayed feedback control (DFC) has been applied to a large variety of systems in physics, biology, medicine, and engineering [63, 64], in purely temporal dynamics as well as in spatially extended systems ([65] and references therein). DFC has also been used to control purely noise-induced oscillations and pattern formation in steady-state regimes in chemical systems [65, 66], electrochemical cells [67], neural systems [68], laser diodes [69], and semiconductor nanostructures [70].

DFC control comprises various aspects such as the stabilization of unstable periodic orbits embedded in a deterministic chaotic attractor, stabilization of unstable fixed points (steady states), or control of the coherence, and timescales of stochastic motion. Moreover, it has been shown to be applicable also to noise-induced oscillations and patterns. DFC has been widely used to control chaos in discharge plasmas. Several interesting reports can be found in the literature [51, 52, 65, 67]. Among them, for instance, Fukuyama et al. [52] applied the method to study the nonlinear periodic regime associated with the current-driven ion acoustic instability (IAI) in a double-discharge plasma system. The discharge chamber was divided at the center into a driver region and a target region by a separating grid that was kept at a floating potential. A DC potential was applied to the other grid in order to excite the instability. The current-driven IAI appeared when the DC potential reached a threshold. Once the instability is established (the threshold was 23 V), the dynamics of the floating potential shows a variety of behaviors. First, a limit cycle appears and persists for values of the DC potential between 23 and 35 V. The amplitude of the limit cycle switches stochastically between two values for the DC potential between 36 and 39 V, and a larger limit cycle appears for 40 V. After this, as the DC potential increases, the system gradually falls into disorder. Then the chaotic state appears at 54 V when the system is completely disturbed. When the potential is 67 V, the instability disappears with a decrease in the noise level. DFC is applied to the nonlinear periodic regime for values of the DC potential between 40 and 45 V. The feedback perturbation signal $F(t)$ applied to the nonlinear system was proportionally adjusted to the difference between two successive values of a dynamic variable $x(t)$, i.e.,

$$F(t) = k\left[x(t-\tau) - x(t)\right] \tag{23}$$

where τ is the time delay and k a proportionality constant. The behavior of the system was investigated by changing the values of τ and k when time-delayed feedback is applied to the nonlinear periodic regime. Intermittency and chaos were observed.

In the remaining of this section, we examine the effects of DFC method on a special periodic regime in discharge plasmas which are often observed in experiments. This can be seen in Figure 10, where the autonomous dynamics as reflected by the time evolution of the positive ion density when the discharge voltage (V_D) has been applied is shown. The gas is supposed to be in a parallel electrode device (Figure 4). One of the electrodes is grounded, and the other is at V_D. Here, we test a DFC method on periodic oscillations with relaxation predicted by the FM for the discharge plasma given by Eqs. (18)– (21). The graph in Figure 10 is obtained by using the deterministic form of Eqs. (18)– (21) with an additional term in continuity equations through which the control is applied as suggested by Pyragas [53]. The continuity equation for the positive ion density is then modified as follows:

$$\frac{\partial n_{\mathrm{p}}}{\partial t} + \nabla \cdot \vec{J}_{\mathrm{p}} = S_{\mathrm{p}} + F(x,t) \tag{24}$$

where $F(x, t)$ is the parameter control given by the formula

$$F(x,t) = k\left[n_{\mathrm{p}}(x,t-\tau) - n_{\mathrm{p}}(t) \right] \tag{25}$$

with k a constant. As it is well known, Eq. (25) for the control term ensures that the system will not drive to a regime where the target dynamics were naturally stable. Equations (24) and (19)–(21) are solved by means of the spectral Chebyshev collocation numerical method described in Section 4.3. The value of the constant k is chosen to optimize the effect of the feedback on the system's oscillatory regime. This process is made by trial and error. The value of the delay time τ is directly related with the period of the rapid oscillations observed between the oscillations in Figure 10.

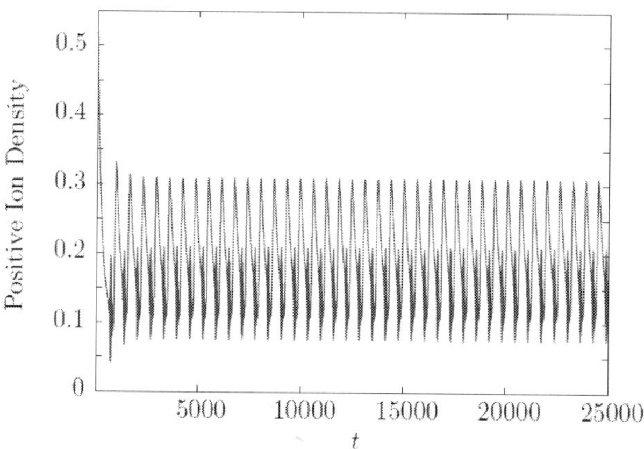

Figure 10. Time evolution of the positive ion density. The dynamics, obtained by numerical solution of Eqs. (18)– (21), shows a periodic behavior with relaxation. The discharge voltage is fixed in 460 V.

The observed periodic oscillation behavior with relaxation in Figure 10 was experimentally found by, for instance, Nurujjaman and Sekar Iyengar [71] in the cathode sheath in a cylindrical DC discharge device by measuring the floating voltage in the plasma which is directly related with the charge density (see also [42] and [72]). It can be seen in Figure 11 that the power spectrum and the phase space plot were included.

Figure 12 shows the result of the application of the control method sketched in Eqs. (24) and (25) to the oscillatory regime between $t=10000$ and $t=20000$. As observed, the stabilization of

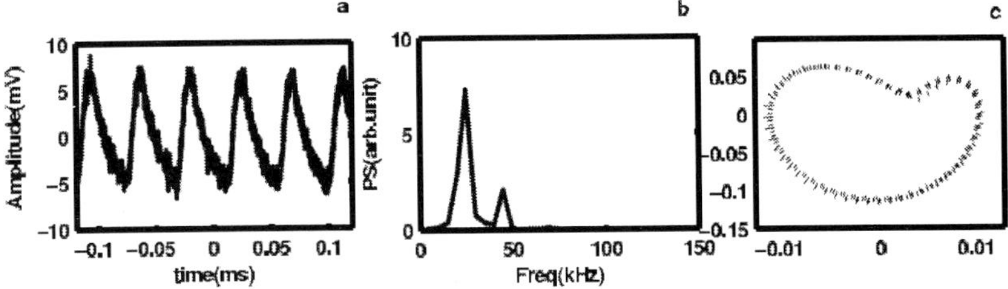

Figure 11. a) Measured floating potential in a discharge plasma at 292 V, b) power spectrum, and c) phase space plot. PS, power spectrum. From Nurujjaman and Sekar Iyengar [71].

the dynamics is easily achieved with great efficiency. The optimum value of the constant k used to obtain Figure 12 is 0.1, while other constants are $\tau = 5$. We leave, for future work, implementing the control method to an experimental plasma. In the laboratory, the density $n_p(x, t)$ should be replaced by an accessible parameter of the system, e.g., the floating potential and the control should be externally applied on the system.

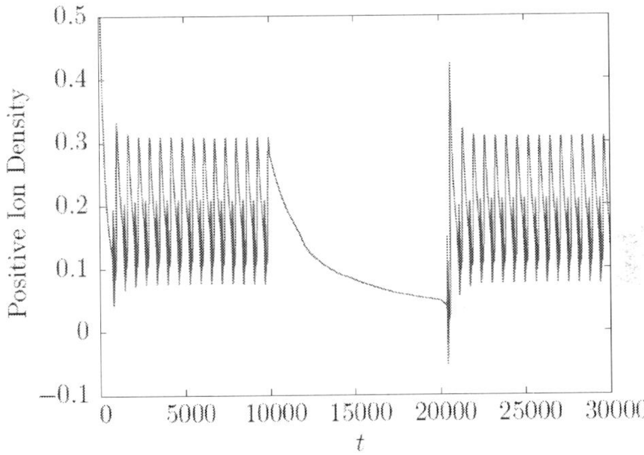

Figure 12. Delayed feedback control is applied to the time series of positive ion density from Figure 10. The control is turned on at $t = 10000$ and turned off at $t = 20000$. Time is in a.u.

6. Conclusion

Low-temperature (LT) plasmas have a substantial role in diverse scientific areas and modern technologies. Industrial applications of LT plasmas form a very important part of the productive infrastructure of advanced economies all over the world. In LT plasmas, a large number and diversity of reactive species are generated that activate physical and chemical processes hard to obtain in ordinary chemical environments. Their stochastic, nonlinear dynamics, and chemical kinetics strongly determine the efficiency and effectiveness of LT plasma-based procedures. Typical applications are etching, spectrochemical analysis, deposition of thin films on substrates, and others. Much is known on low-temperature plasmas by fluid model-based investigations (Sections 2 and 3). Fluid models have been used to study the fundamental properties and dynamic behavior of several gases: argon, oxygen, hydrogen, and helium in a diversity of configurations and its results confirmed by experiments. In Section 4, we analyzed the diffusive and wave-like transport of particles in LT plasmas (glow discharges). To this end, we used the well-known relationship between the random fluctuations of physical observables and the transport processes occurring in the system. We concluded that both the diffusive transport and the wave-like transport of particles coexist in the numerical simulations and experiments that were carried out. Section 5 was devoted to analyzing the state of the art of the nonlinear dynamics of LT plasmas; in this section, besides discussing the wide variety of nonlinear behaviors shown by LT plasmas, we illustrated the case of particle density oscillation control through a time-delayed feedback technique. The understanding of the many complex chaotic behaviors commonly observed in glow discharges still offers an exciting field for scientific research.

Author details

Aldo Figueroa, Raúl Salgado-García, Jannet Rodríguez, Farook Yousif Bashir, Marco A. Rivera and Federico Vázquez*

*Address all correspondence to: vazquez@uaem.mx

Centro de Investigación en Ciencias, Universidad Autónoma del Estado de Morelos, México

References

[1] Bogaerts A, Neyts E, Gijbels R, van der Mullen J. Gas discharge plasmas and their applications. Spectrochimica Acta Part B 2002; 57: 609–658.

[2] Anghel SD, Frentiu T, Simon A. Atmospheric pressure plasmas in resonant circuits. The Open Plasma Physics Journal 2009; 2: 8–16.

[3] Merche D, Vandencasteele N, Reniers F. Atmospheric plasmas for thin film deposition: a critical review. Thin Solid Films 2912; 520: 4219–4236.

[4] Otto C, Zhan S, Rost F. Physical methods for cleaning and desinfection of surfaces. Food Engineering Reviews 2011; 3: 171–188.

[5] Pappas D. Status and potential of atmospheric plasma processing of materials. Journal of Vacuum Science and Technology A 2011; 29: 020801.

[6] Rafatov I, Bogdanov EA, Kudryavtsev AA. On the accuracy and reliability of different fluid models of the direct current glow discharge. Physics of Plasmas 2012; 19: 033502.

[7] Kushner MJ. Hybrid modelling of low temperature plasmas for fundamental investigations and equipment design. Journal of Physics D 2009; 42: 194013.

[8] Vázquez F, del Río JA. A non-equilibrium variational principle for the time evolution of an ionized gas. Physical Review E 1993; 47: 178–183.

[9] Jou D, Lebon G, Casas-Vázquez J. Extended Irreversible Thermodynamics, 4th Ed. Netherlands: Springer, 2010.

[10] Kolobov VI, Tsendin LD. Analytic model of the cathode region of a short glow discharge in light gases. Physical Review A 1992; 46: 7837–7852.

[11] Boeuf JP, Pitchford LC. Two-dimensional model of a capacitively coupled rf discharge and comparisons with experiments in the Gaseous Electronic Conference reference reactor. Physical Review E 1995; 51: 1376–1390.

[12] Bruhn B, May B, Marbach J. Hopf-bifurcations, global coupling and hysteresis in dc-driven oxigen discharges. Contributions to Plasma Physics 2010; 51: 650–671.

[13] Hayashi T. Mixed-mode oscillations and chaos in a glow discharge. Physical Review Letters 2000; 84: 3334–3337.

[14] Ghosh S, Shaw PK, Saha D, Janaki MS, Sekar Iyengar AN. Irregular-regular-irregular mixed mode oscillations in a glow discharge plasma. Physics of Plasmas 2015; 22: 052304.

[15] Wang Y, Wang D. Influence of impurities on the uniform atmospheric-pressure discharge in helium. Physics of Plasmas 2005; 12: 023503.

[16] Wang YH, Zhang YT, Wang DZ, Kong MG. Period multiplication and chaotic phenomena in atmospheric dielectric-barrier glow discharges. Applied Physics Letters 2007; 90: 071501.

[17] Wang Y, Shi H, Sun J, Wang D. Period-two discharge characteristics in argon atmospheric dielectric-barrier discharges. Physics of Plasmas 2009; 16: 063507.

[18] Zhang J, Wang Y, Wang D. Numerical study of period multiplication and chaotic phenomena in an atmospheric radio-frequency discharge. Physics of Plasmas 2010; 17: 043507.

[19] Grabec I, Mandelj S. Desynchronization of striations in the development of ionization turblence. Physics Letters A 2001; 287: 105–110.

[20] Chirokov A, Khot SN, Gangoli SP, Fridman A, Henderson P, Gutsol AF, and Dolgopolski A. Numerical and experimental investigation of the stability of radio-frequen-

cy (RF) discharges at atmospheric pressure. Plasma Source Science Technology 2009; 18: 025025.

[21] Raizer YP. Gas Discharge Physics. Berlin: Springer, 1991.

[22] Bruhn B, Richter A, May B. On the stability of a dc-driven oxygen discharge in cylindrical geometry. Physics of Plasmas 2008; 15: 053505.

[23] Ferreira CM, Gousset G, Touzeau M. Quasi-neutral theory of positive columns in electronegative gases. Journal of Physics D 1988; 21: 1403–1413.

[24] Franklin RN, Snell J. Modelling discharges in electronegative gases. Journal of Physics D 1999; 32: 2190–2203.

[25] Lampee M, Manheimer WM, Fernsler RF. The physical and mathematical basis of stratification in electronegative plasmas. Plasma Sources Science Technology 2004; 13: 15–26.

[26] Nighan WL, Wiegand WJ. Influence of negative-ion processes on steady-state properties and striations in molecular gas discharges. Physical Review A 1974; 10: 922–945.

[27] Bruhn B, Richter A, May B. On the stability of a dc-driven oxygen discharge in cylindrical geometry. Physics of Plasmas 2008; 15: 053505.

[28] Testrich H, Pasedag D, Naurátil Z, May B, Reimer R, Wilke C, Wagner H-E. Investigations on the stability of the low pressure positive column in oxygen. Journal of Applied Physics D. 2009; 42: 145207.

[29] Marconi UMB, Puglisi A, Rondoni L, Vulpiani A. Fluctuation–dissipation: response theory in statistical physics. Physics Reports 2008; 461: 111–195.

[30] Li B, Hazeltine RD. Applications of noise theory to plasma fluctuations. Physical Review E 2006; 73: 065402.

[31] Kimiagar S, Movahed MS, Khorram S, Tabar MRR. Markov Properties of Electrical Discharge Current Fluctuations in Plasma. Journal of Statistical Physics 2011; 143: 148–167.

[32] Vázquez F, Márkus F. Size scaling effects on the particle density fluctuations in confined plasmas. Physics of Plasmas 2009; 16: 112303.

[33] Vázquez F, Márkus F. Non-Fickian particle diffusion in confined plasmas and the transition from diffusive transport to density waves propagation. Physics of Plasmas 2010; 17: 042111.

[34] Graves DB, Jensen KF. A continuum model of DC and RF discharges. IEEE Transactions on Plasma Science 1986; PS-14: 78–91.

[35] Grmela M, Öttinger HC. Dynamics and thermodynamics of complex fluids. I. Development of a general formalism. Physical Review E 1997a; 56: 6620–6632.

[36] Öttinger HC, Grmela M. Dynamics and thermodynamics of complex fluids. II. Illustrations of a general formalism. Physical Review E 1997b; 56: 6636–6655.

[37] Zhao S, Yedlin MJ. A new iterative Chebyshev spectral method for solving the elliptic equation $\nabla \cdot (\sigma \nabla u) = f$. Journal of Computational Physics 1994; 113: 215–223.

[38] Peyret, R. Spectral Methods for Incompressible Viscous Flow; 5th ed. New York: Springer-Verlag, 2002.

[39] Salgado-García R. Resonant response in nonequilibrium steady states. Physical Review E 2012; 85: 051130.

[40] Pugliese E, Meucci R, Euzzor S, Freire JG, Gallas JAC. Complex dynamics of a dc glow discharge tube: experimental modeling and stability diagrams. Scientific Reports 2015; 5: 8447.

[41] Letellier C, Dinklage A, El-Naggar H, Wilke C, Bonhomme G. Experimental evidence for a torus breakdown in a glow discharge. Physical Review E 2000; 63: 7219–7226.

[42] Nurujjaman MD, Narayanan R, Sekar Iyengar AN. Parametric investigation of nonlinear fluctuations in a dc glow discharge plasma. Chaos 2007; 17: 043121.

[43] Wiggins S. Global Bifurcations and Chaos Analytical Methods. New York: Springer-Verlag, 1984.

[44] Nurujjaman M, Iyengar ANS, Parmananda P. Noise-invoked resonances near a homoclinic bifurcation in the glow discharge plasma. Physical Review E 2008; 78: 026406.

[45] Braun T, Lisboa JA, Gallas JAC. Evidence of homoclinic chaos in the plasma of a glow discharge. Physical Review Letters 1992; 68: 2770–2773.

[46] Cheung PY, Wong AY. Chaotic behavior and period doubling in plasmas. Physical Review Letters 1987; 59: 551.

[47] Rosa E, Pardo W, Ticos CM, Walkenstein JA, Monti M. Phase synchronization of chaos in a discharge tube. International Journal of Bifurcation Chaos 2000; 10: 2551–2564.

[48] Ticos CM, Rosa E, Pardo W, Walkenstein JA, Monti M. Experimental real-time phase synchronization of a paced chaotic plasma discharge. Physical Review Letters 2000; 85: 2929–2932.

[49] Davis MS, Nutter NG, Rosa E. Driving phase synchronous plasma discharges with superimposed signals. International Journal of Bifurcation Chaos 2007; 17: 3513–3518.

[50] Dinklage A, Wilke C, Bonhomme G, Atipo A. Internally driven spatiotemporal irregularity in a dc glow discharge. Physical Review E 2001; 62: 042702.

[51] Klinger T, Schröder C, Block D, Greiner F, Piel A, Bonhomme G, Naulin V. Chaos control and taming of turbulence in plasma devices. Physics of Plasmas 2001; 8: 1961–1968.

[52] Fukuyama T, Watanabe Y, Taniguchi K, Shirahama H, Kawai Y. Dynamical behavior of the motions associated with the nonlinear periodic regime in a laboratory plasma subject to delayed feedback. Physical Review E 2006; 74: 016401.

[53] Pyragas K. Continuous control of chaos by self-controlling feedback. Physics Letters A 1992; 170: 421.

[54] Fradkov AL, Evans RJ. Control of chaos: methods and applications in engineering. Annual Reviews in Control 2005; 29: 33–56.

[55] El-Dessoky MM. Global stabilization for novel chaotic dynamical system. Life Science Journal 2014; 11: 157–162.

[56] Ott E, Grebogi C, Yorke J. Controlling chaos. Physical Review Letters 1990; 64: 1196–1199.

[57] Boccaletti S, Grebogi C, Lai Y-C, Mancini H, Maza D. The control of chaos: theory and applications. Physics Reports 2000; 329: 103–197.

[58] El-Dessoky MM, Agiza HN. Global stabilization of some chaotic dynamical systems. Chaos, Solitons & Fractals 2009; 42: 1584–1598.

[59] Ramesh M, Narayanan S. Chaos control by non-feedback methods in the presence of noise. Chaos, Solitons & Fractals 1999; 10: 1473–1489.

[60] Wang T, Wang X, Wang M. Chaotic control of Hénon map with feedback and non-feedback methods. Communications in Nonlinear Science and Numerical Simulation 2011; 16: 3367–3374.

[61] Pyragas K, Pyragas V, Kiss IZ, Hudson JL. Adaptive control of unknown unstable steady states of dynamical systems. Physical Review E 2004; 70: 026215.

[62] Chen G. Controlling chaotic and hyperchaotic systems via a simple adaptive feedback controller. Computers and Mathematics with Applications 2011; 61: 2031–2034.

[63] Pyragas K. Delayed feedback control of chaos. Philosophical Transactions of the Royal Society 2006; A 364: 2309.

[64] Schöll E., and Schuster, H. G. (Editors). Handbook of Chaos Control. Weinheim: Wiley-VCH, 2008.

[65] Schöll E, Hövel P, Flunkert V, Dahlem MA. Time-delayed feedback control: from simple models to lasers and neural systems. In: Complex time-delay systems. Berlin-Heildeberg: Springer, 2010, pp. 85–150.

[66] Balanov AG, Beato V, Janson NB, Engel H, Schöll E. Delayed feedback control of noise-induced patterns in excitable media. Physical Review E 2006; 74: 016214.

[67] Parmananda P, Madrigal R, Rivera M, Nyikos L, Kiss IZ, Gaspar V. Stabilization of unstable steady states and periodic orbits in an electrochemical system using delayed feedback control. Physical Review E 1999; 59: 5266–5271.

[68] Hauschildt B, Janson NB, Balanov AG, Schöll E. Noise-induced cooperative dynamics and its control in coupled neuron models. Physical Review E 2006; 74: 051906.

[69] Flunkert V, Schöll E. Suppressing noise-induced intensity pulsations in semiconductor lasers by means of time-delayed feedback. Physical Review E 2007; 76: 066202.

[70] Hizanidis J, Schöll E. Control of coherence resonance in semiconductor superlattices. Physical Review E 2008; 78: 066205.

[71] Nurujjaman MD, Sekar Iyengar AN. Chaotic-to-ordered state transition of cathode-sheath instabilities in DC glow discharge plasmas. PRAMANA 2006; 67: 299–304.

[72] Nurujjaman MD, Narayanan R, Sekar Iyengar AN. Continuous wavelet transform based time-scale and multifractal analysis of the nonlinear oscillations in a hollow cathode glow discharge plasma. Physics of Plasmas 2009; 16: 102307.